PHYSIOL,
endocrinology

THE OVARY

SECOND EDITION

Volume III

Regulation of Oogenesis and Steroidogenesis

Contributors

D. T. Baird
T. G. Baker
Janet E. Booth
Jennifer H. Dorrington
C. A. Finn
R. B. Heap
J. S. M. Hutchinson
Doreen V. Illingworth
P. Neal
P. J. Sharp

THE OVARY

SECOND EDITION

Volume III

Regulation of Oogenesis and Steroidogenesis

EDITED BY **Professor Lord Zuckerman**

Zoological Society of London
London, England
and
University of East Anglia
Norwich, Norfolk
England

AND **Barbara J. Weir**

Wellcome Institute of Comparative Physiology
Zoological Society of London
London, England
and
Journal of Reproduction and Fertility
Cambridge, England

ACADEMIC PRESS *New York San Francisco London 1977*

A Subsidiary of Harcourt Brace Jovanovich, Publishers

ACADEMIC PRESS, INC.
111 Fifth Avenue, New York, New York 10003

United Kingdom Edition published by
ACADEMIC PRESS, INC. (LONDON) LTD.
24/28 Oval Road, London NW1

Library of Congress Cataloging in Publication Data

Zuckerman, Professor Lord,   Date   ed.
    The ovary.

    Includes bibliographies and index.
    1.   Ovaries.   I.   Title.   [DNLM: 1.   Ovary.   QL876
096]
QP261.Z8 1976   599'.01'66   76-13955
ISBN 0–12–782603–3

Contents

List of Contributors

Numbers in parentheses indicate the pages on which the authors' contributions begin.

D. T. BAIRD (305) M. R. C. Unit of Reproductive Biology, Department of Obstetrics and Gynaecology, University of Edinburgh, Edinburgh, Scotland

T. G. BAKER (1), Hormone Laboratory, Department of Obstetrics and Gynaecology, University of Edinburgh, Edinburgh, Scotland

JANET E. BOOTH* (151), Department of Physiology, Royal Veterinary College, University of London, London, England

JENNIFER H. DORRINGTON (359), Banting and Best Department of Medical Research, University of Toronto, Toronto, Ontario, Canada

C. A. FINN† (151), Department of Physiology, Royal Veterinary College, University of London, London, England

R. B. HEAP (59), Department of Physiology, A. R. C. Institute of Animal Physiology, Babraham, Cambridge, England

J. S. M. HUTCHINSON‡ (227), Department of Chemical Pathology, St. Thomas's Hospital Medical School, London, England

DOREEN V. ILLINGWORTH§ (59), Department of Physiology, A. R. C. Institute of Animal Physiology, Babraham, Cambridge, England

P. NEAL (1), Hormone Laboratory, Department of Obstetrics and Gynaecology, University of Edinburgh, Edinburgh, Scotland

P. J. SHARP (227), Agricultural Research Council's Poultry Research Centre, Edinburgh, Scotland

---

* Present address: Department of Physiology, Charing Cross Hospital Medical School, London, England.

† Present address: Department of Veterinary Clinical Studies, University of Liverpool, Liverpool, England.

‡ Present address: Department of Developmental Biology, University of Aberdeen, Aberdeen, Scotland.

§ Present address: Division of Steroid Endocrinology, Department of Chemical Pathology, University of Leeds, Leeds, England.

Preface

Insofar as its speed is necessarily that of its slowest member, the completion of a book by several hands is somewhat like the voyage of a convoy of ships. This does not, however, make the enterprise any the less valuable. The first edition of "The Ovary" appeared over fifteen years ago, and it is time the reviews it incorporated were brought up-to-date. The slowness with which some chapters arrived was not, however, the only reason for the delay in the appearance of this new edition. The editors could perhaps have tried to be more demanding of their contributors than they were. As it turned out, however, illness held up some chapters, and one, which was all but completed, had to be restarted when its main author was killed in an air crash. In 1961 I found myself constrained to apologize for the tardy appearance of the original edition, and I therefore do so again for this new one, not only to those authors who were first with their contributions and who have had to wait longest to see them in print, but also to our patient publishers, Academic Press, and to the scientific public for whom the work is designed.

Like any book of reviews, it is obvious that the chapters of this new edition will fail to mention some papers that may have appeared in the past three to four years. This, however, detracts little from their value. Immediately before the texts were sent to the publisher every author was invited to update, given she or he so wished, what had been written. No doubt there is a lesson to be drawn from the fact that few felt this necessary. A review of a subject does not gain in value if it merely catalogues the names of authors who have written about it, with a brief reference to the summaries of their published papers. For a review is nothing if it is not critical, and has little merit if its purpose is not to focus on such generalizations as are justified by the pieces of information on which it is based. In the ideal, the updating of a review should be an exercise in which the validity of any general proposition that has already been defined is examined in the light of new experimental data and in which new hypotheses are formulated where they are called for by new findings. This, of course, is the personal view of one editor who knows full well that it is not necessarily accepted universally.

I have heard it said that with the enormous growth of the world's scientific effort over the past two to three decades budding scientists are often advised that when they survey the literature which bears on the problems they are investigating there is little point in going back more than ten years, so rapidly is what is already known overlaid by new observation. This is something which I am sure many who belong to an older generation of scientists greatly regret. Sometimes it results in what is already established being "rediscovered." Often new findings are treated out of all proportion to the major generalizations to which they relate. What is more—and this applies not only to the fields of science with which "The Ovary" deals—the vast expansion of scientific activity over recent years has inevitably resulted in resources becoming available not only for what is called "big science" but for a wider range of enquiry in "small science" than would ever have been possible in more penurious times. When editing a work such as the present one, it is difficult to avoid the impression that on occasion an experiment or set of experiments has been undertaken merely because a professor or supervisor has had to provide a theme for a postgraduate's thesis. One also finds that new techniques have defined topics for experiment without the kind of critical preliminary evaluation of the limitations of new methods of enquiry in relation to the central questions which it is hoped they will help elucidate. With the vast growth in experimental work—and inevitably, therefore, the occasional dilution of its quality—"controls" also sometimes appear inadequate. In spite of these general observations, I have, however, no hesitation in saying that there has been a considerable increase in our knowledge of the ovary over the period since the appearance of the first edition.

Dr. Anita Mandl and Professor Peter Eckstein collaborated with me in the editing of the first edition but were unable to help in this, Dr. Mandl because she had retired from academic life and Professor Eckstein because of the heavy load of other work which he has since assumed. Fortunately I was able to recruit as coeditor Dr. Barbara Weir, to whom my thanks, as well as that of the contributors, are due, as they also are to Academic Press.

S. Zuckerman

Acknowledgments

The help of the librarians and their staff of the Royal Society of Medicine (Mr. R. Wade), Wellcome History of Science Library (Mr. L. Symons), and of the Zoological Society of London (Mr. R. A. Fish) in checking references is greatly appreciated.

Preface to the First Edition

To the best of my knowledge no book on the normal ovary has appeared in English since the publication, in 1929, of Professor A. S. Parkes' monograph entitled "The Internal Secretion of the Ovary". The greater length of the present work, the object of which is to provide a detailed account of the principal aspects of ovarian development, structure and function as understood today, reflects the vigour with which researches on these subjects have been pursued over the past thirty years.

An almost unlimited number of topics could have been regarded as falling within the scope of the review. Since there was a necessary limit to size, the two volumes it constitutes cannot, accordingly, be claimed to exhaust the subject they were designed to cover. To whatever extent arbitrariness marks the fields dealt with, the treatise also partakes of a characteristic common to all scientific reviews, and one which reflects the fact that the content, pattern and emphasis of different fields of knowledge are always in a state of change.

The original intention was to publish the work in a single volume. When it became necessary to allocate the material to two, some rearrangement of chapters was called for, and the original sequence of topics which had been planned was changed. In the main, those chapters which relate to what might be called the natural history of the ovary are now included in Volume I, while information derived from more experimental and chemical studies is assembled in Volume II. The two volumes overlap to some extent, as do certain topics, but so far as possible this has been dealt with by means of cross-references.

I am deeply grateful to the many contributors for their generous assent to my invitation to participate in what has proved a lengthier and more arduous task than I, and perhaps they anticipated at the outset. The authors of the various chapters are of course individually responsible for the content and bibliographic references as well as the style and accuracy of their contributions.

When manuscripts started to arrive, I had to turn to two of my colleagues, Dr. Anita Mandl and Dr. Peter Eckstein, for assistance in the work of editing, and of arranging for the translations of those chapters which were submitted in French. I am deeply grateful for their help, as I am also to the Academic Press for its tolerance during the long period in which this review has been in train. My thanks are also due to Miss Heather Paterson for her able help in preparing manuscripts and checking proofs, to Mr. L. T. Morton for compiling the Subject Index, and to the Academic Press for constructing the Author Index.

The delays which are inevitably associated with the production of a lengthy treatise have meant that a number of contributions appear less up-to-date in print than they did in typescript. Even though references to papers published in last year's scientific journals may be lacking, I nonetheless believe that at the moment the two volumes provide a more comprehensive picture of the whole subject than can be found in any other single work.

August, 1961   *S. Zuckerman*

Contents of Other Volumes

THE OVARY

SECOND EDITION

Volume III

Regulation of Oogenesis and Steroidogenesis

1

Action of Ionizing Radiations on the Mammalian Ovary

T. G. Baker and P. Neal

I. INTRODUCTION

During the 15 years that have elapsed since the first edition of this book appeared, the pace of research on the radiobiology of the gonads has accelerated greatly, only to decline during the past 7 years, which have witnessed a sharp fall in the number of papers published on the subject. Besides being important in their own right, studies of the radiosensitivity of ovarian tissues have also prompted a whole series of fundamental studies of aspects of reproductive biology which might otherwise have been neglected. Thus, as a direct outcome of radiobiological studies, knowledge of oogenesis, reproductive capacity, and spontaneous mutation rates has been considerably enhanced. Furthermore, ionizing radiations have provided the biologist with a powerful "tool" with which to damage or destroy selected groups of cells. The development of microbeam techniques has made possi-

TABLE I

Criteria for Assessing the Radiosensitivity of Germ Cells[a]

Biochemical effects (including endocrine changes)
Morphological changes
   light microscope studies
   electron microscope studies
Cell population studies (cell death)
Functional changes
   fertility and fecundity
   reproductive capacity
Genetic effects

---

[a] From Mandl (1964).

ble studies of the role of regions of cells and even specific organelles in cellular integrity and biochemistry.

Most advances in the radiobiology of the ovary have been related to gametogenesis—with which the greater part of this review is concerned. No attempt is made to repeat information contained in the review by Lacassagne *et al.* (1962) in the first edition. Instead, we follow the style proposed by Mandl (1964) and consider the biological effects of exposure to radiation in terms of (1) the various criteria used to assess the effect (see Table I); (2) variations in response to age, species, and strain; and (3) stage of the mitotic or meiotic cycle.

II. GENERAL PRINCIPLES OF RADIOBIOLOGY

The nature and mode of action of ionizing radiations and their general effects on cells and tissues have been the subject of many reviews and books (see Lea, 1955; Alexander, 1965; Casarett, 1968; Lawrence, 1971). Since a useful summary of the physical properties of the various types of radiation has been provided by Lacassagne *et al.* (1962), only a few general principles need be stated here.

The majority of radiobiological studies involving the ovary have employed either X or γ rays. These electromagnetic radiations have similar properties and are highly penetrating. They ionize by eliminating electrons from stable molecules and not directly as do charged corpuscular rays. The penetration of the secondary radiation (electrons or β rays) is a function of the photon energy of the electromagnetic radiation with which it is produced. Gamma rays (together with α and β rays) arise from the disintegration of naturally occurring or artificially produced unstable radioisotopes (e.g., cobalt-60; strontium-90; radium-226); X rays are produced under certain electrical conditions.

The only other ionizing radiations that have been used in studies on the ovary are neutrons and β rays. The former have occasionally been employed to assess radiation-induced mutagenesis (see Section III,E) and in radiotherapy, while β rays have been studied indirectly because the radio-isotopes used for autoradiographic purposes sometimes cause extensive damage to ovarian tissues (germinal and/or somatic; see Section IV). Beta rays are electrons which are negatively charged particles of low mass and rarely have sufficient energy to penetrate more than a few millimeters of tissue. By contrast, neutrons are large heavy particles with no electrical charge: they are produced artificially in atomic reactors, and are highly penetrating and densely ionizing over large distances.

All of these radiations have the common property of imparting their energy to the tissues through which they pass. The process occurs very rapidly (probably within 10^{-13} seconds: Lawrence, 1971) and is sometimes referred to as the *physical stage* in the production of the radiolesion. This is followed by the *chemical stage* during which the free radicals, produced by the elimination of electrons during ionization, exert their effect within the tissues before they again become stable. The free radicals probably have a very short life and are effective over minute distances (probably of the order of a nanometer or less). The final *physiological stage* is concerned with the consequences of the chemical changes, which may be metabolic or structural and may result in the repair of the radiolesion or the death of the cell. This stage is of long duration and the damage (particularly to the nuclear chromatin) may remain latent for many years.

The interactions between ionizing radiations and biological systems occur at random over distances which differ according to the type of radiation employed. Thus, X and γ rays induce ionizations at fairly wide intervals over long distances. On the other hand, neutrons are densely ionizing and highly penetrating, while electrons have relatively short tracks. Different types of radiation can thus be compared in terms of either the number of ionizations per unit length of track, which is called the linear energy transfer (LET), or in terms of their relative biological efficiency (RBE). The RBE varies according to the system being studied, is largely dependent on LET, and is a comparison of the dose of each type of radiation required to induce the same quantitative change in the cell type or organ under investigation. For the purposes of the present discussion, LET is low for X and γ rays and highest for such fission products as neutrons. Thus, a particular effect is induced by lower doses of neutrons than of X or γ rays.

The physical conditions of the exposure (such as dose rate and degree of fractionation) and various environmental factors (temperature, atmospheric pressure, oxygen tension, etc.) greatly influence the response of tissues to irradiation. These factors have been reviewed by Mandl (1964) and will be considered only briefly in the present text. These physical conditions should clearly be stated in order to allow valid comparisons of results from studies using different species or types of radiations. When such details are not provided it is usually assumed that the irradiation was acute and nonfractionated at normal temperature and pressure in air.

Doses of radiation are expressed either in terms of the exposure actually delivered from the source (roentgen, "R," or "r" units) or preferably as the dose actually absorbed by the target tissue (measured in rads). For soft tissues, such as the ovary exposed to X or γ rays, 1 roentgen unit is equivalent to some 0.93 to 0.98 rads. Bearing in mind the limitations of radiation dosimetry, and also the errors inherent in quantitative biological procedures, it is sufficient for our purposes to regard roentgen (R) and rad units as having the same magnitude. It must be emphasized, however, that this relationship does not hold for exposures of hard tissues (bone or cartilage) or when fission products are used.

A proportion of the ionizations produced by a given dose of radiation will occur within cells, while the remainder will affect extracellular materials and tissue fluids. Although difficult to study directly, the irradiation of supporting fluids may result in the degradation of essential nutrients (especially proteins) or in damage to molecules which in their changed state may be toxic to the cells. So far the only evidence for such effects of exposure to radiation derives from studies of cells in tissue culture for which irradiated culture medium is said to be an inadequate support for normal function in unirradiated cells (Levinson, 1966; Szumiel *et al.,* 1971).

Radiation effects within the cells themselves are largely governed by their form and function; thus, lymphocytes are very sensitive while cells in the central nervous system of adult mammals are highly resistant. Radiation effects are usually studied by histochemical and autoradiographic techniques: morphological features alone may give little indication of cellular radiosensitivity. For example, oocytes at the diplotene stage of meiotic prophase undergo marked variations in their susceptibility while only showing minor changes in appearance (see Section III,B).

In general terms, cells undergoing division are highly sensitive during metaphase and also at the S phase of interphase (when DNA synthesis is occurring). That this is due largely to effects within the chromatin has been shown from studies in which localized areas of oocytes are exposed to a microbeam of radiation: exposure of the nucleus results in death or damage to the cell but has little effect on metabolic processes. Very large doses of X rays are required to damage or kill cells in which only the cytoplasm is exposed (Puck, 1960; Smith, 1964; Miller *et al.,* 1965).

III. RADIOSENSITIVITY OF GERM CELLS

A. Biochemical Changes

Irradiation-induced biochemical changes in the mammalian ovary are difficult to interpret since the germ cells are intimately associated with a variety of types of somatic cells which have different structures and functions and which are largely concerned with the secretion of steroid hormones. In the adult female the production of ovarian steroids is partly dependent upon the integrity of germ cells since the normal development of oocytes and follicles is interrelated and interdependent. However, some estrogen is secreted by the ovaries of rodents, at least for a time after the complete elimination of germ cells (Lacassagne *et al.,* 1962; Matsumoto, 1972: see Section V).

Studies in which the various cellular components of the ovary are separated using enzyme techniques have rarely been attempted and may be of limited value. Somatic cells in monolayer cultures undergo morphological changes; therefore, their biochemistry and radiosensitivity may differ from the situation in the intact gonad. Similarly, small oocytes isolated from their follicular cells usually undergo degeneration, while oocytes from antral and preantral follicles spontaneously resume meiosis (progressing from diplotene to metaphase I or II) when removed from the follicle into chemically defined media (see Edwards, 1966). This meiotic maturation in the absence of the usual inhibition imposed by the follicle prior to the so-called "LH surge" often leads to abnormalities in, or death of, the oocyte; only a small proportion of such oocytes develop normally after subsequent fertilization *in vitro* or *in vivo* (see Schuetz, 1969).

Following sublethal irradiation, normal biochemical function may be restored as the result of repair to enzyme systems or replacement of such damaged organelles as mitochondria (see Section III,B,1). There is also evidence for the repair of DNA in oocytes undergoing meiotic prophase even though DNA synthesis would not normally be possible. The use of autoradiographic techniques following the injection of [³H]thymidine has shown that "unscheduled" DNA synthesis occurs in irradiated oocytes of guinea pigs (diplotene stage: Crone and Peters, 1968; Crone, 1970) and in irradiated human and mouse spermatocytes at pachytene and diplotene (Kofman-Alfaro and Chandley, 1971; Chandley and Kofman-Alfaro, 1971).

In terms of radiation damage, it is important to consider effects within individual cell types as well as those affecting the organ as a whole. Thus, the estrogen-secreting capacity of the ovary is initially only slightly affected by doses of radiation which in rats and mice destroy all the germ cells. It would seem that estrogen is synthesized by the stroma and/or interstitial cells as well as by components of the follicle (see Westman, 1958; Lacassagne *et al.*, 1962). Furthermore, the secretion of progesterone during the luteal phase of the cycle in sheep is unaffected by X irradiation (Ichikawa *et al.*, 1968; see also Lacassagne *et al.*, 1962). The fine structure of the corpus luteum in the rat is also unaffected by X irradiation (Flaks and Bresloff, 1966).

It is clearly essential to distinguish between the direct effects of ionizing radiations and indirect effects mediated by other endocrine glands. When the whole body is exposed to large doses of radiation the pituitary gland and adrenals may be affected. Changes in the output of adrenal steroids are an obvious effect but the action of irradiation on the pituitary is less clear cut. Some workers have claimed that the pituitary is so resistant to irradiation that gonadal morphology and function are unaltered [see Hartman and Smith (1938) and Van Eck (1959) on the monkey], although recent studies on rats show that exposure of the head leads to changes in ovarian structure associated with a modification of estrous cycles (Matsumoto, 1972). There can be little doubt that injections of exogenous gonadotropins increase the rate of atresia* of oocytes in the irradiated hypophysectomized rat (Beaumont, 1969) and thus changes in the output of protein hormones following irradiation of the pituitary could also modify the rate of spontaneous atresia of germ cells.

B. Morphological Changes

1. Light Microscopy

The earliest morphological changes in X-irradiated germ cells seem to affect the nucleus and involve condensation of the chromosomes and wrinkling or disruption of the nuclear envelope (see Murray, 1931; Lacassagne, 1937; Beaumont, 1965; Baker, 1966a). Subsequently the nucleus becomes pyknotic and the cytoplasm eosinophilic, the latter probably being due to damage within the mitochondria and membrane systems (Goldfeder, 1965). The appearance of germ cells undergoing radiation-induced degeneration superficially resembles the process of spontaneous atresia which affects germ cells at all stages in their development, but the actual processes involved may well be very different (Ingram, 1962). Damaged germ cells are either eliminated from the ovary by phagocytosis or rapidly undergo repair to become indistinguishable from their normal counterparts (Mandl, 1959,

---

* In this chapter the term atresia is used in its loosest and most generally used sense as being the process of degeneration affecting oocytes and follicles at all stages in their development (but see also Volume I, Chapters 2 and 6).

1963). On the basis of differential counts of normal and degenerating germ cells, the proportion that is damaged following exposure to X rays is very high (Baker, 1966a). This observation indicates that either a large proportion of the damaged cells survive after repair of the radiolesion, or that the criteria used to identify atretic cells (condensation of chromatin, pyknosis, eosinophilia, etc.) are inadequate. That the latter statement is at least partly true is shown from studies in which different methods of fixation are employed: the number of atretic germ cells (whether radiation-induced or spontaneously occurring) is greatest when such harsh fixatives as Carnoy and Zenker are used, moderate with Bouin's aqueous fluid, and minimal following the well buffered procedures of fixation used for electron microscopy (osmium tetroxide or glutaraldehyde: Baker and Franchi, 1972b, and unpublished observations). It is also generally appreciated that the personal bias and mood of the investigator have a profound affect on the assessment of atresia since the criteria employed are highly subjective.

Radiation-induced changes in morphology affect germ cells at all stages in their development. Thus oogonia undergo degeneration either at interphase ("pyknotic oogonia") or during an ensuing mitotic division ("atretic divisions": see Beaumont and Mandl, 1962; Beaumont, 1962; Mandl, 1964; Baker and Beaumont, 1967). Similarly, radiation-induced atresia affects oocytes at leptotene, zygotene, or pachytene, although it is most common at the pachytene stage. Such cells become grossly pyknotic and their nuclear envelopes disrupt or disappear (the so-called "Z" cells first described by Beaumont and Mandl, 1962; see also Beaumont, 1962; Baker and Beaumont, 1967).

The earliest signs of damage in oocytes at the diplotene/dictyate (so-called "resting") stage of meiotic prophase are condensation of the nuclear chromatin and wrinkling of the nuclear envelope. The cytoplasm subsequently becomes eosinophilic and the whole cell shrinks (Murray, 1931; Lacassagne, 1937; Mandl, 1959; Baker, 1966a). The damaged germ cells either undergo repair or are eliminated from the ovary by somatic phagocytes within 0.5 to 2 days (mouse and rat: Murray, 1931; Mandl, 1959) or over a period of days or weeks (guinea pig and monkey ovaries: Baker, 1966a; Ioannou, 1969). The time at which degenerative changes occur appears to be independent of dose, the latter only influencing the proportion of oocytes which are affected (Mandl, 1959, 1964).

Morphological changes in follicular oocytes vary considerably with the stage of growth attained by the oocyte and its follicle and with the species. Primordial oocytes in rats and mice are very sensitive to radiation and doses which destroy the germ cells have little effect on the granulosa cells (Mandl, 1964; Peters, 1969; Baker and Neal, 1969; see also Section III,C). By contrast, primordial oocytes at diplotene in monkeys and guinea pigs are

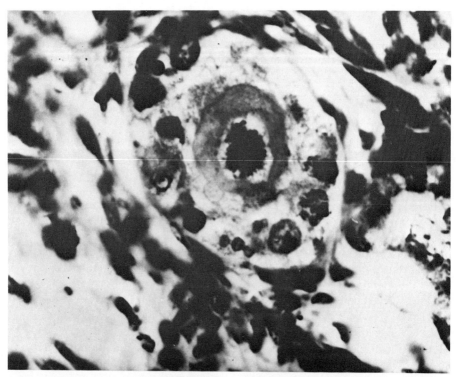

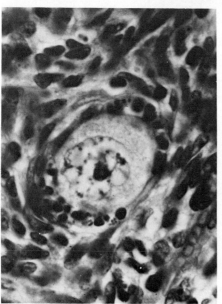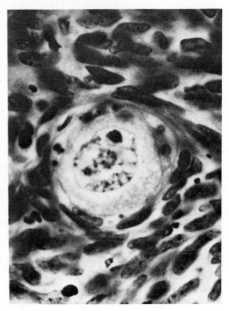

8

so resistant to X irradiation that the granulosa cells (which probably have a similar sensitivity to those in rats), are destroyed long before the oocyte (Baker, 1966a; Ioannou, 1969). Damage to the follicular epithelium is sometimes repaired by mitotic divisions in surviving cells, but those which fail to undergo repair rapidly become pyknotic or karyolytic (see Figs. 1–3), and the oocyte often appears to be too large for the follicle in which it resides (Baker, 1966a). When the dose of radiation is sufficient to cause the death of the oocyte, changes occur which superficially resemble those occurring spontaneously during each reproductive cycle when the oocyte becomes pyknotic, eosinophilic, and hyalinized.

The radiosensitivity of multilayered and Graafian follicles is in general much the same in all the species which have been studied, although mice are twice as sensitive as rats, guinea pigs, or monkeys (about 2000 R, cf. 4000 to 5000 R, respectively; see Mandl, 1964; Baker, 1971a). Cells within the membrana granulosa rapidly become pyknotic and may be shed into the follicular fluid. Furthermore, the cytoplasmic processes arising from both the oocyte and the granulosa cells (and which interdigitate within the zona pellucida) are withdrawn or damaged and the oocyte becomes eosinophilic and hyalinized (Mandl, 1964).

2. Electron Microscope Studies

Studies of the ultrastructural events associated with radiation damage have largely been confined to oocytes at the diplotene and dictyate stages of meiotic prophase and have been hampered by difficulties in distinguishing degenerative changes occurring spontaneously in oocytes from those induced by X rays. The few studies that have been reported have been confined to mice and rhesus monkeys (Parsons, 1962; Bojadjieva-Mihailova, 1964; Baker and Franchi, 1972b).

Parsons (1962) studied the effects on the oocytes of 4-day-old mice of 7 to 200 R whole-body X irradiation. He detected radiation damage within 7 hours of exposure to as little as 7 R, and the effect became more marked as the dose and postirradiation interval were extended. The earliest change was seemingly a transitory reduction in the number of mitochondria, after which the "normal" number was restored within a short time. The first permanent

---

Fig. 1. Monkey oocyte at stage II of follicular growth at an advanced stage of degeneration. The chromosomes are thickened and contracted to form a "knot," the cytoplasm is eosinophilic, and the granulosa cells are undergoing pyknosis. (5000 R; 7 days after irradiation). ×1275.

Figs. 2 and 3. Seemingly normal monkey oocytes surrounded by primordial follicles; the granulosa cells show various degrees of degeneration. (5000 R; 7 days after irradiation). ×850.

changes were detected in the nucleus and confirmed previous observations with the light microscope: the nuclear envelope became irregular in outline and the chromatin condensed. Marked shrinkage occurred in all oocytes, although some subsequently expanded and became karyolytic while others remained small and became pyknotic. Half of the population of "expanded" oocytes subsequently became normal in appearance within 24 hours of exposure, the remainder being eliminated from the ovary. In contrast, follicle cells appeared resistant even to the highest doses of X rays used (200 R) and no degenerative changes were observed in ovaries fixed 24 hours after exposure (Parsons, 1962). Bojadjieva-Mihailova (1964) found that the response of mouse oocytes to 600–1200 R varied considerably; some cells were grossly affected while others appeared normal. Mitochondria survive even in highly abnormal oocytes, but the proportion of the cytoplasm occupied by mitochondria and the endoplasmic reticulum increases considerably during the 24 hours following exposure to 1200 R X rays. The follicle cells were again shown to be unaffected by the doses of radiation used (Bojadjieva-Mihailova *et al.*, 1968).

The results of these studies of ultrastructural changes in X-irradiated mouse oocytes are difficult to interpret and merely confirm the marked radiosensitivity of germ cells in this species. Baker and Franchi (1972b), therefore, studied structural changes induced in oocytes of the rhesus monkey, a species in which germ cells are known to be highly resistant to X rays (Baker, 1966a; Baker and Beaumont, 1967). Biopsy specimens of ovary were removed at intervals from 7 hours to 30 days after exposure to 1000 to 12,000 R in an attempt to compare radiation-induced and spontaneous atresia. It was found that the two processes are essentially similar up to 15 hours after exposure, although a higher proportion of the irradiated oocytes undergo atresia. The changes consist of swelling of the endoplasmic reticulum, loss of cristae from mitochondria, and disruption or breakage of cellular membranes. However, radiation-induced degeneration differs from spontaneous atresia in that damage within the oocyte is usually preceded by changes within the surrounding follicle cells (especially with doses above 3000 R): in controls, the follicle cells are either affected concurrently or remain unaffected. The degree of damage to granulosa cells becomes more marked with increasing doses; 28 days after exposure to 5000 R relatively few follicles are seen, and the surviving oocytes sometimes appear to be abnormally large compared with the follicular remnants that surround them. These "giant oocytes" (Baker, 1966a; Baker and Franchi, 1972b) are often the size of cells which would normally be found in multilayered or Graafian follicles, although the follicle consists of only a single incomplete layer of flattened cells. These observations indicate that the oocyte can survive after exposure to large doses of radiation provided that a few granulosa cells remain undamaged. Sometimes the surviving follicle cells may partly restore the epithelial covering by repeated mitotic divisions. It is unlikely, however, that such follicles can complete this growth process to a point where ovulation could occur: large follicles are rarely found in ovaries within 30 days of exposure to doses in excess of 1000 R (Baker, 1966a).

Parkin (1970) has shown that the pattern of radiation-induced atresia observed in rat oocytes is very like that in the rhesus monkey, but that the doses of X rays required are very much lower (up to 300 R; see Mandl, 1964). However, the follicle cells remain unaffected and survive to phagocytose the remains of the oocyte. Phagocytosis by granulosa cells is also the process whereby spontaneously degenerating oocytes appear to be eliminated from the fetal and juvenile ovary (Franchi and Mandl, 1962; Baker, 1963, 1966b; Baker and Franchi, 1969, 1972a). Damaged oocytes in irradiated ovaries of monkeys are probably eliminated in a similar fashion, although the degeneration of the majority of the granulosa cells may prevent their direct participation in the process (Baker and Franchi, 1972b).

It was previously mentioned (Section III,A) that the initial radiolesion in germ cells is believed to occur in the nucleus and to involve chromosomal breaks and biochemical changes. The aim of most electron microscope studies of radiation-induced atresia has been to detect morphological changes in the chromosomes consistent with current hypotheses of the mechanism of action of ionizing radiations (see Section III,F). Unfortunately, this has not been possible, and only what appear to be relatively advanced stages of pyknosis have been detected. It seems probable that the electron microscope studies alone will not be effective in elucidating these problems, and that new developments in electron cytochemistry and autoradiography will be required.

C. Cell Population Studies

1. Embryonic Period

It was shown early in the history of radiobiology that exposure of embryonic and fetal gonads to ionizing radiations results in a reduction in the total population of germ cells (e.g., Brambell *et al.,* 1927a,b). It is now well established (Lacassagne *et al.,* 1962; Mandl, 1964) that the susceptibility of germ cells to radiation-induced cell death is dependent upon the developmental stage of the cell at the time of exposure and that it varies with the species. Primordial germ cells are relatively sensitive to X irradiation, some two-thirds of them in embryonic rats being destroyed within 5 days of exposure to 100 R (Beaumont, 1962, 1965, 1966). In the mouse, however, the sensitivity of primordial germ cells is low during migration but increases once the cells have colonized the developing gonad (Mintz, 1959; Henricson and Nilsson, 1970). The reduction in the population of primordial germ cells results from their death during migration, their failure to reach the developing gonad, or from a reduction in their rate of mitotic division (Baker, 1971c). There can be little doubt, however, that such germ cells are capable of mitosis after X irradiation. Beaumont (1962, 1965, 1966) has shown that exposure of rat embryos on the tenth day of gestation to 100 R results in a 70% reduction in the germ cell population by day 15.5 after conception but, by the eighteenth day of intrauterine life, the population of cells has been partly restored to some 50% of the number found in coeval controls. It would seem, therefore, that primordial germ cells which survive exposure to radiation are capable of undergoing more mitotic divisions than their normal counterparts in unirradiated controls (but see below for oogonia).

Little is known of the radiosensitivity of primordial germ cells in species other than mice and rats. Preliminary results for bovine embryos indicate that the cells are more resistant than those in rodents; large numbers survive after whole-body exposure of the mother to 400 R γ rays (Erickson, 1965, 1966).

Oogonia are most sensitive to X irradiation when their mitotic index is at a peak. The majority of the affected cells die at metaphase or anaphase of an ensuing division ("atretic divisions"), although some cells degenerate at interphase ("pyknotic oogonia": see Baker and Beaumont, 1967). Pyknotic oogonia are relatively rare in the ovaries of irradiated rats but are common in the rhesus monkey. There is some evidence that oogonia undergoing their final (premeiotic) division are more sensitive than during earlier mitoses (Beaumont, 1962, 1965, 1966; Mandl, 1964). Unlike primordial germ cells, destruction of a proportion of the population of oogonia is followed by only a partial restoration in numbers by mitosis. Beaumont (1965, 1966), for example, has shown that exposure of rats to 100 R on day 15.5 of gestation (when the majority of germ cells are oogonia) results in a reduction in the number of cells which is not restored with increasing age. There appears to be a finite number of mitotic divisions through which oogonia pass before becoming oocytes, and consequently the total number of oocytes is determined by the size of the oogonial stem-cell population.

Oogonia in the ovaries of fetal calves (Erickson, 1965, 1966), rhesus monkeys, and humans (Baker and Beaumont, 1967; Baker, 1969; Baker and Neal, 1969) are far more resistant than the corresponding cells in rodents. For example, Baker and Beaumont (1967) showed that exposure of fetal rhesus monkeys, aged 4–4.5 months, to 350 to 600 R X rays caused a 7% reduction in the population of oogonia: when the postirradiation interval was extended to 10–14 days the depletion of germ cells was only 4%, reflecting the survival and mitotic division of oogonia. The proportion of germ cells destroyed by exposure to X rays is dose-dependent: by increasing the dose to 1000 R the population of oogonia was reduced to 43% of that in coeval controls (Baker and Beaumont, 1967).

Baker and Neal (1969) showed that the radiosensitivity of oogonia in the ovaries of mice, rats, and rhesus monkeys remains virtually the same whether they are exposed to X rays in organ culture or *in vivo*. Furthermore, they showed that doses of 2000 to 4000 R eliminate only a proportion (80% during the sixth month of gestation) of the oogonia in human ovaries *in vitro* (Baker, 1969), and thus human oogonia may be even more resistant than those in the rhesus monkey.

2. Oocytes in Fetal Ovaries

The incidence of radiation-induced cell death in oocytes is known to be dependent upon the stage of meiotic prophase attained by the cell at the time of exposure (Mandl, 1964; Baker, 1971a,c). In the fetal rat, radiosensitivity decreases progressively between leptotene and the onset of the diplotene stage, after which it increases as cells progress through the early part of the dictyate stage (Beaumont, 1962, 1965). A similar response has been demonstrated for oocytes in mice, although in this species there is some confusion as to whether oocytes at the pachytene stage are sensitive or resistant. Peters (1961) and Peters and Borum (1961) claim that pachytene is a highly sensitive stage. They observed that on the day of birth untreated animals possess 45% of their "stock" of oocytes at pachytene with a further 5% undergoing spontaneous degeneration. Animals irradiated at this time and killed 1 day later contain no oocytes at pachytene although those at diplotene were unaffected. By contrast, W. L. Russell *et al.* (1959b) have demonstrated that the fertility of mice exposed to 300 R X rays on the day of birth is not markedly reduced, indicating that a large proportion of the oocytes at pachytene and diplotene survive to progress to the dictyate stage. The apparent discrepancy in the results of various authors could be explained if different strains of mice were used, since the timing of meiotic prophase and the radiosensitivity of germ cells is known to be strain-dependent (see Peters, 1969; Baker, 1971a,c, and Section III,F). Furthermore, the pachytene stage is of relatively long duration. Exposure to X rays towards the end of the period, when large numbers of oocytes are undergoing spontaneous atresia, may speed up the process of degeneration to such an extent that the oocytes would appear to be more sensitive. If, however, irradiation were to be carried out during early pachytene, when large numbers of cells would be expected to progress normally to diplotene, their radiosensitivity would probably be low (Baker, 1971a,c,).

Oocytes in calf, monkey, and human fetuses are far more resistant to radiation-induced cell death than are those in mice and rats. Exposure of pregnant cows to 400 R γ radiation has little effect on the population of oocytes during the first 119 days of intrauterine life when a variety of germ cell stages (primordial germ cells, oogonia, and meiotic prophase stages from leptotene to zygotene) coexist within the ovary (Erickson, 1965, 1966). Irradiation during the period 119 to 154 days *post coitum,* however, results in a 68% reduction in the population of germ cells. This highly sensitive period corresponds to a time when many oocytes are progressing from zygotene to pachytene, a transition during which there is a high incidence of spontaneous atresia in controls. The high sensitivity may thus represent merely a more rapid elimination of germ cells which would have been lost anyway, although the number of cells eliminated is probably increased.

The few reports on the radiosensitivity of oocytes in fetal rhesus monkeys show that doses of 200 to 400 R have no effect on the population of oocytes (Ozzello and Rugh, 1965). In a more extensive study, Baker and Beaumont (1967) found that none of the fetuses aged 2 months to full term (6 months) was completely devoid of germ cells after exposure to 350 to 600 R X rays and that a severe reduction in the number of germ cells occurred only after doses of 1000 R or more. The most marked effect on the population of germ cells was found after exposure during the fifth month of gestation, a period associated with a high rate of spontaneous atresia and with the cessation of mitotic activity in oogonia (Baker, 1966b; Baker and Beaumont, 1967).

The response of fetal human oocytes to X irradiation is strikingly similar to that in the rhesus monkey and calf. The only "direct" studies have been made by Meyer *et al.* (1969) who studied the effects of radiation in women who were themselves exposed when they were fetuses *in utero*: the results are considered in Section III,E. A second approach involved the irradiation of gonads from human fetuses aged 2 to 7 months *post coitum.* For each fetus, one ovary was exposed to X rays while the contralateral ovary served as control: both gonads were subsequently maintained in organ culture (Baker and Neal, 1969; Baker, 1969). A comparison was made of the structure of irradiated and control ovaries at various intervals after the onset of culture. It was found that the procedures for organ culture had little effect on the morphological appearance of the ovary; the number, appearance, and distribution of germ cells were relatively unchanged. Exposure to 2000–4000 R affected germ cells at all stages in their development, although oogonia undergoing mitosis appeared to be the most sensitive. Quantitative studies were confined to a fetus aged 6 months *post coitum* in which exposure to 4000 R caused a 65% reduction in the number of germ cells within 7 days of treatment. A similar response is achieved in fetal rhesus monkeys by exposure to 2000 R and confirms that human oogonia and oocytes are resistant to radiation and are among the most radioresistant germ cells known (Baker, 1969).

3. Oocytes in Juvenile Mammals

In most mammals, oocytes enter the diplotene or dictyate stages of meiotic prophase about the time of birth (see Baker, 1972, and Volume I, Chapter 2). This so-called "resting stage" (period of arrested development) persists throughout follicular growth and ends about 36 hours before ovulation (see Edwards, 1966). It is characterized by the presence within the nucleus of lampbrush chromosomes, which actively produce ribonucleoprotein (see Section III,F). The period of arrested development is thus of great importance to the radiobiologist since the oocytes constitute a uniform population of cells which are not capable of cell division and persist for a prolonged period of time (up to 45 years in women).

In mice and rats, the radiosensitivity of oocytes changes dramatically within the first few weeks after birth; sensitivity increases gradually to a peak value which varies between inbred strains, after which it decreases again. Thus, exposure of newly born mice of the Street strain to 20 R X rays induces the destruction of 50% of the oocyte population (mainly pachytene and diplotene), while treatment 1 week later eliminates 86% of oocytes at the dictyate stage (Peters, 1969). By the third week, sensitivity has increased to a maximum and only 6% of the oocytes survive a dose of 20 rads. Thereafter the oocytes become increasingly resistant to X irradiation, about 12% surviving for 24 hours after exposure during the seventh week (see Fig. 4). In the Bagg strain of mice, peak sensitivity is reached 1 week earlier (14 days *post partum*: Rugh and Wohlfromm, 1964). It is noteworthy that the rate of ovarian development also varies between these strains of mice. Thus, the peak number of medium-sized follicles in normal Bagg mice occurs during the second week whereas the comparable time in Street mice is about 21 days (Peters, 1969).

Biphasic age changes in the radiosensitivity of oocytes have also been reported for the rat (Sladić-Simić *et al.,* 1963; Hahn and Birrell, 1969; Matsumoto, 1971). Oocytes in the 10-day-old rat are much more sensitive to an exposure of 190 rads than are those in the day-old-rat or day-18 *post partum* rat (Matsumoto, 1971). It is not clear whether these changes also occur in other mammals, such as calves, rhesus monkeys, or man, although cursory observations indicate that they may (Erickson, 1965, 1966, 1967; Baker, 1966a).

The only other species for which data are available on the sensitivity of oocytes during the juvenile period are the cow and pig. In the calf, a dose of 300 to 600 R γ rays has little effect on the number of oocytes but exposure to 900 R eliminates half the population within a few days (Erickson, 1967).

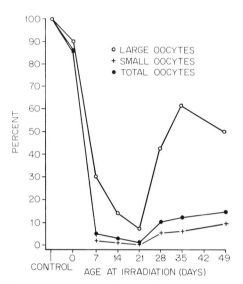

Fig. 4. Survival of oocytes in Street mice 7 weeks after irradiation (20 R) at different ages in early life. From Peters (1966).

Oocytes in prepubertal pigs seem to be more sensitive to γ radiation, 400 R eliminating 22% of the cells, while 600 R kills 80% (Erickson, 1967). Thus, the LD_{50} for pig oocytes is probably of the order of 500 R compared with 900 R for the cow (see Table II).

4. Sexually Mature Mammals

After the onset of sexual maturity, the response of oocytes to X irradiation becomes more stable and is influenced mainly by the stage of growth attained by the follicle, although species variation remains marked.

In mice and rats, oocytes within primordial follicles (primordial oocytes) are markedly more radiosensitive than those in multilayered follicles. The population of primordial oocytes is reduced exponentially in rats exposed to 29–4000 R X rays, the majority being destroyed within 18 hours of exposure to 300 R (Mandl, 1959; Mole and Papworth, 1966). But oocytes which have started to grow and have become surrounded by a complete layer of cuboidal granulosa cells (stage II of follicular growth) are far more resistant, some 26% surviving for 4 days after treatment with 4000 R (Mandl, 1959). By contrast, oocytes in "growing" and Graafian follicles normally account for only 8 to 10% of the population of germ cells in control rats: after exposure to 1000 R and 5000 R the proportion rose to 78% and 93%, respectively (Mandl, 1959; Mandl and Zuckerman, 1956a). Similar results were obtained by Kitaeva (1960) who exposed the ovaries of rats and mice to fractionated doses of X irradiation. She found that the depletion of primordial oocytes is roughly comparable in rats treated with 100 R and mice treated with 15 R, thus confirming the marked differences in the response to radiation of oocytes in the two species. These doses appear to be the LD_{50} values although variations due to age, strain, and fractionation of dose are known to occur (see also Karmysheva, 1963; Bocharov *et al.*, 1964; Oakberg and Clark, 1964; Mandl, 1964). The long term effects of X irradiation on the ovaries in rats were studied by Ingram (1958). Several months after exposure to 31 R and 630 R the population of oocytes had declined to 83% and 1%, respectively, of that in coeval controls.

It is almost impossible to plot dose–response curves for other mammalian species since variation between the number of oocytes/animal is very high. The best that can be done is to compare the response to X or γ rays of one ovary using the unexposed contralateral ovary as a control. However, the normal variation between the population of oocytes in the right and left ovaries of the untreated animal can be quite large, e.g., up to 20% in the rhesus monkey (Green and Zuckerman, 1951, 1954; Baker, 1966a), a point which is often overlooked in radiobiological studies (e.g., Van Eck, 1956).

TABLE II

Doses of Radiation (in R) Required to Kill All Oocytes[a]

Species	Primordial follicles[b]		"Growing" and Graafian follicles
	Juvenile	Adult	
Mouse[c]	15	50	2,000
	(7)	(10–15)	—
Rat[c]	100	315	4,400
	—	(100)	—
Guinea pig	—	15,000	—
	—	(500)	—
Pig	—	—	—
	(500)	(>500)	—
Cow	—	—	—
	(900)	(>900)	—
Monkey	?2,000	7,000	~5,000
	—	(5,000)	—
Man[d]	—	?5,000	—
	—	(?2,000)	—

[a] For references and time taken to eliminate germ cells, see text. (From Baker 1971a).

[b] Figures in parentheses are the LD_{50} doses.

[c] Varies with age, strain, and radiation procedure (see text).

[d] Depends on authority and on fractionated doses.

Reifferscheid (1910, 1914) was probably the first author to appreciate the marked differences in the response to radiation of the ovaries in different mammals. Although his studies were carried out before the advent of precise radiation dosimetry, he clearly showed that oocytes of mice are far more sensitive to radiation-induced cell death than those in rats, rabbits, and dogs, which in turn are more sensitive than those in primates. Many years were to elapse before his pioneering studies were confirmed (e.g., rat: Mandl, 1959; rabbit: Desaive, 1954; dog: Andersen *et al.*, 1961; primates: Baker, 1966a, 1971a; Baker and Franchi, 1972b; see also Table II).

It has been repeatedly shown that the primordial oocytes in the rhesus monkey are highly resistant to X irradiation; they are at least fifteen times less sensitive than those in the rat (Bocharov *et al.*, 1964). Van Eck (1956) confirmed Reifferscheid's (1910, 1914) observation that oocytes in primordial follicles are more resistant than those in multilayered follicles, although the design of her experiments, and some of the conclusions drawn from them, are open to criticism. She failed to consider variations in the size of the population of oocytes with age and believed that those oocytes which were destroyed by irradiation were replaced from the so-called "germinal epithelium," a conclusion which is no longer tenable (see Volume I, Chapter 2). In a more extensive study, Baker (1966a) exposed rhesus monkeys to unilateral ovarian X irradiation (doses of 1000 to 12,000 R), the contralateral ovary serving as control. It was found that a dose of 5000 R destroyed 50% of the oocytes within 30 days of exposure, while doses in excess of 7000 R were required to destroy *all* of the oocytes within this period. The minimum time taken to eliminate 50% of the germ cells appeared to be 4 days with a dose of 9000 R, while total elimination of the oocytes, in the same amount of time, needed 12,000 R. Oocytes surrounded by more than four layers of granulosa cells (multilayered or "growing" follicles) were sensitive to lower doses of radiation, only a small proportion surviving for 4 days after exposure to 4000 to 5000 R (Baker, 1966a).

Most of the other mammalian species for which quantitative data are available are intermediate in response between the mouse and the monkey. The $LD_{50/30}$ for primordial oocytes in the guinea pig is about 500 R, although 15,000 R is required to destroy *all* the oocytes (Ioannou, 1969; see also Oakberg and Clark, 1964). However, the response of primordial oocytes in this species is complicated by the fact that there are two distinct populations of cells differing in their morphology and radiosensitivity. The "large" type of primordial oocyte has a similar appearance to comparable cells at diplotene in the primordial follicles of other species, such as monkey and man (Ioannou, 1964; Baker, 1963, 1966a,b). But in the other ("contracted") type of primordial oocyte the chromosomes appear to be condensed toward the center of the nucleus. "Contracted" oocytes are the most radiosensitive: they are either killed by exposure to X rays or are transformed into the "large" type (Ioannou, 1969). This observation may explain the marked difference between the dose of radiation required to kill all the oocytes and that which destroys only half the population.

In general only qualitative data are available for the rabbit and hamster. Oocytes in these species seem to have radiosensitivities comparable to those in the rat (Lacassagne, 1913; Brooksby *et al.,* 1964; Moawad *et al.,* 1965). Thus, doses of less than 700 R destroy the majority of primordial follicles and also a proportion of those that have completed their growth phase (rabbit: Desaive, 1954; hamster: Brooksby *et al.,* 1964). The LD_{50} for oocytes in adult cows and pigs would be expected to be at least equivalent to that during prepubertal life (i.e., 900 R and 500 R, respectively; see p. 16 and Erickson, 1967). Preliminary results indicate that oocytes in the beagle may have a radiosensitivity comparable to that in the prepubertal pig (Andersen *et al.,* 1961; Baker, 1971a).

Oocytes at diplotene within multilayered follicles show a less variable response to ionizing radiations than do those at earlier meiotic stages. In the mouse (Peters, 1969; Desaive, 1957), rabbit (Desaive, 1954; Lacassagne *et al.,* 1962), and possibly the hamster (Brooksby *et al.,* 1964), such oocytes are destroyed by exposure to about 2000 R, while those of the other species for which data are available require 4000 to 5000 R (rat: Mandl, 1959, 1963; monkey: Baker, 1966a). According to Reifferscheid (1910, 1914) human oocytes may be expected to be as radioresistant as those of the monkey. However, few reliable data are available for human oocytes at any stage of follicular growth, and these relate mostly to radiotherapeutic exposures of women who are of menopausal age and generally show various degrees of pathological change within the ovaries (see Baker, 1971b).

Preovulatory maturation of oocytes within Graafian follicles commences with the onset of the so-called "LH surge" which occurs, for example, about 13 hours before ovulation in the mouse, and 36 hours in man (see Volume I, Chapter 2). Thereafter, radiosensitivity varies according to the stage of meiosis attained by the oocyte and increases by a factor of ten as the cell passes from the dictyate stage to metaphase I (Mandl, 1963). It is, however, difficult to distinguish between the effects of radiation on the oocyte itself and those affecting somatic components of the follicle. Effects within the oocyte are best studied by means of the induction of dominant lethals (see Section III,E).

D. Functional Changes

The production of viable young involves such factors as fertility, fecundity, sterility, and litter size, which are known to change with increasing age due largely to the reduction in the total "stock" of oocytes by atresia.

Ideally, studies of functional changes after exposure to X rays should involve counts of the total number of offspring produced in the entire reproductive life-span of the animal, and should take account of the loss caused by embryonic mortality (see Mole, 1959; Mandl, 1964). In practice, however, these studies can only be made for such laboratory rodents as mice and rats. In other species (e.g., man, monkey, cow, sheep), a number of factors such as longevity, variation between individuals, small litter size, and sporadic breeding preclude a full study of reproductive capacity. Nevertheless, the results of investigations into various facets of reproductive function (e.g., cycle length, incidence of ovulation, and conception) have provided useful comparative data (see below).

1. Induction of Sterility

It is well established that large doses of X rays lead to early sterility by exhausting the supply of ovarian follicles (see Brambell and Parkes, 1927; Murray, 1931; L. B. Russell et al., 1959; Lacassagne et al., 1962). But the effect of small doses is less clear, since animals continue to breed so long as viable oocytes are ovulated. In mice, exposure to doses of 20 to 160 R result in total sterility, although the time at which the effect occurs varies between strains and is related to the age of the animal at the time of exposure (Furth, 1949; W. L. Russell et al., 1959b; Peters and Levy, 1963; Rugh and Wohlfromm, 1964; Peters, 1969; Nash, 1969). Nash (1969) found that exposure of pregnant mice to 160 R on the sixth, tenth, or fourteenth days of gestation led to the birth of sterile offspring, while treatment on the seventeenth day after conception merely impaired fertility, as indicated by an increase in the number of sterile matings with males of proven fertility.

Radiation-induced changes in the reproductive performance of postnatal mice closely parallels changes in sensitivity in terms of cell death. Thus, 25% of CF_1 mice exposed to 30 R on day 14 after birth became sterile compared with only 10% when the dose was administered 1 week later (Rugh and Wohlfromm, 1964). Similarly, exposure of Street mice on days 14 or 21 post partum results in 54% and 75% sterility, compared with 0% on the day of birth, 13% on day 7, and 5% on day 28 (Peters and Levy, 1964). Peters (1969) has shown that Bagg mice are more resistant to the induction of sterility by X rays than are those of the Street and CF_1 strains, a dose of 20 rads having no effect between days 1 and 50, apart from day 14 when 5% of the animals are affected.

One of the most obvious signs of sterility in irradiated mammals is the cessation of estrous cycles and atrophy of the vagina (e.g., rabbit:

Lacassagne, 1913; Lacassagne *et al.,* 1962). When the effects of irradiation are less extreme, the changes are reversible and estrous cycles recommence after the period of sexual inactivity during which the animals will not accept the male. A similar situation exists in the guinea pig, although the occurrence of estrous cycles is not necessarily a sign of fertility (Genther, 1934). But the mouse is exceptional in that animals whose irradiated ovaries contain no oocytes, follicles, or corpora lutea, continue to experience estrous cycles of approximately the same duration as those in control females (Parkes, 1928). The question thus arises as to which cellular component in the irradiated mouse ovary produces estrogen in a cyclical fashion to permit the continuation of estrous cycles. Shortly after exposure to X rays, cords of epithelial tissue arise within the ovary (seemingly from the coelomic epithelium covering the organ: Brambell *et al.,* 1927a,b). These authors were able to show that when the epithelioid tissue was healthy and glandular, estrous cycles persisted; but when the cells were small, compressed and vacuolated, irregular cycles or persistent vaginal cornification occurred. In some animals the epithelioid tissue resembled lutein cells and the cycles ceased altogether. Brambell and his coworkers therefore concluded that the cells derived from proliferations of the coelomic (or "germinal") epithelium produce estrogen in response to pituitary hormones and so maintain the cycle. Their views have largely been supported by numerous other investigations (see Mandl and Zuckerman, 1956b; Lacassagne *et al.,* 1962). As would be expected from this hypothesis, the cyclical changes cease immediately after bilateral ovariectomy and the mice become permanently anestrous.

The ovaries of rats exposed to 4000 R are devoid of oocytes but the animals experience a series of more or less normal estrous cycles for up to 40 days before entering a phase of prolonged estrus which may last for 14 weeks (Mandl and Zuckerman, 1956a). The animals then become anestrous. The results of a series of experiments, including ovariectomy and hypophysectomy, led Mandl and Zuckerman to conclude that estrogen is produced by the ovaries of X-ray sterilized rats in the same way that it is in mice. More recently, Matsumoto (1971, 1972) has shown that exposure of the rat ovary to 190 R X rays on the day of birth or on day 18 *post partum* has little effect on the duration of estrous cycles occurring within 80 days of treatment, but that exposure on day 10 induces considerable variations in the cycle and sometimes continuous estrus. It is likely, however, that the animals used in these experiments were not completely devoid of ovarian follicles and that the follicular cells, as well as the nodules of epithelioid tissue within the ovary, may have been producing estrogen. It is interesting to note, however, that Matsumoto (1971, 1972) found that adult rats exposed to 190 R whole-body X irradiation often showed persistent cornification of the vaginal epithelium, while others, whose ovaries alone were exposed, exhibited normal estrous cycles. This would indicate that exposure of the pituitary and/or other organs of the body can in some way mediate the estrogen effect.

2. Fertility Span

The interval between the birth of the first and last litters (fertility span) varies between different strains of mice (Jones and Krohn, 1961; Ehling, 1964). Irradiation of the ovary results in a shortening of the fertility span, which also varies with the age at exposure. In mice of the Street strain exposed to 20 R X rays, sensitivity increases from the day of birth (40% reduction) to day 17 (72%) and then decreases to day 49 when the reduction in fertility span is 18% (Peters and Levy, 1964). Similar results have been obtained for the Bagg (Peters, 1969) and CF_1 (Rugh and Wohlfromm, 1964) strains, although the effect in Bagg mice was less marked than in the other two strains. The CF_1 and Street mice remained fertile for only 6 to 12 weeks after irradiation while Bagg mice were able to produce viable young for an average of 20 weeks (see Peters, 1969).

3. Average Number of Litters

Exposure of juvenile mice to low doses of X rays induces a reduction in the number of litters produced. Mice of the Street strain exposed to 20 R on day 14 *post partum* produce fewer litters (0.8) than animals exposed on day 49 (7: Peters and Levy, 1964). Similarly, L. B. Russell *et al.* (1959) showed that exposure to 85 R during the second week induced sterility, although animals treated as adults gave birth to an average of 14.2 litters. Adult mice are largely unaffected by doses of less than 20 R although 30 R causes a 50% reduction in the number of litters (Peters and Levy, 1964; Lindop *et al.*, 1966; see Fig. 5).

4. Litter Size

The number of offspring in a litter depends on the number of eggs which are ovulated, fertilized, and, after fertilization, do not undergo early embryonic death. The possibility of early embryonic death is greatly increased following X irradiation owing to the induction of dominant lethals (see page 28). Litter size in unirradiated controls also becomes reduced with increasing age (e.g., Biggers *et al.*, 1962a,b), and thus comparisons of irradiated and control animals must be made at identical ages. In juvenile mice, irradiation generally results in a reduction in litter size (Ingram, 1958; Oakberg, 1966), although Lindop *et al.* (1966) have reported than 1- to 4-

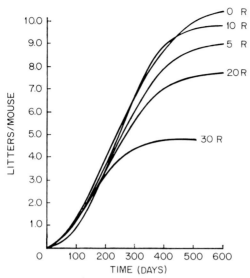

Fig. 5.   Relationship between different doses of X rays given to 28-day-old SAS/4 mice and the number of litters born. From Lindop *et al.* (1966).

week-old mice treated with 5 to 10 R produce larger litters than controls. There can be little doubt that exposure of adult mammals to small doses of X rays (up to 50 R, depending on species) results in a transitory increase in the number of ovulations (e.g., mouse: Russell and Russell, 1956; Karmysheva, 1963; Hahn and Morales, 1964; rat: Mandl, 1965; Hahn and Birrell, 1969; rabbit: Moawad *et al.*, 1965; Feingold and Hahn, 1972; hamster: Hahn, 1972). This radiation-induced superovulation is similar to the now discredited practice of inducing ovulation in infertile women with low-dosage irradiation of the ovary (see Israel, 1952, 1966; Kaplan, 1958, 1959).

Maximum litter size is reached sooner, and the onset in the decline in the number of offspring with increasing age occurs earlier, in irradiated animals than in controls (Brown *et al.*, 1964; Oakberg, 1966; Byskov and Peters, 1972). Gowen and Stadler (1964) irradiated male mice on the forty-sixth day after birth and showed that 20 R induced a pronounced reduction in litter size in the female offspring which continued throughout the twelve deliveries that occurred in some of the animals. Mice exposed to 80 R at the same age had even fewer young in the first three litters and were sterile thereafter. Peters and Levy (1964) showed that a single dose of 20 R administered at various ages from 35 to 49 days had little effect on the litter size of Street mice until after the fourth litter when a gradual reduction occurred. However, animals exposed between 7 and 28 days showed a reduction in the number of young in all litters.

Peters (1969) has briefly reviewed the factors responsible for controlling litter size in normal and irradiated mice. In young adults, the prenatal loss of embryos is normally negligible (McLaren and Michie, 1959) but subsequently increases with advancing age due to changes within the uterus (e.g., Biggers *et al.*, 1962a,b; Finn, 1963; Talbert and Krohn, 1966). Litter size during the early period of sexual maturity thus represents the total number of eggs ovulated, fertilized, and implanted. This view confirms the observations of Jones and Krohn (1961), who showed that the number of large follicles and of ovulated eggs is not reduced with increasing age. By contrast, the decrease in litter size in irradiated mice is due to a reduction in the number of preovulatory follicles (Oakberg, 1966; Lindop *et al.*, 1966). It is unlikely that the uterine factors which affect controls would have an influence on litter size in irradiated animals, since the latter cease to breed long before uterine aging becomes evident (Lindop *et al.*, 1966; Peters, 1969). This view has been confirmed by means of experiments involving egg transfer; oocytes show the same response to irradiation in irradiated uteri as they do in nonirradiated uteri (Lin and Glass, 1962a,b; Glass and McLure, 1964). Other evidence of radiation affecting the oocyte rather than the uterus or oviduct derives from the work of Brent and Bolden (1967) who showed that only ovarian irradiation resulted in embryonic mortality. Exposure of the uterus and oviducts of rats had little if any effect. Conversely, Chang and Harvey (1964) have shown that in rabbits exposure of the uterus to radiation results in a cumulative effect, although the most marked effect occurs in the egg and is expressed in terms of the early death of embryos (Chang and Hunt, 1960). Valentini and Hahn (1971) exposed the right uterine horn of rats to 200–400 R and showed that the survival of embryos within this horn was improved if the left lower quadrant of the body was shielded from the beam. They postulated that at least a proportion of the embryonic deaths that occurred in the irradiated uterus resulted from "maternal dysfunction" rather than as a direct effect of the radiation on the eggs (but see Gibbons and Chang, 1973). It is suggested that the maternal dysfunction may arise from defects in the uterus, the ovaries (e.g., reduced steroid output), or even intestinal damage (toxic factors from extensive ulceration of the mucosa: see Valentini and Hahn, 1971).

5. Total Reproductive Capacity

Rugh and Jackson (1958) irradiated fetal mice on the sixteenth day of gestation and found that 100 rads reduced the number of young born over a 9 month period to about half that in controls. The greatest reduction in reproductive capacity occurred in animals exposed on day 13.5 *postcoitum,* a time when the majority of germ cells are oogonia undergoing mitosis

(Russell *et al.,* 1960). In the Street and Bagg strains, irradiation between days 1 and 35 *postpartum* results in a decrease in reproductive capacity (Peters and Levy, 1964), but Street mice produce fewer offspring and thus show the greater effect. As with other criteria of reproductive capacity, the age of the animal has a marked effect on the observed response. In juvenile mice of the Street strain the maximum decline occurs during the third week after birth whereas in Bagg mice this occurs 1 week earlier (Peters, 1969; see Fig. 6).

When fertility is expressed as the total number of young produced, exposure of mice aged 8–10 weeks to 25 R X rays has no effect during the first 3 months after treatment. However, the animals subsequently become sterile at an age corresponding to two-thirds of the reproductive life-span in controls (Mole, 1959). The total number of offspring born to the irradiated mice was half that of the controls, although fertility during the first 3 months was not impaired.

Changes in reproductive capacity are largely governed by the size of the population of oocytes within the ovary. Ingram (1958) correlated the number of oocytes/ovary with the decline in fertility and concluded that the two phenomena were directly related. Peters (1969) has suggested that this view may be only partly true since in young animals a better correlation is found between total reproductive performance and the number of oocytes in multilayered follicles within the ovary. A minimum number of these follicles has to be present at maturity to permit offspring to be born: virtually

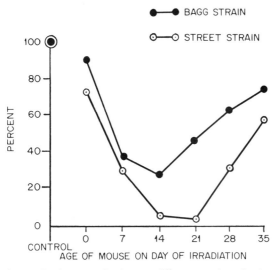

Fig. 6. Total reproductive capacity in two different strains of mice after irradiation with 20 R in early life. From Peters (1968).

no young are born when the number of large follicles is less than 15% of the total population, compared with about three-quarters of the possible maximum of young when 60 to 90% of the follicles are enlarged (Peters, 1966). It has yet to be established, however, that a similar relationship holds with advancing age.

6. Reproductive Potential in Mammals Other than Mice and Rats

Reports on the effects of ionizing radiations on reproductive potential in other mammals are limited. Brooksby *et al.* (1964) have indicated that exposure of hamsters to 100–300 R results in a decrease in the average litter size from eight to ten young/litter in controls to two in irradiated females (see also Hahn, 1972). In contrast, the number of young produced by female beagles exposed to 300 R (which may be the $LD_{50/30}$ dose) was apparently unaffected over a 4-year period of study (Andersen, 1964a,b; Andersen *et al.*, 1961). Whole-body exposure of sheep to 300–500 R X rays resulted in a sharp decline in reproductive capacity in those sheep which survived the treatment (Terry *et al.*, 1964). The ewes were left for 2 months after irradiation before mating was attempted. There were eight ewes/ group, and of the survivors (four at 450 R, six at 400 R, six at 350 R, and seven at 300 R) only two produced viable young, but the authors do not report whether the irradiated animals bred on more than one occasion. However, fractionation of the exposure (200 R given as 25 R at 8 R/min every 28 days) did not adversely affect the first breeding (Terry *et al.*, 1964). (See also Addendum, p. 45.)

Studies of reproductive function in irradiated primates have generally been confined to investigations of the menstrual cycle and the incidence of pregnancy. Exposure of the whole reproductive tract (van Wagenen and Gardner, 1960; Yakovleva and Novikova, 1963), or of the ovaries alone (Baker *et al.*, 1969b), has little effect on the duration of menstrual cycles unless the dose exceeds 4000 R. Amenorrhoea rapidly ensues when the population of oocytes is reduced by doses of 5000–7000 R to below 50% of the control level (Baker, 1966a; Baker *et al.*, 1969b). Irradiated monkeys ovulate eggs of normal appearance and the animals become pregnant when mated with untreated males. The incidence of live births is low if the radiation is given as a whole-body exposure (Yakovleva and Novikova, 1963), or to the entire reproductive tract (van Wagenen and Gardner, 1960), but approaches the normal value for captive animals when only the ovaries are exposed to X rays (Baker *et al.*, 1969b). Thus of twenty-four pregnancies occurring in monkeys exposed to 500–750 R X or γ rays, only ten live offspring were born (Yakovleva and Novikova, 1963). The remainder consisted of eleven abortions and three stillbirths, a far higher incidence than occurred in control monkeys in their colony. Similarly, van Wagenen and Gardner (1960) obtained only one viable offspring from twenty-one pregnancies in animals whose pelvic region was exposed to 600 R. Abortions occurred most frequently between days 39 and 49 of gestation. The authors suggested that the high incidence of abortions resulted from exposure of the oviducts and uterus, although the data of Brent and Bolden (1967) indicate that, at least in the rat, these structures are resistant to radiation (see page 24). This suggestion clearly warrants further attention, since irradiation of the ovaries alone with 500 to 2000 R had little affect on the incidence of pregnancy in the experiments of Baker *et al.* (1969b): three animals treated with 500, 1000, and 2000 R became pregnant, while two monkeys exposed to 4000 R aborted.

It is well established that therapeutic exposures of the pelvic region of women results in amenorrhea which may be permanent if the total dose (fractionated) exceeds 400 R. The duration of the period of amenorrhea varies with the dose of radiation and the age of the patient at the time of exposure. Most of the women who are subjected to radiotherapy are over the age of 40 years, and it is thus not surprising that amenorrhea and sterility rapidly ensue. Irradiation possibly accelerates the rate of spontaneous atresia of oocytes and thus causes the menopause to occur a few years earlier than normal (see Baker, 1971c). Treatment of young women (under 35 years of age) with 400 to 750 R X rays (fractionated) is followed by amenorrhea of relatively short duration, and pregnancy may ensue before the period of X-ray induced amenorrhea has ended (e.g., up to 10 years after exposure to doses of as much as 2000 R; Gans *et al.*, 1963; Vuksanovic, 1966; see also Baker, 1971c). These fractionated doses of radiation used for the treatment of various disorders in women are difficult to equate with the acute doses used in experiments on animals. It is clear, however, that the human ovary is more resistant than that of the mouse or rat, its response corresponding more closely to that of the rhesus monkey.

E. Genetic Effects

Ionizing radiations induce structural and biochemical changes in germ cells which can result in mutations or chromosomal aberrations. Most of the mutations are deleterious and recessive and only affect the phenotype of the individual when occurring on both chromosomes of a given pair. However, dominant mutations are often lethal and result in the death and resorption of the embryo about the time of implantation ("dominant lethals": see below).

Studies of radiation genetics are complex and require large numbers of animals. For this reason such studies in mammals have largely been confined to mice for which inbred strains and adequate genetic markers are available. However, some of the results obtained are at variance with those from studies in which *Drosophila* and other insects were used (see Searle, 1962, 1966; Sobels, 1963). One of the main problems arising in experiments involving radiation genetics in mice is that doses of radiation which induce a detectable rate of mutations also kill most of the oocyte population, and sterility ensues after the production of only one or two litters (Russell, 1963). One possible solution to this problem involves the use of wild mice which are three times more resistant to the induction of reciprocal translocations than are the usual laboratory strains (Searle *et al.*, 1970).

1. Dominant Lethals

Irradiation of gametes results in the death of a proportion of the embryos shortly before or immediately after implantation. It is generally accepted that this is the result of dominant mutations and chromosomal aberrations, which are loosely grouped as "dominant lethals." The results of a number of studies on male and female gametes have shown that the majority of dominant lethals are due to the unbalanced products of translocation (Russell, 1964; Ford *et al.*, 1969; UNSCEAR, 1966, 1971), which arise from faulty recombination following the breakage of chromosomes.

The proportion of embryos that die increases with the dose of radiation employed (Russell *et al.*, 1954; Bateman, 1958). It has been suggested that the duration of embryonic survival is inversely related to the number of chromosomal breaks induced in the gamete (Lea, 1955). This explanation is supported by Bateman (1958) who showed that all embryos derived from gametes carrying a single chromosomal break, and 25% of those with multiple breaks, are capable of forming deciduomata and therefore represent postimplantation deaths.

The incidence of radiation-induced dominant lethality is dependent upon the stage of meiosis attained by the oocyte at the time of exposure: the highest frequency is found after treatment at metaphase I, while the lowest occurs at the dictyate stage (L. B. Russell and Russell, 1956; W. L. Russell and Russell, 1959; Mandl, 1963; Edwards and Searle, 1963; Harvey and Chang, 1963; Chang and Harvey, 1964). The frequency of dominant lethals also varies between species, although the effect is complicated by the stage of germ cell differentiation at which irradiation takes place, and also by sex. In male mammals, exposure to 500 rads X rays results in a lower yield of dominant lethals in guinea pigs, hamsters, and rabbits than in the mouse, although the four species show a comparable pattern of sensitivity relative to stages of germ cell differentiation (Shapiro *et al.*, 1961; Searle, 1962; UNSCEAR, 1971). The mature diplotene oocytes of the hamster and guinea pig, however, are more sensitive to the induction of dominant lethals than are mice (Lyon and Smith, 1971). More data are needed to assess the significance of these variations with sex and species.

2. Translocations

Little reliable information is available about the induction of translocations in mammalian oocytes apart from that which results in the death of the embryo (see above). Léonard and Deknudt (1967) studied the induction of chromosomal rearrangements by irradiating mouse embryos with 100 R during the preimplantation period (pronuclear stage). The effects of irradiation were assessed by studying chromosome preparations of the testes of the male offspring subsequently born to the males and females previously irradiated *in utero*. None of the sons of these irradiated females showed any signs of chromosomal abnormality, whereas the offspring of the irradiated male parents had translocations in spermatocytes.

Only two studies have so far been reported in which the induction of translocations has been examined in the oocytes of adult mice. Russell and Wickham (1957) describe a slight decrease in the fertility of male mice whose mothers had been exposed to 400 R X rays: one male out of 320 experimental animals was "semisterile" (a term used by geneticists to imply reduced litter size in offspring of irradiated mice; see Lüning and Searle, 1971), and produced "semisterile" offspring. The authors concluded that this male offspring (and its progeny) was heterozygous for a reciprocal translocation. More recently, Searle (1971) and A. G. Searle and C. V. Beechey (unpublished observations, 1971, cited in UNSCEAR, 1971) have carried out a large-scale study involving the irradiation of late dictyate oocytes with either fast neutrons (100 to 200 rads) or X rays (300 rads). There was no evidence of inherited "semisterility" in the neutron series or with male progeny in the X-ray series. However, eight of the 293 daughters of the mice treated with X rays were judged "semisterile" on the criterion of litter size, and four of these showed definite evidence of being translocation carriers. There is no explanation as yet for the low induction rate of translocations following exposure to irradiation, but it may be that the tight pairing of the chromosomes and their high efficiency for repair are important factors.

3. Chromosome Deletion

Besides inducing the breakage of chromosomes, exposure to ionizing radiations may result in the loss of whole chromosomes or of fragments during an ensuing cell division. The loss of an autosome is probably lethal owing to the resulting loss of a large number of genes. However, the loss of a Y chromosome need have no adverse effect since it carries very few genes. Similarly, in female offspring, the loss of one X chromosome may have few consequences because one of the two X chromosomes in female mammals is normally inactivated (Lyon, 1961). Thus, XO mice (but not YO) are viable and fertile (Russell, 1964), although the XO condition in man leads to Turner's syndrome characterized by sterility due to the elimination of the germ cells shortly after birth (Singh and Carr, 1966, 1967; see also Volume 2, Chapter 5).

Russell (1964) and Searle (1971) have shown that mice are highly sensitive to the induction of X chromosome deletion, the effect being proportional to dose and increasing with dose rate. Russell (1964) exposed mice to 400 R X rays (at 80 R/minute) or γ rays (0.6 R/minute) at a time when the oocytes in multilayered follicles were at the dictyate stage. The mice were subsequently mated with untreated males whose X chromosomes carried the dominant sex-linked gene *Greasy* (*Gs*). Of the female progeny, twenty-one of 6674 in the high-dose group (0.31%) and fifty out of 7576 (0.66%) in the low dose-rate group were found to be XO, compared with a spontaneous occurrence of 0.05%. Similarly, Searle (1971), who exposed mouse dictyate oocytes to 200 rads fast neutrons, obtained one definite and one presumptive case of XO out of thirty-seven animals tested.

The X chromosome may be eliminated from the oocyte by two possible mechanisms: nondisjunction, in which case one of the cells resulting from a meiotic division has two X chromosomes while the other has none, or by breakage of chromosomes and elimination of fragments at the next division. Sobels (1968) regards chromosomal breaks as the main source of XO offspring, since only one instance of XXY (which with nondisjunction would be in equal numbers with XO) has been recorded for the mouse.

4. Induction of Point Mutations

It is now generally accepted that the frequency of spontaneously occurring mutations in laboratory strains of mice is of the order of 1.4×10^{-6} per locus per gamete (see Schlager and Dickie, 1966, 1967, 1971; Batchelor *et al.*, 1969; Russell, 1971; UNSCEAR, 1971), and estimates are available of the doses of radiation required to induce a comparable yield of mutations and to double this rate (e.g., Lüning and Searle, 1971).

In irradiated mice, the induction of mutations is lower after exposure to fractionated doses than after a single acute exposure (W. L. Russell *et al.*, 1958, 1959a; Russell, 1963, 1965; Russell and Kelly, 1966). This dose-rate effect holds true for male and female germ cells, although the response is most pronounced with oocytes (Sobels, 1963). Russell (1967a,b,c) and Russell (1968) have discussed the mechanism of the dose-rate effect and have suggested that the results can be explained in terms of "one-hit" mutational events (see UNSCEAR, 1971). The rate of mutation after exposure to an acute dose of 50 R is only one-third of that following 400 R; but when 400 R is given as eight fractions of 50 R the effect is only half that with the acute exposure to 400 R. Furthermore, exposure to fast neutrons is almost twenty times more effective in inducing specific locus mutations than is chronic γ radiation (Batchelor *et al.*, 1969; Searle and Phillips, 1968, 1971). Thus, no mutations were recorded in either the control series of mice or in those exposed to 412 rad γ rays over a 12-week period, and only one mutation was scored among the progeny (32,221) of mice exposed to 79.7 rads fast neutrons (with 57.8 rads γ-ray contamination). The mutation rate has been estimated to be 0.3×10^{-7} per locus per rad per gamete (dictyate oocytes) for neutron irradiation, compared with a figure for γ rays which is not significantly different from that of controls (over 100,000 progeny tested; see Batchelor *et al.*, 1969; UNSCEAR, 1971). These results confirm the work of Russell (1969) who has found only three specific locus mutations amongst the 258,400 progeny of mice exposed to 412 R chronic γ radiation. This figure falls well within the limits of the spontaneous mutation frequency and confirms the supposition that mouse oocytes at the dictyate stage are highly radioresistant by this criterion (but compare with cell death: see page 16). These findings clearly have an important bearing on the assessment of the genetic consequences of radiation in the human populations for which the majority of exposures are fractionated.

It has been established that the yield of mutations increases more or less linearly with dose and hence that there is no "safe" dose below which mutations will not occur. Nevertheless, the incidence of radiation-induced mutations has proved to be much lower than was originally expected and depends to a large extent on the interval between irradiation of the oocyte and the time of fertilization (see UNSCEAR, 1966, 1971). For example, Russell (1957, 1965, 1967a,b,c), working with mice, found no mutations in 71,324 progeny conceived more than 7 weeks after irradiation, whereas eleven mutations were found among 169,325 offspring born within the 7-week interval. The former progeny were derived from oocytes which were in primordial follicles at the time of exposure, and the latter from oocytes already involved in follicular growth. Russell (1971) has suggested three possible reasons for the low yield of mutations in primordial oocytes: (1) they may be resistant to the induction of mutations; (2) they have a high intrinsic efficiency for repair or selection; and (3) the damaged oocytes are rapidly eliminated before or during maturation. The third possibility would seem to be highly probable since mouse oocytes are eliminated within 18 hours of exposure to X rays (see Section III,C) and only a small proportion of the total population is actually ovulated (e.g., about 400 in women). It may well be that the damaged oocytes are eliminated during a process of "germinal selection" in the period of follicular growth (Searle, 1962). This may account for the observation that fewer mutations are found in the germ cells of fetal mice than in adults (Searle, 1962; see also below. It is also known that irradiated mice make far better use of their "stock" of oocytes after irradiation than do controls (see Ingram, 1958; Oakberg, 1966).

The results of recent autoradiographic studies by Oakberg (1967, 1968) suggest that mutational frequency may be related to the metabolic activity of the oocyte. In terms of the incorporation of [^{3}H]uridine into ribonucleic acid (RNA), the small oocytes at diplotene are metabolically very active but have a low mutational frequency, whereas the large oocytes are unlabeled and yet have a high rate of mutation induction. Oakberg concludes that because the capacity for repair is probably closely correlated with metabolic activity, the change in mutational frequency with oocyte size and stage of follicular growth may represent a change in the ability to repair radiolesions. While such a relationship may hold true for the oocyte at diplotene, it certainly is not a general phenomenon. Thus, high metabolic activity is maintained during early embryonic development, including the phase of migration and active proliferation of primordial germ cells, which nevertheless show a high level of mutational activity (Searle and Phillips, 1971). These authors exposed mouse embryos *in utero* to neutron irradiation at a dose of 108.5 rad (plus 20.5 rad γ ray contamination) at 0.011 rad/minute. The mice were subsequently mated at 8 weeks of age to a suitably "marked" test stock and the progeny were scored for mutations at specific loci. Searle and Phillips (1971) found clusters of mutations in the gonads of the survivors which permitted the mutation frequency per locus to be calculated at 5.4×10^{-5} in male and 6.4×10^{-5} in female primordial germ cells. If one allows for dose attenuation owing to the depth of the fetal germ cells in relation to maternal plus fetal tissues, the rates are one-third higher: in either event they do not differ significantly between the sexes. The rates are less than in mature oocytes irradiated at 0.17 rad/minute but much higher than after chronic exposures (total of 79.9 rad at 0.0007 rad/ minute).

Carter (1960) investigated the effect of acute X irradiation (300 R) on mouse fetuses aged 12.5 to 15.5 days *postcoitum* and found a mutation frequency of 1.02×10^{-7} per locus: this is higher than after low-dose γ irradiation of oocytes in adult mice. In a further study, Carter *et al.* (1960) found that exposure to 200 R X rays at the pachytene stage (day 17.5) induced a mutation frequency of only 0.7×10^{-7} per locus, which is lower than that obtained in adults (4.2×10^{-7} for X rays). It would therefore seem that dictyate oocytes in multilayered follicles are more sensitive than oogonia and oocytes at pachytene, which in turn are more sensitive than oocytes in primordial follicles (diplotene/dictyate stage). It is also becoming increasingly clear that the immature oocyte at the dictyate stage (in a follicle consisting of a single layer of granulosa cells) is the only germ cell stage which is insensitive to the induction of mutations.

5. Genetic Effects of Exposure to Irradiation in Man

A substantial proportion of spontaneous abortions in women result from chromosomal aberrations. According to one estimate, about one-quarter of all abortions are due to this cause and 1% of all livebirths suffer from the severe effects of chromosomal disorders (UNSCEAR, 1966). It has therefore been suggested that radiation-induced translocations may be more common in man than in the mouse and that small doses of X rays may increase the frequency of chromosomal disorders and congenital defects. At the present time, however, there is virtually no information on the genetic effects of irradiation on human germ cells. The studies of the survivors of the nuclear explosions at Hiroshima and Nagasaki were designed to detect only dominant traits as evidence of gene mutations (Neel and Schull, 1956; Neel, 1963). As Stevenson (1968) has pointed out, if these studies *had* revealed a significant increase in genetic damage, we would have known that man was highly sensitive to genetic changes induced by radiation. In fact the only detectable change was a shift in the sex ratio in favor of males (see also Mondorf and Faber, 1968; Meyer *et al.*, 1969): such results are in marked contrast to those from experiments with mice in which fewer males than females are born to irradiated parents (see Zuckerman, 1965). However, studies of human populations exposed to nuclear hazards, and of patients subjected to diagnostic or therapeutic irradiation, have shown that the rate of stillbirths, abortions, congenital malformations, and leukemia are all increased (see Rugh, 1960; de Bellefeuille, 1961a,b; Conrad *et al.*, 1966; Meyer *et al.*, 1969; Stewart, 1969). It remains to be established whether these changes represent genetic changes in the oocyte or are the consequences of environmental factors within the uterus.

Sobels (1968, 1969) has recently considered the potential genetic hazards arising from treatment of thyroid tumors in women with [131]I. It has been calculated that the gonadal exposure resulting from radiotherapy with iodine is of the order of 0.01 rad/minute, giving a total dose to the ovary of some 80 to 160 rad (Henneman and Mellink, 1968). This is certainly a large dose compared with the usual diagnostic X-ray film, for which the dose rarely exceeds 0.5 rad. Nevertheless, for a variety of reasons, including data for mice, Sobels concludes that the risk arising from ^{131}I therapy is small. It is clear that women exposed to doses of 500 R can produce viable offspring who show few, if any, signs of radiation-induced chromosomal aberrations and mutations (see Baker, 1971c). Such changes would, however, be difficult to detect unless the abortuses were also studied. The incidence of recessive mutations will probably never be known since they would appear in a relatively small proportion of consanguineous marriages, which themselves are very rare. On the basis of data derived from studies on mice, one may perhaps surmise that the incidence of mutations in man may be low since most of the exposures are chronic or fractionated. Dahl-Iversen and Hamburger (1948) have suggested that the dose of radiation required to double the spontaneous mutation frequency is 30 to 60 R for the mouse and 30 to 80 R for man, but Neel (1963) believes that these estimates could well be revised upwards in view of the available data (but also see Lüning and Searle, 1971).

F. Factors Underlying the Differential Radiosensitivity of Germ Cells

It has been shown in the preceding sections of this chapter that, irrespective of the criterion chosen to assess the effect, radiosensitivity varies considerably with phases in the development and differentiation of germ cells. The factors underlying this differential radiosensitivity remain poorly understood, although a number of general conclusions and hypotheses can be suggested. Thus, one would expect germinal and somatic cells at comparable stages of development to show similar responses to X irradiation. With the exception of oocytes at the diplotene and dictyate stages of meiotic prophase, this statement is generally true although species differences have been recorded. Mitotically active "stem cells" within the body (including primordial germ cells and oogonia) are readily killed by ionizing radiations, either by "interphase death" or "mitotic delay and death" (see Lawrence, 1971). Oocytes and spermatocytes in many animals become increasingly resistant to X irradiation as they progress from leptotene to pachytene. It may be postulated that this change in radiosensitivity is due to pairing and condensation of the chromosomes and possibly to the onset of the "lampbrush" stage (see below).

It has often been suggested that changes in radiosensitivity of oocytes during the diplotene/dictyate stage of meiotic prophase are due to variations in the form and metabolic activity of the chromosomes (see Baker, 1963; Mandl, 1964; Oakberg and Clark, 1964). The results of studies of the fine structure of chromosomes have added considerable support to this view. Mammalian chromosomes at diplotene pass through a stage in which they resemble the lampbrush chromosomes characteristic of oocytes in lower vertebrates. Thus, the primordial follicles of the rabbit, cow, rhesus monkey, and man contain oocytes in which the chromosomal threads consist of a bifid axis (presumably the chromatids of one chromosome) with loops of chromosomal material at intervals along their length (Baker, 1963, 1966a, 1971b; Baker and Franchi, 1967a,b,c; Zybina, 1969). These loops bear clusters of granules (especially at their distal ends) which correspond in size to those of ribonucleoprotein, and thus the chromosome is similar to, although smaller than, the lampbrush chromosomes in the newt (see Callan and Lloyd, 1960a,b). Lampbrush-type chromosomes are also found in oocytes within multilayered and Graafian follicles in all the species which have so far been examined (see Baker and Franchi, 1967a,b,c). By contrast, the small primordial oocytes of mice and rats contain what appears to be a mass of fine threads which superficially resembles a reticulum (Beaumont and Mandl, 1962). These chromosomes are difficult to identify with the electron microscope and the granules are widely separated (Franchi and Mandl, 1962; Tsuda, 1965).

Autoradiographic studies of the uptake of [^{3}H]uridine (an RNA precursor) into the oocytes of mice, rats, and monkeys have revealed a high rate of specific incorporation of the isotope along the length of the chromosome. In the monkey, the label is found in the loop matrix of lampbrush chromosomes within 30 minutes of the time of injection: some 2 hours later it moves into the cytoplasm of the oocyte, where it may have a number of important functions (see Baker, 1971b). A similar rate of incorporation of the isotope is found in mice and rats, although the precise localization of silver grains over chromosomal threads can only be detected at pachytene and diplotene. At the dictyate stage the chromosomes are initially too indistinct to permit adequate visualization of the threads, which only become well defined after the onset of follicular growth (stage II of Mandl and Zuckerman, 1951).

It has been suggested that oocytes, in which the loops on lampbrush chromosomes are relatively condensed and are surrounded by a thick sheath of ribonucleoprotein granules, are resistant to radiation damage, while those with highly tortuous ("spun-out") loops and a thin diffuse matrix of ribonucleoprotein are radiosensitive (see Baker and Franchi, 1967b; Baker *et al.*, 1969a; Baker, 1971a,b; and Fig. 7). The condensed type of chromosome is found in primordial oocytes (those in primordial follicles) in such species as the guinea pig, cow, monkey, and man, while oocytes in multilayered follicles are generally slightly less condensed. It is only the small primordial oocytes in some rodents (and particularly the mouse and

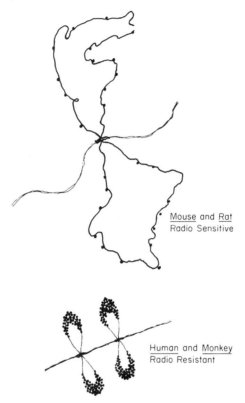

Mouse and Rat
Radio Sensitive

Human and Monkey
Radio Resistant

Fig. 7. Diagram to show how chromosomal configuration may influence the radiosensitivity of primordial oocytes. From Baker (1971a).

rat) which have highly diffuse chromosomes (true dictyate stage) and which are highly radiosensitive. Thus, variations in radiosensitivity, both between and within species, could be due to fluctuations in the activity of lampbrush chromosomes, expressed in terms of loop size and metabolic activity.

On the basis of this hypothesis, changes in the radiosensitivity of the oocyte during the juvenile period (in terms of cell death and reproductive capacity: see Sections III,C and D) can be explained as being due to (1) the transition from diplotene (virtually no loops) to dictyate stage (progressively enlarging loops and increasing radiosensitivity) and (2) increasing metabolic activity and RNA synthesis, which may impart some resistance (see below). Similarly, the change from the highly radiosensitive small primordial oocyte stage (stage I) in the rat and mouse to the resistant stage with a cuboidal granulosa cell covering (stage II) is correlated with a reduction in loop size (such that the chromosomes become clearly visible with light and electron microscopes), and with an increase in the thickness of the sheath investing the chromosome. The diplotene/dictyate stage is therefore maintained throughout follicular growth to the onset of preovulatory maturation, but the morphology of the chromosomes changes frequently, as does radiosensitivity.

It has also been suggested that the sheath of ribonucleoprotein may act as a "splint" by holding the broken ends of chromosomes together until a break is repaired. Miller *et al.* (1965) showed that few breaks could be detected in the lampbrush chromosomes of newt oocytes (which are highly radioresistant) even after exposure to 50,000 R X rays. However, treatment with proteolytic enzymes and ribonuclease dispersed the granular matrix surrounding the chromosomes and breaks were then detected. If a "splinting action" is involved in holding the zone of breakage together until repair is completed, the extreme radiosensitivity of oocytes in rats and mice may be due to the sheath being too thin to approximate and hold the broken ends of the chromosome before repair can be initiated. Moreover, the precursors of the ribonucleoprotein synthesis by lampbrush chromosomes may be suitable "building blocks" for utilization during the repair process (Oakberg, 1968).

There can be little doubt that chromosomes damaged by ionizing radiations are capable of undergoing repair. Studies of somatic cells in tissue culture have shown that repair involves "unscheduled" DNA synthesis (Rasmussen and Painter, 1966; Hill, 1967; Evans and Norman, 1968; Painter and Cleaver, 1969; Richold and Arlett, 1972). It has been mentioned previously (Section III,A) that "unscheduled" DNA synthesis also occurs in oocytes (Crone, 1970) and spermatocytes (Kofman-Alfaro and Chandley, 1971; Chandley and Kofman-Alfaro, 1971) following exposure to X or ultraviolet irradiation. Crone (1970) showed by autoradiography that [³H]thymidine (which is normally taken up only at the "S" phase of interphase or at preleptotene) is incorporated into the dictyate oocytes of 1-day-old guinea pigs within 1 hour of exposure to 5000 R X rays. However, the radioisotope was not incorporated by mouse oocytes at the dictyate stage after exposure to 200 R X rays. Crone (1970) therefore concluded that the marked resistance of primordial oocytes in guinea pigs to radiation (see Section III,C) is due to their ability to repair the initial damage, an ability lacking in the small radiosensitive oocytes of mice.

It remains to be established whether hypophysectomy and/or treatment with various hormones can induce changes in the morphology and metabolic activity of oocytes, since these factors seem to have a profound affect on the response of oocytes to X irradiation. It is hoped that further studies will be made of the fine structure of chromosomes in the oocytes of a variety of mammalian species to test the validity of hypotheses relating to differential radiosensitivity.

IV. IRRADIATION FROM INJECTED RADIOISOTOPES

Radioisotopes have been widely used as "tracers" in autoradiographic studies of various biochemical pathways within the ovary. Furthermore, other isotopes (injected for radiotherapeutic or diagnostic purposes in organs some distance from the ovary) may exert an effect while in the blood stream, or as the result of nonspecific uptake within the gland. It has long been known that radioisotopes can have a deleterious effect on ovarian tissue, the effect being proportional to the amount of the isotope incorporated and with the type and energy spectrum of the radiation emitted. Isotopes emitting low energy radiations, such as β rays from tritium, exert their effects over very small distances. However, the high energy radiation from such isotopes as strontium-90, cobalt-60, and polonium-208 or -210 can penetrate great distances and act as a source external to the ovary. For example, ^{90}Sr is actively incorporated into the long bones where it acts as a source of high energy radiation affecting other organs. Pregnant rats injected with 20 μCi of ^{90}Sr on days 11 or 16 *post coitum* produce offspring in which the population of oocytes is reduced by 50% on day 2 after birth (Henricson and Nilsson, 1965; Samuels, 1966; Nilsson and Henricson, 1969). The decrease in the number of primordial follicles is not followed by a reduction in the population of Graafian follicles produced at each reproductive cycle, although the incidence of corpora lutea and atretic follicles is greater than in controls (Nilsson and Henricson, 1969). However, such high-energy emitters as ^{90}Sr are rarely used in studies involving the ovary. It is therefore important to consider potential effects from the isotopes commonly used in autoradiographic studies.

1. Radioactive Iodine

Iodine-131 has a half-life of about 8 days and produces both β and γ rays (see Lacassagne *et al.*, 1962). It is not incorporated generally within the ovary but may be selectively taken up by the walls of large follicles (Godwin *et al.*, 1951; Kurland *et al.*, 1951; Bengtsson *et al.*, 1963). Injection of 200 μCi of ^{131}I into pregnant mice on the seventeenth day of gestation eliminates all the oocytes from a proportion of the embryos (Speert *et al.*, 1951). Similarly, 5-day-old females which receive the isotope in their mother's milk (300 μCi injected on day 5 of lactation) show a 65% reduction in fertility, although their brothers are seemingly unaffected (Rugh, 1953).

The response of adult mice to ^{131}I injections appears to vary with dose. Thus, 0.5 to 50 μCi is said to induce only temporary irregularities in the estrous cycle, after which normal litters are produced (Duffy *et al.*, 1950). By contrast, Gorbman (1950) found that mice became sterile after injection with 2 μCi ^{131}I/mg of thyroid. In the rat, exposure to the radiation from 20 to 200 μCi also results in irregularities in the estrous cycle, the duration of which is proportional to the dose of isotope administered (Tillotson *et al.*, 1953). Ovarian function is subsequently restored although the rats produce smaller litters.

As with X and γ rays, exposure to incorporated radioactive iodine results in the death of germ cells and in a reduction of ovarian volume. Rugh (1953) injected mice once a week over a 6-month period with 4 μCi and reported that the volume of the ovary was reduced, corpora lutea were absent, and the remaining follicles showed signs of degeneration. Reddi (1971a,b, 1973) has largely confirmed these observations for the ovaries of mice and cows in which the oocyte population is markedly reduced. He has suggested that the accumulation of the isotope in the walls of multilayered follicles (see also Bengtsson *et al.*, 1963) not only results in the elimination of these follicles, but also acts as a source of irradiation which destroys the oocytes and/or granulosa cells in small follicles.

2. Radioactive Sulfur

Autoradiographic studies have demonstrated that sulfur-35 is incorporated into mucopolysaccharides. It rapidly appears in theca and granulosa cells, and in the follicular fluid (Odeblad, 1952a,b; Odeblad and Boström, 1953; Gothié, 1954; Moricard and Gothié, 1955). It would seem that ^{35}S causes the death of many germ cells when injected during fetal life (Penchansky and Gonzalez Cueto, 1968). Whether destruction of oocytes and the consequential reduction in fertility occurs during adult life remains to be established.

3. Tritium

This isotope of hydrogen, with a half-life of 12.26 years, emits low energy β rays and is widely used as a "tracer" in autoradiographic studies of ovarian function, as [^{3}H]thymidine, [^{3}H]uridine, or as precursors of various proteins or steroids. It has often been suggested that the radiation from tritium is so weak (and thus the penetration so low) that the radioisotope is unlikely to damage ovarian cells. However, the results of numerous studies have shown that this view is no longer tenable, at least for germ cells in embryonic mice and rats. For example, [^{3}H]thymidine so severely depletes the population of germ cells during prenatal life, that the subsequent reproductive life-span of the animal is reduced (Callebaut, 1968; Haas and Fliedner, 1971; Ghosal *et al.*, 1972; Baker and McLaren, 1973). For example, Baker and McLaren (1973) gave twice-daily injections of [^{3}H]thymidine (total of seven injections, each of 4, 40, or 400 μCi) to mice aged between 13 and 16 days *post coitum.* It was found that the population of oocytes declined markedly with increasing age after exposure, and that the effect was proportional to the dose of isotope used. The size and weight of the ovary, and also reproductive capacity, were markedly reduced. These observations fully confirm the earlier studies of Callebaut (1968), but are in conflict with those of Borum (1967; see also Lima de Faria and Borum, 1962) who claims that 80 to 100% of the germ cells are labeled for up to 50 days after birth. It is conceivable that these apparently conflicting results merely represent genetic differences between inbred strains of mice (see Section III,C).

The information that is now available makes it difficult to calculate the dose of radiation that germ cells receive from incorporated [³H]thymidine. By considering the number of disintegrations occurring in each cell, it has been estimated that rat spermatocytes receive a dose of approximately 300 rad from an injection of 50 μCi/gm body weight (Kiselieski *et al.,* 1964). Samuels (1966) believes that mouse oocytes receive a similar dose from polonium-210. If these figures are correct and can be applied to [³H]thymidine injections in the female, it is not surprising that the germ cells are destroyed since they are highly sensitive to radiation-induced death (see Section III,C).

4. Radioactive Phosphorus

Phosphorus-32 is incorporated into nucleic acids and is rapidly concentrated into the granulosa cells of growing follicles (Leblond *et al.,* 1948; Gothié, 1954). Injections of 25 to 250 μCi of the isotope into adult mice results in a reduction in the size of the ovary, atresia in many of the follicles, and the absence of corpora lutea (Warren *et al.,* 1950). The first radiolesions occur in the granulosa cells, which undergo degeneration long before changes occur in the oocyte at doses of 1 to 3 μCi/gm of rat or mouse (Odeblad, 1952a,b; Reither and Lang, 1955; T. G. Baker and P. J. Ford, unpublished observations). It has recently been shown that the cuboidal granulosa cells surrounding oocytes at stage II of follicular growth are more sensitive to destruction by ³²P (0.05 μCi injected into adult mice) than are the flattened granulosa cells around stage I oocytes (Reddi *et al.,* 1968). This finding probably reflects the high uptake of radioactive phosphate into nucleic acids in the actively dividing granulosa cells that have commenced follicular growth (stage II). A similar observation was made by Oakberg (1968) who injected 75 μCi of [³H]uridine into mice: the granulosa cells which were heavily labeled showed marked degenerative changes.

This brief resumé of the literature relating to the use of radioisotopes as "tracers" in autoradiographic studies clearly indicates that the dose chosen should always be kept as low as possible. Even then, studies in which cells are damaged by irradiation from an internal source are of dubious value, and are unlikely to provide meaningful data on normal biochemical pathways within the ovary.

V. EFFECTS OF RADIATION ON THE SOMATIC COMPONENTS OF THE OVARY

Exposure to ionizing radiations induces hyperemia (Van Eck, 1959; Borovkov, 1968; Hahn and Ward, 1971) and destruction of the follicles (see Lacassagne *et al.,* 1962). The ovary subsequently decreases in size and weight and becomes greyish in color. These changes become more marked with increasing dose and interval between exposure to radiation and autopsy.

Changes in the size and weight of the ovary are largely the result of the elimination of the follicles, although other tissues also regress. By 3 to 4 weeks after exposure to X rays, all that remains of the large follicles in the rabbit ovary are the theca and the zona pellucida that once surrounded the oocyte. The follicular fluid often contains pyknotic debris that was once the membrana granulosa, which degenerates completely by the tenth day after exposure. In some cases the hyalinized remains of the oocyte can be seen within the follicles. Blood vessels within the theca sometimes rupture, and the resulting hemorrhagic follicles may persist for several months (Lacassagne, 1913, 1936, 1937). Similar changes have been reported for the X-irradiated ovaries of other species, although their magnitude and timing vary in extent (mouse: Murray, 1931; rat: Mandl, 1959; guinea pig: Genther, 1931; monkey and man: Reifferscheid, 1910, 1914).

The coelomic ("germinal") epithelium covering the ovary shows little signs of damage for a period of weeks following exposure to irradiation, after which some of the cells undergo degeneration. The morphological appearance of the remaining cells changes from cuboidal to a flattened epithelium, the effect only becoming fully apparent in animals whose ovaries are fully devoid of follicles (see Lacassagne *et al.,* 1962). In the mouse and rat, epithelial cords (which seem to be derived from the coelomic epithelium) grow into the ovary and may secrete the estrogen that maintains the estrous cycles in these species (Van Eck and Freud, 1949a,b; Mandl and Zuckerman, 1956a,b; Westman, 1958; Slate and Bradburg, 1962: see also page 21). Thus, in these rodents, the coelomic epithelium would seem to be resistant to the doses of radiation usually applied to the ovary, and its proliferation may sometimes compensate for the loss in weight caused by the destruction of the follicles (Levine and Witschi, 1933; Borovkov, 1968; Matsumoto, 1971; Baker and McLaren, 1973). The proliferation can become extensive, leading to the formation of adenomata, while multiplication of surviving granulosa cells results in the formation of granulosa cell tumors (Furth and Butterworth, 1936; Furth and Furth, 1936; Butterworth, 1937; Furth and Boon, 1947; see Lacassagne *et al.,* 1962).

There are few reports on the effect of ionizing radiations on the coelomic epithelium in other species, although this apparent deficiency in the literature may imply radioresistance. In those species in which the oocytes are highly resistant to death by X rays (e.g., monkey), the granulosa cells are severely depleted in number (Baker and Franchi, 1972b). The surviving cells appear to restore, at least partly, the normal population of follicle cells around oocytes, and cases of radiation-induced granulosa cell tumors have been reported for the guinea pig (Schmidt, 1939) and ferret (Parkes *et al.,* 1932).

Ovarian interstitial tissue may be refractory to X irradiation (Bouin *et al.,* 1906), although some authors claim that it undergoes marked changes (Specht, 1906; Fellner and Neumann, 1907). In the rabbit, the gland is certainly reduced in size from the second to the sixth month after exposure, but subsequently grows again. Lacassagne (1913) believes that these changes in the size of the interstitial gland are secondary rather than the direct effect of X rays. He concludes that interstitial tissue usually arises as glandular nodules from the theca interna of atretic follicles. Once these follicles have been destroyed by irradiation, the extent of the interstitial gland would thus gradually decline until (according to Lacassagne, 1913) new tissue is reconstituted from the stroma.

No morphological or functional changes occur in the fully differentiated corpus luteum (mouse: Parkes, 1929; rabbit: Lacassagne, 1936, 1937; guinea pig: Genther, 1931). Furthermore, Regaud and Lacassagne (1913) showed that corpora lutea form normally after the irradiation of Graafian follicles, although they subsequently degenerate prematurely on the eighth day after formation. When rabbits are exposed to 1300 R or more between 2 and 24 hours after copulation, the secretory capacity of the corpus luteum is reduced and deciduomata are not formed in response to traumatization of the uterus. The effect is dose dependent, however, in that the secretory capacity of the gland is seemingly unaffected by exposures of 500 to 1300 R, and deciduomata are formed in response to stimulation of the uterus. With the eventual involution of the corpora lutea formed before or immediately after irradiation, ovaries subjected to sterilizing doses do not again form corpora lutea. But if sufficient follicles remain for ovulation to continue, secretory corpora lutea are formed (Bouin *et al.,* 1906). Ichikawa *et al.* (1968) have recently shown that the corpus luteum in sheep ovaries autotransplanted to the neck continues to secrete steroids (progesterone, 20α-dihydroprogesterone, estrone, and estradiol-17β) in a normal fashion after exposure to 1000–2000 R X irradiation. Similar results have been obtained for rat ovaries exposed to 2000 R (MacDonald *et al.*, 1969; Christiansen *et al.*, 1970). This finding is in keeping with the results of Flaks and Bresloff (1966) who showed that the fine structure of the corpus luteum in rats is not changed following exposure to 4000 R, a dose which is known to destroy all the follicles (Mandl, 1959).

VI. SUMMARY AND CONCLUSIONS

It is clear that the effects of radiation on the mammalian ovary are complex and highly variable depending on the particular end point that is selected to assess the results. It may therefore be useful to provide a summary of the main points that can be drawn from this review. The main conclusion is that no matter which criterion is chosen to assess the effect of a particular dose of X rays, the response varies considerably between species and between inbred and outbred strains within a species. Consequently, the practice of extrapolating data from laboratory rodents to man may be of doubtful value. Furthermore, it is no longer meaningful to use the term "radiosensitivity" unless the particular criterion chosen to assess the effect is clearly defined and the physical conditions of the exposure (e.g., dose rate, acute or chronic exposure, temperature, oxygen tension) are stated.

The response of germ cells to ionizing radiations varies considerably with age and with the stage of mitosis or meiosis at the time of exposure. In terms of cell death and reproductive capacity, primordial germ cells are more resistant than oogonia, which in turn are more sensitive than oocytes at early stages of meiotic prophase. Oocytes become increasingly more resistant as they progress from leptotene to pachytene, but gradually become more sensitive (in rodents, but possibly not in other species) during the early part of the dictyate stage. Sensitivity of the oocyte also changes with the phases of follicular growth. In general terms, multilayered follicles show a comparable radiosensitivity irrespective of species, but the small primordial follicles vary from extremely sensitive (mouse and rat) to highly resistant (monkey).

Using the criterion of genetic effects, the small dictyate oocytes of mice and rats are resistant to the induction of mutations and translocations, but moderately sensitive to X-chromosome deletion. This is in marked contrast to their sensitivity in terms of cell death; the small dictyate oocytes of these rodents are amongst the most radiosensitive cells known. Oocytes in multilayered follicles are resistant to cell death, sensitive to the induction of

mutations, and vary in terms of translocations and dominant lethals according to the chromosomal configuration of the oocyte (i.e., stage of preovulatory maturation). But whichever genetic criterion is used to assess
radiosensitivity, the effect is considerably reduced by splitting the total dose
into a series of fractions and also by lowering the dose rate. This is
important in any consideration of genetic effects in irradiated human
patients since radiotherapeutic exposures are generally fractionated.

Little is known of the effects of ovarian irradiation in women, although
there can be little doubt that the sensitivity of the germ cells, in terms of the
criterion of cell death, is far more comparable to that of the rhesus monkey
than to that of laboratory rodents. It has been shown that oogonia and
oocytes in the fetus, and also oocytes in the young adult, survive relatively
high doses of X and γ rays, and that ovulation and conception can subsequently occur. Progeny with severe genetic lesions are likely to be
resorbed about the time of implantation, or to be aborted or stillborn,
along with the large number of individuals carrying "spontaneous" genetic
damage. The level of recessive mutations may never be known, since these
are virtually impossible to detect in human populations. Thus, in the 27
years that have elapsed since Hiroshima and Nagasaki were destroyed by
nuclear bombs, little reliable information about genetic changes has
emerged.

The marked fluctuations in radiosensitivity that occur between species
and with age cannot be explained at present. The most important factors
appear to be the chromosomal configuration and metabolic activity of the
oocyte, and an hypothesis is outlined which accounts for most of the experimental results that have emerged from studies of cell death and reproductive capacity.

The somatic components of the ovary are more uniform in their response
to radiation. Hyperemia and loss of weight are said to occur even at low
doses, but the effect probably differs widely between species. Such ovarian
structures as the interstitial gland and corpus luteum are refractory and
continue to secrete steroids after exposure to high doses of X rays. But most
of the estrogen secreted by the ovary is normally produced by the multilayered follicles, which are destroyed by doses of 2000 to 4000 R X rays. In
mice and rats, estrogen is secreted even after the elimination of all the follicles, and cyclical changes in the uterus and vagina persist for prolonged
periods. It would seem that, in the absence of follicles, the interstitial gland
and the epithelial cords produced by the coelomic epithelium of the ovary
take over the role of steroid secretion.

The effect of irradiation from internal emitters (radioisotopes) is complicated by the different energies and sites of incorporation of the substances.
It is clear, however, that even those producing low energy β radiation (e.g.,

tritium) can cause depletion of germ cells, particularly in the embryo. Consequently, the amount of radioisotope used in autoradiographic studies should be kept to the minimum.

VII. ADDENDUM

A number of important papers have appeared in the literature during the interval between the preparation and publication of this chapter. However, none of these articles warrants a change in the views expressed above. The most significant study was that by Erickson *et al.* (1976) and is worthy of special mention. Erickson and his coworkers describe an experiment in which the long-term (up to 14 years) reproductive performance of a group of 280 Hereford cows was studied. The animals were exposed at an age of 15 to 18 months to γ rays at doses of 0 to 600 R (dose rate 0.69 to 0.78 R/minute). Although a few of the animals died during the postirradiation interval, many survived to permit a maximum of 13 calves to be sired by unirradiated bulls. Irrespective of the dose of irradiation administered, the number of oocytes, proportion of viable matings, pregnancies reaching full term, and number of viable calves weaned were all within the normal range for the controls. This data provides much support for the conclusions on page 26, based on limited results of studies in rhesus monkeys and women, that reproductive capacity is unlikely to be affected by radiation unless the dose employed is around the LD_{50} level.

REFERENCES

Alexander, P. (1965). "Atomic Radiation and Life." Penguin Books, London.
Andersen, A. C. (1964a). Reproductive ability of the aging X-irradiated and sham-treated female beagle. *In* "Effects of Ionizing Radiation on the Female Reproductive System" (W. D. Carlson and F. X. Gassner, eds.), pp. 323–331. Pergamon, Oxford.
Andersen, A. C. (1964b). Syndromes affecting reproduction in normal and X-irradiated female beagles. *In* "Effects of Ionizing Radiation on the Reproductive System" (W. D. Carlson and F. X. Gassner, eds.), pp. 377–392. Pergamon, Oxford.
Andersen, A. C., Schultz, F. T., and Hage, T. J. (1961). The effect of total body X-irradiation on reproduction of the female beagle to 4 years of age. *Radiat. Res.* **15,** 745–753.
Baker, T. G. (1963). A quantitative and cytological study of germ cells in human ovaries. *Proc. R. Soc. London, Ser. B* **158,** 417–433.
Baker, T. G. (1966a). The sensitivity of oocytes in post-natal rhesus monkeys to X-irradiation. *J. Reprod. Fertil.* **12,** 183–192.

Baker, T. G. (1966b). A quantitative and cytological study of oogenesis in the rhesus monkey. *J. Anat.* **100**, 761–776.

Baker, T. G. (1969). The sensitivity of rat, monkey and human oocytes to X-irradiation in organ culture. *In* "Radiation Biology of the Fetal and Juvenile Mammal" (M. R. Sikov and D. D. Mahlum, eds.), CONF-690501, pp. 955–961. U.S. At. Energy Comm., Washington, D.C.

Baker, T. G. (1971a). Comparative aspects of the effects of radiation during oogenesis. *Mutat. Res.* **11**, 9–22.

Baker, T. G. (1971b). Electron microscopy of the primary and secondary oocyte. *Adv. Biosci.* **6**, 7–23.

Baker, T. G. (1971c). Radiosensitivity of mammalian oocytes with particular reference to the human female. *Am. J. Obstet. Gynecol.* **110**, 746–761.

Baker, T. G. (1972). Oogenesis and ovarian development. *In* "Reproductive Biology" (H. Balin and S. R. Glasser, eds.), pp. 398–437. Excerpta Med. Found., Amsterdam.

Baker, T. G., and Beaumont, H. M. (1967). Radiosensitivity of oogonia and oocytes in the foetal and neonatal monkey. *Nature (London)* **214**, 981–983.

Baker, T. G., and Franchi, L. L. (1967a). The fine structure of oogonia and oocytes in human ovaries. *J. Cell Sci.* **2**, 213–224.

Baker, T. G., and Franchi, L. L. (1967b). The structure of the chromosomes in human primordial oocytes. *Chromosoma* **22**, 358–377.

Baker, T. G., and Franchi, L. L. (1967c). The fine structure of chromosomes in bovine primordial oocytes. *J. Reprod. Fertil.* **14**, 511–513.

Baker, T. G., and Franchi, L. L. (1969). The origin of cytoplasmic inclusions from the nuclear envelope of mammalian oocytes. *Z. Zellforsch. Mikrosk. Anat.* **93**, 45–55.

Baker, T. G., and Franchi, L. L. (1972a). The fine structure of oogonia and oocytes in the rhesus monkey (*Macaca mulatta*). *Z. Zellforsch. Mikrosk. Anat.* **126**, 53–74.

Baker, T. G., and Franchi, L. L. (1972b). Electron microscope studies of radiation-induced degeneration in oocytes of the sexually mature rhesus monkey. *Z. Zellforsch. Mikrosk. Anat.* **133**, 435–454.

Baker, T. G., Beaumont, H. M., and Franchi, L. L. (1969a). The uptake of tritiated uridine and phenylalanine by the ovaries of rats and monkeys. *J. Cell Sci.* **4**, 655–675.

Baker, T. G., and McLaren, A. (1973). The effect of tritiated thymidine on the developing oocytes of mice. *J. Reprod. Fertil.* **34**, 121–130.

Baker, T. G., and Neal, P. (1969). The effects of X-irradiation on mammalian oocytes in organ culture. *Biophysik* **6**, 39–45.

Baker, T. G., Kelly, W. A., and Marston, J. H. (1969b). The reproductive capacity of rhesus monkeys following X-irradiation of the ovaries. *Strahlentherapie* **138**, 355–360.

Batchelor, A. L., Phillips, R. J. S., and Searle, A. G. (1969). The ineffectiveness of chronic irradiation with neutrons and gamma rays in inducing mutations in female mice. *Br. J. Radiol.* **42**, 448–451.

Bateman, A. J. (1958). Mutagenic sensitivity of maturing germ cells in the male mouse. *Heredity* **12**, 213–232.

Beaumont, H. M. (1962). The radiosensitivity of germ cells at various stages of ovarian development. *Int. J. Radiat. Biol.* **4**, 581–590.

Beaumont, H. M. (1965). The short-term effects of acute X-irradiation on oogonia and oocytes. *Proc. R. Soc. London, Ser. B* **161**, 550–570.

Beaumont, H. M. (1966). The effects of acute X-irradiation on primordial germ cells in the female rat. *Int. J. Radiat. Biol.* **10**, 17–28.

Beaumont, H. M. (1969). Effect of hormonal environment on the radiosensitivity of oocytes. *In* "Radiation Biology of the Fetal and Juvenile Mammal" (M. R. Sikov and D. D. Mahlum, eds.), CONF-690501, pp. 943–953. U.S. At. Energy Comm., Washington, D.C.

Beaumont, H. M., and Mandl, A. M. (1962). A quantitative and cytological study of oogonia and oocytes in the foetal and neonatal rat. *Proc. R. Soc. London, Ser. B* **155**, 557–579.

Bengtsson, G., Ewaldsson, B., Hansson, E., and Ullberg, S. (1963). Distribution and fate of ^{131}I in the mammalian ovary. *Acta Endocrinol. (Copenhagen)* **42**, 122–128.

Biggers, J. D., Finn, C. A., and McLaren, A. (1962a). Long-term reproductive performance of female mice. 1. Effect of removing one ovary. *J. Reprod. Fertil.* **3**, 303–312.

Biggers, J. D., Finn, C. A., and McLaren, A. (1962b). Long-term reproductive performance of female mice. 2. Variation of litter size with parity. *J. Reprod. Fertil.* **3**, 313–330.

Bocharov, Yu.-S., Bocharov, E. V., and Mickleyeva, G. A. (1964). The comparative radiosensitivity of ovaries in monkeys (*Macaca mulatta*) and mice under X-ray irradiation. (Inst. Biol. Thys. Acad. Sci. U.S.S.R.) *Biol. Abstr.* **45**, No. 36063.

Bojadjieva-Mihailova, A. (1964). Electron microscopical studies of the ovaries of embryos and newborn white mice under the influence of roentgen rays. *Izv. Inst. Morfol., Bulg. Akad. Nauk.* **9–10**, 161–165, cited in *Excerpta Med., Sect.* 14, **19**, (Abstr. No. 2511) (1965).

Bojadjieva-Mihailova, A., Boneva, L., and Hadjioloff, A. (1968). Application of the mathematical-statistical method for the evaluation of the ultrastructural dimensions of ovocytic mitochondria after X-ray irradiation. *Proc. European Reg. Conf. Electron Microsc. 4th, 1968* Vol. 1, pp. 605–606.

Borovkov, A. I. (1968). Morphology of the ovaries in mice after general single dose of X-ray irradiation. *Arkh. Patol.* **30**, 31–36.

Borum, K. (1967). Oogenesis in the mouse: A study on the origin of the mature ova. *Exp. Cell Res.* **45**, 39–47.

Bouin, P., Ancel, P., and Villemin, F. (1906). Sur la physiologie du corps jaune de l'ovaire. Recherches faites á l'aide des rayons X. *C. R. Soc. Biol.* **61**, 417–419.

Brambell, F. W. R., and Parkes, A. S. (1927). Changes in the ovary of the mouse following exposure to X-rays. Part III. Irradiation of the non-parous adult. *Proc. R. Soc. London, Ser. B* **101**, 316–328.

Brambell, F. W. R., Parkes, A. S., and Fielding, U. (1927a). Changes in the ovary of the mouse following exposure to X-rays. Part I. Irradiation at three weeks old. *Proc. R. Soc. London, Ser. B* **101**, 29–56.

Brambell, F. W. R., Parkes, A. S., and Fielding, U. (1927b). Changes in the ovary of the mouse following exposure to X-rays. Part 2. Irradiation at or before birth. *Proc. R. Soc. London, Ser. B* **101**, 95–114.

Brent, R. L., and Bolden, B. T. (1967). The indirect effect of irradiation on embryonic development. III. The contribution of ovarian irradiation, uterine irradiation, oviduct irradiation and zygote irradiation to fetal mortality and growth retardation in the rat. *Radiat. Res.* **30**, 759–773.

Brooksby, G., Sahinin, F., and Soderwall, A. L. (1964). X-irradiation effects on reproductive function in the female hamster. *In* "Effects of Ionizing Radiation on the Reproductive System" (W. D. Carlson and F. X. Gassner, eds.), pp. 349–360. Pergamon, Oxford.

Brown, S. O., Krise, G. M., Pace, H. B., and De Boer, J. (1964). Effect of continuous radiation on reproductive capacity and fertility of the albino rat and mouse. *In* "Effects of Ionizing Radiation on the Reproductive System" (W. D. Carlson and F. X. Gassner, eds.), pp. 103–110. Pergamon, Oxford.

Butterworth, J. S. (1937). Observations on the histogenesis of ovarian tumors produced in mice by X-rays. *Am. J. Cancer* **31**, 86–99.

Byskov, A. G., and Peters, H. (1972). Ovarian development and fertility of the progeny of mice irradiated before mating. *Radiat. Res.* **49**, 351–358.

Callan, H. G., and Lloyd, L. (1960a). Lampbrush chromosomes. *In* "New Approaches in Cell Biology" (P. M. B. Walker, ed.), pp. 23–46. Academic Press, New York.

Callan, H. G., and Lloyd, L. (1960b). Lampbrush chromosomes of crested newts *Triturus cristatus* (Laurenti). *Philos. Trans. R. Soc. London, Ser. B* **243**, 135–219.

Callebaut, M. (1968). Development of the gonads of mice after intense incorporation of tritiated thymidine during the period of oogenesis. *Experientia* **24**, 828–829.

Carter, T. C. (1960). Mutations induced in germ cells of the foetal female mouse. *Genet. Res.* **1**, 59–61.

Carter, T. C., Lyon, M. F., and Phillips, R. J. S. (1960). The genetic sensitivity to X-rays of mouse foetal gonads. *Genet. Res.* **1**, 351–355.

Casarett, A. P. (1968). "Radiation Biology." Prentice-Hall, Englewood Cliffs, New Jersey.

Chandley, A. C., and Kofman-Alfaro, S. (1971). "Unscheduled" DNA synthesis in human germ cells following UV irradiation. *Exp. Cell Res.* **69**, 45–48.

Chang, M. C., and Harvey, E. B. (1964). Effects of ionizing radiation of gametes and zygotes on the embryonic development of rabbits and hamsters. *In* "Effects of Ionizing Radiation on the Reproductive System" (W. D. Carlson and F. X. Gassner, eds.), pp. 73–89. Pergamon, Oxford.

Chang, M. C., and Hunt, D. M. (1960). Effects of *in vitro* radiocobalt irradiation of rabbit ova on subsequent development *in vivo* with special reference to the irradiation of the maternal organism. *Anat. Rec.* **137**, 511–519.

Christiansen, J. M., Keyes, P. L., and Armstrong, D. T. (1970). X-irradiation of the rat ovary luteinized by exogenous gonadotropins: influence on steroidogenesis. *Biol. Reprod.* **3**, 135–139.

Conrad, R. A., Lowrey, A., Maynard, T. K., and Scott, W. A. (1966). Ageing studies in a Marshallese population exposed to radioactive fall out in 1954. *In* "Radiation and Ageing" (P. J. Lindop and G. A. Sacher, eds.), pp. 345–360. Taylor & Francis, London.

Crone, M. (1970). Radiation stimulated incorporation of [³H]thymidine into diplotene oocytes of the guinea pig. *Nature (London)* **228**, 460.

Crone, M., and Peters, H. (1968). Unusual incorporation of tritiated thymidine into early diplotene oocytes of mice. *Exp. Cell Res.* **50**, 664–667.

Dahl-Iversen, E., and Hamburger, C. (1948). Experimental studies on cystic glandular hyperplasia of the endometrium. II. Inability of X-ray treatment of the ovaries to produce cystic glandular hyperplasia of the endometrium in rhesus monkeys. *Acta Obstet. Gynecol. Scand.* **27**, 317–326.

de Bellefeuille, P. (1961a). Genetic hazards of radiation to man. Part 1. *Acta Radiol.* **56**, 65–80.

de Bellefeuille, P. (1961b). Genetic hazards of radiation to man. Part 2. *Acta Radiol.* **56**, 149–159.

Desaive, P. (1954). Influence du mode d'irradiation, de l'hypophysectomie, des hormones gonadotropes et des radio-protecteurs chimiques sur la résponse de l'ovaire de lapine aux rayons Röntgen. *Acta Radiol.* **41**, 545–557.

Desaive, P. (1957). Restoration of primordial follicles in the irradiated ovary. *In* "Advances in Radiobiology" (C. G. de Hevesy, A. G. Forssberg and J. B. Abbat, eds.), pp. 274–280. Oliver & Boyd, Edinburgh.

Duffy, B. J., Godwin, J. T., Trunnel, J. B., and Rawson, R. W. (1950). Radioiodine (^{131}I) and gonadal function: an experimental and clinical study. *J. Clin. Endocrinol. Metab.* **10**, 810–811.

Edwards, R. G. (1966). Mammalian eggs in the laboratory. *Sci. Am.* **215**, 72–80.

Edwards, R. G., and Searle, A. G. (1963). Genetic radiosensitivity of specific post-dictyate stages in mouse oocytes. *Genet. Res.* **4**, 389–398.

Ehling, U. H. (1964). Strain variation in reproductive capacity and radiation response of female mice. *Radiat. Res.* **23**, 603–610.

Erickson, B. H. (1965). Radiation effects on gonadal development in farm animals. *J. Anim. Sci.* **24**, 568–583.

Erickson, B. H. (1966). Development and radio-response of the prenatal bovine ovary. *J. Reprod. Fertil.* **11**, 97–105.

Erickson, B. H. (1967). Effect of gamma radiation on the prepuberal bovine ovary. *Radiat. Res.* **31**, 441–451.

Erickson, B. H., Reynolds, R. A., and Murphree, R. L. (1976). Late effects of ^{60}Co radiation in the bovine oocyte as reflected by oocyte survival, follicular development and reproductive performance. *Radiat. Res.* **68**, 132–137.

Evans, R. G., and Norman, A. (1968). Radiation stimulated incorporation of thymidine into the DNA of human lymphocytes. *Nature (London)* **217**, 455–456.

Feingold, S. M., and Hahn, E. W. (1972). Postconception development of rat ova following X-ray induced superovulation. *Radiat. Res.* **51**, 110–120.

Fellner, O., and Neumann, F. (1907). Der Einfluss der Röntgenstrahlen auf die Eirstöcke trächtiger Kaninchen und auf die Trächtigheit. *Z. Heilkd.* **28**, 162–170.

Finn, C. A. (1963). Reproductive capacity and litter size in mice: effect of age and environment. *J. Reprod. Fertil.* **6**, 205–214.

Flaks, B., and Bresloff, P. (1966). Some observations on the fine structure of the lutein cells of X-irradiated rat ovary. *J. Cell Biol.* **30**, 227–238.

Ford, C. E., Searle, A. G., Evans, E. P., and West, B. J. (1969). Differential transmission of translocations induced in spermatogonia of mice by irradiation. *Cytogenetics* **8**, 447–470.

Franchi, L. L., and Mandl, A. M. (1962). The ultrastructure of oogonia and oocytes in the foetal and neonatal rat. *Proc. R. Soc. London, Ser. B* **157**, 99–114.

Furth, J. (1949). Relation of pregnancies to induction of ovarian tumors in mice by X-rays. *Proc. Soc. Exp. Biol. Med.* **71**, 274–277.

Furth, J., and Boon, M. C. (1947). Induction of ovarian tumors in mice by X-rays. *Cancer Res.* **7**, 241–245.

Furth, J., and Butterworth, J. S. (1936). Neoplastic diseases occurring among mice subjected to general irradiation with X-rays. II. Ovarian tumors and associated lesions. *Am. J. Cancer* **28,** 66–95.

Furth, J., and Furth, O. B. (1936). Neoplastic diseases produced in mice by general irradiation with X-rays. I. Incidence and types of neoplasms. *Am. J. Cancer* **28,** 54–65.

Gans, B., Bahary, C., and Levie, B. (1963). Ovarian regeneration and pregnancy following massive radiotherapy for dysgerminoma. *Obstet. Gynecol.* **22,** 596–600.

Genther, I. T. (1931). Irradiation of the ovaries of guinea pigs and its effect on the oestrous cycle. *Am. J. Anat.* **48,** 99–137.

Genther, I. T. (1934). X-irradiation of the ovaries of guinea pigs and its effect on subsequent pregnancies. *Am. J. Anat.* **55,** 1–45.

Ghosal, S. K., Beauregard, L. J., Lamarche, P., DeLuca, F. G., and Parmar, C. P. (1972). Uridine-^{3}H uptake by human oocytes. *Can. J. Physiol. Pharmacol.* **50,** 285–288.

Gibbons, A. F. E., and Chang, M. C. (1973). Indirect effects of X-irradiation on embryonic development: irradiation of the exteriorized rat uterus. *Biol. Reprod.* **9,** 133–141.

Glass, L. E., and McLure, T. R. (1964). Equivalence of X-irradiation *in vivo* or *in vitro* on mouse oocyte survival. *J. Cell. Comp. Physiol.* **64,** 347–354.

Godwin, J. T., Duffy, B. J., Fitzgerald, P. J., Trunnel, J. B., and Rawson, R. W. (1951). Pathological effects and radioautographic localization of ^{131}I in humans. *Cancer* **4,** 936–951.

Goldfeder, A. (1965). The role of mitochondrial integrity in cellular radiosensitivity. *J. Cell Biol.* **27,** 34A.

Gorbman, A. (1950). Functional and structural changes consequent to high dosage of radioactive iodine. *J. Clin. Endocrinol. Metab.* **10,** 1177–1191.

Gothié, S. (1954). Etude comparée de la répartition du ^{32}P et du ^{35}S dans l'organisme de lapin, spécialement dans l'ovaire. *C. R. Soc. Biol.* **148,** 1210–1213.

Gowen, J. W., and Stadler, J. (1964). Acute irradiation effects on reproductivity of different strains of mice. *In* "Effects of Ionizing Radiation on the Reproductive System" (W. D. Carlson and F. X. Gassner, eds.), pp. 45–58. Pergamon, Oxford.

Green, S. H., and Zuckerman, S. (1951). The number of oocytes in the mature rhesus monkey (*Macaca mulatta*). *J. Endocrinol.* **7,** 194–202.

Green, S. H., and Zuckerman, S. (1954). Further observations on oocyte numbers in mature rhesus monkeys (*Macaca mulatta*). *J. Endocrinol.* **10,** 284–290.

Haas, R. J. and Fliedner, T. M. (1971). The effect of tritiated thymidine on the oocytes of foetal rats following maternal infusion in pregnancy. *Int. J. Radiat. Biol.* **19,** 197–200.

Hahn, E. W. (1972). Litter size increases in the X-irradiated hamster and its relationship to dose. *J. Anim. Sci.* **22,** 649–651.

Hahn, E. W., and Birrell, J. D. (1969). Differences in response of Long-Evans, Sprague-Dawley, Wistar and Sherman rats to X-ray induced increases in litter size. *Lab. Anim. Care* **19,** 720–723.

Hahn, E. W., and Morales, R. L. (1964). Superpregnancy following pre-fertilization X-irradiation of the rat. *J. Reprod. Fertil.* **7,** 73–78.

Hahn, E. W., and Ward, W. F. (1971). Changes in ovarian intravascular compartment prior to superovulation in X-irradiated rats. *Radiat. Res.* **46**, 192–198.

Hartman, C. G., and Smith C. (1938). Non-effect of irradiation of the hypophysis in sterile monkey females. *Proc. Soc. Exp. Biol. Med.* **39**, 330–332.

Harvey, E. B., and Chang, M. C. (1963). Effects of X-irradiation of ovarian ova on the morphology of fertilized ova and development of embryos. *J. Cell. Comp. Physiol.* **61**, 133–143.

Henneman, G., and Mellink, J. H. (1968). The chances of genetic damage in the progeny of a woman treated with [131]I. I. Clinical problem and estimation of radiation dose to the gonads. *Folia Med. Neerl.* **11**, 40–47.

Henricson, B., and Nilsson, A. (1965). Effect of radiostrontium on oocytes and follicles of adult mice. *Acta Radiol.* **3**, 296–304.

Henricson, B., and Nilsson, A. (1970). Roentgen ray effects on the ovaries of foetal mice. *Acta Radiol.* **9**, 443–448.

Hill, M. (1967). Non-S-phase incorporation of ³H-thymidine into DNA of X-irradiated mammalian cells. *Int. J. Radiat. Biol.* **13**, 199–203.

Ichikawa, Y., McCracken, J. A., Baird, D. T., and Uno, A. (1968). Effect of X-ray irradiation on ovarian steroid secretion in the sheep. *Proc. Int. Congr. Endocrinol., 3rd, 1968* Excerpta Med. Found. Int. Congr. Ser. No. 157, Abstract 210.

Ingram, D. L. (1958). Fertility and oocyte numbers after X-irradiation of the ovary. *J. Endocrinol.* **17**, 81–90.

Ingram, D. L. (1962). Atresia. *In* "The Ovary" (S. Zuckerman, ed.), 1st ed., Vol. 1, pp. 247–273. Academic Press, New York.

Ioannou, J. M. (1964). Oogenesis in the guinea pig. *J. Embryol. Exp. Morphol.* **12**, 673–691.

Ioannou, J. M. (1969). Radiosensitivity of oocytes in post-natal guinea pigs. *J. Reprod. Fertil.* **18**, 287–295.

Israel, S. L. (1952). The empiric use of low dosage irradiation in amenorrhea. *Am. J. Obstet. Gynecol.* **64**, 971–983.

Israel, S. L. (1966). Radiation therapy and ovulation. *In* "Ovulation" (R. B. Greenblatt, ed.), pp. 91–97. Lippincott, Philadelphia, Pennsylvania.

Jones, E. C., and Krohn, P. L. (1961). The relationship between age and numbers of oocytes and fertility in virgin and multiparous mice. *J. Endocrinol.* **21**, 469–495.

Kaplan, I. I. (1958). The treatment of female sterility with X-ray therapy directed to the pituitary and ovaries. *Am. J. Obstet. Gynecol.* **76**, 447–453.

Kaplan, I. I. (1959). Genetic effects in children and grandchildren of women treated for infertility and sterility by roentgen therapy. *Radiology* **72**, 518–521.

Karmysheva, Y. (1963). Oocyte maturation division in the mammalian ovary during various phases of the estrous cycle and in acute radiation. *Fed. Proc. Fed. Am. Soc. Exp. Biol., Suppl.* **22**, T314–T316.

Kiselieski, W. E., Samuels, L. D., and Hiley, P. C. (1964). Dose-effect measurements of radiation following administration of tritiated thymidine. *Nature (London)* **202**, 458–459.

Kitaeva, O. N. (1960). Cited by Mandl (1964).

Kofman-Alfaro, S., and Chandley, A. C. (1971). Radiation induced DNA synthesis in spermatogenic cells of the mouse. *Exp. Cell Res.* **69**, 33–44.

Kurland, G. S., Freedberg, A. S., and McManus, M. J. (1951). Distribution of I[131] in tissue obtained at necropsy or at surgical operation in man. *J. Clin. Endocrinol. Metab.* **11**, 843–856.

Lacassagne, A. (1913). "Etude histologique et physiologique des effets produits sur l'ovaire par les rayons X." Thèse Médecine, Lyon.

Lacassagne, A. (1936). Untersuchungen über die Radiosensibilität des Corpus luteum und der Uterusschleimhaut mit Hilfe eines küntslich erzengten Deziduoms beim Kaninchen. *Strahlentherapie* **56**, 621–625.

Lacassagne, A. (1937). Etude de la radiosensibilité de corps jaune et de la muqueuse utérine au moyen de déciduome artificiel chez la lapine. *Radiophysiol. Radiother.* **3**, 315–322.

Lacassagne, A., Duplan, J. F., Marcovich, H., and Raynaud, A. (1962). The action of ionizing radiations on the mammalian ovary. *In* "The Ovary" (S. Zuckerman, ed.), 1st ed., Vol. 2, pp. 463–532. Academic Press, New York.

Lawrence, C. W. (1971). "Cellular Radiobiology." Arnold, London.

Lea, D. E. (1955). "Action of Radiations on Living Cells." Cambridge Univ. Press, London and New York.

Leblond, C. P., Stevens, C. E., and Bogoroch, R. (1948). Histological localization of newly-formed desoxyribonucleic acid. *Science* **108**, 531–533.

Léonard, A., and Deknudt, G. (1967). Relation between the X-ray dose and the rate of chromosome rearrangements in spermatogonia of mice. *Radiat. Res.* **32**, 35–41.

Levine, W. T., and Witschi, E. (1933). Endocrine reactions in female rats after X-ray treatment of the ovaries. *Proc. Soc. Exp. Biol. Med.* **30**, 1152–1153.

Levinson, W. (1966). Toxic effect of X-irradiated medium on chick embryo cells. *Exp. Cell Res.* **43**, 398–413.

Lima de Faria, A., and Borum, K. (1962). The period of DNA synthesis prior to meiosis in the mouse. *J. Cell Biol.* **14**, 381–388.

Lin, T. P., and Glass, L. E. (1962a). Effects of *in vitro* X-irradiation on the survival of mouse eggs. *Radiat. Res.* **16**, 736–745.

Lin, T. P., and Glass, L. E. (1962b). Cause of pre-implantation death of mouse oocytes X-irradiated *in vitro*. *Anat. Rec.* **142**, 253.

Lindop, P. J., Rotblat, J., and Vatistas, S. (1966). The effect of age and hypoxia on the long-term response of the ovary to radiation. *In* "Radiation and Ageing" (P. J. Lindop and G. A. Sacher, eds.), pp. 307–324. Taylor & Francis, London.

Lüning, K. G., and Searle, A. G. (1971). Estimates of the genetic risk from ionizing irradiation. *Mutat. Res.* **12**, 291–304.

Lyon, M. F. (1961). Gene action in the X-chromosome of the mouse (*Mus musculus*, L). *Nature (London)* **190**, 372–373.

Lyon, M. F., and Smith, B. D. (1971). Species comparisons concerning radiation-induced dominant lethals and chromosome aberrations. *Mutat. Res.* **11**, 45–58.

MacDonald, G. J., Keyes, P. L., and Greep, R. O. (1969). Steroid secreting capacity of X-irradiated rat ovaries. *Endocrinology* **84**, 1004–1008.

McLaren, A., and Michie, D. (1959). Superpregnancy in the mouse. I. Implantation and foetal mortality after induced superovulation in females at various ages. *J. Exp. Biol.* **36**, 281–300.

Mandl, A. M. (1959). A quantitative study of the sensitivity of oocytes to X-irradiation. *Proc. R. Soc. London, Ser. B* **150**, 53–71.

Mandl, A. M. (1963). The radiosensitivity of oocytes at different stages of maturation. *Proc. R. Soc. London, Ser. B* **158**, 119–141.

Mandl, A. M. (1964). The radiosensitivity of germ cells. *Biol. Rev. Cambridge Philos. Soc.* **39**, 288–371.

Mandl, A. M. (1965). Superovulation following ovarian X-irradiation. *J. Reprod. Fertil.* **8**, 375–396.

Mandl, A. M., and Zuckerman, S. (1951). The relation of age to numbers of oocytes. *J. Endocrinol.* **7**, 190–193.

Mandl, A. M., and Zuckerman, S. (1956a). The reactivity of the X-irradiated ovary of the rat. *J. Endocrinol.* **13**, 243–261.

Mandl, A. M., and Zuckerman, S. (1956b). Changes in the mouse ovary after X-ray sterilization. *J. Endocrinol.* **13**, 262–268.

Matsumoto, A. (1971). Effects of perinatal X-ray irradiation on subsequent development of ovaries in rats. *Annot. Zool. Jpn.* **44**, 99–104.

Matsumoto, A. (1972). Effects of X-ray irradiation in early postnatal rats on subsequent development of ovaries. *Annot. Zool. Jpn.* **45**, 80–87.

Meyer, M. B., Mertz, T., and Diamond, E. L. (1969). Investigation of the effects of prenatal X-ray exposure of human oogonia and oocytes as measured by later reproductive performance. *Am. J. Epidemiol.* **89**, 619–635.

Miller, O. L., Carrier, R. F., and von Borstel, R. C. (1965). *In situ* and *in vitro* breakage of lampbrush chromosomes by X-irradiation. *Nature (London)* **206**, 905–908.

Mintz, B. (1959). Continuity of the female germ cell line from embryo to adult. *Arch. Anat. Microsc. Morphol. Exp.* **48**, 155–172.

Moawad, A. H., Rakoff, A. E., and Kramer, S. (1965). A histologic study of the effects of low-dosage irradiation of rabbit ovaries. *Fertil. Steril.* **16**, 370–383.

Mole, R. H. (1959). Impairment of fertility by whole body irradiation. *Int. J. Radiat. Biol.* **1**, 107–114.

Mole, R. H., and Papworth, D. G. (1966). The sensitivity of rat oocytes to X-rays. *Int. J. Radiat. Biol.* **10**, 609–615.

Mondorf, L., and Faber, M. (1968). The influence of radiation on human fertility. *J. Reprod. Fertil.* **15**, 165–169.

Moricard, R., and Gothié, S. (1955). Etude de la répartition du S^{35} dans les cellules folliculeuses périovocytaires au cours de l'ovogénèse et de la terminaison de la première mitose de maturation chez la lapine adulte. *C. R. Soc. Biol.* **149**, 1918–1922.

Murray, J. M. (1931). A study of the histological structure of mouse ovaries following exposure to Roentgen irradiation. *Am. J. Roentgenol. Radium Ther.* [N. S.] **25**, 1–46.

Nash, D. J. (1969). Lifetime reproductive performance of mice exposed as embryos to X-irradiation. *Biol. Bull. (Woods Hole, Mass.)* **137**, 189–201.

Neel, J. V. (1963). "Changing Perspectives on the Genetic Effects of Radiation." Thomas, Springfield, Illinois.

Neel, J. V., and Schull, W. J. (1956). Studies on the potential genetic effects of the atomic bomb. *Acta Genet.* **6**, 183–196.

Nilsson, A., and Henricson, B. (1969). The effect of ^{90}Sr on the ovaries of the fetal mouse. *In* "Radiation Biology of the Fetal and Juvenile Mammal" (M. R. Sikov and D. D. Mahlum, eds.), CONF-690501, pp. 313–322. U.S. At. Energy Comm., Washington, D.C.

Oakberg, E. F. (1966). Effect of 25r of X-rays at 10 days of age on oocyte numbers and fertility of female mice. *In* "Radiation and Ageing" (P. J. Lindop and G. A. Sacher, eds.), pp. 293–306. Taylor & Francis, London.

Oakberg, E. F. (1967). ^{3}H-uridine labeling of mouse oocytes. *Arch. Anat. Microsc. Morphol. Exp.* **56**, Suppl. 3–4, 171–184.

Oakberg, E. F. (1968). Relationship between stage of follicular development and RNA synthesis in the mouse oocyte. *Mutat. Res.* **6**, 155–165.

Oakberg, E. F., and Clark, E. (1964). Species comparisons of radiation response of the gonads. *In* "Effects of Ionizing Radiation on the Reproductive System" (W. D. Carlson and F. X. Gassner, eds.), pp. 11–24. Pergamon, Oxford.

Odeblad, E. (1952a). A study of the short-time effects of the mouse ovary of internal irradiation with P^{32}. *Acta Radiol.* **38**, 33–40.

Odeblad, E. (1952b). Studies on the effects on the ovarian follicles in the mouse of the internal irradiation with P^{32} with special reference to the quantitative evaluation. *Acta Radiol.* **38**, 375–381.

Odeblad, E., and Boström, H. (1953). A time-picture relation study with autoradiography on the uptake of labelled sulphate in the Graafian follicles of the rabbit. *Acta Radiol.* **39**, 137–140.

Ozzello, L., and Rugh, R. (1965). Acute pathologic alteration in X-irradiated primate fetuses. *Am. J. Roentgenol., Radium Ther. Nucl. Med.* [N. S.] **93**, 209–221.

Painter, R. B., and Cleaver, J. B. (1969). Repair replication, unscheduled DNA synthesis and the repair of mammalian DNA. *Radiat. Res.* **37**, 451–466.

Parkes, A. S. (1928). On the occurrence of the oestrous cycle after X-ray sterilisation. Part IV. Irradiation of the adult during pregnancy and lactation; a general summary. *Proc. R. Soc. London, Ser. B* **102**, 51–62.

Parkes, A. S. (1929). The function of the corpus luteum. I. The mechanism of oestrus inhibition. *Proc. R. Soc. London, Ser. B* **104**, 171–182.

Parkes, A. S., Rowlands, I. W., and Brambell, F. W. R. (1932). Effect of X-ray sterilisation on oestrus in the ferret. *Proc. R. Soc. London, Ser. B* **109**, 425–434.

Parkin, P. (1970). "The effects of X-irradiation on primordial oocytes in the rat." Ph.D. Thesis, University of Birmingham, Birmingham, England.

Parsons, D. F. (1962). An electron microscope study of radiation damage in the mouse oocyte. *J. Cell Biol.* **14**, 31–48.

Penchansky, L., and Gonzalez Cueto, D. (1968). Histological study of mice gonads treated with ^{32}P during their fetal life. *Medicina (Mexico City)* **28**, Supp. 1, 161–164.

Peters, H. (1961). Radiation sensitivity of oocytes at different stages of development in the immature mouse. *Radiat. Res.* **15**, 582–593.

Peters, H. (1966). Age dependence of radiosensitivity of mouse oocytes. *In* "Radiation and Ageing" (P. J. Lindop and G. A. Sacher, eds.), pp. 281–292. Taylor & Francis, London.

Peters, H. (1968). Late effects of irradiation in infancy on the mouse ovary. *In* "Symposium on the Postnatal Development of Phenotype", Liblice, p. 197. Academia, Prague.

Peters, H. (1969). The effect of radiation in early life on the morphology and reproductive function of the mouse ovary. *Adv. Reprod. Physiol.* **4**, 149–185.

Peters, H., and Borum, K. (1961). The development of mouse ovaries after low dose irradiation at birth. *Int. J. Radiat. Biol.* **3**, 1–6.

Peters, H., and Levy, E. (1963). The effect of irradiation in infancy on the fertility of female mice. *Radiat. Res.* **18**, 421–428.

Peters, H., and Levy, E. (1964). Effect of irradiation in infancy on the mouse ovary. A quantitative study of oocyte sensitivity. *J. Reprod. Fertil.* **7**, 37–45.

Puck, T. T. (1960). Radiation and the human cell. *Sci. Am.* **202**, 142–153.

Rasmussen, R. E., and Painter, R. B. (1966). Radiation-stimulated DNA synthesis in cultured mammalian cells. *J. Cell Biol.* **29**, 11-19.

Reddi, O. S. (1971a). Effect of ^{131}I on spermatogonia and oocytes in mice. *Indian J. Med. Res.* **59**, 494-498.

Reddi, O. S. (1971b). Effect of ^{131}I on the survival of oocytes in cows. *Int. J. Radiat. Biol.* **19**, 497-498.

Reddi, O. S. (1973). Follicular damage to bovine ovary from radioiodine. *Indian J. Med. Sci.* **42**, 648-650.

Reddi, O. S., Vasudevan, B., Reddy, P. P., Rani, K. P., Kumari, I. R., and Vijayalaxmi, C. (1968). Effect of ^{32}P on the survival of spermatogonia and oocytes in mice. *Nature (London)* **219**, 5161.

Regaud, C., and Lacassagne, A. (1913). Sur les conditions de la stérilisation des ovaires par les rayons X. *C. R. Soc. Biol.* **74**, 783-786.

Reifferscheid, K. (1910). Histologische Untersuchungen über die Beeinflussung menschlicher und tierischer Ovarien durch Röntgenstrahlen. *Z. Roentgenkol.* **12**, 233-254.

Reifferscheid, K. (1914). Die Einwirkung der Röntgenstrahlen auf tierische und menschliche Eierstöcke. *Strahlentherapie* **5**, 407-426.

Reither, K. H., and Lang, G. (1955). Zur Frage der schädigenden Wirkung von P[32] auf das Ovargewebe. *Strahlentherapie* **98**, 453-463.

Richold, M., and Arlett, C. F. (1972). Repair replication and unscheduled DNA synthesis in synchronous mammalian cells. *Int. J. Radiat. Biol.* **21**, 127-136.

Rugh, R. (1953). Selective sterilization of young female mice by radioiodine transmitted through the mother's milk. *Proc. Soc. Exp. Biol. Med.* **83**, 762-764.

Rugh, R. (1960). General biology: gametes, the developing embryo and cellular differentiation. *In* "Mechanisms in Radiobiology" (M. Errera and A. Forssberg, eds.), Vol. 2, pp. 1-94. Academic Press, New York.

Rugh, R., and Jackson, S. (1958). Effects of fetal X-irradiation upon the subsequent fertility of the offspring. *J. Exp. Zool.* **138**, 209-221.

Rugh, R., and Wohlfromm, M. (1964). X-irradiation sterilization of the pre-mature female mouse. *Atompraxis* **10**, 511-518.

Russell, L. B. (1964). Experimental studies on mammalian chromosome aberrations. *In* "Mammalian Cytogenetics and Related Problems in Radiobiology" (C. Pavan *et al.,* eds.), pp. 61-86. Pergamon, Oxford.

Russell, L. B. (1968). The use of X-chromosome anomalies for measuring radiation effects in different germ cell stages of the mouse. *In* "Effects of Radiation on Meiotic Systems," pp. 27-40. IAEA, Vienna.

Russell, L. B. (1971). Definition of functional units in a small chromosomal segment of the mouse and its use in interpreting the nature of radiation-induced mutations. *Mutat. Res.* **11**, 107-123.

Russell, L. B., and Russell, W. L. (1956). The sensitivity of different stages of oogenesis to the radiation induction of dominant lethals and other changes in the mouse. *In* "Progress in Radiobiology" (J. S. Mitchell, B. E. Holmes, and C. L. Smith, eds.), pp. 187-195. Oliver and Boyd, Edinburgh.

Russell, L. B., and Wickham, I. (1957). The incidence of disturbed fertility among male mice conceived at various intervals after irradiation of the mother. *Genetics* **42**, 392-393.

Russell, L. B., Stelzner, K. F., and Russell, W. L. (1959). Influence of dose rate on radiation effect on fertility of female mice. *Proc. Soc. Exp. Biol. Med.* **102**, 471-479.

Russell, L. B., Badgett, S. K., and Saylors, C. L. (1960). Comparison of the effects of acute, continuous and fractionated irradiation during embryonic development. *Int. J. Radiat. Biol. Suppl.* pp. 343–359.

Russell, W. L. (1957). Shortening of life in the offspring of male mice exposed to neutron radiation from the atomic bomb. *Proc. Natl. Acad. Sci. USA* **43**, 324–329.

Russell, W. L. (1963). The effect of radiation dose rate and fractionation on mutation in mice. *In* "Repair from Genetic Radiation Damage and Differential Radiosensitivity of Germ Cells" (F. H. Sobels, ed.), pp. 205–217. Pergamon, Oxford.

Russell, W. L. (1965). Effect of the interval between irradiation and conception on the mutational frequency in female mice. *Proc. Natl. Acad. Sci. USA* **54**, 1552–1557.

Russell, W. L. (1967a). Factors that affect the radiation induction of mutations in the mouse. *An. Acad. Bras. Cienc.* **39**, Suppl., 66–75.

Russell, W. L. (1967b). Recent studies on the genetic effects of radiation in mice. *In* "Biological Interpretations of Dose from Accelerator-Produced Radiation" (R. Wallace, ed.), CONF-670305, pp. 81–87. U.S. At. Energy Comm., Washington, D.C.

Russell, W. L. (1967c). Repair mechanisms in radiation mutation induction in the mouse. *Brookhaven Symp. Biol.* **20**, 179–189.

Russell, W. L. (1969). Observed mutation frequency in mice and the chain processes affecting it. *In* "Mutation as a Cellular Process," Ciba Found. Symp., 1969 pp. 216–228.

Russell, W. L., and Kelly, E. M. (1966). Mutation frequency in female mice exposed to high intensity X-irradiation delivered in small fractions. *Science* **154**, 427–428.

Russell, W. L., and Russell, L. B. (1959). The genetic and phenotypic characteristics of radiation-induced mutations in mice. *Radiat. Res., Suppl.* **1**, 296–305.

Russell, W. L., Russell, L. B., and Kimball, A. W. (1954). The relative effectiveness of neutrons from a nuclear detonation and from a cyclotron in inducing dominant lethals in the mouse. *Am. Nat.* **88**, 269–286.

Russell, W. L., Russell, L. B., and Kelly, E. M. (1958). Radiation dose rate and mutation frequency. *Science* **128**, 1546–1550.

Russell, W. L., Russell, L. B., and Cupp, M. B. (1959a). Dependence of mutation frequency on radiation dose-rate in female mice. *Proc. Natl. Acad. Sci. USA* **45**, 18–23.

Russell, W. L., Russell, L. B., Steele, M. H., and Phipps, E. L. (1959b). Extreme sensitivity of an immature stage of the mouse ovary to sterilization by X-irradiation. *Science* **129**, 1288.

Samuels, L. D. (1966). Effects of polonium-210 on mouse ovaries. *Int. J. Radiat. Biol.* **11**, 117–129.

Schlager, G., and Dickie, M. M. (1966). Spontaneous mutation rates at five coat color loci in mice. *Science* **151**, 205–206.

Schlager, G., and Dickie, M. M. (1967). Spontaneous mutations and mutation rates in the house mouse. *Genetics* **57**, 319–330.

Schlager, G., and Dickie, M. M. (1971). Natural mutation rates in the house mouse: estimates for five specific loci and dominant mutations. *Mutat. Res.* **11**, 89–96.

Schmidt, I. G. (1939). Changes in the genital tracts of guinea pigs associated with cystic and 'interstitial gland' ovaries of long duration. *Endocrinology* **24**, 69–81.

Schuetz, A. W. (1969). Oogenesis: processes and their regulation. *Adv. Reprod. Physiol.* **4**, 99–148.

Searle, A. G. (1962). A review of the factors affecting the incidence of radiation-induced lethals in mammals. *Strahlentherapie, Suppl.* **51**, 215–222.

Searle, A. G. (1966). Progress in mammalian radiation genetics. In "Radiation Research" (G. Silini, ed.), pp. 469–481. North-Holland Publ., Amsterdam.

Searle, A. G. (1971). Attempts to induce translocations in female mice. *Proc. Int. Congr. Radiat. Res., 4th, 1970* p. 194.

Searle, A. G., and Phillips, R. (1968). Genetic insensitivity of the mouse dictyate oocyte to chronic irradiation. In "Effects of Radiation on Meiotic Systems," pp. 17–23. IAEA, Vienna.

Searle, A. G., and Phillips, R. (1971). The mutagenic effectiveness of fast neutrons in male and female mice. *Mutat. Res.* **11**, 97–105.

Searle, A. G., Berry, R. J., and Beechey, C. V. (1970). Cytogenetic radiosensitivity and chiasma frequency in wild-living male mice. *Mutat. Res.* **9**, 137–140.

Shapiro, N. I., Plotnikova, Ye. D., Strashnenko, S. I., and Sulikova, V. I. (1961). Relative genetic radiosensitivity of various species of mammals. *Radiobiologiya* **1**, 93–103.

Singh, R. P., and Carr, D. H. (1966). The anatomy and histology of XO human embryos and fetuses. *Anat. Rec.* **155**, 369–384.

Singh, R. P., and Carr, D. H. (1967). Anatomic findings in human abortions of known chromosomal constitution. *Obstet. Gynecol.* **29**, 806–818.

Sladić-Simić, D., Zivkovic, N., Pavić, D., and Martinovitch, P. N. (1963). The effects of total-body X-irradiation on the reproductive glands of infant female rats. In "Cellular Basis and Aetiology of Late Somatic Effects of Ionizing Radiation" (R. J. C. Harris, ed.), pp. 327–338. Academic Press, New York.

Slate, W. G., and Bradburg, J. T. (1962). Ovarian function and histology after X-ray irradiation in albino rats. *Endocrinology* **70**, 1–6.

Smith, C. L. (1964). Microbeam and partial cell irradiation. *Int. Rev. Cytol.* **16**, 133–153.

Sobels, F. H., ed. (1963). "Repair from Genetic Radiation Damage and Differential Radiosensitivity of Germ Cells." Pergamon, Oxford.

Sobels, F. H. (1968). The chances of genetic damage in the progeny of a woman treated with [131]I. II. Evaluation of the genetic risk. *Folia Med. Neerl.* **11**, 48–52.

Sobels, F. H. (1969). Estimation of the genetic risk resulting from treatment of women with [131]I. *Strahlentherapie* **138**, 172–177.

Specht, O. (1906). Mikroskopische Befunde an röntgenisierten Kaninchenovarien. *Arch. Gynakol.* **78**, 458–472.

Speert, E. H., Quimby, E. H., and Werner, S. C. (1951). Radioiodine uptake by the fetal mouse thyroid and resultant effect on late life. *Surg., Gynecol. Obstet.* **93**, 230–242.

Stevenson, A. C. (1968). The timing of necessary diagnostic and therapeutic exposures in women—the risks in perspective. *Br. J. Radiol.* **41**, 720–721.

Stewart, A. M. (1969). Radiogenic cancers of children. In "Radiation Biology of the Fetal and Juvenile Mammal" (M. R. Sikov and D. D. Mahlum, eds.), CONF-69051, pp. 681–692. U.S. At. Energy Comm., Washington, D.C.

Szumiel, I., Ziemba-Zak, B., Rosiek, O., Sabliński, J., and Beer, J. Z. (1971). Harmful effects of an irradiated cell culture medium. *Int. J. Radiat. Biol.* **20**, 153–161.

Talbert, G. B., and Krohn, P. L. (1966). Effect of maternal age on viability of ova and uterine support of pregnancy in mice. *J. Reprod. Fertil.* **11**, 399–406.

Terry, J. L., Damewood, L. A., and De Boer, J. (1964). The effects of chronic LD_{50} X-ray doses on the reproductive potential of sheep. In "Effects of Ionizing Radiation on the Reproductive System" (W. D. Carlson and F. X. Gassner, eds.), pp. 333–336. Pergamon, Oxford.

Tillotson, F. W., Rose, R. G., and Warren, S. (1953). A physiopathologic study of the effect of radioactive iodine on the ovary. *Am. J. Roentgenol., Radium Ther. Nucl. Med.* [N. S.] **70**, 599–604.

Tsuda, H. (1965). An electron microscope study on the oogenesis of the mouse, with special reference to the behaviours of oogonia and oocytes at meiotic prophase. *Arch. Histol. Jpn.* **25**, 533–555.

UNSCEAR (1966). "Report of the United Nations Scientific Committee on the Effects of Atomic Radiation," General Assembly, 21st Session, Official Records, Suppl. No. 14 (A/6314). United Nations, New York.

UNSCEAR (1971). "Report of the Scientific Committee on the Effects of Atomic Radiation to the General Assembly, Genetic Effects of Ionizing Radiation," A/ AC.82/R.256/Rev.2. United Nations, New York. Also published as: "Ionizing Radiations: Levels and Effects," U.N. (1972).

Valentini, E. J., and Hahn, E. W. (1971). The indirect effect of radiation on embryonic mortality. *Int. J. Radiat. Biol.* **20**, 259–267.

Van Eck, G. J. V. (1956). Neo-ovogenesis in the adult monkey. Consequences of atresia of oocytes. *Anat. Rec.* **125**, 207–224.

Van Eck, G. J. V. (1959). Effect of low dosage X-irradiation upon pituitary gland and ovaries of the rhesus monkey. *Fertil. Steril.* **10**, 190–202.

Van Eck, G. J. V., and Freud, J. (1949a). I. Structure and function of mouse ovaries after X-raying. *Arch. Int. Pharmacodyn. Ther.* **78**, 49–62.

Van Eck, G. J. V., and Freud, J. (1949b). II. Action of gonadotrophins and of oestrogens on X-rayed mouse ovaries. *Arch. Int. Pharmacodyn. Ther.* **78**, 67–68.

Van Wagenen, G., and Gardner, W. U. (1960). X-irradiation of the ovary in the monkey (*Macaca mulatta*). *Fertil. Steril.* **11**, 291–302.

Vuksanovic, M. M. (1966). Pregnancy following ovarian irradiation. *Am. J. Roentgenol., Radium Ther. Nucl. Med.* [N. S.] **97**, 951–956.

Warren, S., MacMillan, J. C., and Dixon, F. J. (1950). Effects of internal irradiation of mice with P^{32}. Part II. Gonads, kidneys, adrenal glands, digestive tract, spinal cord, lungs and liver. *Radiology* **55**, 557–570.

Westman, A. (1958). The influence of X-irradiation on the hormonal function of the ovary. *Acta Endocrinol. (Copenhagen)* **29**, 334–346.

Yakovleva, L. A., and Novikova, M. I. (1963). Late disturbances in the reproductive functions of the monkey following acute radiation sickness. *Fed. Proc. Fed. Am. Soc. Exp. Biol., Suppl.* **22**, T833–835.

Zuckerman, S. (1965). The sensitivity of the gonads to radiation. *Clin. Radiol.* **16**, 1–15.

Zybina, E. V. (1969). Behaviour of the chromosomal nucleolar apparatus during the growth period of the rabbit oocytes. *Excerpta Med., Sect. I* **23**, 711.

2

The Mechanism of Action of Estrogens and Progesterone

R. B. Heap and Doreen V. Illingworth

I. INTRODUCTION

The steroid nucleus (see Fig. 1) is distributed widely in nature. It has been identified in compounds found in many organisms ranging from algae and higher plants to insects and mammals. Examples of its bizarre distribution include: substances such as antheridiol, a sterol of the fungus *Achlya,* which stimulates the male strain to secrete a hormone that induces the genesis of sex organs in the female (Barksdale, 1969); the sterol diosgenin, the product of a plant indigenous to Mexico that has contributed in no small way to the development of the modern steroid industry because of its suitability as a starting material in the synthesis of progesterone, androgens, and estrogens; and the discovery of appreciable amounts of cholesterol and progesterone in the seeds of Red Delicious apples (Gawienowski and Gibbs, 1968). The steroid nucleus is found in compounds differing as widely in their biological properties as bile acids that aid lipid absorption from the intestine to plant glycosides that are used in cardiac therapy and in the making of arrow poisons, and ecdysone which is concerned with molting in insects. Thus, in the synthesis and secretion of steroid hormones, mammalian and nonmammalian species are making use of a basic molecular structure that is commonly found in nature to regulate their reproductive processes.

The synthesis of the ovarian steroid hormones, estrogens and progesterone, is in no sense a unique feature of the ovary, for other endocrine organs, such as the placenta and the adrenals in the female and the testes and the adrenals in the male, possess similar steroidogenic properties. The role of estrogens and progesterone in reproduction became prominent during the evolution of viviparity with the specialization of fetal and maternal tissues in the formation of the placenta, the retention of young within the female genital tract, the preparation of the mammary gland for lactation, and the subsequent parental care of offspring (Amoroso, 1968). Ovarian steroids, together with the hormones secreted by the pituitary and placenta, play a major role in the regulation of these events. In this chapter we shall be concerned with how ovarian steroid hormones are adapted to participate as regulators of cell function, and how they interact with one another to produce their complex and distinctive effects. Special consideration will be given to the actions of estradiol-17β and progesterone (Fig. 1), since these ovarian steroids are the most biologically active representatives of their groups so far found in nature, and to the target cells of the oviduct and uterus whose response to the action of these hormones has been studied in detail.

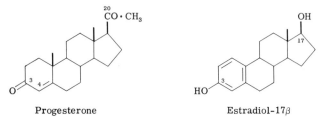

Progesterone   Estradiol-17β

Fig. 1.   The structure of progesterone and estradiol-17β.

II. MECHANISMS OF HORMONE ACTION

It has been observed that "ideas about hormone action, like styles in women's clothing, are subject to changes in fashion" (Hechter and Halkerston, 1964). A brief survey of some of the more important theories of hormone action will show that the analogy may be even more accurate than at first appears, for theories, like fashions, can be found to regain lost favor. Recent attempts to discover the primary action of steroids have emphasized the important role of hormone receptors located in target cells (Bush, 1965; Baulieu *et al.*, 1971). Yet, it is many years since Ehrlich adopted the "lock and key" hypothesis of Emil Fischer to describe how drugs and dyes interact with receptors of specific and complementary configuration and since Clark (1937) proposed that the action of drugs was proportional to the fraction of receptors they occupied, reaching a maximum effect when the available receptor sites were saturated. These early ideas of tissue receptors, though modified in the light of modern findings, were rapidly extended to cover all classes of humorally active agents, and are supported by the results of present investigations into the primary action of estrogens and progesterone.

Much is known about the physiology of steroid secretion and the biochemistry of steroid synthesis. The study of the biological functions of steroids has naturally lead to increased interest in the actions of these ubiquitous compounds at a cellular and molecular level. Grant (1969) has discussed the experimental approaches that have been used to study the mechanism of steroid action; the analytical approach, whereby the early events that occur in target cells after exposure to hormone treatment have been subjected to a form of biochemical dissection, and the inductive approach, whereby a theoretical model of hormone action is proposed and tested by an investigation of the interaction between the steroid and a particular cell component, with the aim that a cell's response to hormone treatment can be reconstructed. From these investigations there is growing evidence that the mechanisms by which ovarian steroids act are similar to those discovered for other hormones.

An early view of hormone action proposed that steroids functioned as regulators of cell metabolism through their action on specific enzymes. Green (1941) suggested that substances functioning in small amounts in biological systems did so by influencing enzyme systems. If this were true for estrogens and progesterone, the control of key enzymes that act as "pacemakers" regulating important metabolic pathways could provide the essential link between the steroid and its known biological effect (Tepperman and Tepperman, 1960). Even though it has been demonstrated that estrogens act as co-enzymes for transhydrogenation *in vitro* (Villee *et al.,* 1960; Talalay and Williams-Ashman, 1960), this theory of enzyme regulation has fallen into disrepute, and a detailed appraisal by Hechter and Halkerston in 1964 produced little evidence that steroids act directly on enzyme systems in eukaryotes. There is ample evidence in bacteria (prokaryotes), however, that exposure to steroids can effectively induce steroid dehydrogenase enzymes (Δ^5-3-ketosteroid isomerase of *Pseudomonas testeroni*; Talalay and Wang, 1955), though it should be noted that steroids themselves are absent from bacteria.

A different mechanism of action has been proposed by those who consider that hormones regulate the permeability of the cell membrane. In this way steroids such as estrogens and progesterone may influence membrane permeability to small molecules and ions, as at the onset of myometrial contractions during labor (Coutinho, 1965; Kao, 1967; Carsten, 1968). Kroeger (1967) has argued that changes in the intracellular concentrations of certain ions *modify* gene activity and that solutions of high ionic strength mimic the actions of androgen- and estrogen-induced RNA synthesis. Although the effects of ions and hormones are not additive he suggests a common point of attack for both agents. But although steroid hormones are known to influence the transport of certain compounds (e.g., amino acids, RNA precursors) into cells, it cannot be said with certainty that this is due either to a direct steroid–membrane interaction or some other steroid-induced event.

A dominant theme of recent studies has been the finding that a number of growth and developmental hormones regulate the biochemical machinery that is responsible for protein synthesis in target cells (Korner, 1964; Williams–Ashman *et al.,* 1964; Tata, 1968, 1970; Mueller, 1971). There are many metabolic actions of hormones that are not involved with protein synthesis (Hechter and Halkerston, 1964); however, those actions which influence growth and development require the synthesis of new proteins (Tata, 1970). The analytical approach has revealed a great deal about the biochemistry of target cells when influenced by steroid hormones, but as yet it has not shown precisely how steroids stimulate protein synthesis. Two explanations are topical. One theory is that hormones regulate protein synthesis by a derepression of gene activity which influences the transcription of DNA (Karlson, 1963; Sekeris, 1967). This explanation has been derived from the classical studies of Jacob and Monod (1961) who demonstrated that microbial protein synthesis is regulated by messenger RNA (mRNA) synthesis. However, this explanation of hormone action has been criticized because protein synthesis in eukaryotic cells may be regulated in ways that are not directly associated with mRNA transcriptions (Tomkins and Ames, 1967; Tomkins and Thompson, 1967; Tata, 1970; Ohno, 1971) and because the evidence for a direct interaction between steroid hormones and genes (or repressors) is so far inconclusive. A second explanation of hormone-induced protein synthesis contends that steroids affect the translation of the RNA template. According to this view, specific hormones block either the action or the production of some hypothetical "translation repressor." The removal of repressor molecules allows the translation of specific mRNA to proceed (Tomkins *et al.*, 1969).

Because of the multiplicity of hormone actions, it would be presumptuous to suggest that any single explanation adequately explains how steroids act to stimulate protein synthesis. The search for the earliest detectable change in a responsive cell has brought us nearer to an understanding of the primary action of individual hormones and, with this in mind, the actions of estrogens and progesterone on the rat uterus and chick oviduct have been intensively studied. Whereas the biochemist is intent on defining this critical event in detail, the physiologist and endocrinologist is also concerned with knowing how these primary events result in the uterus eventually becoming receptive to an implanting blastocyst and, later, ejecting a mature fetus— two of many processes intimately associated with the long-term or chronic effects of estrogen and progesterone action.

III. DYNAMICS OF STEROID METABOLISM

If estrogens and progestagens are to function as regulators of cell metabolism in target tissues and influence the formation of specific proteins, it is to be expected that they must undergo a rapid turnover to fulfill this role effectively. In the absence of rapid metabolism, gross accumulation at target sites would quickly diminish their regulatory efficacy and, in view of their high potency, would give rise to undesirable toxic effects. The pathways of catabolism include hydroxylation, hydrogenation, and subsequent conjugation of progesterone and oxidation, hydroxylation, and conjugation of estradiol-17β. The pathways of their metabolism have been extensively reviewed by Dorfman and Ungar (1965).

The physicochemical properties of estradiol-17β, estrone, and progesterone are consistent with their regulatory role (Table I). Both estrone and estradiol-17β (C_{19} steroids) as well as progesterone (C_{21} steroid) have high lipid solubilities and high lipid–water distribution coefficients which aid ready diffusion into a lipid phase. Distribution coefficients (K) are considerably greater than those of other biologically active steroids such as testosterone, cortisol, and aldosterone but are similar to that of pregnenolone, the precursor of progesterone (Heap *et al.*, 1970). The rate at which estradiol, estrone, and progesterone are removed from blood is high and is

TABLE I

Aqueous and Lipid (Phosphatidylcholine) Solubilities and Distribution Coefficients of
Estradiol-17β, Estrone, and Progesterone[a]

	Progesterone	Estradiol-17β	Estrone
Aqueous solubility in buffered KCl at 37°C, μmoles/liter	48.0	15.0	8.2
Lipid solubility at 37°C, nmoles/ μmole lipid	99.3	26.0	33.6
Distribution coefficient (K)[b] at 37°C	2601	2314	4428

[a] Data from Heap et al. (1970).

[b] $K = \dfrac{\text{Solubility in lipid (nmoles/}\mu\text{moles lipid)}}{\text{Solubility in KCl (nmoles/ml Tris-buffered KCl)}}$

related approximately to the K value of the steroid in those species so far
investigated. Thus, the metabolic clearance rate (MCR) in the human—the
amount of whole blood completely and irreversibly cleared of steroid in unit
time (Tait and Burstein, 1964)—is higher for progesterone, pregnenolone,
estrone, and estradiol-17β (K, 2314 to 4428) than for cortisol, aldosterone,
testosterone, and dehydroepiandrosterone sulfate (K, 45 to 514). Androstenedione, however, is anomalous for it has a high MCR from blood in man
but a low distribution coefficient (Horton and Tait, 1966; Baird et al., 1969;
Heap et al., 1970). The high clearance of progesterone frequently exceeds
hepatic blood flow, so that even if the liver extracts all the progesterone it
receives, extrahepatic metabolism must take place. In man it has been calculated that the splanchnic clearance and extrasplanchnic clearance of
progesterone are approximately the same (Little et al., 1966). A similar
conclusion has been reached in studies of conscious sheep in which
splanchnic extraction of progesterone is about 90% and accounts for
approximately one-half the total clearance from blood (Bedford et al.,
1974). Among extrahepatic tissues, appreciable amounts of progesterone
are removed by the head and viscera (and probably kidney) in sheep (Bedford et al.,1972b;1973).In addition there is considerable progesterone uptake
by the goat mammary gland—a target organ for the steroid (Heap and
Linzell, 1966; Heap et al., 1969; Slotin et al., 1970).

The rapid removal of steroid from blood has been studied by two techniques; either by a single intravenous injection or by a constant infusion of
labeled steroid until a steady state is reached. In the latter instance the
removal of steroid from blood after stopping the infusion has been measured
in the same way as after a single intravenous injection. The rate of removal

can be described by a double exponential function as would be expected if the compound were distributed within the body between two exchangeable compartments. The relative volumes of the two compartments indicate that progesterone is rapidly removed from blood rather than rapidly distributed in a large volume throughout the body. In ovariectomized women, the half-life for the rate of removal from the "inner" compartment (volume equal to 12 liters) is short (0.96 minutes) while that of the "outer" compartment (15–27 liters) is longer (10.7 minutes) (Little et al., 1966). In pregnant sheep (Slotin et al., 1971) and guinea pigs (Illingworth, 1970) the first exponent has a short half-life (2.29 and 2.4 minutes) while that of the second exponent is longer (49 and 28 minutes, respectively).

Another feature of ovarian steroids is their relatively low concentration in arterial blood (Grant, 1969). On an average, the blood concentration of progesterone is 10^{-9} to 10^{-7} M, but there are wide variations depending on the species and the time of the reproductive cycle (see Langecker and Damrosch, 1968; Heap et al., 1972). The absolute blood level of a particular hormone can be misleading, for it is the product of a dynamic equilibrium of the hormone involving its rate of production, its biological half-life and its distribution within the body. A more important consideration is the amount of steroid presented to the cell membrane of target tissues and its concentration at intracellular receptor sites. This raises the fundamental question of the mechanism of cell recognition and the availability of a hormone to target cells. Ovarian steroid hormones are transported from their site of synthesis to their site of action in association with plasma proteins (Tait and Burstein, 1964; Sandberg et al., 1966; Westphal, 1971). This association is believed to affect the availability of steroids to target cells. A large proportion of steroid in blood is bound to albumin and is readily available for tissue metabolism. Since the association constant is low, the protein–steroid complex rapidly dissociates. Proteins such as corticosteroid-binding globulin (CBG, transcortin; Sandberg et al., 1966; Seal and Doe, 1966) and sex hormone-binding globulin (SHBG; Mercier-Bodard et al., 1970) possess a high affinity but a low capacity for progesterone and estradiol-17β, respectively, and probably provide a reservoir of steroid in circulation that dissociates only slowly. These proteins may serve to protect the steroid against metabolism and prevent the accumulation of biologically active compounds in an unbound and active form in extravascular spaces where their effects could be deleterious (Tait and Burstein, 1964; Baird et al., 1969; Westphal, 1971).

Investigations have suggested that when a steroid, such as cortisol, is bound to a high-affinity protein (CBG) it is biologically inert (Sandberg et al., 1966). Cortisol clearance from blood is much reduced when the concentration of CBG is increased in pregnancy (Seal and Doe, 1966). With respect to progesterone, a notable example of this phenomenon is found in the guinea pig when, during pregnancy, the MCR decreases by about 90% and the absolute level of progesterone in blood increases about one hundred-fold. There is only about a tenfold increase in the actual production rate of the hormone (Illingworth *et al.,* 1970; Challis *et al.,* 1971). This "progesterone-conserving" mechanism occurs after about the fifteenth day of gestation and it is associated with an increased production of a plasma protein with a high affinity for progesterone (progesterone-binding protein, PBP: Heap, 1969; Diamond *et al.,* 1969; Milgrom *et al.,* 1970a; progesterone-binding globulin, PBG: Westphal, 1971). Association of progesterone with PBG presumably reduces its metabolism in the liver (and possibly extrahepatic tissues). The presence of a "progesterone-conserving" mechanism has been found in other hystricomorph rodents, including the coypu (Illingworth and Heap, 1971; Heap and Illingworth, 1974), and it may be that this mechanism represents one of several ways in which the progesterone requirements of pregnancy are met. At one extreme the production rate of progesterone increases, as in women; at the other extreme the rate of clearance is reduced, as in guinea pigs and coypu (Fig. 2).

The rates of clearance of estradiol-17β, estrone, and progesterone from blood are summarized in Table II where MCR has been expressed, for comparative purposes, relative to surface area ($W^{0.75}$; Kleiber, 1965). It is notable that in the species studied so far, the MCR ($1/\text{minute}/W^{0.75}$) of all three steroids (and of cortisol; Paterson and Harrison, 1968) is greater in sheep than in women, rats, or guinea pigs. In man, the MCR of estradiol-17β corrected for surface area is similar to that of rats, whereas the MCR of estrone is somewhat higher; the MCR of progesterone in man is slightly higher than that of nonpregnant guinea pigs, but this may be because the guinea pigs were lightly anesthetized. An explanation for the high values in sheep may be the absence of appreciable concentrations of high-affinity binding proteins such as CBG (Paterson and Hills, 1967) or of other plasma proteins with a high affinity for estrogens and progesterone.

Whereas the availability of ovarian steroids to target cells and their rate of removal from blood may be modified by high-affinity plasma proteins, recent findings raise some doubt as to whether steroids bound in this way are biologically inactive. Milgrom and Baulieu (1970a) have reported the intracellular identification of a progesterone-binding protein in the rat uterus which so far cannot be distinguished from plasma CBG. In addition, experiments *in vitro* have shown that rat CBG in the presence of corticosterone forms a complex with a sedimentation coefficient of 3.45 S; when the steroid is absent, a reversible polymerization produces di-, tetra-, and octameric, as well as the monomeric forms (Chader and Westphal, 1968;

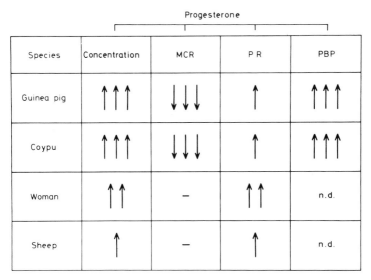

Species	Concentration	MCR	P R	PBP
Guinea pig	↑↑↑	↓↓↓	↑	↑↑↑
Coypu	↑↑↑	↓↓↓	↑	↑↑↑
Woman	↑↑	−	↑↑	n.d.
Sheep	↑	−	↑	n.d.

Fig. 2. Summary of the comparative changes in the peripheral plasma concentration, the metabolic clearance rate (MCR), and the production rate (PR) of progesterone and the plasma concentration of progesterone binding protein (PBP) during pregnancy. The direction of the arrow indicates an increase or decrease in the pregnant compared with the nonpregnant condition; the number of arrows indicates the order of change; a dash, no change; and n.d., not detected. Sources of information: guinea pig, Challis *et al.* (1971), Illingworth *et al.* (1970), Heap (1969), Burton *et al.* (1971); coypu, Rowlands and Heap (1964), Illingworth and Heap (1971); woman, Short (1961), Little *et al.* (1967), Little and Billiar (1969); sheep, Bassett *et al.* (1969), Bedford *et al.* (1972b).

Westphal, 1971). Whether or not the conformation of other plasma transport proteins is affected by their association with steroids is not yet known; however, if this is the case, the functions of these proteins may not be limited to vascular transport and steroid conservation. The finding that the rate constant for the dissociation of cortisol from human transcortin is appreciably faster (Paterson and Harrison, 1972) than previously anticipated also throws doubt on the view that steroids bound by high-affinity plasma proteins are rendered biologically inactive.

IV. ESTROGENS

A. Physiological Effects

Detailed descriptions of the physiological effects of estrogens will be found in the reviews of Parkes and Deanesly (1966), Deanesly (1966), Emmens (1967), and Briggs and Brotherton (1970) (see also Volume III,

TABLE II

The Metabolic Clearance Rate of Progesterone, Estradiol-17β, and Estrone from Blood in Man, Sheep, Guinea Pigs, and Rats

	Metabolic clearance rate		
	liters/ minute	liters/ minute/ $W^{0.75}$	Source
Progesterone			
Man[a]			
male	2.563	0.0979	Little *et al.* (1966)
ovariectomized female	2.565	0.1106	
Guinea pig[b]			
nonpregnant	0.067	0.0757	Illingworth *et al.* (1970)
pregnant (after day 20 *post coitum*)	0.005	0.0056	
Sheep			
nonpregnant (normal cycle)	2.752	0.1938	Bedford *et al.* (1972a,b)
pregnant	3.483	0.2039	
Estrone			
Human being			
male	2.694	0.1251	Longcope *et al.* (1968)
female	2.222	0.1032	
Rat			
nonpregnant adult females	0.015	0.0622	DeHertogh *et al.* (1970)
Sheep			
pregnant and nonpregnant	3.435	0.1887	Challis *et al.* (1972)
Estradiol-17β			
Human being			
male	2.125	0.0986	Longcope *et al.* (1968)
female	1.583	0.0735	
Rat			
nonpregnant adult females	0.018	0.0747	DeHertogh *et al.* (1970)
Guinea pig			
pregnant	0.111	0.1082	Challis *et al.* (1971)
Sheep			
pregnant and nonpregnant	2.449	0.1303	Challis *et al.* (1972)

[a] Figures calculated on a basis of MCR/plasma/blood ratio of 1.7.
[b] Figures calculated assuming an hematocrit of 40 and negligible binding of [^{3}H]progesterone to maternal erythrocytes. Values for MCR were obtained under light pentobarbitone anesthesia.

Chapter 3). Estrogens stimulate the growth of the female reproductive tract and mammary glands by stimulating protein synthesis and the rate of mitosis, though their mitogenic action is not confined entirely to tissues of the Müllerian duct system since epithelial cells such as those of the mammary, nasal, and buccal mucosa are also affected. Estrogens play a major part in the expression of behavioral estrus and the hormonal events of the estrous cycle in nonprimate female mammals and of the menstrual cycle in primates (see Volume II, Chapter 9). In late pregnancy estrogens stimulate myometrial activity and influence the physicochemical properties of the ground substance of the cervix and pubic symphysis which assists pelvic relaxation at the time of parturition. In addition they promote a wide variety of extragenital changes including the development of secondary sexual characteristics in females, such as the localized deposition of fat, the distribution of hair, and the texture of skin. Estrogens have mildly anabolic properties and may increase the retention of nitrogen and sodium. An important feature of the biological properties of estrogens is their interaction with other hormones. They antagonize or enhance the actions of progestagens according to the proportions of the two steroids present (Deanesly, 1966); they influence the secretion of pituitary protein hormones by either stimulation (e.g., prolactin, Pasteels, 1970) or inhibition (e.g., FSH; Motta et al., 1970); and they oppose the actions of androgens on the female genitalia, though extragenital effects may be synergistic.

B. Early Effects of Estrogens

A suitable model system for the study of the primary effects of estrogen action should be highly responsive and readily available. The uterus of the rat has proved invaluable for *in vitro* and *in vivo* investigations due to the fact that it is rapidly transformed from the undeveloped condition of immature animals to an actively growing tissue after the single subcutaneous injection of 10 μg estradiol-17β. However, it should be noted that investigators do not always define whether the observed effects of estradiol-17β are attributable to changes in the endometrium or myometrium. In some instances unphysiological doses of estrogens have been used so that the responses may not resemble the normal reaction in the intact animal.

Little is known of the way in which estrogens are transported from blood across the outer cell membrane to intracellular receptor sites; but target organ responses may be observed soon after hormone treatment (Table III). Within minutes of injection, estradiol-17β (0.5 μg/100 gm body weight) causes a rapid uterine hyperemia in ovariectomized rats. This effect may be observed as soon as 25 seconds after an intravenous injection (mean value, 44.8 $\pm$ 3.9 seconds). Together with a marked increase in 3′5′-cyclic AMP (cyclic AMP), uterine hyperemia is one of the earliest results of estrogen treatment. Szego (1965) suggests that uterine hyperemia can be attributed to a depletion of histamine in the rat uterus, caused presumably by a reduction of tissue stores rather than by an increase in histamine synthesis.

TABLE III

Summary of Early Changes in the Rat Uterus after Estrogen Injection

Time of change	Response to estrogen treatment	Source
Within 15 minutes	Increased concentration of $3':5'$ cyclic AMP	Szego and Davis (1967)
	Histamine mobilization	Szego (1965)
	Uptake of estrogen after local injection (mouse)	Stone and Martin (1964)
	Estradiol-17β bound to receptors	Stumpf (1968)
	Uterine hyperemia	Szego (1965)
	Uptake of [^{3}H]uridine into nuclear RNA	Means and Hamilton (1966)
	Increased chromatin template activity (rabbit)	Church and McCarthy (1970)
Within 30 minutes	[^{3}H]Uridine uptake	Hamilton et al. (1965)
	Uterine hyperemia (rabbit)	Markee (1932)
	Increased chromatin template activity	Teng and Hamilton (1968)
Within 1 hour	RNA synthesis and increased nuclear RNA	Gorski and Nicolette (1963)
	Increased RNA polymerase activity	Hamilton et al. (1965)
	Induced protein detected	Barnea and Gorski (1970)
	Increased glucose metabolism	Nicolette and Gorski (1964)
	Increased uptake of amino acids in vitro	Riggs et al. (1968)
	Increase in nuclear phospholipids	Gorski and Nicolette (1963)
Within 2 hours	Accumulation of water	Astwood (1938)
	Increase in glycogen content	Bitman et al. (1965)
	Increase in light RNA synthesis	Wilson (1963)
	Increased RNA polymerase activity	Gorski and Nicolette (1963)
	Nuclear RNA transported to cytoplasm	Hamilton et al. (1968a)
	Increase in content and synthesis of phospholipid	Aizawa and Mueller (1961) Aizawa and Nishigori (1968)
Within 4 hours	Accumulation of serum albumin, oxygen, urea, glucose, protein, sucrose, and NaCl	see Segal and Scher (1967)
Later than 4 hours	Accumulation of peak water	Astwood (1938)
	Increase in glycogen content	Walaas (1952)
	Increase in uterine weight	Cole (1950)
	Accumulation of α-aminoisobutyrate	Halkerston et al. (1960)
	DNA synthesis (later than 5 hours, mouse; later than 24 hours, rat)	Martin and Finn (1971) Aizawa and Mueller (1961)

Histamine increases the amount of blood supplied to the uterus, expands the microcirculation of the target organ, and enhances the amount of essential precursors for cell synthesis (Szego, 1965; Szego and Davis, 1967). When antihistamines are given locally, acute hyperemia and edema are blocked; when histamine liberation is defective, as in alloxan diabetes, the uterine response to estradiol-17β is impaired. The acute effects of estrogen on the uterus can be mimicked by histamine, histamine liberators or histidine (a direct precursor of histamine), and to a moderate degree by 5-hydroxytryptamine. Histamine, as with estradiol-17β, increases the labeling of uterine protein or lipid by the incorporation of [U-^{14}C]glycine given locally or parenterally (Szego, 1965). Objections to the attractive hypothesis that histamine release is an important component of early estrogen action were prompted by the large amounts of histamine required to mimic hormone action and by the high doses of antihistamines used to inhibit its effects. Inhibition may have resulted as much from the toxic effects of antihistamines as from their specific antagonism of the release and action of histamine (Martin, 1962).

Another early event induced by estrogen in the rat uterus is a transient increase in cyclic AMP. Within 15 seconds of an intravenous injection of estradiol-17β, uterine cyclic AMP approximately doubles in concentration and then falls to its original value within the next hour. Cyclic AMP values in the uterus decline after ovariectomy, but low doses of estradiol-17β (0.06 μg/100 gm body weight) evoke a marked increase. Diethylstilbestrol, but not estradiol-17α, produces a similar response (Szego and Davis, 1967). Cyclic AMP has been proposed as a universal "second messenger" transferring the stimulus of a "primary messenger" (the hormone) to an intracellular site of action (Sutherland *et al.*, 1969; Robison *et al.*, 1971). It seems improbable that cyclic AMP serves a similar function in the action of estrogens on the mammalian uterus since, among other reasons, it fails to mimic their effects. Szego and Davis (1967) propose that the acute effects of estrogen on uterine cyclic AMP reflects an interaction between membrane-bound adenylcyclase and the hormone.

Within an hour of estrogen treatment, uterine hyperemia (rabbit: Markee, 1932; MacLeod and Reynolds, 1938) is followed by water imbibition. Water accumulation follows an unusual pattern for it reaches a first peak about 6 hours after injection and a second peak about 24 hours later (Astwood, 1938). The first peak is closely followed by an increase in uterine dry weight (Cole, 1950), and the second peak is associated with a more striking increase in the rate of uterine growth (Telfer, 1953; Mueller *et al.*, 1958; Spaziani and Szego, 1959).

During the early stages of estrogen action, and preceding the time when significant increases in the growth of the tissue can be detected, there are

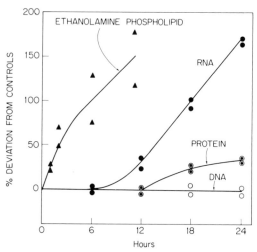

Fig. 3. Alterations in the composition of the uterus of rats after a single injection of estradiol-17β (10 μg) at zero time. The results are expressed as the % deviation from control animals during the first 24 hours after hormone treatment. (From Aizawa and Mueller, 1961.)

other important changes in tissue composition (Fig. 3). Phospholipid, RNA and protein components increase in concentration within a few hours, whereas DNA concentration (Mueller *et al.*, 1958; Aizawa and Mueller, 1961) remains constant for about 36 hours after estradiol treatment. In contrast to the rapidity of these early changes, gene replication and cell division show a considerable time lag (Mueller, 1971) and the mechanisms that regulate these basic processes of cell growth is a fascinating question that awaits elucidation. The application of refined techniques of separation and identification has shown that the increased synthesis of nuclear RNA (Means and Hamilton, 1966; Hamilton, 1968), the induction of a specific protein (Barnea and Gorski, 1970), and the formation of new mRNA species (DeAngelo and Gorski, 1970) are among other early events in the action of estrogens on the rat uterus, and in some instances changes have been noted within minutes of hormone treatment (Table III). Experiments with inhibitors of RNA and protein synthesis (actinomycin D and puromycin, respectively) have shown that the early responses depend on the synthesis of both new RNA and protein (Mueller *et al.*, 1961; Ui and Mueller, 1963; Gorski and Axman, 1964). The dangers inherent in the use of these inhibitors that are relatively toxic have been frequently discussed (Revel *et al.*, 1964; Honig and Rabinovitz, 1965; Laszlo *et al.*, 1966). It is notable that the increased uptake of water by the rat uterus induced by estrogen administration is inhibited by only about 70% after massive doses of actinomycin D which reduce RNA synthesis below that of untreated control animals (Ui and Mueller, 1963).

C. The Localization of Estrogens in Target Cells

Whereas estradiol-17β is rapidly metabolized within the body, it is retained within target organs like the uterus (Glascock and Hoekstra, 1959; Jensen and Jacobson, 1962; Stone, 1963; Stone *et al.*, 1963; Fig. 4) because of the properties of highly specialized receptor mechanisms. The pioneer investigations of Jensen and Jacobson (1962) revealed that after an injection of estradiol-17β *in vivo,* unchanged [³H]estradiol-17β could be isolated from the rat uterus at least 2 hours after treatment (Fig. 4: Jensen *et al.,* 1966). Of the labeled estradiol retained within the uterus, approximately half was associated with a nuclear fraction and up to 30% with a high-speed supernatant fraction (cytosol) (Noteboom and Gorski, 1965). By autoradiographic techniques, Stumpf (1968) and others have confirmed that [³H]estradiol is localized within two major sites of intracellular binding,

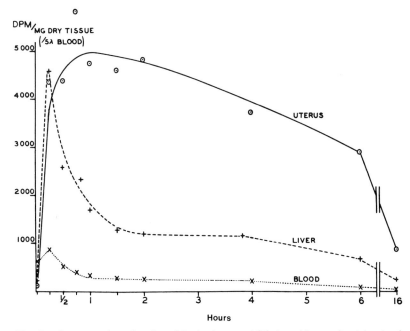

Fig. 4. Concentration of radioactivity in tissues of 24-day-old rats after injection (s.c., 0.1 μg, 11.8 μCi) of [6,7-¹H]estradiol-17β. Animals were pretreated with 0.1 μg nonradioactive estradiol on days 22 and 23. (From Jensen and Jacobson, 1962.)

namely the nucleus and cytoplasm. High-affinity binding proteins have now been reported in a wide variety of estrogen-sensitive tissues in many species (see Table 1 of Mueller *et al.*, 1972, for summary and references; see King, 1972, for detailed bibliography). In addition to those found in the uterus of the rat, estrogen receptors have been reported in the mammary gland, hormone-dependent mammary tumors, hypothalamus, and pituitary of the rat, the uterus and vagina of the mouse, the oviduct and liver of the chick, and the corpus luteum of the rabbit.

D. Characteristics of Receptors

1. Specificity

Estrogen receptor mechanisms are highly specific for biologically active, natural estrogens such as estradiol-17β and synthetic estrogens such as hexestrol, 17α-ethynylestradiol and 17α-methylestradiol. Steroids of low estrogenic potency like testosterone, cortisol, and the 17α-epimer of estradiol do not competitively inhibit the binding of [^{3}H]estradiol-17β (Noteboom and Gorski, 1965; Jensen *et al.*, 1966). Uterine receptors have a lower affinity for estrone than for estradiol-17β (Korenman, 1969), except that some of the injected dose is reduced to estradiol-17β which is then incorporated within the rat uterus (Jensen and Jacobson, 1962). Similarly, a synthetic estrogen, mestranol (3-methyl ether of 17α-ethynylestradiol), is not bound by the uterine receptor mechanism, but demethylation in the liver results in the formation of 17α-ethynylestradiol which can be bound by the receptors (Jensen *et al.*, 1966). When estradiol is linked to the large molecular chain of polystyrene by a 17α-propyl side chain, its binding by cytosol receptors is not reduced (Vonderhaar and Mueller, 1969). Thus, the minimum steric requirements for receptor recognition of biologically active estrogen molecules appear to be a phenolic group at C_3 of ring A and an alcohol of β-orientation at C_{17}. The 17α-substituent seems to have no part in the estrogen receptor interaction.

The binding of estradiol-17β to cytosol or nuclear receptors is not inhibited by antibiotics that block the synthesis of RNA and proteins (Jensen, 1965; Jensen *et al.*, 1966). Antiestrogens such as Clomiphene and Nafoxidine block the uterotrophic action of estrogens and inhibit their early uptake by uterine receptors. The effect is dose dependent and the amounts of [^{3}H]estradiol-17β bound in both the nucleus and cytosol are reduced by about the same extent (Callentine *et al.*, 1966; Jensen *et al.*, 1966, 1967a,b; Wyss *et al.*, 1968).

A distinction can be made between the uptake of a hormone and its retention within responsive cells. In rats injected with physiological doses (10 to 100 ng) of hormone, the quantity of [³H]estradiol present in the uterus 6 hours after injection increases proportionally with the injected dose. Above these levels, although uptake is related to dose, the amount retained for a period of 6 hours is about 25 pg/mg dry weight (Jensen et al., 1967b). Thus, retention in the uterus and the binding to a specific receptor mechanism appear to be primary events of estrogen action in the rat uterus and not merely passive phenomena unrelated to hormone action.

2. Physicochemical Properties

Estrogen receptors comprise a family of macromolecular complexes formed of different subunits (see Table III of Mueller et al., 1972, for summary and reference). The major part of estradiol in the uterus is bound to the nuclear sediment apparently in association with nucleochromatin (King et al., 1966; Maurer and Chalkley, 1967). When the nuclear receptor is released from isolated uterine nuclei by treatment with 0.3 to 0.4 M KCl or with pancreatic DNase at 0°C, its sedimentation coefficient is 5 S in a sucrose gradient containing 0.4 M KCl (Gorski et al., 1968), though it aggregates in a medium of low ionic strength to give a series of heterogeneous complexes that sediment rapidly (Mueller et al., 1972). The nature of this close association between the nuclear estrogen receptor and chromatin (Toft, 1972) remains to be elucidated. In contrast to nuclear receptor molecules, another receptor isolated from the cytosol fraction of rat uterine homogenates sediments in a sucrose density gradient near to yeast alcohol dehydrogenase and has a coefficient of 8 S (Erdos, 1968; Rochefort and Baulieu, 1968; Jensen et al., 1969a,b) or 9.5S (Toft and Gorski, 1966; Toft et al., 1967). This cytosol receptor isolated from the uterus of the rat and other species has a high dissociation constant (10^9 to 10^{11} M^{-1}: Erdos et al., 1968; Gorski et al., 1968; Talwar et al., 1968; Puca and Bresciani, 1969; Hähnel, 1971), and probably only one type of binding site (Toft and Gorski, 1966; Gorski et al., 1968; Giannopoulos and Gorski, 1971a). The concentration of receptor molecules in the rat uterus ranges from 4 to 5 × 10^{-14} moles/uterus (Noteboom and Gorski, 1965; Toft and Gorski, 1966). The noncovalent binding of the high-affinity estrogen-receptor complex isolated from the uterine cytosol (8–9 S) dissociates reversibly in 0.3 to 0.4 M KCl to yield two 4 S subunits. Recently it has been found that when this happens only one of the two subunits, the A subunit (4 S), binds estradiol-17β. The other subunit (B, 4–5 S) can be dissociated into even smaller components (B′), neither of which bind estradiol-17β

(Mueller, 1971; Mueller *et al.,* 1972). These conclusions, based on the results of several experimental procedures, are reviewed and discussed in detail by Mueller *et al.* (1972).

The physical character of these heterogeneous cytosol receptors is markedly affected by temperature, pH, and sulfhydryl reacting agents (Toft *et al.,* 1967; Vonderhaar *et al.,* 1970a,b). When labeled estradiol-17β is incubated with cytosol receptor at 0° and 23°C, and centrifuged through a sucrose gradient of low ionic strength at 0–4°C, the steroid is associated mainly with 8–9 S receptor. At 37°C, however, the estrogen–receptor complex is found as a 4 S subunit and the 8–9 S receptor is largely absent. This 4 S receptor is no longer capable of binding with "native" 4 S subunits dissociated in salt concentrations of 0.3 to 0.4 M KCl. Its formation depends on incubation at 37°C and on the presence of the steroid, for when incubations are performed in the absence of estradiol-17β, 9 S macromolecules are still present (Vonderhaar *et al.,* 1970a,b). The formation of this temperature-sensitive receptor is influenced by the pH during temperature treatment. At pH 8 a 5 S-estradiol binding receptor is produced, while at pH 7 the receptor is predominantly 4 S in nature (Brecher *et al.,* 1970; Mueller *et al.,* 1972). But since intracellular pH is probably less than pH 7.0, the alteration of receptor structure occurs far from its optimum pH if a similar mechanism exists *in vivo.* Other factors, including ageing and prolonged exposure to salt treatment, also alter the character of the receptor *in vitro* so that reaggregation of subunits is no longer possible. These 4 S (and 5 S) estrogen complexes formed at 37°C, or at pH 8, or during salt treatment of receptor preparations, fail to associate with freshly prepared "native" subunits because of the formation of some surface factor (C unit). Evidence obtained so far suggests that this active agent is apparently a small, dialyzable, ether-soluble component able to prevent the reaggregation of dissociated cytosol–estradiol subunits. This factor does not cause a release of estradiol once the steroid has been bound to the cytosol receptor (Mueller *et al.,* 1972).

Estrogen receptor molecules interact with a number of biologically important substances. The 8–9 S cytosol receptor combines with RNase to form a recognizably different moiety (Jensen *et al.,* 1969b). The 8 S receptor can be precipitated by protamines, histones, or polylysine (King *et al.,* 1969), and when exposed to purine and pyrimidine nucleotides its binding properties are enhanced by compounds characterized by the imidazole ring, namely ATP, GTP, their nucleotides, and even free bases (Vonderhaar *et al.,* 1970a). The increased binding with imidazole compounds is observed only when [^{3}H]estradiol is also present. It is possible that these imidazole compounds somehow alter the conformation of receptors to permit more efficient binding. Perhaps these compounds contribute in a significant way to the mechanism of action of estrogens by facilitating recognition by the receptor proteins of nucleotide bases of DNA, RNA, or even bases in catalytic regulatory proteins (see Mueller, 1971).

An important aspect of the study of estrogen-receptor molecules concerns their isolation and purification. Treatment with Ca^{2+} of the dissociated 8 S complex results in a stable subunit (4 S) which does not aggregate during classical purification procedures. When purified from calf uterus, the Ca^{2+} stabilized cytosol receptor has a molecular weight of 75,000, pI 6.4 (DeSombre et al., 1969, 1971). A recent partial purification by Puca et al. (1971) of the 8 S receptor from the calf uterus reveals a molecular weight of 288,000, pI 6.2; the 4.5 S subunit dissociates into two dissimilar proteins, with molecular weights of approximately 61,000, pI 6.6 and 6.8. Knowledge of the nuclear receptor complex, however, is less detailed than that of the cytoplasmic receptor, though the two appear to be related as they have immunological (Jungblut et al., 1970), chemical (King et al., 1969), and physical (Puca and Bresciani, 1969) properties in common. Whereas the 8 S receptor of the cytosol is readily demonstrable in vitro by adding estradiol-17β to a high speed supernatant fraction of rat uterus homogenates, the 5 S-estradiol complex is produced predominantly in the nucleus when the incubation mixture contains a cytosol fraction or a whole uterine homogenate. The 5 S receptor complex is not formed with a "pure" washed nuclear fraction alone, nor when the cytosol receptor is destroyed by heating at 45°C (Brecher et al., 1967; Jensen et al., 1967a, 1968, 1969a).

E. Two-Step Mechanism of Estrogen Action

The fate of estrogen molecules when they enter the responsive cells of the rat uterus is summarized in Fig. 5 (Jensen et al., 1969a, 1971a,b; Mueller et al., 1972). Jensen and co-workers propose that upon entering the cell the estradiol-17β molecule combines with a soluble receptor in the cytosol of the rat uterus (A subunit, 4 S) by a temperature-insensitive process. The transfer, or translocation, of estradiol-17β from the cytosol to the nucleus comprises a second step. This step is temperature sensitive since it occurs at 37°, but not at 0°C, and it is thought to consist of the transport of the 4 S steroid–receptor complex into the nucleus (Jensen et al., 1969a). More recently it has been proposed that the 4 S steroid–receptor complex is converted into a 5 S component before translocation into the nucleus (Jensen et al., 1971b). Translocation is a process that is unaffected by inhibitors of RNA (actinomycin D) or protein synthesis (cycloheximide or puromycin). It depends upon the initial formation of the 4 S estrogen–receptor complex within the uterine cytosol and upon some special feature of

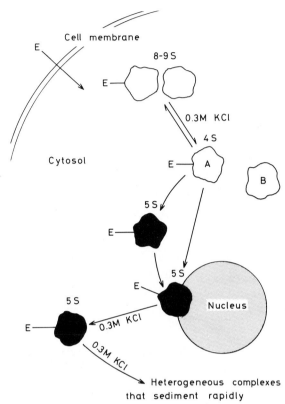

Fig. 5. Two-step mechanism of estrogen action (see text for sources of information).

nuclei isolated from target cells. This unidentified property of target cell nuclei contributes to the transfer of the hormone from cytosol to nucleus, for when nuclei from diaphragm (nontarget tissue) are substituted for those from the uterus, nuclear binding of estradiol-17β is not found *in vitro* (Jensen *et al.,* 1969a). During the process of translocation the 4 S estrogen–receptor complex within the cytosol is depleted (Jensen *et al.,* 1968; 1971a,b; Giannopoulos and Gorski, 1971a,b). The quantity of nuclear receptor isolated after estrogen treatment is less than the quantity of 8 S (two 4 S subunits) receptor that disappears from the cytosol. Thus, either the rates of turnover of cytosol and nuclear receptors differ, or the estradiol–8 S receptor complex acts on some cytoplasmic target. If estradiol-17β is to fulfill a regulatory role, it must be transported not only from the cytoplasm to the nucleus, but also in a reverse manner from the nucleus to the cytoplasm and eventually into blood. It may be that cytosol receptor protein facilitates this process, but whether estrogen is released from the cell in a free or protein-bound form is not known.

F. Estrogens and the Synthesis of Nucleic Acids and Proteins

Simultaneously with studies of the primary action of estrogens on responsive cells, several groups of workers have been trying to determine how
estrogens stimulate cell growth and activity. Mueller and coworkers are
among those (Mueller *et al.,* 1958, 1972; Talwar and Segal, 1964; Trachewsky and Segal, 1968; O'Malley *et al.,* 1969; Raynaud-Jammet and
Baulieu, 1969; Hamilton, 1971a,b) who support the view that estrogens
regulate gene expression or transcription. They suggest that this is achieved
by a modification of the availability or the conformation of specific proteins
which play a vital role in the control of the activity of genes (Mueller *et al.,*
1972). On the other hand, Tata (1968, 1970), in a comprehensive treatment
of the subject of hormone action, has concluded that "most growth and
developmental hormones probably alter the biosynthetic activity of cells at
multiple steps of both the transcription and translation processes." The evidence available lends strong support to the idea that estrogens affect
transcription, but the probability that they also regulate the translation of
pre-existing mRNA templates by the inhibition of a "translation repressor,"
or some other mechanism, has not been disproved.

It is commonly held that RNA is the vehicle for the transfer of information from the genes to the cytoplasm of the cell and that DNA is the template on which this RNA is formed before it is transported to the cytoplasm
where it in turn acts as a template for protein synthesis. Before considering
the effects of estrogens on RNA synthesis, it is important to recall that
RNA may be formed within the nucleus and may be engaged in the regulation of genetic activity without transfer to the cytoplasm (rapidly turning
over nuclear RNA); RNA may be synthesized on DNA without becoming a
template of protein synthesis; or it may act as a template for protein synthesis even though it is not synthesized on the DNA in the cell nucleus
(Harris, 1970).

It is well established that estrogens produce an early stimulation of RNA
synthesis in the rat uterus (Ui and Mueller, 1963; Gorski and Nicolette,
1963; Wilson, 1963). Within 2 minutes of treatment, estradiol-17β given to
adult ovariectomized rats increases the incorporation of [^{3}H]uridine into
rapidly labeled nuclear RNA (Means and Hamilton, 1966). The change in
the specific activity of labeled RNA isolated in subcellular fractions suggests that part of the newly synthesized nuclear RNA is transported to the
cytoplasm by a process that is accelerated by the presence of the hormone
(Hamilton *et al.,* 1968b; Hamilton, 1971a,b; Fig. 6). Since estrogens
probably stimulate the increased synthesis of all classes of RNA, a problem
arises in the characterization of these different forms, particularly the identification of small amounts of hormone-specific mRNA produced in the

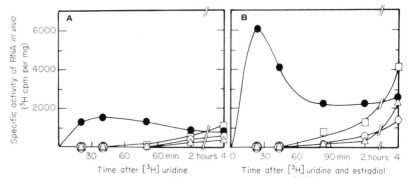

Fig. 6. The effect of estradiol-17β on the incorporation of [³H]uridine into subcellular RNA *in vivo* in the uterus of ovariectomized rats. At time zero animals received (i.p.) 100 μCi of [³H]uridine (left), or 100 μCi [³H]uridine and 10 μg of estradiol-17β (right). ● Nuclear RNA; ○, mitochondrial RNA; □, microsomal RNA; △, supernatant RNA. (From Hamilton *et al.*, 1968b.)

midst of relatively large amounts of rRNA and transfer RNA (tRNA). The identification of new species of mRNA is essential proof of the theory that estrogens regulate gene expression, though the task of identifying these species has been compared with that of finding a needle in a haystack! The pattern of synthesis of RNA species *in vivo* has been examined in uterine homogenates of immature rats after estradiol-17β injection (1 μg intraperitoneally). Among the RNA species formed, the earliest change occurred in Q_1-RNA, believed to be a precursor of rRNA, which increased to 54% above the control value within 30 minutes of hormone treatment. This rise was followed by increases in the radioactivity of rRNA, 7 S RNA (possibly derived from rRNA during extraction) and tRNA. There was a considerably slower response in the DNA-like RNA fraction. The results indicated that one of the primary effects of estrogen in the uterus is to promote a massive synthesis of rRNA, followed later by an increase in tRNA synthesis and by a relatively small change in DNA-like RNA. These studies failed to detect any appreciable increase in the synthesis of RNA species associated with mRNA (Billing *et al.*, 1969a,b). However, in later studies, Knowler and Smellie (1971) found that the incorporation of labeled precursors into RNA species of very high molecular weight (>45 S) preceded their incorporation into rRNA and 4 S RNA. The authors suggested that the increased synthesis of these species could reflect the formation of new mRNA and that translation could be a prerequisite of increased rRNA synthesis.

Whereas these and other observations (Luck and Hamilton, 1972) provide important information about the nature of RNA species in uterine homogenates after hormonal stimulation *in vivo*, they do not illuminate the intracellular location of these changes. Experiments in Mueller's laboratory, mentioned earlier (p. 72), suggested that estrogens probably acted on the transcription of genetic information since actinomycin D blocks the estradiol-stimulated incorporation of [^{3}H]uridine into new RNA species and [2-^{14}C]glycine into new protein (Ui and Mueller, 1963). Furthermore, hormone activity in the uterus is strikingly dependent on the synthesis of new protein due to the fact that when puromycin is injected to block protein synthesis, the estradiol-induced synthesis of uterine RNA and phospholipid is prevented (Mueller *et al.*, 1961). As a consequence of these findings the purpose of other studies has been to examine the effects of estrogens on the activity of RNA polymerase (the enzyme responsible for DNA transcription) on the template activity of uterine chromatin and on the character of RNA that hybridizes with DNA *in vitro* in an attempt to identify the formation of new species of mRNA.

1. DNA-Dependent RNA Polymerase

Estradiol-17β stimulates the activity of DNA-dependent RNA polymerase in uterine nuclei of rats (Gorski, 1964; Nicolette and Mueller, 1966a,b; Hamilton *et al.*, 1965, 1968a,b). The enhanced activity of the enzyme depends on an increased synthesis of proteins that normally accompanies the hormonal stimulation of the rat uterus. Puromycin or cycloheximide, given at doses that block uterine protein synthesis, inhibit RNA polymerase activity, though the inhibition with cycloheximide is reversible (Gorski *et al.*, 1965; Gorski and Morgan, 1967). The dependence of an enhanced enzyme activity on continuing protein synthesis has been amply confirmed. Thus, in prolonged incubations of uterine tissue in culture medium RNA polymerase activity increases even if the rats have not been treated with estrogen. Enzyme activity is only raised when protein synthesis is raised (Nicolette, 1969).

Further studies on the stimulation of RNA polymerase activity by estrogens has revealed an unexpected property in respect of the sensitivity of the enzyme to temperature. The increased activity of this enzyme in the uterus of rats treated with estrogens can be sustained *in vitro* at 0° and 37°C, though not at 23°C. An explanation of this observation may lie in the suggestion that at 23°C those mechanisms concerned with enzyme inactivation exceed those of activation (Nicolette *et al.*, 1968). The phenomenon has also been studied by the use of MPB [1-(β-4-pyridiethyl)benzimidazole], an inhibitor of RNA synthesis. When uteri are treated with MPB in tissue culture, RNA polymerase activity may be protected from the inhibitory effects of cycloheximide so that RNA synthesis can proceed. The protective role of MPB, however, is completely ineffective at 23°C. It appears, therefore, that the increase in uterine RNA polymerase activity induced by estrogens requires the continued synthesis of some protein utilized in RNA synthesis and that the supply or activation of such protein(s) to the catalytic site is temperature-dependent. This protein may be concerned with the processing or the transport of newly synthesized RNA from the polymerase site. Attempts to identify this putative protein have shown that uterine nuclei contain a labile substance of high molecular weight which is essential for polymerase activity and which is removed by 0.1 M ammonium sulfate precipitation. Soluble fractions of uteri from rats and other species have been reported to contain a factor that stimulates RNA polymerase activity (Mueller *et al.*, 1972).

Determinations of DNA-dependent RNA polymerase show that two types may be distinguished. One type is activated by Mg^{2+} and its products are similar to rRNA. Another type is activated by Mn^{2+} and ammonium sulfate and its products resemble DNA and are of mRNA type (Hamilton *et al.*, 1965, 1968b; Hamilton, 1971a,b; Fig. 6). Autoradiography associates the Mg^{2+}-activated type with the nucleolus (the site of rRNA synthesis) of the rat uterus and the Mn^{2+}-activated form with an extranuclear site (Laguens, 1964). In this context it is interesting to note that changes in nucleolar structure (increased heterogeneity) have been noted within 2 hours of the local administration of estradiol-17β into the vagina of ovariectomized mice (Pollard, 1970). Thus, there is substantial evidence that estradiol-17β stimulates DNA-dependent RNA polymerase activity of the rat uterus. Increased activity, however, does not conclusively show that estrogens activate transcription processes since a stimulation of activity may derive instead from the removal of a translation inhibitor (Blatti *et al.*, 1970; Barry and Gorski, 1971).

2. Uterine Chromatin

The template activity of uterine chromatin is increased within 30 minutes of injecting rats with 10 μg estradiol-17β and reaches a maximum after about 8 hours (Teng and Hamilton, 1968). An even more pronounced effect has been found with chromatin isolated from the endometrium of ovariectomized rabbits (Church and McCarthy, 1970). The association of basic and acidic proteins with nucleochromatin is well known but the role of these proteins has not been clearly defined. Stedman and Stedman (1950) have suggested that histones (basic proteins) regulate the activity of genes and the question has been frequently raised whether or not steroids (and other hormones) act by a derepression of specific genes, namely by the removal or inactivation of gene-regulating components such as histones (DeLange and Smith, 1971). Teng and Hamilton (1968) have found in rats that the

RNA:DNA ratio of uterine chromatin increases by about 17% after estrogen administration. Moreover, DNA-dependent RNA synthesis may be inhibited by histones to an extent that is inversely correlated with the ratio of lysine:arginine in the basic protein (Teng and Hamilton, 1969). Since the concentration of histones in uterine chromatin decreases within 8 hours after estrogen treatment (Teng and Hamilton, 1968), it is tempting to infer that these basic proteins, normally associated with DNA, no longer shield specific genetic information from transcription. However, further studies using [³H]tryptophan (which is absent from histones) and [¹⁴C]amino acids (which are common to histones) show that estradiol-17β stimulates the incorporation of these labeled precursors into an acidic, nonhistone protein present in the nucleus (Teng and Hamilton, 1970; Hamilton, 1971b). Whether it is this protein which affects the transcriptional process, and whether it is the same protein as that isolated from chromatin and which binds estradiol-17β (King et al., 1969) are, as yet, unanswered questions. So far as histones are concerned there is some indication that they repress the transcription of DNA (de Rueck and Knight, 1966), but there is little convincing evidence that they bind steroids in any specific manner (Sluyser, 1971). What is more to the point is whether estrogen–receptor complexes and chromatin-repressor substances play a part in the activation of specific areas of the genome. King et al. (1971) proposed that in rats a chromatin acceptor of the estrogen–receptor complex in uterine nuclei is a histone and that its properties as an acceptor molecule are dependent on ovarian stimulation since they are much reduced after ovariectomy.

3. Messenger RNA

A key to the action of estradiol-17β would be the conclusive demonstration that new mRNA molecules are produced in response to an organ-specific hormonal stimulation. Church and McCarthy (1970) found that new populations of DNA-like RNA molecules appear in the uterus of ovariectomized rabbits 1 hour after estrogen treatment. However, the ability of rapidly labeled nuclear RNA, formed under the influence of estrogens, to anneal with homologous DNA does not provide conclusive evidence of new RNA serving as a template for protein synthesis in the cytoplasm, for it is also essential to know if the newly synthesized RNA leaves the nucleus (Grant, 1969). Other evidence has been reported for the synthesis of new mRNA in the uterus of rats treated with estradiol, though the analytical problems involved in identification of new species of mRNA have also been emphasized yet again (Trachewsky and Segal, 1968).

Another way of looking at this problem is to see whether the formation of new mRNA species induced by estradiol-17β treatment stimulates the synthesis of a small number of specific protein molecules, which then give rise to a general increase of ribosomes, protein synthesis, and finally, cell growth. Gorski and colleagues found that labeled amino acids were incorporated into a specific soluble protein when incubated *in vitro* with homogenates of uterine tissue from rats previously injected with estradiol-17β (Notides and Gorski, 1966). The specific protein induced by hormone treatment (induced protein) was detectable 40 minutes after treatment and reached its maximum rate of formation between 5 and 80 minutes later (Barnea and Gorski, 1970; Fig. 7). Actinomycin D (8 mg/kg) prevented the induction of this specific uterine protein when given before the estrogen treatment. The synthesis of what may be the mRNA of this induced protein occurred almost instantaneously, was complete within about 60 minutes, and was inhibited by actinomycin D. Unlike other early effects of estrogens, it was apparently not dependent on prior protein synthesis (DeAngelo and Gorski, 1970; Fig. 7). The synthesis of this specific uterine protein, and its mRNA, is one of the earliest macromolecular events that occur after estrogen treatment. It can be stimulated *in vitro* by the incubation of immature rat uteri with estradiol-17β (10^{-9} M). It can be stimulated also by other estrogens including diethylstilbestrol, estriol, and estradiol-17α (in decreasing order of magnitude), but it cannot be stimulated with cyclic AMP or its butyryl analogue (Katzenellenbogen and Gorski, 1971).

The results so far are in keeping with the concept of a very early derepression or activation of the genome by estradiol in tissue-responsive cells. The increased synthesis of a few specific RNA and protein molecules is accompanied by an increased synthesis of RNA (mainly rRNA) and protein in the nucleus which, together with mRNA molecules, are transported in some selective way to the cytoplasm. Some form of selective transport is indicated by the discoveries that estradiol treatment causes a relatively large increase in tRNA of oviduct nuclei in the chick, but a smaller increase in cytoplasmic tRNA (Dingman *et al.*, 1969). Similarly, when actinomycin D is given to ovariectomized rats 30 minutes after hormone treatment, the transport of newly formed RNA from the nucleus to the microsomal fraction is blocked. When the inhibitor is given 2 hours after hormone treatment, the transport is not blocked, although the synthesis of nuclear RNA is much reduced (Hamilton, 1971a,b).

4. Ribosome Formation

The appearance of newly formed ribosomes in the cytoplasm is an essential component of enhanced protein synthesis. In the immature rat, the uterus contains relatively few cytoplasmic ribosomes but after estrogen treatment *in vivo* the rate of synthesis and the cytoplasmic concentration of

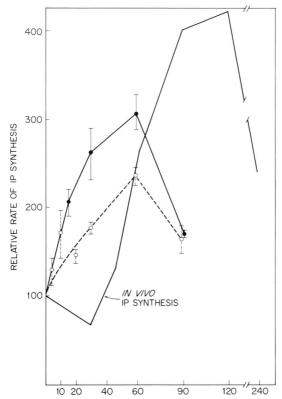

Fig. 7. The induction of new uterine protein (induced protein, IP) by estradiol treatment (10 μg) in ovariectomized rats (Barnea and Gorski, 1970), and the accumulation of its messenger RNA (IP-mRNA) in immature and ovariectomized rats (from DeAngelo and Gorski, 1970). For the demonstration of IP-mRNA accumulation, rats were treated with estradiol-17β and killed 5 to 10 minutes later. Uteri were incubated with actinomycin D (20 μg) and [³H]leucine (estrogen treated) or [¹⁴C]leucine (controls) for 2 hours. The induced protein synthesized was identified after electrophoretic separation. The ratio of ³H : ¹⁴C was used to calculate the relative rate of IP synthesis. Actinomycin D blocked mRNA synthesis during incubation so that IP synthesis *in vitro* depended on the quantity of mRNA formed *in vivo* in response to estrogen treatment before the animals were killed.

monomeric ribosomes are increased. When cell-free systems are used to study the incorporation of amino acids by microsomal or ribosomal preparations, it has been found that the number, activity, and stability of cytoplasmic polyribosomes *in vivo* are influenced by estrogens (Hamilton, 1971a). Tata (1968, 1970) has promoted the idea that increased protein synthesis may be related to the ability of ribosomes to translate information from mRNA. Hence, a redistribution process may occur that involves the segregation and selection of polyribosomes and the formation of "rough" endoplasmic reticulum, a site of active protein synthesis. In the light of this proposal the findings that estrogens produce an early stimulation of phospholipid synthesis (Aizawa and Mueller, 1961), and that they affect phospholipid metabolism in the uterus (Aizawa and Nishigori, 1968), gain special importance if hormone action is associated with the formation of new membranes. A fascinating possibility is that steroid hormones, such as estradiol-17β, may contribute to the association of polysomes and the smooth endoplasmic reticulum of hormone-specific cells. When polysomes from the liver of male rats are presented to smooth membranes prepared by fractionation of the same tissue, rough membranes form if estradiol or corticosterone (50 μg/ml) are present in the medium. These steroids, unlike cortisol, estrone, and cholesterol, protect "rough" membranes against the action of the carcinogen, aflotoxin B, which removes polysomes from them (James *et al.*, 1969; Williams and Rabin, 1969; Sunshine *et al.*, 1971; Blyth *et al.*, 1971). In spite of the relatively high concentrations of estradiol-17β used, it would be interesting to know whether or not a similar phenomenon occurs in other tissues such as the uterus, for it could provide evidence for a direct effect of the steroid on the assembly process of protein synthesis.

5. DNA Synthesis

Whereas uterine RNA synthesis is rapidly increased and the activity of uterine chromatin is also raised by estradiol-17β administration, the total RNA content of the uterus shows little change until 6 to 8 hours after hormone treatment (Aizawa and Mueller, 1961; Billing *et al.*, 1969b). The DNA content of the uterus of ovariectomized or immature rats has an even longer latent period and increases about 36 hours after estradiol administration (Mueller *et al.*, 1958). The precise way in which estrogens cause mitosis is not known, despite the fact that the mitogenic action of estrogens is one of their most characteristic properties. Galand and Leroy (1971) consider that estrogens accelerate generative cycles already in progress in tissue-responsive cells rather than activate new cycles in otherwise dormant cells. Estrogen treatment appears to shorten the G_1 (the reorganization of nuclear material) and S phases (the time of DNA replication) of prophase. Examples of these actions are to be seen in the mammary gland of mice where the numbers of cells undergoing DNA replication at any one time increase. Bresciani (1971) suggests that high-affinity tissue receptors in the mammary gland appear to inhibit the rate of cell replication, for in mammary tumors where the rate of replication is high, the concentration of binding receptors is low. He suggests that the interaction between estrogens and their receptor sites removes a local inhibition and thereby enhances cell replication. The proliferative action of estrogens on cell division in the uterus is discussed elsewhere (Martin and Finn, 1971; also Volume III, Chapter 3).

G. Estrogens and Transhydrogenation

The pronounced effects of estrogen upon the growth of target tissues result in radical alterations of enzyme activities. Little proof has come to light of specific enzyme activation by estrogen (except for the stimulation of RNA polymerase in uterine nuclei) or of specific enzyme induction, though it has been proposed that estrogens may act as cofactors essential for the function of already existing enzyme systems. An example is the estradiol-17β stimulation of a transhydrogenase in human placenta.

Estradiol-17β has been shown to stimulate the transfer of hydrogen from NADPH to NAD^+ and to function as a coenzyme in transhydrogenation so that estradiol-17β is cyclically reduced and oxidized to estrone (Villee and Hagerman, 1953; Talalay and Williams-Ashman, 1958, 1960; Villee *et al.*, 1960).

$$\text{NADPH} \quad \text{Estrone} \quad \text{NADH}$$
$$\text{NADP}^+ \quad \text{Estradiol-17}\beta \quad \text{NAD}^+$$

A bone of contention for many years has been the question whether estradiol stimulates transhydrogenation by an activation of a single 17β-hydroxysteroid dehydrogenase (Talalay, 1957; Talalay and Williams-Ashman, 1958, 1960) or by a stimulation of enzymes that are concerned with both transhydrogenation and dehydrogenation (Villee and Hagerman, 1958; Villee *et al.*, 1960). The problem has been largely resolved by the demonstration of two transhydrogenase enzyme systems in human placental extracts. Only one of these transhydrogenases is involved in the cyclic oxidation and reduction of estradiol-17β and estrone, respectively, while the other transhydrogenase appears to transfer hydrogen atoms directly from NADPH to NAD^+ (Karavolas and Engel, 1966).

The idea that estradiol may act by the regulation of a rate-limiting step in energy metabolism is an appealing one, especially because of the growth-promoting properties of this hormone. Unfortunately the physiological importance of these findings is far from clear. The uterus, as distinct from the placenta, converts little estrone to estradiol-17β (Williams-Ashman and Liao, 1964) and contains little, if any, estrogen-sensitive transhydrogenation (Jensen *et al.*, 1966; Grant, 1969). It is also unclear what part these transhydrogenation reactions play in the economy of placental function since estrogen has no effect on transhydrogenase activity in the placenta of many other species (see Grant, 1969). Karavolas has proposed that estradiol-17β-activated transhydrogenase may control the rate of estradiol-17β biosynthesis by the human placenta (see Engel, 1970).

H. Estrogen Action in the Domestic Chicken

The domestic chicken has been used by several investigators to study how estrogens act on hormonally sensitive tissues. The two model systems that have been examined in greatest detail are the synthesis of the egg-yolk protein, phosvitin, by the chick liver and the synthesis of the egg-white protein, ovalbumin, by the chick oviduct.

The precursors of egg yolk are formed in the liver and transported in plasma to the ovary. Estrogen treatment of roosters and nonlaying hens results in an increased level of all the plasma lipids and phosphoproteins present in normal, laying hens (see Gilbert, 1967). A large proportion (75%) of the phosphoproteins in blood can be accounted for by phosvitin and the remaining 25% by lipovitellin. Phosvitin, which comprises 7 to 8% of the yolk proteins, is transported in plasma as a very stable complex tightly bound to lipovitellin. The complex can be separated from plasma proteins by DEAE-cellulose chromatography and phosvitin isolated by incubation at 41°C, pH 8.0, with pancreatic lipase (Beuving and Gruber, 1971a). Estrogen-stimulated phosvitin is not stored in the liver but is released into the bloodstream, since the blood concentration of labeled phosvitin formed from radioactive precursors is increased, whereas hepatic concentration is not (Greengard *et al.*, 1965). In roosters and nonlaying hens, phosvitin (an egg-yolk phosphoprotein) first appears in the blood 20 hours after an injection of estradiol-17β or stilbestrol (Greengard *et al.*, 1964; Beuving and Gruber, 1971b); the amount of phosvitin produced and the duration of the increased synthesis are dose-dependent. At the highest doses administered (20 mg single injection of estradiol), phosvitin reaches peak concentration 5 days after injection and then declines 3 to 5 days later. As in the estradiol-induced synthesis of proteins in mammals, phosvitin synthesis in chicks can be blocked by actinomycin D and by cycloheximide (Beuving and Gruber, 1971b).

If a second dose of estradiol-17β is given to chicks when blood phosvitin levels are falling, an increased synthesis of phosvitin is obtained, but with a much shorter time-lag of about 6 hours instead of 24 hours. This second response can be reproduced for up to 3 weeks after the decline of hormonally stimulated phosvitin levels. It appears to consist of two phases; an initial phase 5 to 10 hours after treatment which is resistant to actinomycin D, and a second phase which is susceptible to actinomycin D inhibition. The results have been interpreted as evidence that estradiol has a dual effect on the induction of phosvitin synthesis; namely, the stimulation of new mRNA molecules, and the facilitation of translation processes (Beuving and Gruber, 1971b).

Phosvitin is an unusual protein since it consists of 50% serine residues, so that identification of hormonally stimulated tRNA (seryl-tRNA) is made easier. Liver seryl-tRNA is higher in birds during phosvitin production (Bernfield, 1971), a finding consistent with the view that estradiol stimulates RNA synthesis. Despite the lag in phosvitin production it is probable that RNA synthesis increases rapidly after hormone treatment (Beuving and Gruber, 1971b). Much of the 20-hour time lag is presumably taken up with the organization of new ribosomes since the response to a second hormone injection is much more rapid.

A second model used in the study of estrogen action is the hormonally sensitive chick oviduct. Estrogen converts the immature chick oviduct into an actively growing tissue that secretes proteins which are deposited around the yolk of the egg. Relatively large doses of estrogen given over several days cause differentiation of the thin epithelium of the chick oviduct into three cell types; tubular gland cells which secrete the protein, ovalbumin; goblet cells which secrete avidin after progesterone administration; and ciliated cells which are concerned with fluid transport (Brant and Nalbandov, 1956; O'Malley et al., 1969; Yu et al., 1971). Oviduct synthesis of ovalbumin is not seen for about 6 days during treatment with diethylstilbestrol, even though cytodifferentiation and hyperplasia are far advanced. Within 24 hours of treatment stromal edema is marked, mononuclear cells migrate into oviduct tissues, and ribosomes aggregate prior to an increasing development of the rough endoplasmic reticulum. After 4 days of treatment tubular glands are formed by epithelial invagination into stromal tissue, and about 2 days later granules of ovalbumin can be seen within the glandular cells and in the acinar lumen of the glands. If estrogen treatment is continued there is a gradual increase in ovalbumin synthesis but, if stopped, ovalbumin synthesis falls though it can be restimulated if treatment is resumed (O'Malley et al., 1969).

The occurrence of estrogen–receptor mechanisms in the chick oviduct which possess a high degree of specificity was indicated by early studies (Jonsson and Terenius, 1965; Terenius, 1969). The nature of the radioactivity retained by the chick oviduct was not identified in these studies. From the pattern of hormonally induced events in the chick oviduct, it has been inferred yet again that estrogens regulate gene expression in respect of cytodifferentiation and the synthesis of new proteins (ovalbumin, lysozyme) (Fig. 8). The activity of the DNA-chromatin template, studied by nearest-neighbor frequency analysis of dinucleotide pairs, is qualitatively different after 3 days

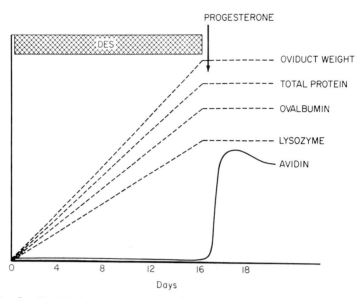

Fig. 8. The biochemical responses of the immature chick oviduct after treatment with steroid hormones. DES, diethylstilbestrol. (From O'Malley *et al.*, 1969.)

of diethylstilbestrol treatment and even more so after 6 days. A single injection of 5 mg diethylstilbestrol causes a prolonged stimulation of RNA polymerase and changes the pattern of high molecular weight RNA and nuclear RNA (4 S) species (McGuire and O'Malley, 1968; Dingman *et al.*, 1969). Hormone stimulation gives rise to new species of RNA as shown by DNA/RNA hybridization techniques; 20 days of estrogen treatment produces a marked increase in new species of RNA actively competing for DNA binding sites. Yet these experiments provide no conclusive evidence for the production of new mRNA since they do not distinguish between rapidly synthesized nuclear RNA (an early event induced by estrogen in the rat uterus) and new mRNA; both are species that should hybridize with DNA (O'Malley *et al.*, 1969).

Primary stimulation of cell differentiation and protein synthesis by estradiol-17β is arrested if hormone treatment ceases; the oviductal synthesis of ovalbumin declines, the concentration of ribosomes falls, and polysomes become monosomes. This sequence of events can be reversed if estradiol-17β treatment is repeated (second response). This provides an opportunity to study the synthesis of ovalbumin in differentiated tissue without the complication of those factors that are concerned with cytodifferentiation. During this secondary response, labeled precursors are more rapidly incorporated into RNA and protein (especially ovalbumin which comprises at least 50% of the soluble proteins) in oviduct magnum explants. Actinomycin

D (5 mg/kg) will inhibit the incorporation of precursors into RNA and protein, though at a low dose (2 mg/kg) it blocks only the incorporation into RNA and not that into protein. It has been concluded from these studies that the estrogen-induced secondary response of ovalbumin synthesis, in the absence of simultaneous cytodifferentiation, does not require rRNA synthesis (Palmiter *et al.*, 1971).

In monolayer cell cultures estrogens given *in vitro* will stimulate some, but not all, of the changes observed *in vivo*. After stilbestrol treatment, a mainly fibroblast culture will become epithelioid in association with an increased DNA synthesis and with the induction of ovalbumin synthesis (Kohler and O'Malley, 1967). Ovalbumin synthesis, however, cannot be stimulated *in vitro* (probably because of the lack of some specific factor such as ovalbumin mRNA), but once synthesis has begun *in vivo*, it can be sustained *in vitro* for some hours (Palmiter *et al.*, 1971). Despite the failure to induce ovalbumin synthesis *in vitro*, such techniques have been profitably employed to study other aspects of the mechanism of action of estrogens on the chick oviduct (Kohler and O'Malley, 1967; O'Malley *et al.*, 1969; Palmiter *et al.*, 1971; Means *et al.*, 1971; Means and O'Malley, 1971a). Newly formed polyribosomes stimulated by estrogen treatment are found to have an increased capacity to synthesize proteins in a cell free system. The proteins formed differ qualitatively from those in the unstimulated oviduct. Qualitative changes in the synthesized protein suggest that there are also qualitative changes in the mRNA population. Polysomal protein synthesis *in vitro* reaches a maximum after 4 days of estrogen treatment *in vivo* which precedes the time when cytodifferentiation is complete and ovalbumin synthesis begins (day 6). Peak synthesis of polysomes corresponds with the time of striking increases in ribosome synthesis and the conversion of monosomes to polysomes.

The results obtained with the chick oviduct as a model for the study of the mechanisms of estrogen action point to control systems that operate at a transcriptional level of cell activity. But the findings hitherto do not preclude the possibility that estrogens act at multiple sites in target cells and that other events, such as the observed alteration in the number of tRNA molecules, could influence translational processes, and lead to a stimulation of protein synthesis and growth.

V. PROGESTERONE

A. Physiological Effects

The biological effects of progesterone have been extensively reviewed (Boscott, 1962a; Fotherby, 1964; Deanesly, 1966; Zarrow, 1961, 1965;

Gibian and Unger, 1968; Briggs and Brotherton, 1970; Volume III, Chapter 3). Briefly speaking, progesterone plays an essential part in the maintenance of pregnancy in all mammalian species. As "the hormone of pregnancy" it is concerned with the stimulation of endometrial proliferation and prepares the uterus to receive the implanting blastocyst. It inhibits myometrial activity and serves to maintain the developing conceptus. Yet this is not all, for among a variety of other effects, progesterone influences the rate of egg transport through the Fallopian tube, it causes mucification of the vagina, and it stimulates the lobuloalveolar growth of the mammary gland. Progesterone is also known to inhibit ovulation during the normal cycle and in pregnancy by blocking the release of pituitary gonadotropins. In late pregnancy the delivery of the young is frequently, though not invariably, associated with a decreased secretion rate of progesterone and in consequence it has been implied that the hormone has an important regulatory role in the onset of parturition and lactation, as well as in the maintenance of gestation.

Among other diverse actions, progesterone is thermogenic and causes a rise in body temperature of 0.5 to 1°F in women. It promotes water excretion, influences electrolyte balance, and is reputed to be natriuretic and diabetogenic in human pregnancy. Large doses of progesterone, in common with some other steroids, produce hypnotic effects. One of the most striking features of progesterone is its interaction with estrogen. Often complicated by a dependence on the prior sensitization of target tissues by estrogens, the precise mechanism by which progestagens act has been obscure and difficult to study. Recent findings suggest that the mechanisms of progestagen and estrogen actions may have much in common.

B. Localization of Progesterone in Target Cells

1. Mammals

Early events in the action of progesterone on target cells of the uterus have not been as clearly defined as those of estrogens. This is partly because it is difficult to detect distinctive changes in cell function immediately after progesterone has been administered alone, and partly because the typical progestational responses such as the proliferation of endometrial glands in the rabbit occur most successfully when tissues have already responded to estrogen stimulation. Rapid metabolism of progesterone by target organs such as the uterus and mammary gland has precluded the use of autoradiography for the localization of the steroid in intracellular sites because it is impossible to distinguish between ^{3}H-labeled radiometabolites and [^{3}H]-progesterone itself.

An early indication that low concentrations of progesterone are retained in certain organs was obtained by Lawson and Pearlman (1964) in experiments on rats. After [7α-^{3}H]progesterone had been continuously infused into conscious rats (18 to 19 days pregnant) for up to 21 hours, progesterone and 20α-dihydroprogesterone (20α-diHP) were isolated and characterized in samples of plasma, mammary gland, uterus, and adipose tissue. Concentrations were greater in the mammary gland and adipose tissue than in plasma. In the uterus the concentration of labeled progesterone was slightly greater than that of plasma, though in only one of two experiments, while labeled 20α-diHP could not be detected. The molar concentrations were 10^{-6} to 10^{-7} mol/kg in the mammary gland and even lower (10^{-9} mol/kg) in the uterus (Lawson and Pearlman, 1964). Since tissue samples were taken shortly after the infusion of isotope had been stopped, the isotope concentrations reported are probably underestimates of those at steady-state. Nonetheless, they emphasize the low concentrations found in target organs and the relative absence of long-term retention, a characteristic feature of estrogen action. Similar results of a low progesterone uptake by the uterus were previously reported by Riegel *et al.* (1950), Berliner and Wiest (1956), and Wiest (1963b), while the accumulation of progesterone in adipose tissue was noted by Zarrow *et al.* (1954), Kaufmann and Zander (1956), and Plotz and Davis (1957).

From measurements of endogenous progesterone levels in pregnant rats, uterine concentrations were found to be often greater than those in plasma or placentae. After ovariectomy on the fifteenth day of gestation and daily progesterone injections, the concentration recorded in the uterus was greater (6.2 μg/100 gm) than that of placentae (1.9 μg/100 gm) or peripheral plasma (2.6 μg/100 ml). The authors (Csapo and Wiest, 1969) suggested that the higher uterine concentrations may be attributable to a local synthesis of placental progesterone, but they may also reflect a capacity of the pregnant uterus, as distinct from the placenta, to retain progesterone.

The results of these and similar studies, provided no direct proof of the presence of progesterone receptors in target organs since the amounts of steroid detected might be explained by the lipid solubility of progesterone and its high lipid–water partition coefficient (Table I). Evidence has now come to light for the existence of a specific receptor complex within the uterus of the rat (Milgrom and Baulieu, 1970a,b) and guinea pig (Falk and Bardin, 1970; Milgrom *et al.,* 1970a,b). In the rat, progesterone binding protein has been detected in uterine homogenates after sucrose gradient centrifugation of a high-speed supernatant fraction. Surprisingly, this binding component (3.6 S) could not be distinguished from the plasma transport protein, corticosteroid binding globulin (CBG). The concentration of CBG-like protein was increased by estradiol treatment. Progesterone and cortisol are bound strongly when incubated with a high-speed supernatant (105,000 *g*) of a uterine homogenate, but when incubated with intact uterine horns, only progesterone is bound. A similar difference in uterine binding has been demonstrated after an injection of ^{3}H-labeled steroids *in vivo* followed by sucrose gradient separation of the binding components. In these experiments [^{3}H]progesterone was bound to a 4 S component whereas [^{3}H]-cortisol was not (Table IV: after Milgrom and Baulieu, 1970a,b).

Another progesterone binding component has been found in the guinea pig uterus (uterine PBP). Labeled progesterone of high specific activity injected into nonpregnant ovariectomized guinea pigs was retained longer in the uterus than in the heart or diaphragm. This uptake, which was enhanced by pretreatment with estrogen, was notable in that even after 1 hour almost 90% of the total radioactivity in the uterus was still in the form of [^{3}H]progesterone. Moreover, the uptake was largely specific for progesterone; there was no uptake of [^{3}H]cortisol and the uptake of progesterone was unaffected by the competition of testosterone, cortisol or estradiol-17β (Fig. 9; Falk and Bardin, 1970).

Uterine PBP in guinea pigs, as in rats, can be located in the 105,000 *g* supernatant fraction of homogenates, and it possesses a number of interesting physical properties. After sucrose gradient centrifugation the sedimentation coefficient is 6.7 S in Tris buffer (10 m*M*) and 4.3 S in 0.3 *M* KCl, indicating some dissociation of the protein into subunits. The possibility that sulfhydryl (SH) groups are implicated in the binding reaction has been shown by *p*-hydroxymercuribenzoate treatment which inhibits progesterone

TABLE IV

Properties of Progesterone Binding Protein (CBG-like protein) in Rat Uterus[a]

Binding properties greatly reduced by heating 105,000 *g* supernatant of uterine homogenate at 60°C for 20 minutes.

CBG-like protein present in 105,000 *g* supernatant prepared from castrated rats treated with three daily injections of 0.2 μg of estradiol-17β; sucrose gradient centrifugation revealed a sedimentation coefficient of about 4 S.

Competitive inhibition studies showed the following order of decreasing affinity; cortisol >progesterone > 17-ethinyl-19-nortestosterone > chlormadinone acetate.

Association constant of CBG-like protein in plasma $1.75 \pm 0.28 \times 10^8\ M^{-1}$; in uterus $1.54 \pm 0.29 \times 10^8\ M^{-1}$.

Uterine binding abolished after incubation with pronase (protease) or trypsin; not significantly altered by deoxyribonuclease or ribonuclease.

Diaphragm and kidney contained no similar binding component.

Concentration of specific binding sites per milligram of total uterine protein lower in prepubertal rats.

Uterine binding protein could not be washed out from extravascular sites.

---

[a] Data from Milgrom and Baulieu (1970a,b).

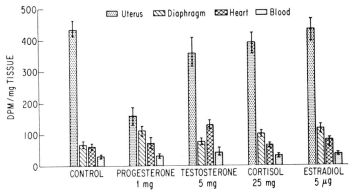

Fig. 9. The tritium concentration of tissues from guinea pigs sensitized with estrogens and treated with [³H]progesterone (means ± S.E.). Fifteen minutes before labeled progesterone was given (i.v.), progesterone, testosterone, cortisol, and estradiol were administered. Note the marked decrease in the uterine concentration of tritium when guinea pigs were injected with 1 mg of unlabeled progesterone. (From Falk and Bardin, 1970.)

binding by uterine PBP (Milgrom *et al.*, 1970b). The similarity between the uterine and plasma proteins that bind progesterone in the rat does not seem to apply in the guinea pig. Thus, uterine PBP found in nonpregnant guinea pigs differs markedly from plasma progesterone binding globulin (PBG) found in high concentration in pregnant guinea pigs (Heap, 1969; Milgrom *et al.*, 1970a; Burton *et al.*, 1971). The chief differences lie in their sedimentation coefficients (4.5 S for plasma PBG and 6.7 S for uterine PBP, Milgrom *et al.*, 1970b) and in their relative temperature stabilities; plasma PBG is thermostable, (Heap, 1969; Milgrom *et al.*, 1970a; Burton *et al.*, 1971), whereas uterine PBP is comparatively thermolabile (Milgrom *et al.*, 1970b). Apparently uterine PBP can be distinguished from plasma CBG in guinea pigs since uterine PBP will bind progesterone but not cortisol, whereas plasma CBG will bind both steroids (Rosenthal *et al.*, 1969).

Other examples of progesterone-binding macromolecules in mammalian tissues have been reported, including a component which has been located in the cytoplasm of the rat prostate that binds pregnenolone and progesterone (Karsznia *et al.*, 1969) and a component that has been found to bind progesterone in the corpus luteum of the cow (Leymarie and Gueriguian, 1969). In both instances the functional significance of these macromolecules is obscure, since neither organ is regarded as a target for progesterone. The macromolecules isolated from the prostate possess properties that differ from those of other tissue receptors in being thermostable and showing an increased affinity for steroids at 60°C. Another site of progesterone uptake is the mammalian brain, where the retention of radioactivity after treatment with labeled progesterone (Laumas and Farooq, 1966; Laumas, 1967) is a finding of special interest in view of its probable influence on progesterone-sensitive areas of the brain (see Volume III, Chapter 3).

Some of the difficulties encountered in identifying the uptake and localization of progesterone in target tissues have arisen because of the use of labeled progesterone with a low specific activity and because of the long time interval between administration of the dose and the measurement of radioactivity in tissues. In some experiments the endogenous concentration of progesterone in blood has not been known, so a single injection of tracer into an anesthetized rat would have little chance of demarcating specific receptors already saturated with progesterone secreted from other steroidogenic organs such as the adrenals (Feder *et al.*, 1968; Feder and Ruf, 1969). Even topical application in the mouse uterus and vagina *in vivo* has produced little evidence of local retention (Stone and Baggett, 1965).

2. Vertebrates Other Than Mammals

Convincing evidence for the presence of specific receptor mechanisms for progesterone within target cells has been presented by O'Malley and his co-workers in their studies of the chick oviduct. Although the progesterone binding components found in the rat uterus have not been distinguished from the plasma binding protein CBG, it has been demonstrated that progesterone receptors in the chick oviduct differ in several respects from proteins found in chick plasma (Sherman *et al.*, 1970; O'Malley *et al.*, 1971; Schrader and O'Malley, 1972). Within 15 minutes of injection, progesterone is bound to a soluble macromolecule in the chick oviduct which can be identified by gel-filtration on Sephadex G-200. Even 1 minute after injection [^{3}H]progesterone has been found attached to this macromolecule, reaching its maximum concentration 24 minutes later. The proportion found within oviductal nuclei increases progessively during this period (O'Malley *et al.*, 1969). The properties of the cytoplasmic receptor for progesterone in the chick oviduct are summarized in Table V. Functional tests have revealed that compounds which actively induce avidin synthesis are also the most active competitors of progesterone binding.

Since progesterone induces the synthesis of a new protein (avidin) by the chick oviduct, and presumably alters gene expression in a selective manner to effect new RNA transcription, experiments have been performed to see if a macromolecular receptor exists within the nucleus of the target epithelium. The studies showed that the two-step mechanism of estrogen binding in the rat uterus (p. 78; Jensen *et al.*, 1968; Gorski *et al.*, 1968) and of 5α-

TABLE V

Characteristics of Progesterone Receptor in the 105,000 g Supernatant of Oviduct
Homogenates from Chicks Treated with Estrogens[a]

Present in supernatant at 105,000 g; heat sensitive, nondialyzable and precipitated by
  ammonium sulfate; isoelectric point, pI 4.0.
Sedimentation coefficient in the absence of KCl, $s_{20,w}^{\circ}$ of about 5 S and 8 S; molecular
  weights (MW) of about 1.0 and 5.6 $\times$ 10^5. Plasma CBG in chicks has $s_{20,w}^{\circ}$ of
  approximately 3.75 and MW of about 6.0 $\times$ 10^4.
In solutions of 0.3 M KCl, the cytoplasmic binding macromolecule and CBG have a
  similar sedimentation coefficient, 3.8 S.
Cytoplasmic macromolecules from oviduct resolved from CBG on Agarose A-0.5 m gel
  filtration.
Dissociation constant of cytoplasmic receptor, 8 $\times$ 10^{-10} M in 0.3 M KCl at 1°C.
Binding reduced by mild heating.
Pronase and p-hydroxymercuribenzoate (10^{-3} M), but not ribo- or deoxyribonucleases,
  destroyed the complex implicating protein in the cytoplasmic binding receptor.
Specificity tests indicate the following order of binding; progesterone > testosterone
  > 20α-diHP > estradiol-17β > cortisol > estrone > androstenedione.
Found only in progesterone-responsive target tissues (oviduct).
Concentration increased tenfold by estrogen pretreatment.

---

[a] Data from: O'Malley *et al.* (1969, 1970); Sherman *et al.* (1970); and Schrader and
O'Malley (1972).

dihydrotestosterone binding in the rat ventral prostate (Fang *et al.*, 1969)
apparently applies to progesterone in the chick oviduct (O'Malley *et al.*,
1970). Specific progesterone binding molecules were identified in the
nucleus of the oviduct of estrogen-treated chicks. They were indistinguishable from, though they appeared later than, those receptors found in the
cytoplasmic supernatant. A hormone-dependent transfer of the progesterone–receptor complex from the cytoplasm into the nucleus could
be detected and monitored. As the concentration of cytoplasmic receptor
protein decreased, that of the nuclear receptor increased, a process which
took place within 20 minutes of [³H]progesterone injection (Fig. 10). There
was a twofold specificity in these reactions; namely, oviduct cytosol and
oviduct nuclei were required for the nuclear transfer and retention of the
steroid–receptor complex. If nuclei or cytosol were taken from a nontarget
tissue, the reactions failed to occur.

Further information is accumulating supporting the theory that the hor-
mone–receptor complex appears to interact with the genome at specific
DNA sites which are determined by chromatin acidic, nonhistone proteins.
It has been speculated that these may be activator sites in the operons regu-
lating the functional response of target cells to progesterone stimulation
(Spelsberg *et al.*, 1971; O'Malley *et al.*, 1972b).

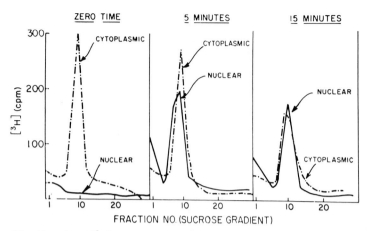

Fig. 10. Transfer of [³H]progesterone-receptor complex of the chick oviduct from the cytosol to the nucleus *in vitro* (37°C). Oviduct segments were incubated with [³H]progesterone (5 minutes at 0°C). At zero time the steroid had formed a complex with the cytosol receptor. Subsequent incubations at 37°C led to a progressive increase in the proportion of steroid found in the nucleus. (From O'Malley *et al.*, 1970.)

C. Progesterone and the Synthesis of Proteins and Nucleic Acids

1. Avidin

Estrogen stimulation of the uterus (Gorski *et al.*, 1965), androgen-promoted growth of male accessory sex glands (Williams-Ashman *et al.*, 1964), and glucocorticoid induction of liver enzymes (Garren *et al.*, 1964; Kenney *et al.*, 1965) are examples of ways in which steroids may initiate and regulate protein synthesis in target organs. Another example is provided by the progesterone-induced synthesis of avidin, an egg-white protein produced in the chick oviduct. As early as 1924 Boas noted that a 20% egg-white diet produced a nutritional deficiency in rats. Subsequently the factor concerned was named avidin by Eakin *et al.* (1941) because of its remarkable avidity for biotin, resulting in the condition known as "egg white deficiency." The mechanism of induction of avidin synthesis by the goblet cells has a specific requirement for progesterone, certain other Δ⁴-3-ketosteroids, and for progestationally active steroids such as 17-ethynyl-19-nortestosterone, 19-norprogesterone, and 17-methyl nortestosterone. Studies of the structure–function requirements indicate that the only progestationally inactive compounds found to stimulate avidin synthesis include substance S (4-pregnene-17α,21-diol-3,20-dione) and allopregnan-21-ol-3,20-dione. It is also possible that antiestrogenic activity may be an important requirement (O'Malley *et al.*, 1969). Progesterone itself has little effect on the biosynthesis of other oviductal proteins that are induced by estrogen treatment, namely ovalbumin and lysozyme. Interactions between estrogens and progesterone, however, affect the stimulation of avidin whose synthesis is greater when both steroids are administered (Korenman and O'Malley, 1968). Nevertheless, immature chicks treated *in vivo* with progesterone only (estrogen pretreatment omitted) produce avidin 10 to 20 hours later (O'Malley *et al.*, 1969).

How progesterone stimulates avidin synthesis has not been finally resolved. Induction appears not to depend on new DNA synthesis since progesterone fails to stimulate [³H]thymidine incorporation into DNA (O'Malley and McGuire, 1968), and neither hydroxyurea inhibition of DNA synthesis nor colchicine treatment interfere with avidin production. Actinomycin D, however, does cause a 90% inhibition of avidin synthesis in tissue cultures of chick oviduct even when general protein synthesis is not blocked, so that DNA-dependent RNA polymerase appears to be implicated in the induction mechanism. Cycloheximide inhibits avidin synthesis almost completely so that induction seems to involve the transfer of amino acids from soluble RNA to ribosomal polypeptides (O'Malley *et al.*, 1969).

Experiments on nuclear RNA and on RNA polymerase activity indicate that progesterone acts initially on the nucleus. In chicks pretreated with estrogen and given a single injection of progesterone, RNA polymerase activity falls transiently and then rises prior to avidin synthesis 6 to 8 hours later (Fig. 11). Similar results with rapidly labeled nuclear RNA have confirmed the early involvement of the nucleus in the progesterone-induced synthesis of avidin in cultured oviduct tissue (O'Malley *et al.*, 1969). Thus, newly synthesized nuclear RNA was found to possess a greater percentage of high molecular weight RNA species after estrogen and progesterone treatment, and there was a pronounced increase in nuclear transfer RNA (Dingman *et al.*, 1969).

Further support for the hypothesis that progesterone acts at a nuclear level of protein synthesis has been derived from studies of RNA synthesized before avidin induction. A significant change in the dinucleotide composition of the RNA after progesterone treatment (nearest-neighbor frequency analysis and hybridization-competition studies) suggests that progesterone induces a new species of nuclear RNA (Hahn *et al.*, 1968; O'Malley *et al.*, 1969). It is probable that this new RNA may contain mRNA specific for avidin synthesis by the ribosomes (Means and O'Malley, 1971b), but definitive evidence is awaited.

O'Malley (1971) summarizes the actions of progesterone in the chick oviduct in the following way: uptake of progesterone by the target cell and binding to specific macromolecules in the cytoplasm; transport of the progesterone-receptor complex to the nucleus; binding of the complex to "acceptor" sites on the genome and activation of transcription processes

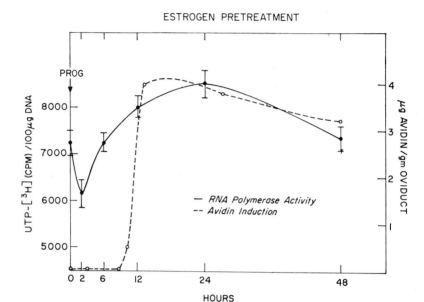

Fig. 11. The stimulation of nuclear RNA polymerase activity and of avidin synthesis in the oviduct of estrogen-pretreated chicks after a single injection of progesterone (5 mg, s.c.). (From O'Malley et al., 1969.)

that result in new RNA species; transport of hormone-mediated RNA to cytoplasm; and ribosomal synthesis of new proteins (avidin) (Fig. 8). Whereas this scheme was undoubtedly an oversimplification, it emphasized that the mechanism by which progesterone acts may be "a qualitative progesterone-induced change in nuclear gene transcription." Although unphysiological doses of estrogens (stilbestrol, 5 mg/day for 6 to 10 days) and progesterone (5 mg, single injection) have been used in experiments on immature chicks, the sequence of biochemical changes closely resembles that produced by the endogenous hormones of the mature chicken.

The proposal that cyclic AMP may have a widespread function as a "second messenger" of hormone action (Robison et al., 1971) prompted a further study of whether it played any part in progesterone-induced avidin synthesis. Adenyl cyclase activity in chick oviduct homogenates was not affected by estrogens, but progesterone caused a 344% increase over controls by 24 hours after treatment in vivo. Tissue levels of cyclic AMP were raised 6 and 24 hours after a progesterone injection (5 mg), though the rate of increase suggested that it played only a secondary role in avidin induction. The progesterone-stimulated increase in adenyl cyclase activity appears to be organ-specific, since it is not seen in chick liver or lung. It could not be accounted for by catecholamine release, or be reproduced in *vitro*. Furthermore, the stimulation of avidin synthesis by progesterone treatment could not be mimicked by cyclic AMP or N^6-2′-*O*-dibutyryl cyclic 3′5′-AMP. So it seems improbable that cyclic AMP acts as a "second messenger" of progesterone action in the chick oviduct (Kissel *et al.*, 1970; Rosenfeld *et al.*, 1970).

2. Carbonic Anhydrase

Progesterone is reputed to affect the activity of many enzymes in mammalian tissues. These effects have been frequently described in animals sensitized with estrogens or inferred from changes observed in target tissues during estrous or menstrual cycles or during pregnancy when the corpus luteum is active. Consequently the precise role of each hormone is difficult to assess in these circumstances. Typical examples of instances where progesterone is believed to affect enzyme activities in mammalian tissues have been discussed elsewhere (Boscott, 1962a; Fotherby, 1964).

Carbonic anhydrase is a zinc–protein enzyme widely distributed in mammalian tissues. In the female reproductive tract the enzyme is located in the endometrium, placenta, and the Fallopian tube. In the rabbit, its activity is closely related to the progestational state. Progesterone, and to a lesser extent ethisterone and methyltestosterone, produce conspicuous increases in endometrial carbonic anhydrase activity, but only after the rabbits have been sensitized by endogenous or exogenous estrogens. In other species the distribution of carbonic anhydrase is much more variable. The endometrium of rats, hamsters, guinea pigs, and cows is devoid of activity, while the uterine mucosa of sheep is particularly rich in the enzyme. In sheep the enzyme activity seems to be independent of endocrine influences since it is demonstrable in prepubertal and ovariectomized animals (Lutwak-Mann, 1955). Similarly, progesterone treatment has no effect on uterine carbonic anhydrase in mice (Madjerek and Van der Vies, 1961). The placenta of most species studied (Lutwak-Mann, 1955) contains carbonic anhydrase activity that is located mainly in the maternal placental tissue.

Carbonic anhydrase activity in the rabbit endometrium has been used in the bioassay of progestagens (Lutwak-Mann and Adams, 1957; Miyake, 1962). Biochemically, the enzyme catalyses the reversible reaction:

$$H_2CO_3 \rightleftharpoons H_2O + CO_2$$

but its physiological significance in the reproductive tract remains something of a mystery (Lutwak-Mann, 1955; Miyake and Pincus, 1958, 1959). When blocked by acetazolamide, there is no inhibition of the profuse endometrial proliferation of estrogen-sensitized rabbits treated with progesterone (Knudsen *et al.*, 1969) and no immediate effect on embryo development in pregnant rabbits (Lutwak-Mann, 1955). Hager and Gilbert (1966) suggest that progesterone stimulates carbonic anhydrase activity in the rabbit endometrium by an activation of gene expression since actinomycin D given before (but not after) steroid injections reduces the levels of enzyme activity. Although the hypothesis that progesterone regulates transcriptional processes is fashionable, the lack of detailed information about control animals in the abstract published by Hager and Gilbert (1966) suggests that it should be treated with caution. More recently a close correlation has been reported between the overall patterns of the total uterine zinc content and carbonic anhydrase levels in the pregnant rabbit. This is surprising in view of the fact that not more than 1% to 2% of total uterine zinc is associated with carbonic anhydrase. The finding that progestagens secreted during pregnancy may mobilize zinc to the uterus as well as activate carbonic anhydrase (Lutwak-Mann and McIntosh, 1969) is an interesting possibility that merits further study.

3. Lactose Synthetase

The hormonal control of lactose synthesis by the mammary gland of certain species illustrates a different action of progesterone on target cells, namely an inhibition of protein synthesis. Turkington and Hill's (1969) experiments have shown that progesterone inhibits the synthesis of the enzyme lactose synthetase. The activity of this enzyme is associated with the interaction of two proteins, whey protein or α-lactalbumin (B protein) and galactosyl-transferase (A protein) which specifies glucose as the substrate for galactose transfer in the formation of lactose (Brodbeck and Ebner, 1966; Brodbeck et al., 1967). When α-lactalbumin, probably a rate-limiting subunit of lactose synthetase, is absent, then lactose is formed at a very low rate (Brew, 1969, 1970; Palmiter, 1969a; but see Palmiter, 1969b). Special attention has been directed towards those factors that regulate the synthesis of these two proteins. Mammary explants from mice in midpregnancy have been cultured in the presence of prolactin, insulin, cortisol, and human placental lactogen (HPL). Lactose synthetase activity, which was stimulated by prolactin or HPL treatment, could be blocked by physiological doses (2×10^{-6} M) of progesterone. Progesterone inhibited the synthesis of α-lactalbumin but permitted an increased galactosyl transferase synthesis in culture. At higher concentrations (5×10^{-4} M) the hormonally induced synthesis of both proteins was reduced. Similar effects have been found in pregnant mice injected with 1 mg progesterone (Turkington and Hill, 1969), and the findings are consistent with the idea that progesterone prevents a rise in mammary lactose so that when the endogenous secretion of progesterone is reduced by ovariectomy or parturition, lactogenesis ensues

(Liu and Davis, 1967; Kuhn, 1969a,b, 1971; Yokoyama *et al.,* 1969; Kuhn and Briley, 1970). Whether the progesterone inhibition of lactose synthetase found in rats and mice will be corroborated in all species is doubtful, since lactose synthesis may begin in some species before the onset of parturition and before the time when progesterone levels fall (Cowie and Tindal, 1971). In the mouse, at any rate, the idea that progesterone inhibits the prolactin-stimulated RNA synthesis and RNA polymerase activity of mammary explants, and that progesterone selectively inhibits the transcription processes associated with α-lactalbumin synthesis (but not those of galactosyl transferase or casein phosphoproteins; Turkington and Hill, 1969), provides a different view of the regulatory properties of this steroid hormone.

Early studies of ovarian hormones led to the conclusion that estrogens are responsible for the growth of ducts and that progesterone, as well, is required for complete lobulo-alveolar development (Turner, 1939; Folley and Malpress, 1948; Folley, 1952; Lyons *et al.,* 1958; Cowie, 1971). Multiple hormones are involved in the differentiation of mammary epithelial cells into secretory alveolar cells *in vivo,* and the endocrine effects are dose dependent. Estradiol-17β at low concentrations (10^{-13} M) inhibits DNA synthesis in epithelial cells, but at higher concentrations the cells proliferate. Progesterone stimulates the budding of the ducts and provides new populations of stem cells for alveolar differentiation (Prop, 1963; Bresciani, 1968, 1971). Unexpectedly, the characteristic effects of estrogens and progesterone on mammary cells have not been reproduced *in vitro.* Among a variety of explanations for this anomaly is the prospect that an essential metabolite of estrogens or progesterone (formed in an extramammary site) is lacking *in vitro.* Chatterton (1971) suggests that since progesterone fails to stimulate cell division in mammary explants of mice or rats, metabolites such as 3β-hydroxy-5α-pregnan-20-one (allopregnanolone) formed *in vivo* outside the gland—or even allopregnanolone sulfate produced within the gland—may have important functions in normal tissue. Another possibility is that ovarian hormones reduce the threshold at which mammary cells respond to circulating levels of insulin, growth hormone, and prolactin (Chatterton, 1971).

4. Synthesis of Other Proteins and Nucleic Acids

Progesterone has multiple effects on enzymatic activities and on the protein and nucleic acid content of target organs such as the uterus and mammary gland, but as yet it is difficult to organize these events into a coherent pattern to describe how the steroid acts. The effects on nucleic acids are numerous. For instance, progesterone causes a twofold increase in the content of endometrial RNA and DNA per wet weight of tissue (Telfer and

Hisaw, 1957); it increases the total content of RNA and DNA of the immature rat uterus though DNA concentration is actually decreased (Lerner *et al.*, 1966); it inhibits a major part of the stimulation by estradiol of RNA/DNA levels, phospholipid synthesis, and certain enzyme activities in the immature rat uterus, but when given alone it promotes short-term increases in these same constituents (Harris *et al.*, 1967). Progesterone treatment of tissue cultures of human endometrium has been reported to produce no increase in the uptake of [³H]uridine into RNA (Wilson and King, 1969); in short-term incubations of normal human endometrium, progesterone may even decrease the uptake of [¹⁴C]uridine into RNA and [³H]thymidine into DNA (Nordquist, 1970). In ovariectomized rats, progesterone induces DNA synthesis in the uterine epithelium (7%) but not in the stroma, though estradiol-17β treatment gives an even greater response (16%: Leroy, 1967). In mice, progesterone (1 mg/day/mouse) evokes DNA synthesis in the cells of the duct epithelium of mammary explants and in those of the terminal buds, even if the tissues have not been treated previously with estradiol-17β (Bresciani, 1971).

Progesterone appears to have little, or in some instances a slight inhibitory, effect on the total protein content of the uterus (Drasher, 1952; Brody and Westman, 1958, Leathem, 1959; Little and Lincoln, 1964). It may depress the incorporation of [¹⁴C]glycine into uterine phospholipids and protein in rats (Mueller, 1953) and reduce nuclear protein synthesis in uterine epithelial cells of mice after stimulation by estrogen (Smith *et al.*, 1969). In addition, it is reported that progesterone increases glycogen deposition *in vitro* (Csermely *et al.*, 1969) in human endometrium but not in the uterus of immature rats (Walaas, 1952) or rabbits (Telfer and Hisaw, 1957).

D. Progesterone Metabolism and Mechanism of Action

Is progesterone metabolized to analogous compounds more active than their parent substance? This question became more pertinent when it was reported that testosterone, another Δ^4-3-ketosteroid, is metabolized in target tissues such as the seminal vesicle and epididymis to 5α-pregnan-3-one-17-ol (5α-dihydrotestosterone; 5α-diHT), a compound more biologically active in some test systems than the parent form (Bruchovsky and Wilson, 1968; Northcutt *et al.*, 1969; Wilson and Walker, 1969). At least twenty-six metabolites may be formed from progesterone by the reduction of the C_4 double bond and the C_3 and C_{20} oxogroups, while reduction of derivatives formed by the introduction of oxygen function could yield nearly 1000 metabolites! Of these theoretical compounds, relatively few have been identified in natural sources (van der Molen and Aakvaag, 1967). In man, progesterone is reduced *in vivo* principally to 5β-metabolites (Ungar *et al.*, 1951; Dorfman, 1954) which are excreted mainly in urine (Davis and Plotz, 1958; Sandberg and Slaunwhite, 1958), though in some species a major proportion of progesterone metabolites appear in bile (Taylor and Scratcherd, 1961, 1967; Taylor, 1965; Stupnicki *et al.*, 1969). In addition to the many 5β-metabolites, compounds with a 5α-orientation have been identified in the urine of man and of other species (see Dorfman and Ungar, 1965; man: Fotherby, 1964; Romanoff *et al.*, 1966; Thijssen and Zander, 1966; sow: Schomberg *et al.*, 1966; sheep: Stupnicki and Williams, 1968). The formation of progesterone metabolites with a 5α-configuration has been reported in the liver (Taylor, 1954; Shirley and Cooke, 1968), myometrium (Wichmann, 1967; Wichmann *et al.*, 1967), endometrium (Bryson and Sweat, 1963, 1967), uterus of normal and eviscerated rats (Wiest, 1963a,b), the kidney of rabbits (Chatterton *et al.*, 1969a), and the mammary glands of rabbits and goats (Chatterton *et al.*, 1969b; Chatterton, 1971).

A description of these compounds and the enzymes concerned in their metabolism will be found in the monograph of Dorfman and Ungar (1965).

In the rat uterus the metabolites of progesterone with a 5α-configuration result from the activities of two enzymes, Δ^4-5α-reductase and 3α-hydroxysteroid dehydrogenase. The former enzyme is probably rate-limiting since 3-keto intermediate compounds are absent, whereas 3α-hydroxylated 5α-pregnane products accumulate in uterine tissue (Wiest, 1969). These two enzymes can be located in a particulate cell fraction of whole uterus homogenates, while another progesterone-metabolizing enzyme, 20α-hydroxysteroid dehydrogenase, is found in the soluble fraction of the myometrium of pregnant rats (Wichmann, 1967). Enzymic activities may differ during the estrous cycle, greater activity being found at estrus than in pseudopregnancy (Wiest, 1969). Similarly, estrogens increase the uterine metabolism of progesterone in rats probably because of an enhanced rate of ring A reduction (Armstrong and King, 1971). In the nonpregnant human uterus the metabolism of progesterone differs quantitatively and qualitatively between the endometrium and myometrium. In the myometrium the rate of progesterone metabolism *in vitro* is only one-sixth that of endometrial incubations. A highly polar dihydroxy compound, a major product of endometrial metabolism, is not found in myometrial incubations. As in the rat, the 5α-configuration is predominant among the reduced compounds, but unlike the rat, the C_3-keto group is not reduced (Bryson and Sweat, 1967, 1969; Fig. 12).

In addition to the formation of progesterone metabolites within target organs, other evidence suggests that these metabolites may be selectively removed from blood by the mammary gland and uterus. Allopregnanolone, a 5α-metabolite of progesterone, appears to accumulate in target organs of

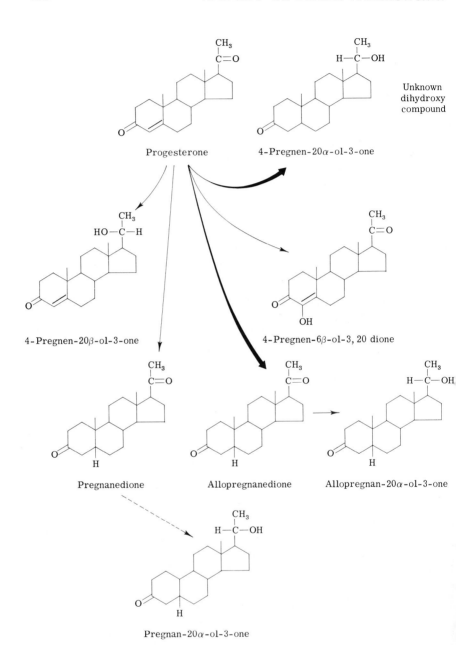

Fig. 12. Metabolism of progesterone in human endometrium *in vitro*. (From Bryson and Sweat, 1967.)

the rat, rabbit, and goat, probably as a result of selective uptake from blood (Wichmann, 1967; Chatterton *et al.,* 1969b; Chatterton, 1971). It is notable that allopregnanolone is selectively retained by mammary tissue in lactating rabbits, whereas 20α-dihydroprogesterone is the predominant metabolite of progesterone found in the mammary glands during gestation (Chatterton, 1971). This finding indicates that the pattern of progesterone metabolism within this target organ may differ according to the reproductive state.

Wiest (1969) has stated that the biological activity of progesterone is a function of the progesterone molecule itself and not of its metabolites. Certainly tests of the biological activities of 5α-orientated and other metabolites provide no conclusive evidence that their activity is greater than, or even equal to, that of progesterone. Such tests are usually performed in several responsive systems; occasionally, they indicate that a metabolite possesses a high potency in one system, but not in another (e.g., 20β-diHP; Zander, 1959). However, the possibility that progesterone may act by its metabolism to a compound with a 5α- configuration, or indeed to some other form, cannot be entirely ruled out. Pregnanolone (5β- configuration), a short-acting, potent, hypnotic steroid (Gyermek, 1967), has a pronounced action on the uterus. Oxytocin-induced contractions of isolated strips of rat uteri, or contractions of the isolated rat duodenum produced by barium chloride, are inhibited to a greater extent by pregnanolone than by progesterone (Gyermek, 1968). Also it should be noted that the progesterone receptor of the uterine cytosol of the guinea pig will bind 5α-pregnanedione. The 5α-metabolite competitively inhibits progesterone binding by about 55% (Milgrom *et al.,* 1970b).

VI. INTERACTIONS OF ESTROGENS AND PROGESTERONE

Basic to our understanding of progesterone action is the question of why many of its biological effects are pronounced only after target tissues have been sensitized by estrogens. The physiological response of these tissues to ovarian hormones is both dose-dependent and ratio-dependent. When the ratio of progesterone:estrogen is large (about 750:1 in rabbits), synergistic responses are commonly found, but when the dose of estrogen is relatively high then the action of progesterone may be wholly, or partially, inhibited. Many examples of the synergistic and antagonistic interactions of these and other hormones have been described; they include such effects as the influence of estrogen–progesterone treatment on endometrial proliferation, pregnancy maintenance, myometrial activity, mammary development, vaginal cornification, uterine growth, and deciduoma formation (see Courrier, 1950; Deanesly, 1966; and Volume II, Chapter 8).

An interaction of estrogen and progesterone has been demonstrated in most of the model systems already discussed; the induction of avidin synthesis by the chick oviduct (O'Malley *et al.*, 1969), progesterone binding in the guinea pig uterus (Milgrom *et al.*, 1970a,b), and the stimulation of carbonic anhydrase activity in the rabbit endometrium (Lutwak-Mann, 1955). Treatment with progesterone alone will induce the formation of avidin by the chick oviduct, but if the chicks are previously treated with estrogen the amount of avidin produced is greatly increased. Continuous administration of diethylstilbestrol has the interesting effect of raising the amount of the progesterone-binding component in the oviduct cytosol, but no avidin is produced in the absence of progesterone. If the estrogen treatment is stopped for 10 days, the density-gradient pattern of the progesterone binding component in the cytosol does not change, but the protein associated with the binding component decreases in concentration (Sherman *et al.*, 1970). Similar results have been reported in ovariectomized guinea pigs in which estradiol-17β treatment stimulates the binding of progesterone by uterine proteins. In untreated guinea pigs, uterine binding proteins sediment in a sucrose density gradient with a composite peak (7 to 4S), but after estrogen treatment there is a rise in the amount of a heavy binding component (6.7 S: Milgrom *et al.*, 1970b). These findings are in keeping with the idea that estrogens stimulate the synthesis of soluble binding components whose high affinity for progesterone protects the steroid from rapid metabolism within the target organ. This ensures a low, but effective concentration of progesterone at its site(s) of action. The fact that progesterone treatment alone will induce avidin synthesis, although only at a low rate, suggests that one effect of the estrogen-stimulated synthesis of progesterone-binding proteins is to reduce the threshold dose of progesterone.

In addition to the view that estrogens reduce the threshold dose of progesterone by stimulating the synthesis of progesterone-binding components, there is some evidence that the formation of other steroid receptor macromolecules may be influenced by hormones. Progesterone and estradiol-17β raise the concentration of estradiol receptors in the uterine cytosol of rats (Steggles *et al.*, 1971), and ovariectomy or hypophysectomy, but not adrenalectomy, reduce their concentration. The nature of the endocrine factors that control the production of this 4 S cytoplasmic receptor, which fails to aggregate at low ionic strength, has not yet been defined (King *et al.*, 1971). In mice, progesterone increases the uptake of [^{3}H]estradiol-17β by stromal, but not epithelial, cells *in vivo* (Smith *et al.*, 1970). Interactions between estradiol-17β and progesterone (or synthetic progestagens) with respect to their uptake and retention by target tissues have been reported in women (Eisenfeld and Axelrod, 1965; Brush *et al.*,

1967), mice (Kraay and Black, 1970), and rats (Eisenfeld and Axelrod, 1965).

In contrast to the synergistic interactions of these ovarian hormones, there are circumstances in which progesterone inhibits the action of estrogens (Oka and Schimke, 1969; Yu *et al.*, 1971). Progesterone (1 mg/day), given to chicks simultaneously with estrogen, antagonizes the estrogen-induced changes of the immature oviduct (growth, differentiation, lysozyme and ovalbumin synthesis). Progesterone appears to be a potent antagonist of estrogen at a dose rate of 1:1, but it does not inhibit the other changes that normally occur in the chick oviduct within 24 hours of estrogen treatment, namely the uptake of α-aminoisobutyric acid (nonmetabolizable amino acid) and water, nor does it affect the rise in total lipid, phospholipid, and phosphoprotein content of serum, or the increased phosvitin synthesis in the liver. Paradoxically, there are other situations where progesterone mimics the action of estrogens on the chick oviduct; for example, it stimulates the tubular gland cells of the chick oviduct to secrete lysozyme and ovalbumin, normally a prerogative of estrogens. Progesterone will produce these effects only if the oviduct has been previously sensitized by estrogens. These results emphasize the importance of defining the specific cell types that are involved in hormonal stimulation and their competence to respond (Oka and Schimke, 1969).

A different explanation for the striking biological effects of estrogen–progesterone treatments may lie in the direct interaction between these two steroid molecules. When a sonicated suspension of equimolar amounts of solid estradiol-17β and progesterone is allowed to equilibrate with steroid-free KCl contained in a dialysis sac, approximately equal amounts of estradiol and progesterone enter the sac (solubility about 35 μmol/liter). When measured separately under the same experimental conditions, the solubilities of estradiol and progesterone are 23.5 and 42.6 μmol/liter, respectively. When dialysis sacs contain phosphatidylcholine liposomes (phospholipid liquid crystals), the uptake of the two steroids from aqueous buffer solutions is again approximately equal (about 30 nmol/μmol lipid), whereas their individual lipid solubilities differ by a factor of three in favor of progesterone (Fig. 13). These observations have been interpreted as evidence that under these experimental conditions estradiol and progesterone are not independent molecules but rather some type of molecular complex, or small mixed micelle, containing approximately equal amounts of the two steroids (Heap *et al.*, 1971).

Theoretically, the structures of these two steroids present several possibilities for intermolecular bonding: hydrogen bonding between the hydroxyl groups of estradiol and the carbonyl groups of progesterone; π-bonding

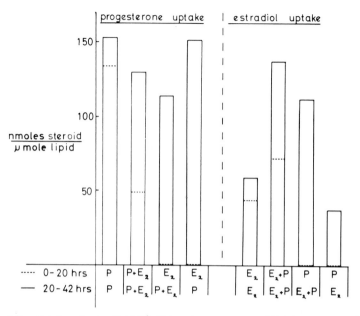

Fig. 13. Uptake of estradiol-17β (E₂) and progesterone (P) (alone or together) by lecithin liposomes suspended in buffered KCl at 37°C and contained within dialysis sacs. Uptake during the first 20 hours shown by dotted line within each column; uptake during 20 to 42 hours shown by the full line at the top of the column. (Data taken from Heap *et al.*, 1971.)

between the aromatic ring of estradiol and the α,β-unsaturated carbonyl of progesterone; and hydrophobic bonding by Van der Waal's forces between the hydrocarbon nucleus of each of the two steroids (Heap *et al.*, 1971). Whether such direct interactions take place *in vivo* is another matter. It has been noted already that the synergistic effects of progesterone and estradiol are most clearly observed when the ratio of the steroid dose is greatly in favor of progesterone (e.g., 750:1; Deanesly, 1966). But more relevant to their mechanism of action is the concentration of these hormones in target cells where the permeability of cell and organelle membranes may be influenced not only by free molecules but also by complexes of these molecules. Indeed, calculations from the results obtained by different workers who infused into rats labeled estradiol and progesterone at a constant rate suggest that the uterine concentrations of these two steroids at steady-state may be rather similar (1 to 7×10^{-9} mol/kg, progesterone isolated specifically; Lawson and Pearlman, 1964; estradiol-17β as total tritium concentration; Laumas, 1967).

VII. MEMBRANE PERMEABILITY

Current theories on the mechanism of action of estrogens and progesterone are frequently interpreted to support the concept that the multiple biological actions of these hormones are derived from "a unique interaction at a single site." Tata (1968) has proposed as a variant of this idea that the early action of steroid hormones involves an adjustment of the permeability barriers of target cells. Specificity could then reside in the regulation of a process that is intracellular, rate-limiting, hormone-specific, and facilitated by a change in the membrane permeability of a cell or of an organelle. It seems improbable that steroid hormones regulate transport mechanisms specifically, since individual hormones may affect the transport of several substrates, and the transport of an individual substrate may be stimulated by several hormones (see Riggs, 1970). But in the light of findings that estrogens influence the entry of substrates (including RNA precursors) into target cells, it is pertinent to enquire how ovarian steroid hormones may affect the permeability properties of cell membranes.

Willmer (1961) advanced the theory that amphipathic steroids penetrate the phospholipid bilayers of cell membranes. He envisaged that this might occur in one of several ways; the lipophilic end of the steroid molecule could associate with the aliphatic end of the phospholipids; the hydrophilic end of the steroid molecule could be orientated in the aqueous phase or towards any polar group in the membranes; or, an end group of the steroid may become attached to a polar group of the phospholipid. He also suggested that double stacking of steroid molecules between aliphatic chains is a theoretical possibility because of the shorter dimensions of steroid molecules compared with those of phospholipids. This double-stacking may be relevant to the various biological effects obtained when estrogens and progesterone are presented in differing order. Accordingly, progesterone preceded by estradiol-17β would lead to the following arrangement of polar groups, from the hydrophobic surface (left) to the hydrophilic surface (right) of the unit layer,

$$CH_3CO \quad O \qquad OH \text{ (alcoholic)} \qquad OH \text{ (phenolic)}$$
$$\text{(progesterone)} \qquad\qquad \text{(estrogen)}$$

while with the reverse treatment the order would be,

$$OH \text{ (alcoholic)} \qquad OH \text{ (phenolic)} \qquad CH_3CO \qquad O$$
$$\text{(estrogen)} \qquad\qquad \text{(progesterone)}$$

and the physiological effect would be very different. If steroid orientation does occur in one of these forms, then the estrogenic properties of two dissimilar molecules like estrone and stilbestrol may be related to the length of the molecules which are almost identical (8.55 Å) (but see Duax et al. 1977). Steroid penetration of phospholipid bilayers would introduce hydrophilic groups into a mainly hydrophobic phospholipid–cholesterol membrane and as a result the phospholipid would acquire special features on its polar surface. Instead of CH_3 groups, the sequence would be broken by a specific pattern of active, polar and more or less hydrophilic groups. Membrane penetration by steroids such as estrogens and progesterone would be expected to change its permeability to water, ions, and possibly other substances. Willmer (1961) suggested that an inability to enter into the phospholipid bilayer may explain the inactivity of steroids with a 5β-configuration compared with those active forms with a 5α-configuration. The bent configuration between ring A and B may prevent penetration and orientation of 5β-steroids between the phospholipid molecules in bimolecular leaflets.

The validity of Willmer's theory has been tested in several ways. Closepacked monolayers of cholesterol and 1-α-(dipalmityl)-lecithin were found to absorb small quantities of progesterone from an aqueous phase, though it was not possible to determine whether the progesterone molecules were arranged at the interface or whether they were packed vertically into the monolayer with one polar group at the interface (Taylor and Haydon, 1965). In other experiments, model membranes formed with liquid crystals of phospholipid (liposomes), and membranes of natural sources (erythrocytes, lysosomes, mitochondria) were used to examine the influence of steroids on membrane permeability (Bangham et al., 1965a,b; Snart and Wilson, 1967; Bangham, 1968). The results obtained with model membranes and lysosomes were substantially similar and showed that steroids such as progesterone and estrogens with polar groups at each end of the molecule clearly altered permeability. In more recent experiments liposomes were allowed to equilibrate with labeled steroids and the following effects on K^+ permeability were noted. Estradiol-17β and estrone produced a predominantly stabilizing effect, whereas progesterone and steroids related to androgens and corticosteroids had a labilizing action (Fig. 13). When the gross effect of a steroid on membrane permeability was not large, after due allowance was made for the amount of steroid actually taken up by the liposomes, the effect of each molecule could be appreciable. It was also demonstrated that the molecular effect of a steroid on K^+ permeability (under equilibrium conditions) was directly proportional to the aqueous solubility of the steroid and inversely proportional to its lipid solubility (Heap et al., 1970, 1971).

Other findings relevant to Willmer's (1961) hypothesis that steroid molecules penetrate phospholipid membranes concern the extent to which interactions between steroids may influence membrane permeability. Incorporation of 0.5% cortisone or cortisol into liposomes before exposure to different steroids has a stabilizing effect on the membrane. It also inhibits the increased permeability normally produced by diethylstilbestrol and etiocholanolone. Similarly, when as little as 0.1% estradiol-17β is preincorporated into liposomes, the release of anions induced by etiocholanolone (but not other steroids) is blocked. These observations have interesting biological implications because an antagonism between glucocorticoids and 5β-H steroids (e.g., etiocholanolone) has been found clinically in the case of fever in man, for which males are more susceptible than females (Weissmann, 1965; Weissmann et al., 1966).

Another example of the effect of steroid interactions on the permeability of a phospholipid model membrane has been described for estradiol-17β and progesterone. When both steroids are presented to liposomes simultaneously, the molecular effects on K^+ permeability are appreciably greater than when they are presented separately. The largest increase in K^+ permeability occurs with liposomes equilibrated for 20 hours with progesterone or estradiol and progesterone, and subsequently transferred to a buffer solution containing both steroids. The results are in keeping with the idea that progesterone and estradiol can form a complex, or small mixed, micelle, in aqueous solution which is orientated in a phospholipid model membrane in such a way as to produce a membrane conformational change similar to molecules of progesterone alone. The steroid-induced effect is one of membrane labilization similar to that caused by progesterone alone rather than a membrane stabilization, as found with estradiol-17β alone (Heap et al., 1971). These studies are therefore consistent with Willmer's (1961) hypothesis insofar as they confirm that steroids will effect changes in the permeability of phospholipid bimolecular leaflets. It is probable, but not yet proven, that this is achieved by steroid penetration between phospholipid molecules to produce conformational and permeability changes, but more rigorous investigations will be required to establish their molecular orientation.

An important site of progesterone action is the smooth muscle cell of the myometrium, and it has been postulated that changes in the ionic permeability of the myometrial cell determined by endocrine factors may underlie its different properties during gestation. The inhibition of myometrial activity in pregnancy allows implantation to proceed normally and prevents the expulsion of fertilized eggs and of the developing conceptus. The inhibition of activity is related in most species to a continuous secretion of progesterone (see Bedford et al., 1972a), which gives rise to what has become known as the "progesterone-block" of myometrial activity (Csapo, 1956, 1969). In the nonpregnant animal, the uterus has a low membrane potential (about 35 to 45 mV) which has been ascribed to a relatively high

Na^+ permeability since a tenfold increase in the external K^+ concentration produces a depolarization of only between 32 and 52 mV, smaller than expected from the potassium diffusion potential (Kuriyama, 1964; Casteels and Kuriyama, 1965). During late pregnancy in the rat, the membrane potential increases to values of about 60 mV and then declines after parturition. Similar changes have been observed in the mouse (Kuriyama, 1961, 1964), guinea pig, and cat (Bülbring et al., 1968). In the rabbit myometrium, however, conflicting results have been reported; some workers claim that estrogen- or progesterone-dominated animals show similar resting potentials (Kao and Nishiyama, 1964), while others have found higher values in progesterone-dominated rabbits. The cells at the placental site were reported to have a higher resting potential (53.4 mV) than those in an interplacental area (42.3 mV; Goto and Csapo, 1959), but local synthesis of progesterone by the rabbit placenta has not yet been conclusively demonstrated.

Intracellular concentrations of ions in the myometrium remain relatively constant throughout pregnancy in several species, apart from an increase in intracellular Cl^- during the early stages of gestation (Bülbring et al., 1968). It is probable, therefore, that changes in ionic permeabilities may lead to the different resting potentials (Kao, 1967). Jones (1970) has shown that uterine muscle becomes more selective toward K^+ than Na^+ after progesterone treatment. In the myometrium of rabbits treated with progesterone, there was an increase in K^+ permeability as measured by the exchange of ^{42}K at steady-state (Jones, 1968). Casteels and Kuriyama (1965) also observed that the resting potential of the myometrium of the pregnant rat followed the changes in K^+ concentration in external solution more closely than in the nonpregnant condition, consistent with the hypothesis that progesterone enhances K^+ selectivity of the myometrial cell.

Estrogens have been found to influence sugar and amino acid transport in uterine tissue (see Riggs, 1970). The injection of rats with metabolic inhibitors (cycloheximide, puromycin, and actinomycin D) before, or at the same time as, estradiol administration will block sugar metabolism and probably sugar transport (Nicolette and Gorski, 1964; Gorski and Morgan, 1967; Roskoski and Steiner, 1967a; Spaziani and Suddick, 1967; Smith and Gorski, 1968) and will also inhibit the uptake of α-aminoisobutyric acid in vitro (Roskoski and Steiner, 1967b). The sugar 2-deoxyglucose is rapidly transported into the rat uterus and metabolized to the 6-phosphate compound by a route similar to that of glucose; both steps are stimulated by estradiol-17β treatment in vivo. The uptake of the sugar 3-O-methylglucose is also stimulated by estradiol administration, but only 2 hours after treatment. From the use of metabolic inhibitors, uptake is apparently related to increased protein synthesis (Smith and Gorski, 1968). The uterine uptake of amino acids is also affected by estrogen treatment (Little and Lincoln,

1964; Segal and Scher, 1967; Riggs, 1970) but as yet it is not known how far these changes are related to the early events of estrogen action, such as the synthesis of the "induced protein" in the rat uterus reported by Barnea and Gorski (1970). Riggs (1970) has discussed the finding that estradiol-17β produces a greater stimulation of the uptake of the short-chain, more polar amino acids (e.g., glycine, L-alanine, L-serine) than of branched chain, and less polar substances (L-leucine, L-valine, L-phenylalanine). Amino acids in the latter group require the presence of extracellular Na^+ for uptake, a process relatively sensitive to metabolic inhibitors. Attention has been drawn to the possibility that two transport systems exist as described for the A and L pathways of multiple neutral amino acid transport in the Ehrlich ascites carcinoma cell (Oxender and Christensen, 1963; Christensen, 1967).

The effects of estrogens and progesterone on membrane permeability may comprise an early action of these steroids on target cells. The rapidity of estrogen action in provoking uterine hyperemia, histamine release, and an increase in cyclic AMP, all membrane-related phenomena, lends strong support for Tata's (1968) proposal that the adjustment of permeability barriers is of prime importance.

VIII. THE RELATIONSHIP BETWEEN CHEMICAL STRUCTURE AND BIOLOGICAL ACTIVITY

Theories of how ovarian hormones induce biochemical changes in target tissues are inadequate if they fail to account for the efficacy of highly potent synthetic compounds that produce estrogenic or progestagenic effects, even when their chemical structure bears little resemblance to that of their endogenous counterparts. This raises the problem of what structural features of steroid (or synthetic) molecules are implicated in receptor recognition—an event believed to be a prerequisite of hormone action. If the physiological effects of these synthetic compounds derive from biochemical events comparable to those produced by endogenous hormones, then a consideration of the relationship between chemical form and biological activity may throw further light on the molecular actions of hormones. A complication lies in the fact that the activity of synthetic compounds may differ according to their route of administration and the species treated. Melengestrol acetate is an example of a synthetic progestagen which is a potent inhibitor of estrus and ovulation when given orally to ruminants, but it is much less effective in nonruminants (Zimbelman et al., 1970).

An analysis of the structural determinants of estrogenicity implicates the following molecular features; an aromatic ring A, an hydroxyl group at C_3

(phenolic), and the absence of an angular methyl group at C_{10} (Shoppee, 1958; Fieser and Fieser, 1959; Boscott, 1962b). However, the variability in the structure of compounds that display estrogenic activity is striking. Compounds range from those secreted by the mammalian ovary and other endocrine glands (e.g., estradiol-17β) to coumestrol, a nonsteroidal estrogen of plant origin; genistein, an active estrogen of subterranean clover; and stilbestrol, a highly potent synthetic estrogen. Although flat aromatic ring structures are common to all compounds, in other respects structures differ greatly (steroids, biphenolics, and aromatic carboxylic acids). Irrespective of their differing structures, it appears that estrogenic activity is closely related to the uptake and retention by high-affinity receptors in target tissues. Highly potent estrogens, including aromatic carboxylic acids, competitively inhibit the uptake of labeled estradiol-17β by the mouse uterus, whereas weakly active compounds, such as estradiol-17α and the racemic isomer of mesostilbestrol are comparatively weak inhibitors (Terenius, 1966). Particularly interesting is the potent estrogen mirestrol, found in a leguminous plant, *Pueraria mirifica,* native to Thailand. This estrogen suppresses the uterine binding of estradiol-17β in immature mice *in vitro,* and has about one-fifth the uterotrophic potency of estradiol (Terenius, 1968). Mirestrol differs markedly from estradiol-17β. It has four hydroxyl groups, compared with the two of estradiol-17β and stilbestrol, and is at least ten times more water-soluble than estradiol-17β (Jones and Pope, 1960). Yet the molecular dimensions of mirestrol and estradiol-17β are remarkably similar. X-ray crystallography reveals that the distances between the C_3 and C_{17} hydroxyl groups of estradiol-17β and between the C_3 and C_{18} hydroxyl groups of mirestrol are almost identical (Taylor *et al.*, 1960).

The major determinants of progestational activity apparently comprise a high electron density between C_4–C_5 and the presence of a $C_{17\beta}$ acetyl radical. Potency is reduced by modification of the C_{17} substituent, removal of the C_{13} methyl group, or the introduction of a double bond at C_{16}. Another essential determinant is the C_2 position where substitution reduces potency. In contrast, the *trans* position of rings A/B, B/C, C/D is not indispensable for when B/C, C/D are in the *cis* form, activity is not abolished. From such results, Miyake and Rooks (1966) have deduced that the most important features of progestational molecules are their attached radicals and the central "core" of the steroid. The entire steroid molecule is only of secondary importance to provide distance between areas of electrical density (C_4 double bond) and proper steric configuration. Thus, the minimum structural components concerned with progestational activity are an unsubstituted C_2, C_4 double bond, $C_{17\beta}$-acetyl group, and $C_{18\beta}$-methyl group (Fig. 14). Activity may be enhanced by substituents at $C_{6\alpha}$, $C_{17\alpha}$, and C_{19} or by an additional double bond at C_6. Testosterone also has a low

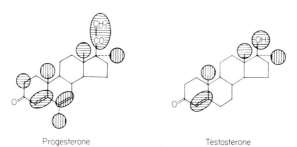

Progesterone   Testosterone

Fig. 14. Structural components concerned with progestational activity. The minimum necessary structural components for the appearance of progestational activity are indicated by horizontal shading. Those positions that are important for increased progestational activity in two Δ^4-3-ketosteroids, progesterone, and testosterone are indicated by vertical shading. (From Miyake and Rooks, 1966.)

degree of progestational activity (0.05 that of progesterone in the Clauberg subcutaneous assay). By a similar analysis of synthetic compounds, it has been deduced that the minimum structural requirements for activity are a double bond at C_4, a $C_{18\beta}$-methyl group (as for progesterone), and a $C_{17\beta}$-hydroxyl group. Progestational activity can be markedly increased if compounds are substituted at $C_{17\alpha}$ and C_{19} positions. More detailed accounts of the biological properties of synthetic progestagens and estrogens will be found in the articles of Boscott (1962a,b), Emmens and Martin (1964), and Junkmann (1968).

In addition to structural determinants, it is probable that the orientation of the steroid molecule with respect to its receptor plays an important part in the mechanism of hormone action. Because of the three-dimensional property of the steroid molecule, the possibilities of configurational associations are numerous. There is strong evidence that the rear, or α-face, of the steroid molecule, rather than the β-face, is involved in the association with proteins and certain other biologically important compounds. Westphal (1971) reviews the evidence for the interaction between Δ^4-3-ketosteroids and proteins such as human serum albumin. The α-face of the steroid molecule has a fairly plane surface, unlike that of the upper, or β-face, which is convex and "turtle-like." The substituents numbered in Fig. 15 are thought to be the points of contact with proteins, being better suited for a close approach to the protein than the curved β-face. A similar conclusion has been reached from studies of the interactions between testosterone and proteins (Marcus and Talalay, 1955; Talalay, 1957), between cortisol and CBG of human plasma (Daughaday, 1958), and between steroids and specific immunoglobulins (Zimmering et al., 1967). The substrate specificity of NAD-linked estradiol dehydrogenase of human placental tissue also supports the conclusion that the α-side of steroid molecule interacts with the

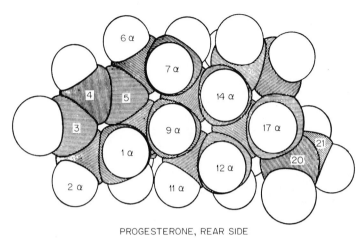

PROGESTERONE, REAR SIDE

Fig. 15. The substituents of the α-face of the progesterone molecule believed to be concerned with steroid–protein interactions. (From Westphal and Ashley, 1959.)

enzyme–coenzyme complex. Reactivity is depressed by α-orientated substituents in rings C or D, with α-orientated angular methyl groups, or with changes at the asymmetric carbon atoms, C_8, C_9, C_{13}, and C_{14}. Another important structural feature of reactive molecules in this system is the planarity of rings A and B, since either or both have to be aromatic in order for the steroid to be utilized as substrate by this enzyme. Interactions of steroids with other compounds, including purine bases, purine nucleosides, and purine-containing nucleotides, also support the conclusion that the α-face of the steroid is closely involved in molecular associations. When substituents are systematically introduced into the steroid molecule, the association constants for adenosine are similar for β-orientated substituents at C_{17} but lower for α-orientated substituents. If substitutions are made at C_{11}, α or β, the association constant tends to be reduced, whereas substituents in rings A or B have little or no effect. From these results it has been concluded that the axial or pseudoaxial hydrogen atoms in rings C and D and part of ring B are involved in the interactions between steroids and purine compounds (Engel *et al.*, 1964).

The interaction between steroid molecules and plasma proteins, such as human serum albumin, may be an inadequate model for steroid interactions with other proteins, such as specific tissue receptors or high-affinity transport proteins such as CBG, PBG, and α_1-acid glycoprotein (AAG or orosomucoid). However, these interactions do share certain features. The "polarity rule" whereby the strength of binding of a steroid to albumin is in general inversely related to the number of polar groups (Eik-Nes *et al.*, 1954), seems to apply to the transport proteins, AAG (Ganguly *et al.*, 1967) and CBG (Westphal, 1971). In respect of AAG and CBG, disulfide bridges are involved in the steroid–protein interaction since cleavage of the S–S bond reduces binding affinity, and blockage by alkylation with iodoacetic acid abolishes binding irreversibly. Similarly, modification of the ϵ-amino group of lysine virtually eliminates the binding properties of the AAG protein molecules. On the basis of these results it has been suggested that the disulfide bridges in the AAG molecule provide a conformational structure that permits an optimal interaction with progesterone and other Δ^4-3-ketosteroids, possibly enveloping the steroid (dimensions $13 \times 7 \times 5$ Å; Fieser and Fieser, 1949; Westphal, 1958) within a protein matrix (Westphal, 1971).

These examples are a few of the many studies in which the physicochemical interaction of steroid and protein molecules have been investigated. A detailed review of this subject is presented in the monograph by Westphal (1971). The question of how estrogens and progesterone act on the responsive cells of target tissues cannot be adequately resolved until more detailed information of the configuration of receptor molecules is obtained.

IX. CONCLUDING REMARKS

Estrogens and progestagens are believed to act on the responsive cells of the reproductive tract in many ways. The regulation of the genetic expression of target cells has become prominent among the theories of the mechanism of action of these ovarian hormones. Yet whether it is accurate to describe these hormones as regulators, a term used both in this chapter and elsewhere in the literature, is a point that should now be questioned. Strictly speaking, neither estrogens nor progesterone can be said to act on target organs in a homeostatic manner, which is a feature of regulatory systems. Nor is there evidence of feedback loops that modify the rate of metabolism and removal of these compounds. On the contrary, the presence of high-affinity tissue receptors causes a retention of these biologically potent molecules that would otherwise be rapidly metabolized and appears to lower their threshold dose. Thus, the presence of these specific receptors in target organs presumably ensures a relatively high concentration of steroid at active sites for a prolonged time—an interpretation hardly consistent with a proposed regulatory role.

A limitation of the theory that estrogens and progesterone influence directly the transcriptional processes of target cells has been the lack of definitive identification of new mRNA molecules synthesized specifically in response to hormone treatment and proof that steroid molecules interact directly with specific regions of nucleochromatin. Because of the technical problems associated with obtaining this information the evidence that these ovarian hormones affect transcriptional processes has only recently been of a direct nature. A different hypothesis, mentioned earlier in this chapter (Section II) and advanced by Tomkins *et al.* (1969) and Ohno (1971), draws attention to a mechanism of action which involves the control of translational processes. These workers have proposed that steroid hormones act as inducers by neutralizing some cytoplasmic repressor, which normally inhibits the translation of specific mRNA and promotes its degradation. In the presence of the steroid the repressor is inactivated, messenger translation occurs, and mRNA degradation is prevented. In this way the concentration of mRNA increases since the transcription of the mRNA is continuous. These workers have postulated that the target cell receptors are allosteric molecules whose conformation is changed by the steroid hormone so that the steroid–receptor complex prevents the synthesis or transport of the repressor or increases its breakdown (Tomkins *et al.*, 1969). Another possibility is that the steroid (inducer) may interact directly with the repressor, transfer it to the nucleus, and so remove a translational inhibitor. This latter proposition has found substantial support in Ohno's (1971) analysis of the action of another gonadal hormone, testosterone. Ohno (1971) contends that the target cells of testosterone contain a protein specified by the wild type allele of the X-linked *Tfm* locus. This protein has a dual role, for when not bound to testosterone it stays in the cytoplasm and represses the translation of certain enzymes; when attached to testosterone the conformation of this allosteric molecule changes, it detaches from mRNA and releases a translation block, and it moves into the nucleus where it activates RNA polymerase I inside the nucleolar region and becomes a nuclear receptor protein. If this single gene product is absent, as in a mutant *Tfm* allele, the testosterone-induced response of target cells is abolished, as in the testicular feminization syndrome.

An obvious limitation of the latter hypothesis is the lack of evidence for a cytoplasmic repressor of translational processes. Nevertheless, it could be envisaged that the migration of an estrogen– or progesterone–receptor complex from the cytosol to nucleus removes repressor protein molecules from their site of translational inhibition. An apparent increase in mRNA molecules soon after hormone treatment could arise from a decrease in degradation of mRNA or from an accumulation of mRNA due to continuous transcription. The finding that the activity of RNA polymerase I in the rat uterus is stimulated by estrogen treatment (Blatti *et al.*, 1970) has been compared by Ohno (1971) with the suggestion that the testosterone–receptor complex activates RNA polymerase I in the nucleolar region.

In respect of the mechanism of action of estrogens and progesterone, it is apparent that there is much support for the view that the induction of transcriptional processes in the nucleus is a dominant feature of their action. However, if future work provides evidence of the presence and identity of repressor molecules, then the possibility that these hormones influence translational processes in the cytoplasm will gain considerable strength. In any event the description of the fundamental mechanisms that are involved in the endocrine control of highly responsive target cells has reached an intriguing stage, and it may turn out that estrogens and progesterone control protein synthesis at several points of the central dogma of transcription and translation.

X. ADDENDUM

Since the above account was completed in 1972, there have been striking advances in the study of how estrogens and progesterone act on target tissues. The following summary contains a selection of recent publications and many of these support the proposition that a dominant, though not exclusive, feature of the action of these steroid hormones is the induction of transcription processes and the stimulation of new protein synthesis.

The nature and distribution of estrogen and progesterone receptors in the cytoplasm and nucleus of target cells have been the subject of reviews and books (Thomas, 1973; O'Malley and Means, 1973; 1974; King and Mainwaring, 1974; Schrader et al., 1974; Liao, 1975; Pasquilini and Scholler, 1975; Baulieu et al., 1975). Techniques for the isolation and purification of receptors, for the determination of receptor concentrations in cytosol and nucleus, and for the study of the translocation of steroid–receptor complexes to the nucleus have been compiled (O'Malley and Hardman, 1975a). Procedures for the investigation of the action of steroid hormones on genome function in eukaryotic cells have also been collated (O'Malley and Hardman, 1975b).

Investigations of the amount of receptors present in target cells have shown that the concentration varies according to the stage of the estrous or menstrual cycle, pregnancy, or lactation (Mester et al., 1974; Chatkoff and Julian, 1973; Robertson et al., 1971; Trams et al., 1973). In the immature rat the cytosol receptor concentration for estradiol rises to a maximum of 20,000–60,000 sites per cell at the age of about 10 days and declines subsequently to values of 15,000–20,000 at 20–22 days. After puberty the values vary between 2000 just after estrus to 20,000 at proestrus. In pregnancy the maximum cytosol receptor level coincides with the time of implantation and reaches a value of 40,000 sites per cell in the endometrium. The value in the myometrium is approximately one-third of that in the endometrium. After ovariectomy, treatment with estradiol has been found to increase the cytosol receptor concentration in the endometrium and myometrium, whereas treatment with progesterone abolishes the rise, though only in myometrial tissue (Mester et al., 1974). The determination of nuclear receptors in target tissues, however, has proved more difficult. Nuclear levels of estrogen receptor in the rat also vary according to the stage of the estrous cycle and they are influenced by hormone treatment of immature animals (Clark et al., 1972; Anderson et al., 1973; Mester and Baulieu, 1975). A correlation has been found between the biological potency of an estrogen and its retention within the nucleus. Uteri from immature rats were incubated with labeled estradiol or estriol in vitro and the translocation of the estrogen receptor was measured. Both estrogens were transferred to the nucleus, though the efficiency of translocation was affected by the presence of albumin in the medium (Anderson et al., 1974). However, the retention of estradiol by the nuclear receptor was appreciably greater than that of estriol. Estriol was found to be as effective as estradiol in stimulating transient water imbibition, glucose metabolism, and RNA polymerase activity, but it failed to produce normal uterine growth (Anderson et al., 1975). Thus, estradiol and estriol (as well as Nafoxidine, Upjohn 11,100A) caused a rapid elevation in the activity of endogenous nuclear RNA polymerase II which was followed by a second rise in rats treated with estradiol or Nafoxidine but not in those treated with estriol (Hardin et al., 1976). These workers suggested that normal uterine growth was related to the long-term nuclear retention of estrogen, to the sustained activation of RNA polymerase II activity, and to the sustained elevation of RNA polymerase I activity (Hardin et al., 1976).

Recent publications on the occurrence of estrogen receptors in various organs and tissues (Fallopian tube, vagina, mammary gland, tumor cells, anterior pituitary and hypothalamus, corpus luteum, testis, epididymis, prostate, pineal, pancreas, and eosinophils) in mammalian and nonmammalian species (including oviduct and liver of chick and liver of amphibians) are cited in the review of Baulieu et al. (1975). The ontogeny of estrogen receptors in embryonic life has been described in the Müllerian duct of the developing chick and it has been demonstrated that the estradiol–receptor complex is translocated into the nucleus (Teng and Teng, 1976).

Substantial advances in the knowledge of the mechanism of progesterone action have been made during the past 4 years. Progesterone receptors have been studied principally in the chick oviduct (see O'Malley and Means, 1974 and Baulieu et al., 1975 for references). It has been demonstrated that their synthesis in the uterus and vagina of the hamster can be induced by guinea pigs, hamsters, rabbits, rats, and women (see O'Malley and Means, 1974 and Baulieu *et al.* 1975 for references). It has been demonstrated that their synthesis in the uterus and vagina of the hamster can be induced by estrogens (Leavitt and Blaha, 1972). In the guinea pig the concentration of uterine cytosol receptors for progesterone is highest during proestrus (about 40,000 binding sites per cell) and it then decreases during estrus and diestrus. At day 7 of pregnancy, when implantation occurs, the concentration is very low (Milgrom *et al.*, 1972a,b). The significance of these findings in relation to the hormonal requirements of implantation in this species (Deanesly, 1960) is not yet understood, but other experiments indicate that estrogens and progesterone have a complex role in the regulation of the number of progesterone receptors (Milgrom *et al.*, 1973). An earlier ambiguity about the role of other binding components in the rat and guinea pig uterus has been partially clarified (Baulieu *et al.*, 1975). Large amounts of CBG-like protein appear to be present in the uterus of the rat (Rosenthal *et al.*, 1974), and the occurrence of significant amounts of progesterone-binding components has also been reported in human endometrium (Kontula *et al.*, 1973; Rao *et al.*, 1974; Haukkamaa and Luukkainen, 1974) so that special precautions are required to distinguish the properties of CBG-like binding from uterine receptor binding.

Among studies of the molecular action of progesterone, the work of O'Malley and co-workers (1972a,b; 1974) and Schimke *et al.* (1975) using the chick oviduct has been prominent. Physicochemical studies have been reported on the cytosol and nuclear receptors (Schrader and O'Malley, 1972; Sherman *et al.*, 1974; Smith *et al.*, 1974; Kuhn *et al.*, 1975; Schrader *et al.*, 1975; Buller *et al.*, 1975a,b; Spelsberg, 1976; Webster *et al.*, 1976; Pikler *et al.*, 1976). These studies reveal that the progesterone receptor in chick oviduct cytosol consists of two 4 S binding components termed A and B. These components are present in near-equal amounts with molecular weights of 110,000 and 117,000, respectively (Kuhn *et al.*, 1975). Each subunit has a binding site for a single molecule of progesterone, and component A binds to DNA whereas component B (and the dimer receptor) binds to purified oviduct chromatin and in particular to an acidic protein in a nonhistone fraction designated AP_3 (Spelsberg *et al.*, 1972). Thus, the "nuclear acceptor hypothesis" advanced by O'Malley and co-workers (1972b, 1974) according to which target cell nuclei should contain acceptor sites with a specific affinity for the receptor molecules has received strong support, and the hormone–receptor complex is now seen as a probable inducer unit that modifies nuclear DNA transcription.

The mechanism of gene activation is being intensively examined, and it is thought to involve the separate interactions of both subunits, with binding between the AP_3 and the B subunit probably determining which genes will be activated. After dissociation of the subunits the A subunit may interact with the DNA enabling a molecule of RNA polymerase to occupy an initiation site. The number of initiation sites for the transcription of particular genes appears to be increased by hormone treatment as shown in studies of ovalbumin synthesis in chicks injected with diethylstilbestrol. The increase in initiation sites was closely correlated with the concentration of nuclear-bound receptors (Kalimi *et al.*, 1976). Current work has been directed toward the demonstration of transcription of the ovalbumin gene *in vitro* (Harris *et al.*, 1976; Tsai *et al.*, 1976), the kinetics of transcription *in vitro* (Hirose *et al.*, 1976), and the effect of estrogen on the nature of poly (A)-containing RNA prepared from chick oviduct (Monahan *et al.*, 1976).

Reservations were expressed in our earlier account about the emphasis given to the description of the action of estrogen and progesterone in terms of transcriptional control (Section IX, this chapter). It can be seen from many of the reports now cited, however, that recent work has largely resolved these doubts, particularly in respect of the hormonal stimulation of specific mRNA synthesis in target cells, as in the chick oviduct. Definitive evidence has been obtained for the synthesis of ovalbumin mRNA after hormone treatment. After estrogen treatment up to 15,000 molecules of ovalbumin mRNA may be present in a single oviduct cell, whereas after hormone treatment ceases the concentration falls to about 10 molecules per cell. Furthermore, studies *in vitro* have shown that chromatin from estrogen-stimulated chick oviducts was capable of supporting ovalbumin mRNA synthesis, whereas that from spleen or from unstimulated chick oviducts failed to produce mRNA sequences in the RNA transcript (Harris *et al.*, 1976). Ovalbumin mRNA has now been purified and its physicochemical properties described (Means *et al.*, 1972; Woo *et al.*, 1975) and used in the partial purification of the coding strand of the ovalbumin gene (Woo *et al.*, 1976).

Parallel studies on how progesterone induces synthesis in the chick oviduct have also provided direct evidence that the appearance of avidin mRNA precedes the accumulation of avidin in the chick oviduct (Chan *et al.*, 1973), that avidin mRNA is present only in the oviduct, and that its intracellular concentration is dependent upon progesterone stimulation (O'Malley *et al.*, 1972). A partial purification of progesterone-induced mRNA for avidin from hen oviduct has been published (Sperry *et al.*, 1976).

The construction of models to describe dynamic biological events is inevitably a hazardous, though popular, exercise. Many of the advances in this field of steroid hormone action can be seen to derive from studies on non-mammalian species, though work on estrogen action in the mammalian uterus has provided many comparable findings. Recently, mammalian systems have been recognized that may contribute to the study of how progesterone acts. For instance, blastokinin (Krishnan and Daniel, 1967; Daniel, 1976) or uteroglobin (Beier, 1966, 1967) is a progesterone-dependent protein synthesized by the uterus of the rabbit. Its synthesis and secretion are modulated by estrogen and progesterone (Bullock and Connell, 1973; Beier, 1976), and the protein itself has the ability to bind progesterone (Beato and Baier, 1975). Uteroglobin mRNA has been prepared from rabbit endometrium after sequential treatment with estrogen and progesterone, and the existence of a precursor protein has been demonstrated (Beato and Nieto, 1976). Studies in the pig have also revealed another progesterone-induced protein, purple phosphatase or lavender protein, which requires a free thiol and one atom of Fe^{3+} for its activity and coloration (Chen et al., 1973; Schlosnagle et al., 1974).

A summary of the major biochemical events of the action of estrogens and progesterone in target cells is given in Fig. 16 which is taken from that proposed by O'Malley and Means (1974). The possible molecular mechanisms involved in a different experimental system, the estrogen-dependent induction of yolk protein synthesis in avian and amphibian liver, have also been summarized by Clemens (1974). It should be noted, however, that Schimke et al. (1975) have questioned a simple model of the effects of

Hormone + receptor

(i) ↓

Hormone–receptor complex (cytoplasm)

(ii) ↓

Hormone–receptor complex (nucleus)

(iii) ↓

Hormone–receptor complex —DNA—AP_3

(iv) ↓

New RNA synthesis (mRNA)

(v) ↓

Protein synthesis

(vi) ↓

Specific target cell response

Fig. 16. Major steps in the biochemistry of steroid hormone action. (i) Hormone enters target cell and binds to specific cytosol receptor; (ii) hormone–receptor complex transformed and translocated to nucleus; (iii) binding of hormone–receptor complex to chromatin DNA and nonhistone–protein fraction AP_3; (iv) synthesis of new species of RNA arising from interaction of receptor subunit A and DNA, stimulation of initiation sites, and synthesis of strand of mRNA that serves as a template for new protein synthesis; (v) transport of hormone-induced RNA to the cytoplasm; (vi) steroid-mediated response to target cell. (From O'Malley and Means, 1974.)

"steroid hormones on gene expression in which each specific receptor complex interacts with a specific chromatin segment to allow transcription of a specific gene." These workers provide numerous examples where this concept is obviously an oversimplification including experiments in which estrogens were found to have multiple effects on ovalbumin mRNA in that they also influenced its rate of degradation as well as its rate of utilization (Schimke *et al.*, 1975). These, and other, findings imply that steroid hormones such as estrogens and progesterone have a variety of mechanisms and that specific gene transcription may be one of several effects.

ACKNOWLEDGMENTS

We are greatly indebted to our colleagues, Drs. D. B. Lindsay, J. R. G. Challis, the late J. L. Linzell, J. S. Perry, and J. C. Watkins for helpful advice and discussions. In particular we thank Dr. G. C. Mueller for an advance copy of his manuscript presented to the Laurentian Hormone Conference, 1971 (Mueller *et al.*, 1972), and Dr. R. T. Chatterton for a copy of his manuscript, "Progesterone and mammary gland development" (Chatterton, 1971). We thank Mrs. Janet Hood and Mrs. Dorothy George, who typed the manuscript, and the Lalor Foundation who supported one of us (D.V.I.).

REFERENCES

Aizawa, Y., and Mueller, G. C. (1961). The effect *in vivo* and *in vitro* of estrogens on lipid synthesis in the rat uterus. *J. Biol. Chem.* **236,** 381–386.

Aizawa, Y., and Nishigori, H. (1968). Action of estrogen on phospholipid metabolism in the uterus. *In* "Biogenesis and Action of Steroid Hormones" (R. Dorfman, K. Yamasaki, and M. Dorfman, eds.), pp. 236–251. Geron-X Inc, Los Altos, California.

Amoroso, E. C. (1968). The evolution of viviparity. *Proc. R. Soc. Med.* **61,** 1188–1200.

Anderson, J. N., Peck, E. J., Jr., and Clark, J. H. (1973). Nuclear receptor-estrogen complex: relationship between concentration and early uterotrophic responses. *Endocrinology* **92,** 1488–1495.

Anderson, J. N., Peck, E. J., Jr., and Clark, J. H. (1974). Nuclear receptor-estrogen complex: *in vivo* and *in vitro* binding of estradiol and estriol as influenced by serum albumin. *J. Steroid Biochem.* **5,** 103–107.

Anderson, J. N., Peck, E. J., Jr., and Clark, J. H. (1975). Estrogen-induced uterine responses and growth: relationship to receptor estrogen binding by uterine nuclei. *Endocrinology* **96,** 160–167.

Armstrong, D. T., and King, E. R. (1971). Uterine progesterone metabolism and progestational response: effects of estrogens and prolactin. *Endocrinology* **89,** 191–197.

Astwood, E. B. (1938). A six-hour assay for the quantitative determination of estrogen. *Endocrinology* **23,** 25–31.

Baird, D. T., Horton, R., Longcope, C., and Tait, J. F. (1969). Steroid dynamics under steady-state conditions. *Recent Prog. Horm. Res.* **25**, 611–656.

Bangham, A. D. (1968). Membrane models with phospholipids. *Prog. Biophys. Biophys. Chem.* **18**, 29–95.

Bangham, A. D., Standish, M. M., and Watkins, J. C. (1965a). Diffusion of univalent ions across the lamellae of swollen phospholipids. *J. Mol. Biol.* **13**, 238–252.

Bangham, A. D., Standish, M. M., and Weissmann, G. (1965b). The action of steroids and streptolysin S on the permeability of phospholipid structures to cations. *J. Mol. Biol.* **13**, 253–259.

Barksdale, A. W. (1969). Sexual hormones of *Achlya* and other fungi. *Science* **166**, 831–837.

Barnea, A., and Gorski, J. (1970). Estrogen-induced protein. Time course of synthesis. *Biochemistry* **9**, 1899–1904.

Barry, J., and Gorski, J. (1971). Uterine ribonucleic acid polymerase. Effect of estrogen on nucleotide incorporation into 3′ chain termini. *Biochemistry* **10**, 2384–2390.

Bassett, J. M., Oxborrow, T. J., Smith, I. D., and Thorburn, G. D. (1969). The concentration of progesterone in the peripheral plasma of the pregnant ewe. *J. Endocrinol.* **45**, 449–457.

Baulieu, E.-E., Alberga, A., Jung, I., Lebeau, M-C., Mercier-Bodard, C., Milgrom, E., Raynaud, J.-P., Raynaud-Jammet, C., Rochefort, H., Truong, H., and Robel, P. (1971). Metabolism and protein binding of sex steroids in target organs: an approach to the mechanism of hormone action. *Recent Prog. Horm. Res.* **27**, 351–412.

Baulieu, E.-E., Atger, M., Best-Belpomme, M., Corvol, P., Courvalin, J.-C., Mester, J., Milgrom, E., Robel, P., Rochefort, H., and de Catalogne, D. (1975). Steroid hormone receptors. *Vitam. Horm.* (*New York*) **33**, 649–736.

Beato, M., and Baier, R. (1975). Binding of progesterone to the proteins of the uterine luminal fluid. Identification of uteroglobin as the binding protein. *Biochim. Biophys. Acta* **392**, 346–356.

Beato, M., and Nieto, A. (1976). Translation of the mRNA for rabbit uteroglobin in cell-free systems. Evidence for a precursor protein. *Eur. J. Biochem.* **64**, 15–25.

Bedford, C. A., Challis, J. R. G., Harrison, F. A., and Heap, R. B. (1972a). The rôle of oestrogens and progesterone in the onset of parturition in various species. *J. Reprod. Fertil. Suppl.* **16**, 1–23.

Bedford, C. A., Harrison, F. A., and Heap, R. B. (1972b). The metabolic clearance rate and production rate of progesterone and the conversion of progesterone to 20α-hydroxypregn-4-en-3-one in the sheep. *J. Endocrinol.* **55**, 105–118.

Bedford, C. A., Harrison, F. A., and Heap, R. B. (1973). The kinetics of metabolism of progesterone in the pregnant sheep. *In* "The Endocrinology of Pregnancy and Parturition". (C. G. Pierrepoint, ed.), pp. 83–93. Alpha Omega Alpha Publ., Cardiff, Wales.

Bedford, C. A., Harrison, F. A., and Heap, R. B. (1974). Splanchnic, uterine, ovarian and adrenal uptake of progesterone and 20α-dihydroprogesterone in the pregnant and non-pregnant sheep. *J. Endocrinol.* **62**, 277–290.

Beier, H. (1966). Das Proteinmilieu in Serum, Uterus und Blastocysten des kaninchens vor der Nidation. *In* "Biochemie der Morphogenese" (W. Beerman, ed.), pp. 1–10. Deutsche Forschungsgemeinschaft, Konstanz.

Beier, H. (1967). Veränderungen am Proteinmuster des Uterus bei dessen

Ernährungsfunktion für die Blastocyste des Kaninchens. *Verh. Deut. Zool. Ges.* **31**, 139–148.

Beier, H. (1976). Uteroglobin and related biochemical changes in the reproductive tract during early pregnancy in the rabbit. *J. Reprod. Fertil., Suppl.* **25**, 53–69.

Berlinger, D. L., and Wiest, W. G. (1956). The extrahepatic metabolism of progesterone in rats. *J. Biol. Chem.* **221**, 449–459.

Bernfield, M. R. (1971). Transfer RNA alterations during estrogen-induced phosvitin synthesis, *Fed. Proc. Fed. Am. Soc. Exp. Biol.* **30**, 1048 (abstr.).

Beuving, G., and Gruber, M. (1971a). Isolation of phosvitin from plasma of estrogenized roosters. *Biochim. Biophys. Acta* **232**, 524–528.

Beuving, G., and Gruber, M. (1971b). Induction of phosvitin synthesis in roosters by estradiol injection. *Biochim. Biophys. Acta* **232**, 529–536.

Billing, R. J., Barbiroli, B., and Smellie, R. M. S. (1969a). The mode of action of oestradiol. I. The transport of RNA precursors into the uterus. *Biochim. Biophys. Acta* **190**, 52–59.

Billing, R. J., Barbiroli, B., and Smellie, R. M. S. (1969b). The mode of action oestradiol. II. The synthesis of RNA. *Biochim. Biophys. Acta* **190**, 60–65.

Bitman, J., Cecil, H. C., Mench, M. L., and Wrenn, T. R. (1965). Kinetics of *in vivo* glycogen synthesis in the estrogen-stimulated rat uterus. *Endocrinology* **76**, 63–69.

Blatti, S. P., Ingles, C. J., Lindell, T. J., Morris, P. W., Weaver, R. F., Weinberg, F., and Rutter, W. J. (1970). Structure and regulatory properties of eucaryotic RNA polymerase. *Cold Spring Harbor Symp. Quant. Biol.* **35**, 649–657.

Blyth, C. A., Freedman, R. B., and Rabin, B. R. (1971). Sex specific binding of steroid hormones to microsomal membranes of rat liver. *Nature (London), New Biol.* **230**, 137–139.

Boscott, R. J. (1962a). The chemistry and biochemistry of progesterone and relaxin. *In* "The Ovary" (S. Zuckerman, ed.), 1st ed., Vol. 2, pp. 47–79. Academic Press, New York.

Boscott, R. J. (1962b). The chemistry and biochemistry of natural and artificial oestrogens. *In* "The Ovary" (S. Zuckerman, ed.), 1st ed., Vol. 2, pp. 1–45. Academic Press, New York.

Brant, J. W. A., and Nalbandov, A. V. (1956). Role of sex hormone in albumen secretion by the oviduct of chickens. *Poult. Sci.* **35**, 692–700.

Brecher, P. I., Vigersky, R., Wotiz, H. S., and Wotiz, H. H. (1967). An *in vitro* system for the binding of oestradiol to rat uterine nuclei. *Steroids* **10**, 635–651.

Brecher, P. I., Numata, M., DeSombre, E. R., and Jensen, E. V. (1970). Conversion of uterine 4S estradiol-receptor complex to 5S complex in a soluble system. *Fed. Proc., Fed. Am. Soc. Exp. Biol.* **29**, 249 (abstr. 14).

Bresciani, F. (1968). Topography of DNA synthesis in the mammary gland. An autoradiographic study. *Cell Tissue Kinet.* **1**, 51–63

Bresciani, F. (1971). Ovarian steroid control of cell proliferation in the mammary gland and cancer. *In* "Basic Actions of Sex Steroids on Target Organs" (P. O. Hubinont, F. Leroy, and P. Galand, eds.), pp. 130–159. Karger, Basel.

Brew, K. (1969). Secretion of α-lactalbumin into milk and its relevance to the organisation and control of lactose synthetase. *Nature (London)* **222**, 671–672.

Brew, K. (1970). Lactose synthetase: evolutionary origins, structure and control. *Essays Biochem.* **6**, 93–118.

Briggs, M. H., and Brotherton, J. (1970). "Steroid Biochemistry and Pharmacology." Academic Press, New York.

Brodbeck, U., and Ebner, K. E. (1966). Resolution of a soluble lactose synthetase into two protein components and solubilization of microsomal lactose synthetase. *J. Biol. Chem.* **241**, 762–764.

Brodbeck, U., Denton, W. L., Tanahashi, N., and Ebner, K. E. (1967). The isolation and identification of the B protein of lactose synthetase as α-lactalbumin. *J. Biol. Chem.* **242**, 1391–1397.

Brody, S., and Westman, A. (1958). Effects of oestradiol and progesterone on the nucleic acid and protein content of the rabbit uterus. *Acta Endocrinol. (Copenhagen)* **27**, 493–498.

Bruchovsky, N., and Wilson, J. D. (1968). The intranuclear binding of testosterone and 5α-androstan-17β-ol-3-one by rat prostate. *J. Biol. Chem.* **243**, 5953–5960.

Brush, M. G., Taylor, R. W., and King, R. J. B. (1967). The uptake of [6,7-^{3}H]-oestradiol by the normal human female reproductive tract. *J. Endocrinol.* **39**, 599–607.

Bryson, M. J., and Sweat, M. L. (1963). Endometrial metabolism of progesterone. *Fed. Proc., Fed. Am. Soc. Exp. Biol.* **22**, 469 (abstr. 1845).

Bryson, M. J., and Sweat, M. L. (1967). Metabolism of progesterone in human proliferative endometrium. *Endocrinology* **81**, 729–734.

Bryson, M. J., and Sweat, M. L. (1969). Metabolism of progesterone in human myometrium. *Endocrinology* **84**, 1071–1075.

Bülbring, E., Casteels, R., and Kuriyama, H. (1968). Membrane potential and ion content in cat and guinea-pig myometrium and the response to adrenaline and nor-adrenaline. *Br. J. Pharmacol.* **34**, 388–407.

Buller, R. E., Schrader, W. T., and O'Malley, B. W. (1975a). Progesterone-binding components of chick oviduct. IX. The kinetics of nuclear binding. *J. Biol. Chem.* **250**, 809–818.

Buller, R. E., Toft, D. O., Schrader, W. T., and O'Malley, B. W. (1975b). Progesterone-binding components of chick oviduct. VIII. Receptor activation and hormone-dependent binding to purified nuclei. *J. Biol. Chem.* **250**, 801–808.

Bullock, D. W., and Connell, K. H. (1973). Occurrence and molecular weight of rabbit uterine "blastokinin." *Biol. Reprod.* **9**, 125–132.

Burton, R. M., Harding, G. B., Rust, N., and Westphal, U. (1971). Steroid–protein interactions. XXIII. Nonidentity of cortisol-binding globulin and progesterone-binding globulin in guinea pig serum. *Steroids* **17**, 1–16.

Bush, I. E. (1965). Hormones and receptors. *Proc. Int. Congr. Endocrinol., 2nd, 1964* Excerpta Med. Found. Int. Congr. Ser. No. 83, pp. 1324–1335.

Callentine, M. R., Humphrey, R. R., Lee, S. L., Windsor, B. L., Schottin, N. H., and O'Brien, O. P. (1966). Action of an estrogen antagonist on reproductive mechanisms in the rat. *Endocrinology* **79**, 153–167.

Carsten, M. E. (1968). Regulation of myometrial composition, growth and activity. *In* "Biology of Gestation" (N. S. Assali, ed.), Vol. 1, pp. 355–425. Academic Press, New York.

Casteels, R., and Kuriyama, H. (1965). Membrane potential and ionic content in pregnant and non-pregnant rat myometrium. *J. Physiol. (London)* **177**, 263–287.

Chader, G. J., and Westphal, U. (1968). Steroid-protein interactions. XVIII. Isolation and observations on the polymeric nature of the corticosteroid-binding globulin of the rat. *Biochemistry* **7**, 4272–4282.

Challis, J. R. G., Heap, R. B., and Illingworth, D. V. (1971). Concentrations of oestrogen and progesterone in the plasma of non-pregnant, pregnant and lactating guinea-pigs. *J. Endocrinol.* **51,** 333–345.

Challis, J. R. G., Harrison, F. A., and Heap, R. B. (1972). The kinetics of oestrogen metabolism in the pregnant sheep. *In* "The Endocrinology of Pregnancy and Parturition" (C. G. Pierrepoint, ed.), pp. 73–82. Alpha Omega Alpha Publ., Cardiff, Wales.

Chan, L., Means, A. R., and O'Malley, B. W. (1973). Rates of induction of specific translatable messenger RNAs for ovalbumin and avidin by steroid hormones. *Proc. Nat. Acad. Sci. USA* **70,** 1870–1874.

Chatkoff, M. L., and Julian, J. A. (1973). Effects of progesterone on the binding of estradiol-receptor to rabbit uterine chromatin. *Biochem. Biophys. Res. Commun.* **51,** 1015–1022.

Chatterton, R. T. (1971). Progesterone and mammary gland development. *In* "The Sex Steroids" (K. W. McKerns, ed.), pp. 345–375. Appleton, New York.

Chatterton, R. T., Chatterton, A. J., and Hellman, L. (1969a). Metabolism of progesterone by the rabbit kidney. *Endocrinology* **84,** 1089–1097.

Chatterton, R. T., Chatterton, A. J., and Hellman, L. (1969b). Metabolism of progesterone by the rabbit mammary gland. *Endocrinology* **85,** 16–24.

Chen, T. T., Bazer, F. W., Cetorelli, J. J., Pollard, W. E., and Roberts, R. M. (1973). Purification and properties of a progesterone-induced basic glycoprotein from the uterine fluid of pigs. *J. Biol. Chem.* **248,** 8560–8566.

Christensen, H. N. (1967). Some transport lessons taught by the organic solute. *Perspect. Biol. Med.* **10,** 471–494.

Church, R. B., and McCarthy, B. J. (1970). Unstable nuclear RNA synthesis following estrogen stimulation. *Biochim. Biophys. Acta* **199,** 103–114.

Clark, A. J. (1937). "The Mode of Action of Drugs on Cells." Williams & Wilkins, Baltimore, Maryland.

Clark, J. H., Anderson, J. N., and Peck, E. J., Jr. (1972) Receptor-estrogen complex in the nuclear fraction of rat uterine cells during the estrous cycle. *Science* **176,** 528–530.

Clemens, M. J. (1974). Regulation of egg yolk protein synthesis by steroid hormones. *Progr. Biophys. Mol. Biol.* **28,** 71–108.

Cole, D. F. (1950). The effects of oestradiol on the rat uterus. *J. Endocrinol.* **7,** 12–23.

Courrier, B. (1950). Interactions between estrogens and progesterone. *Vitam. Horm. (New York)* **8,** 179–214.

Coutinho, E. M. (1965). Hormone induced ionic regulation of labor. *Proc. Int. Congr. Endocrinol. 2nd, 1964* Excerpta Med. Found. Int. Cong. Ser. No. 83, pp. 742–747.

Cowie, A. T. (1971). Influence of hormones on mammary growth and milk secretion. *In* "Lactation" (I. R. Falconer, ed.), pp. 123–140. Butterworth, London.

Cowie, A. T., and Tindal, J. S. (1971). "The Physiology of Lactation," Monogr. Physiol. Soc. No. 22. Arnold, London.

Csapo, A. I. (1956). Progesterone "block." *Am. J. Anat.* **98,** 273–291.

Csapo, A. I. (1969). The four direct regulatory factors of myometrial function. *Ciba Found. Study Group* **34,** 13–55.

Csapo, A. I., and Wiest, W. G. (1969). An examination of the quantitative relationship between progesterone and the maintenance of pregnancy. *Endocrinology* **85,** 735–746.

Csermely, T., Demers, L. M., and Hughes, E. C. (1969). Organ culture of human endometrium: effects of progesterone. *Obstet. Gynecol.* **34**, 252–259.

Daniel, J. C., Jr. (1976). Blastokinin and analogous proteins. *J. Reprod. Feril. Suppl.* **25**, 71–83.

Daughaday, W. H. (1958). Binding of corticosteroids by plasma proteins. III. The binding of corticosteroid and related hormones by human plasma and plasma protein fractions as measured by equilibrium dialysis. *J. Clin. Invest.* **37**, 511–518.

Davis, M. E., and Plotz, E. J. (1958). Hormones in human reproduction. II. Further investigation of steroid metabolism in human pregnancy. *Am. J. Obstet. Gynecol.* **76**, 939–954.

Deanesly, R. (1960). Implantation and early pregnancy in ovariectomized guinea-pigs. *J. Reprod. Fertil.* **1**, 242–248.

Deanesly, R. (1966). The endocrinology of pregnancy and foetal life. *In* "Marshall's Physiology of Reproduction" (A. S. Parkes, ed.), 3rd ed., Vol. 3, pp. 891–1063. Longmans Green, London and New York.

DeAngelo, A. B., and Gorski, J. (1970). Role of RNA synthesis in the estrogen induction of a specific uterine protein. *Proc. Natl. Acad. Sci. USA* **66**, 693–700.

DeHertogh, R., Ekka, E., Vanderheyden, I., and Hoet, J. J. (1970). Metabolic clearance rates and the interconversion factors of estrone and estradiol-17β in the immature and adult female rat. *Endocrinology* **87**, 874–880.

DeLange, R. J., and Smith, E. L. (1971). Histones: structure and function. *Annu. Rev. Biochem.* **40**, 279–314.

de Rueck, A. V. S., and Knight, J., eds. (1966). "Histones. Their role in the Transfer of Genetic Information," Ciba Found. Study Group No. 24. Churchill, London.

DeSombre, E. R., Puca, G. A., and Jensen, E. V. (1969). Purification of an estrophilic protein from calf uterus. *Proc. Natl. Acad. Sci. USA* **64**, 148–154.

DeSombre, E. R., Chabaud, J. P., Puca, G. A., and Jensen, E. V. (1971). Purification and properties of an estrogen-binding protein from calf uterus. *J. Steroid Biochem.* **2**, 95–103.

Diamond, M., Rust, N., and Westphal, U. (1969). High-affinity binding of progesterone, testosterone and cortisol in normal and androgen-treated guinea pigs during various reproductive stages: relationship to masculinization. *Endocrinology* **84**, 1143–1151.

Dingman, C. W., Aronow, A., Bunting, S. L., Peacock, A. C., and O'Malley, B. W. (1969). Changes in chick oviduct ribonucleic acid following hormonal stimulation. *Biochemistry* **8**, 489–495.

Dorfman, R. I. (1954). Neutral steroid hormone metabolites. *Recent Prog. Horm. Res.* **9**, 5–24.

Dorfman, R. I., and Ungar, G. (1965). "Metabolism of Steroid Hormones." Academic Press, New York.

Drasher, M. L. (1952). Morphological and chemical observations on the mouse uterus during oestrous cycle and under hormonal treatment. *J. Exp. Zool.* **119**, 333–353.

Duax, W. L., Weeks, C. M., Rohrer, D. C., and Griffin, J. F. (1977). Crystal and molecular structures of steroids: identification, functional analysis, and drug design. *Proc. Int. Congr. Endocrinol.* 5th Excerpta Med. Found. Int. Congr. Ser. No. 402, **1**, in press.

Eakin, R. E., Snell, E. E., and Williams, R. J. (1941). The concentration and assay of avidin, the injury-producing protein in raw egg white. *J. Biol. Chem.* **140**, 535–543.

Eik-Nes, K., Schellman, J. A., Lumry, R., and Samuels, L. T. (1954). The binding of steroids to protein. I. Solubility determinations. *J. Biol. Chem.* **206**, 411–419.

Eisenfeld, A. S., and Axelrod, J. (1965). Selectivity of estrogen distribution in tissue. *J. Pharmacol. Exp. Ther.* **150**, 469–475.

Emmens, C. W. (1967). The action of oestrogens. *Adv. Reprod. Physiol.* **2**, 213–255.

Emmens, C. W., and Martin, L. (1964). Estrogens. *Methods Horm. Res.* **3**, 1–80.

Engel, L. L. (1970). Eli Lilly Lecture: Estrogen metabolism and action. *Endocrinology* **87**, 827–835.

Engel, L. L., Stoffyn, A. M., and Scott, J. F. (1964). The role of molecular configuration in the interaction of steroid hormones with coenzymes and enzymes. *Horm. Steroids; Biochem., Pharmacol., Ther., Proc. Int. Congr., 1st, 1962* Vol. 1, pp. 291–299.

Erdos, T. (1968). Properties of a uterine oestradiol receptor. *Biochem. Biophys. Res. Commun.* **32**, 338–343.

Erdos, T., Gospodarowicz, D., Bessada, R., and Fries, J. (1968). Propriétés d'un récepteur utérin de l'estradiol. *C. R. Acad. Sci. Ser. D* **266**, 2164–2167.

Falk, R. J., and Bardin, C. W. (1970). Uptake of tritiated progesterone by the uterus of the ovariectomized guinea pig. *Endocrinology* **86**, 1059–1063.

Fang, S., Anderson, K. M., and Liao, S. (1969). Receptor proteins for androgens. On the role of specific proteins in selective retention of 17β-hydroxy-5α-androstan-3-one by rat ventral prostate *in vivo* and *in vitro*. *J. Biol. Chem.* **244**, 6584–6595.

Feder, H. H., and Ruf, K. B. (1969). Stimulation of progesterone release and estrous behavior by ACTH in ovariectomized rodents. *Endocrinology* **84**, 171–174.

Feder, H. H., Resko, J. A., and Goy, R. W. (1968). Progesterone levels in the arterial plasma of pre-ovulatory and ovariectomized rats. *J. Endocrinol.* **41**, 563–569.

Fieser, L. F., and Fieser, M. (1949). "Natural Products Related to Phenanthrene," 3rd ed. Van Nostrand-Reinhold, Princeton, New Jersey.

Fieser, L. F., and Fieser, M. (1959). "Steroids." van Nostrand-Reinhold, Princeton, New Jersey.

Folley, S. J. (1952). Lactation. *In* "Marshall's Physiology of Reproduction" (A. S. Parkes, ed.), 3rd ed., Vol. 2, pp. 525–647. Longmans Green, London and New York.

Folley, S. J., and Malpress, F. H. (1948). Hormonal control of mammary growth. *Hormones* **1**, 695–805.

Fotherby, K. (1964). The biochemistry of progesterone. *Vitam. Horm. (New York)* **22**, 153–204.

Galand, P., and Leroy, F. (1971). Radioautographic evaluation of the effects of oestrogen on cell kinetics in target tissues. *In* "Basic Actions of Sex Steroids on Target Organs" (P. O. Hubinont, F. Leroy, and P. Galand, eds.), pp. 160–171. Karger, Basel.

Ganguly, M., Carnighan, R. H., and Westphal, U. (1967). Steroid-protein interactions. XIV. Interaction between human α_1-acid glycoprotein and progesterone. *Biochemistry* **6**, 2803–2814.

Garren, L. D., Howell, R. R., and Tomkins, G. M. (1964). Mammalian enzyme induction by hydrocortisone: the possible role of RNA. *J. Mol. Biol.* **9,** 100–108.

Gawienowski, A. M., and Gibbs, C. C. (1968). Identification of cholesterol and progesterone in apple seeds. *Steroids* **12,** 545–550.

Giannopoulos, G., and Gorski, J. (1971a). Estrogen receptors. Quantitative studies on transfer of estradiol from cytoplasmic to nuclear binding sites. *J. Biol. Chem.* **246,** 2524–2529.

Giannopoulos, G., and Gorski, J. (1971b). Estrogen binding protein of the rat uterus. Different molecular forms associated with nuclear uptake of estradiol. *J. Biol. Chem.* **246,** 2530–2536.

Gibian, H., and Unger, R. (1968). Wirkungen der Gestagene auf den Stoffwechsel *Handb. Exp. Pharmakol.* **22,** 352–449.

Gilbert, A. B. (1967). Formation of the egg in the domestic chicken. *Adv. Reprod. Physiol.* **2,** 111–180.

Glascock, R. F., and Hoekstra, W. G. (1959). Selective accumulation of tritium-labelled hexoestrol by the reproductive organs of immature female goats and sheep. *Biochem. J.* **72,** 673–682.

Gorski, J. (1964). Early estrogen effects on the activity of uterine ribonucleic acid polymerase. *J. Biol. Chem.* **239,** 889–892.

Gorski, J., and Axman, M. C. (1964). Cycloheximide (actidione) inhibition of protein synthesis and the uterine response to estrogen. *Arch. Biochem. Biophys.* **105,** 517–520.

Gorski, J., and Morgan, M. S. (1967). Estrogen effects on uterine metabolism: reversal by inhibitors of protein synthesis. *Biochim. Biophys. Acta* **149,** 282–287.

Gorski, J., and Nicolette, J. A. (1963). Early estrogen effects on newly synthesized RNA and phospholipid in subcellular fractions of rat uterus. *Arch. Biochem. Biophys.* **103,** 418–423.

Gorski, J., Noteboom, W. D., and Nicolette, J. A. (1965). Estrogen control of the synthesis of RNA and protein in the uterus. *J. Cell. Comp. Physiol.* **66,** 91–100.

Gorski, J., Toft, D.O., Shyamala, G., Smith, D.E., and Notides, A. (1968). Hormone receptors: studies on the interaction of estrogen with the uterus. *Recent Prog. Horm. Res.* **24,** 45–72.

Goto, M., and Csapo, A. I. (1959). The effect of the ovarian steroids on the membrane potential of uterine muscle. *J. Gen. Physiol.* **43,** 455–466.

Grant, J. K. (1969). Actions of steroid hormones at cellular and molecular levels. *Essays Biochem.* **5,** 1–58.

Green, D. E. (1941). Enzymes and trace substances. *Adv. Enzymol.* **1,** 177–198.

Greengard, O., Gordon, M., Smith, M. A., and Acs, G. (1964). Studies on the mechanism of diethylstilboestrol-induced formation of phosphoprotein in male chickens. *J. Biol. Chem.* **239,** 2079–2082.

Greengard, O., Sentenac, A., and Acs, G. (1965). Induced formation of phosphoprotein in tissues of cockerels *in vivo* and *in vitro. J. Biol. Chem.* **240,** 1687–1691.

Gyermek, L. (1967). Pregnanolone: a highly potent, naturally occurring hypnotic-anaesthetic agent. *Proc. Soc. Exp. Biol. Med.* **125,** 1058–1062.

Gyermek, L. (1968). Effects of pregnanolone and progesterone. *Lancet* **2,** 1195.

Hager, C. B., and Gilbert, J. B. (1966). Effect of actinomycin D upon the progesterone-stimulated increase of carbonic anhydrase activity in rabbit endometrium. *Fed. Proc., Fed. Am. Soc. Exp. Biol.* **25,** 760 (Abstr. 3228).

Hahn, W. E., Church, R. B., Gorbman, A., and Wilmot, L. (1968). Estrone- and progesterone-induced synthesis of new RNA species in the chick oviduct. *Gen. Comp. Endocrinol.* **10**, 438–442.

Hähnel, R. (1971). Properties of the estrogen receptor in the soluble fraction of human uterus. *Steroids* **17**, 105–132.

Halkerston, I. D. K., Eichhorn, J., Feinstein, M., Scully, E., and Hechter, O. (1960). Early permeability effects of estradiol on castrate rat uterus. *Proc. Soc. Exp. Biol. Med.* **103**, 796–798.

Hamilton, T. H. (1968). Control by estrogen of genetic transcription and translation. *Science* **161**, 649–661.

Hamilton, T. H. (1971a). Effects of sexual steroid hormones on genetic transcription and translation. *In* "Basic Actions of Sex Steroids on Target Organs" (P. O. Hubinont, F. Leroy, and P. Galand, eds.), pp 56–92. Karger, Basel.

Hamilton, T. H. (1971b). Steroid hormones, ribonucleic acid synthesis and transport, and the regulation of cytoplasmic translation. *Biochem. Soc. Symp.* **32**, 49–84.

Hamilton, T. H., Widnell, C. C., and Tata, J. R. (1965). Sequential stimulations by oestrogen of nuclear RNA synthesis and DNA-dependent RNA polymerase activities in rat uterus. *Biochim. Biophys. Acta* **108**, 168–172.

Hamilton, T. H., Teng, C.-S., and Means, A. R. (1968a). Early estrogen action: nuclear synthesis and accumulation of protein correlated with enhancement of two DNA-dependent RNA polymerase activities. *Proc. Natl. Acad. Sci. U.S.A.* **59**, 1265–1272.

Hamilton, T. H., Widnell, C. C., and Tata, J. R. (1968b). Synthesis of ribonucleic acid during early estrogen action. *J. Biol. Chem.* **243**, 408–417.

Hardin, J. W., Clark, J. H., Glasser, S. R., and Peck, E. J., Jr. (1976). RNA polymerase activity and uterine growth: differential stimulation by estradiol, estriol and nafoxidine. *Biochemistry* **15**, 1370–1374.

Harris, D. N., Lerner, L. J., Hilf, R., Bianchi, A., and Raskin, E. K. (1967). Responses of the immature rat uterus to sex steroids as influenced by time and antagonists. *Fed. Proc., Fed. Am. Soc. Exp. Biol.* **26**, 536 (abstr.).

Harris, H. (1970). "Nucleus and Cytoplasm." 2nd ed. Oxford Univ. Press (Clarendon), London and New York.

Harris, S. E., Schwarz, R. J., Tsai, M.-J., and O'Malley, B. W. (1976). Effect of estrogen on gene expression in the chick oviduct. *In vitro* transcription of the ovalbumin gene in chromatin. *J. Biol. Chem.* **251**, 524–529.

Haukkamaa, M., and Luukkainen, T. (1974). The cytoplasmic progesterone receptor of human endometrium during the menstrual cycle. *J. Steroid Biochem.* **5**, 447–452.

Heap, R. B. (1969). The binding of plasma progesterone in pregnancy. *J. Reprod. Fertil.* **18**, 546–548.

Heap, R. B., and Illingworth, D. V. (1974). The maintenance of gestation in the guinea-pig and other hystricomorph rodents: changes in the dynamics of progesterone metabolism and the occurrence of progesterone-binding globulin (PBG). *Symp. Zool. Soc. London* **34**, 385–415.

Heap, R. B., and Linzell, J. L. (1966). Arterial concentration, ovarian secretion and mammary uptake of progesterone in goats during the reproductive cycle. *J. Endocrinol.* **36**, 389–399.

Heap, R. B., Linzell, J. L., and Slotin, C. A. (1969). Quantitative measurement of progesterone metabolism in the mammary gland of the goat. *J. Physiol. (London)* **200**, 38–40P.

Heap, R. B., Symons, A. M., and Watkins, J. C. (1970). Steroids and their interactions with phospholipids: solubility, distribution coefficient and effect on potassium permeability of liposomes. *Biochim. Biophys. Acta* **218**, 482–495.

Heap, R. B., Symons, A. M., and Watkins, J. C. (1971). An interaction between oestradiol and progesterone in aqueous solutions and in a model membrane system. *Biochim. Biophys. Acta* **233**, 307–314.

Heap, R. B., Perry, J. S., and Challis, J. R. G. (1972). The hormonal maintenance of pregnancy. *In* "Handbook of Physiology" (Am. Physiol. Soc.), (R. O. Greep, ed.) Sect. 7, Vol. II, Part 2, pp. 217–260. Williams & Wilkins, Baltimore, Maryland.

Hechter, O., and Halkerston, I. D. K. (1964). On the action of mammalian hormones. "Hormones" Vol. 5, 697–825.

Hirose, M., Tsai, M.-J., and O'Malley, B. W. (1976). Effect of estrogen on gene expression in the chick oviduct. Kinetics of initiation of *in vitro* transcription on chromatin. *J. Biol. Chem.* **251**, 1137–1146.

Honig, G. R., and Rabinovitz, M. (1965). Actinomycin D: inhibition of protein synthesis unrelated to effect on template RNA synthesis. *Science* **149**, 1504–1505.

Horton, R., and Tait, J. F. (1966). Androstenedione production and interconversion rates measured in peripheral blood and studies on the possible site of its conversion to testosterone. *J. Clin. Invest.* **45**, 301–313.

Illingworth, D. V. (1970). Kinetics of progesterone metabolism in pregnant and nonpregnant guinea pigs. *J. Physiol. (London)* **210**, 99ᴾ–100ᴾ.

Illingworth, D. V., and Heap, R. B. (1971). A decrease in the metabolic clearance rate of progesterone in the coypu during pregnancy. *J. Reprod. Fertil.* **27**, 492–494.

Illingworth, D. V., Heap, R. B., and Perry, J. S. (1970). Changes in the metabolic clearance rate of progesterone in the guinea-pig. *J. Endocrinol.* **48**, 409–417.

Jacob, F., and Monod, J. (1961). Genetic regulatory mechanisms in the synthesis of proteins. *J. Mol. Biol.* **3**, 318–356.

James, D. W., Rabin, B. R., and Williams, D. J. (1969). Role of steroid hormones in the interaction of polysomes with endoplasmic reticulum. *Nature (London)* **224**, 371–372.

Jensen, E. V. (1965). Metabolic fate of sex hormones in target tissues with regard to tissue specificity. *Proc. Int. Congr. Endocrinol., 2nd,* Excerpta Med. Found. Int. Congr. Ser. No. 83, pp. 420–433.

Jensen, E. V., and Jacobson, H. I. (1962). Basic guides to the mechanism of estrogen action. *Recent Prog. Horm. Res.* **18**, 387–408.

Jensen, E. V., Jacobson, H. I., Flesher, J. W., Saha, N. N., Gupta, G. N., Smith, S., Colucci, V., Shiplacoff, D., Neumann, H. G., DeSombre, E. R., and Jungblut, P. W. (1966). Estrogen receptors in target tissues. *In* "Steroid Dynamics" (G. Pincus, T. Nakao, and J. F. Tait, eds.), pp. 133–156. Academic Press, New York.

Jensen, E. V., DeSombre, E. R., Hurst, D. J., Kawashuma, T., and Jungblut, P. W. (1967a). Estrogen-receptor interactions in target tissues. *Arch. Anat. Microsc. Morphol. Exp.* **56**, Suppl., 547–569.

Jensen, E. V., DeSombre, E. R., and Jungblut, P. W. (1967b). Interaction of estrogens with receptor sites *in vivo* and *in vitro*. *Proc. Int. Congr. Horm. Steroids, 2nd, 1966* Excerpta Med. Found. Int. Congr. Ser. No. 132, pp. 492–500.

Jensen, E. V., Suzuki, T., Kawashuma, T., Stumpf, W. E., Jungblut, P. W., and DeSombre, E. R. (1968). A two-step mechanism for the interaction of estradiol with rat uterus. *Proc. Natl. Acad. Sci. U.S.A.* **59**, 632–638.

Jensen, E. V., Suzuki, T., Numata, M., Smith, S., and DeSombre, E. R. (1969a). Estrogen-binding substances of target tissues. *Steroids* **13**, 417–427.

Jensen, E. V., Numata, M., Smith, S., Suzuki, T., Brecher, P. I., and DeSombre, E. R. (1969b). Estrogen-receptor interaction in target tissues. *Symp. Soc. Dev. Biol.* **28**, 151–171.

Jensen, E. V., Numata, M., Smith, S., Suzuki, T., and DeSombre, E. R. (1971a). Estrogen-receptor interaction in target tissues. *In* "The Action of Hormones. Genes to Population" (P. P. Foa, ed.), pp. 20–44. Thomas, Springfield, Illinois.

Jensen, E. V., Numata, M., Brecher, P. I., and DeSombre, E. R. (1971b) Hormone-receptor interaction as a guide to biochemical mechanism. *Biochem. Soc. Symp.* **32**, 31–47.

Jones, A. W. (1968). Influence of oestrogen and progesterone on electrolyte accumulation in the rabbit myometrium. *J. Physiol. (London)* **197**, 19P–20P.

Jones, A. W. (1970). Application of the 'Association-Induction Hypothesis' to ion accumulation and permeability of smooth muscle. *In* "Smooth Muscle" (E. Bülbring *et al.,* eds.), pp. 122–150. Arnold, London.

Jones, H. E. H., and Pope, G. S. (1960). A study of the action of miroestrol and other oestrogens on the reproductive tract of the immature female mouse. *J. Endocrinol.* **20**, 229–235.

Jonsson, C.-E., and Terenius, L. (1965). Uptake of radioactive oestrogens in the chicken oviduct and some other organs. *Acta Endocrinol. (Copenhagen)* **50**, 289–300.

Jungblut, P. W., McCann, S., Görlich, L., Rosenfeld, G. C., and Wagner, R. K. (1970). Binding of steroids by tissue proteins. *In* "Research on Steroids" (M. Finkelstein *et al.,* eds.), pp. 213–232. Pergamon, Oxford.

Junkmann, K. (1968). Chemie der Gestagene. *Hand. Exp. Pharmakol.* **22**, 1–44.

Kalimi, M., Tsai, S. Y., Tsai, M.-J., Clark, J. H., and O'Malley, B. W. (1976). Effect of estrogen on gene expression in the chick oviduct. Correlation between nuclear-bound estrogen receptor and chromatin initiation sites for transcription. *J. Biol. Chem.* **251**, 516–523.

Kao, C. Y. (1967). Ionic basis of electrical activity in uterine smooth muscle. *In* "Cellular Biology of the Uterus" (R. M. Wynn, ed.), pp. 386–448. North-Holland Publ., Amsterdam.

Kao, C. Y., and Nishiyama, A. (1964). Ovarian hormones and resting potential of rabbit uterine smooth muscle. *Am. J. Physiol.* **207**, 793–799.

Karavolas, H. J., and Engel, L. L. (1966). Human placental 17β-estradiol dehydrogenase. III. The separation of a 17β-estradiol-dependent transhydrogenase. *J. Biol. Chem.* **241**, 3454–3456.

Karlson, P. (1963). New concepts on the mode of action of hormones. *Perspect. Biol. Med.* **6**, 203–214.

Karsznia, R., Wyss, R. H., Heinrichs, W. L., and Herrmann, W. L. (1969). Binding of pregnenolone and progesterone by prostatic "receptor" proteins. *Endocrinology* **84**, 1238–1246.

Katzenellenbogen, B. S., and Gorski, J. (1971). *In vitro* induction of the synthesis of a specific uterine protein (IP) by physiological (10^{-9} M) concentration of estradiol-17β (E_2) *Fed. Proc. Fed. Am. Soc. Exp. Biol.* **30**, 1214, Abstr.

Kaufmann, C., and Zander, J. (1956). Progesterone in menschlichem Blut und Gewebe. II. Progesteron im Fettgeweke. *Klin. Wochenschr.* **34**, 7–9.

Kenney, F. T., Wicks, W. D., and Greenman, D. L. (1965). Hydrocortisone stimulation of RNA synthesis in induction of hepatic enzymes. *J. Cell. Comp. Physiol.* **66**, 125–136.

King, R. J. B. (1972). Oestrogen receptor mechanisms. *Bibl. Reprod.* **20**, 1–10 and 159–166.

King, R. J. B., and Mainwaring, W. I. P. (1974). "Steroid–Cell Interactions". Butterworth, London.

King, R. J. B., Gordon, J., Cowan, D. M., and Inman, D. R. (1966). The intranuclear localization of [6,7-³H]oestradiol-17β in dimethylbenzanthracene-induced rat mammary adenocarcinoma and other tissues. *J. Endocrinol.* **36**, 139–150.

King, R. J. B., Gordon, J., and Steggles, A. W. (1969). The properties of a nuclear acidic protein fraction that binds [6,7-³H]oestradiol-17β. *Biochem. J.* **114**, 649–657.

King, R. J. B., Gordon, J., Marx, J., and Steggles, A. W. (1971). Localisation and nature of sex steroids receptors within the cell. *In* "Basic Actions of Sex Steroids on Target Organs" (P. O. Hubinont, F. Leroy, and P. Galand, eds.), pp. 21–43. Karger, Basel.

Kissel, J. H., Rosenfeld, M. G., Chase, L. R., and O'Malley, B. W. (1970). Response of chick oviduct adenyl cyclase to steroid hormones. *Endocrinology* **86**, 1019–1023.

Kleiber, M. (1965). Respiratory exchange and metabolic rate. *In* "Handbook of Physiology" (Am. Physiol. Soc., J. Field, ed.), Sect. 3, Vol. II, pp. 927–938. Williams & Wilkins, Baltimore, Maryland.

Knowler, J. T., and Smellie, R. M. S. (1971). Oestradiol-17β stimulation of ribonucleic acid synthesis in immature rat uterus: the effects of injection route and rat weight. *Biochem. J.* **123**, 33 P.

Knudsen, K. A., Jones, R. C., and Edgren, R. A. (1969). Effect of a carbonic anhydrase inhibitor (diamox) on the progesterone-stimulated rabbit uterus. *Endocrinology* **85**, 1204–1205.

Kohler, P. O., and O'Malley, B. W. (1967). Estrogen-induced morphologic changes in monolayer cultures of immature chick oviduct. *Endocrinology* **81**, 1422–1427.

Kontula, K., Jänne, O., Luukkainen, T., and Vihko, R. (1973). Progesterone-binding protein in human myometrium. Ligand specificity and some physicochemical characteristics. *Biochim. Biophys. Acta* **328**, 145–153.

Korenman, S. G. (1969). Comparative binding affinity of estrogens and its relation to estrogenic potency. *Steroids* **13**, 163–177.

Korenman, S. G., and O'Malley, B. W. (1968). Progesterone action: regulation of avidin biosynthesis by hen oviduct *in vivo* and *in vitro*. *Endocrinology* **83**, 11–17.

Korner, A. (1964). Regulation of the rate of synthesis of messenger ribonucleic acid by growth hormone. *Biochem. J.* **92**, 449–456.

Kraay, R. J.,and Black, L. J. (1971). Influence of progestins on the uptake of radioactive estradiol. *Proc. Int. Congr. Horm. Steroids, 3rd, 1970* Excerpta Med. Found. Int. Congr. Ser. No. 210, p. 150, Abstract No. 307.

Krishnan, R. S., and Daniel, J. C., Jr. (1967). "Blastokinin": inducer and regulator of blastocyst development in the rabbit uterus. *Science* **158**, 490–492.

Kroeger, H. (1967). Hormones, ion balances and gene activity in dipteran chromosomes. *Mem. Soc. Endocrinol.* **15**, 55–65.

Kuhn, N. J. (1969a). Progesterone withdrawal as the lactogenic trigger in the rat. *J. Endocrinol.* **44**, 39–54.

Kuhn, N. J. (1969b). Specificity of progesterone inhibition of lactogenesis. *J. Endocrinol.* **45**, 615–616.

Kuhn, N. J. (1971). Control of lactogenesis and lactose biosynthesis. *In* "Lactation" (I. R. Falconer, ed.), pp. 161–176. Butterworth, London.

Kuhn, N. J., and Briley, M. S. (1970). The roles of pregn-5-ene-$3\beta,20\alpha$-diol and 20α-hydroxy steroid dehydrogenase in the control of progesterone synthesis preceding parturition and lactogenesis in the rat. *Biochem. J.* **117**, 193–201.

Kuhn, R. W., Schrader, W. T., Smith, R. G., and O'Malley, B. W. (1975). Progesterone binding components of chick oviduct. X. Purification by affinity chromatography. *J. Biol. Chem.* **250**, 4220–4228.

Kuriyama, H. (1961). The effect of progesterone and oxytocin on the mouse myometrium. *J. Physiol. (London)* **159**, 26–39.

Kuriyama, H. (1964). Effect of electrolytes on the membrane activity of the uterus. *In* "Pharmacology of Smooth Muscle" (E. Bülbring, ed.), pp. 127–140. Pergamon, Oxford.

Laguens, R. (1964). Effect of estrogen upon the fine structure of the uterine smooth muscle cell of the rat. *J. Ultrastruct. Res.* **10**, 578–584.

Langecker, H., and Damrosch, L. (1968). Der Stoffwechsel des Progesterons. *Hand. Exp. Pharmakol.* **22**, 45–263..

Laszlo, J., Miller, D. S., McCarty, K. S., and Hochstein, P. (1966). Actinomycin D: inhibition of respiration and glycolysis. *Science* **151**, 1007–1009.

Laumas, K. R. (1967). The distribution and localization in tissues of tritium labelled steroid sex hormones. *Proc. Asia Oceania Congr. Endocrinol., 3rd, 1967* Part 1, pp. 124–135.

Laumas, K. R., and Farooq, A. (1966). The uptake *in vivo* of [1,2-³H]-progesterone by the brain and genital tract of the rat. *J. Endocrinol.* **36**, 95–96.

Lawson, D. E. M., and Pearlman, W. H. (1964). The metabolism *in vivo* of progesterone-7-³H; its localization in the mammary gland, uterus and other tissues of the pregnant rat. *J. Biol. Chem.* **239**, 3226–3232.

Leathem, J. H. (1959). Some biochemical aspects of the uterus. *Ann. N.Y. Acad. Sci.* **75**, 463–471.

Leavitt, W. W., and Blaha, G. C. (1972). An estrogen-stimulated, progesterone-binding system in the hamster uterus and vagina. *Steroids* **19**, 263–274.

Lerner, L. J., Hilf, R., Turkheimer, A. R., Michel, I. and Engel, S. L. (1966). Effects of hormone antagonists on morphological and biochemical changes induced by hormonal steroids in the immature rat uterus. *Endocrinology*, **78**, 111–124.

Leroy, F. (1967). Etude histophotométrique de l'endomètre. II. Effets isolés et interactions de l'oestradiol et de la progestérone sur la synthèse d'ADN des noyaux de l'endomètre chez la ratte castrée. *Rev. Fr. Etud. Clin. Biol.* **12**, 902–907.

Leymarie, P., and Gueriguian, J. L. (1969). Progesterone and cortisol binding by the soluble cellular fraction of the corpus luteum of the pregnant cow. *C. R. Acad. Sci. Ser. D* **269**, 1342–1345.

Liao, S. (1975). Cellular receptors and mechanisms of action of steroid hormones. *Int. Rev. Cytol.* **41**, 87–172.

Little, B., and Billiar, R. B. (1969). Progesterone production. *Prog. Endocrinol., Proc. Int. Congr. Endocrinol., 3rd, 1968* pp. 871–879.

Little, B., and Lincoln, E. (1964). Effect of estrogen, progesterone and testosterone on the incorporation of L-valine-1-C^{14} into protein of the rat liver and uterus. *Endocrinology* **74**, 1–8.

Little, B., Tait, J. F., Tait, S. A. S., and Erlenmeyer, F. (1966). The metabolic clearance rate of progesterone in males and ovariectomized females. *J. Clin. Invest.* **45**, 901–912.

Little, B., Tait, J. F., Tait, S. A. S., and Erlenmeyer, F. (1967). Progesterone clearance in non-pregnant and pregnant subjects. *Proc. Int. Congr. Horm. Steroids, 2nd, 1966* Abstract No. 240.

Liu, T. M. Y., and Davis, J. W. (1967). Induction of lactation by ovariectomy of pregnant rats. *Endocrinology* **80**, 1043–1050.

Longcope, C., Layne, D. S., and Tait, J. F. (1968). Metabolic clearance rates and interconversions of estrone and 17β-estradiol in normal males and females. *J. Clin. Invest.* **47**, 93–106.

Luck, D. N., and Hamilton, T. H. (1972). Early estrogen action: stimulation of the metabolism of high molecular weight and ribosomal RNAs. *Proc. Natl. Acad. Sci. U.S.A.* **69**, 157–161.

Lutwak-Mann, C. (1955). Carbonic anhydrase in the female reproductive tract. Occurrence, distribution and hormonal dependence. *J. Endocrinol.* **13**, 26–38.

Lutwak-Mann, C., and Adams, C. E. (1957). Carbonic anhydrase in the female reproductive tract. II. Endometrial carbonic anhydrase as indicator of luteoid potency: correlation with progestational proliferation. *J. Endocrinol.* **15**, 43–55.

Lutwak-Mann, C., and McIntosh, J. E. A. (1969). Zinc and carbonic anhydrase in the rabbit uterus. *Nature (London)* **221**, 1111–1114.

Lyons, W. R., Li, C. H., and Johnson, R. E. (1958). The hormonal control of mammary growth and lactation. *Recent Prog. Horm. Res.* **14**, 219–248.

McGuire, W. L., and O'Malley, B. W. (1968). Ribonucleic acid polymerase activity of the chick oviduct during steroid-induced synthesis of a specific protein. *Biochim. Biophys. Acta* **157**, 187–194.

MacLeod, J., and Reynolds, S. R. M. (1938). Vascular, metabolic and motility responses of uterine tissue following administration of oestrin. *Proc. Soc. Exp. Biol. Med.* **37**, 666–668.

Madjerek, Z., and Van der Vies, J. (1961). Carbonic anhydrase activity in the uteri of mice under various experimental conditions. *Acta Endocrinol. (Copenhagen)* **38**, 315–320.

Marcus, P. I., and Talalay, P. (1955). On the molecular specificity of steroid-enzyme combinations: the kinetics of β-hydroxysteroid dehydrogenase. *Proc. R. Soc. London, Ser. B* **144**, 116–132.

Markee, J. E. (1932). Rhythmic vascular uterine changes. *Am. J. Physiol.* **100**, 32–39.

Martin, L. (1962). The effects of histamine on the vaginal epithelium of the mouse. *J. Endocrinol.* **23**, 329–340.

Martin, L., and Finn, C. A. (1971). Oestrogen-gestagen interactions on mitosis in target tissues. *In* "Basic Actions of Sex Steroids on Target Organs" (P. O. Hubinont, F. Leroy, and P. Galand, eds.), pp. 172–188. Karger, Basel.

Maurer, H. R., and Chalkley, G. R. (1967). Some properties of a nuclear binding site of estradiol. *J. Mol. Biol.* **27**, 431–441.

Means, A. R., and Hamilton, T. H. (1966). Early estrogen action: concomitant stimulations within two minutes of nuclear RNA synthesis and uptake of RNA precursor by the uterus. *Proc. Natl. Acad. Sci. U.S.A.* **56**, 1594–1598.

Means, A. R., and O'Malley, B. W. (1971a). Assessment of sex steroid action *in vitro. Acta Endocrinol. (Copenhagen), Suppl.* **158**, 318–344.

Means, A. R., and O'Malley, B. W. (1971b). Protein biosynthesis on chick oviduct polyribosomes. II. Regulation by progesterone. *Biochemistry* **10**, 1570–1576.

Means, A. R., Abrass, I. B., and O'Malley, B. W. (1971). Protein biosynthesis on chick oviduct polyribosomes. I. Changes during estrogen-mediated tissue differentiation. *Biochemistry* **10**, 1561–1570.

Means, A. R., Comstock, J. P., Rosenfeld, G. C., and O'Malley, B. W. (1972). Ovalbumin messenger RNA of chick oviduct: partial characterization, estrogen dependence, and translation *in vitro. Proc. Natl. Acad. Sci. USA* **69**, 1146–1150.

Mercier-Bodard, C., Alfsen, A., and Baulieu, E. E. (1970). Sex steroid binding plasma protein (SBP). *Acta Endocrinol. (Copenhagen), Suppl.* **147**, 204–224.

Mešter, J., and Baulieu, E. E. (1975). Dynamics of oestrogen-receptor distribution between the cytosol and nuclear fractions of immature rat uterus after oestradiol administration. *Biochem. J.* **146**, 617–623.

Mešter, J., Martel, D., Psychoyos, A., and Baulieu, E. E. (1974). Hormonal control of oestrogen receptor in uterus and receptivity for ovoimplantation in the rat. *Nature (London)* **250**, 776–778.

Milgrom, E., and Baulieu, E. E. (1970a). Progesterone in uterus and plasma. I. Binding in rat uterus 105,000g supernatant. *Endocrinology* **87**, 276–287.

Milgrom, E., and Baulieu, E.-E. (1970b). Progesterone in the uterus and the plasma. II. The role of hormone availability and metabolism on selective binding to uterus protein. *Biochem. Biophys. Res. Commun.* **40**, 723–730.

Milgrom, E., Atger, M., and Baulieu, E.-E. (1970a). Progesterone binding plasma protein (PBP). *Nature (London)* **228**, 1205–1206.

Milgrom, E., Atger, M., and Baulieu, E.-E. (1970b). Progesterone in uterus and plasma. IV. Progesterone receptor(s) in guinea pig uterus cytosol. *Steroids* **16**, 741–754.

Milgrom, E., Perrot, M., Atger, M., and Baulieu, E.-E. (1972a). Progesterone in uterus and plasma. V. An assay of the progesterone cytosol receptor of the guinea pig uterus. *Endocrinology* **90**, 1064–1070.

Milgrom, E., Atger, M., Perrot, M., and Baulieu, E. E. (1972b). Progesterone in uterus and plasma. VI. Uterine progesterone receptors during the estrous cycle and implantation in the guinea pig. *Endocrinology* **90**, 1071–1078.

Milgrom, E., Luu Thi, M., Atger, M., and Baulieu, E. E. (1973). Mechanisms regulating the concentration and the conformation of progesterone receptor(s) in the uterus. *J. Biol. Chem.* **248**, 6366–6374.

Miyake, T. (1962). Progestational substances. *Methods Horm. Res.* **2**, 127–178.

Miyake, T., and Pincus, G. (1958). Anti-progestational activity of estrogens in rabbit endometrium. *Proc. Soc. Exp. Biol. Med.* **99**, 478–482.

Miyake, T., and Pincus, G. (1959). Hormonal influences on the carbonic anhydrase concentration in the accessory reproductive tracts of the rat. *Endocrinology* **65**, 64–72.

Miyake, T., and Rooks, W. H., II. (1966). The relation between the structure and physiological activity of progestational steroids. *Methods Horm. Res.* **5**, 59–145.

Monahan, J. J., Harris, S. E., and O'Malley, B. W. (1976). Effect of estrogen on gene expression in the chick oviduct. Effect of estrogen on the sequence and population complexity of chick oviduct poly (A)-containing RNA. *J. Biol. Chem.* **251**, 3738–3748.

Motta, M., Piva, F., and Martini, L. (1970). The hypothalamus as the center of endocrine feedback mechanisms. *Hypothal., Proc. Workshop Conf., 1969* pp. 463–489.

Mueller, G. C. (1953). Incorporation of glycine-2-C^{14} into protein by surviving uteri from α-estradiol-treated rats. *J. Biol. Chem.* **204**, 77–90.

Mueller, G. C. (1971). Estrogen action: a study of the influence of steroid hormones on genetic expression. *Biochem. Soc. Symp.* **32**, 1–29.

Mueller, G. C., Herranen, A. M., and Jervell, K. F. (1958). Studies on the mechanism of action of estrogens. *Recent Prog. Horm. Res.* **14**, 95–129.

Mueller, G. C., Gorski, J., and Aizawa, Y. (1961). The role of protein synthesis in early estrogen action. *Proc. Natl. Acad. Sci. U.S.A.* **47**, 164–169.

Mueller, G. C., Vonderhaar, B., Kim, U. H., and Le Mahieu, M.A. (1972). Estrogen action: an inroad to cell biology. *Recent Prog. Horm. Res.* **28**, 1–49.

Nicolette, J. A. (1969). The effect of prolonged incubation on RNA polymerase activity from surviving rat uteri. *Arch. Biochem. Biophys.* **135**, 253–258.

Nicolette, J. A., and Gorski, J. (1964). Effect of estradiol on glucose-U-C^{14} metabolism in the rat uterus. *Arch. Biochem. Biophys.* **107**, 279–283.

Nicolette, J. A., and Mueller, G. C. (1966a). Effect of actinomycin D on the estrogen response in uteri of adrenalectomized rats. *Endocrinology* **79**, 1162–1165.

Nicolette, J. A., and Mueller, G. C. (1966b). *In vitro* regulation of RNA polymerase in estrogen-treated uteri. *Biochem. Biophys. Res. Commun.* **24**, 851–857.

Nicolette, J. A., Le Mahieu, M. A., and Mueller, G. C. (1968). A role of estrogens in the regulation of RNA polymerase in surviving rat uteri. *Biochim. Biophys. Acta* **166**: 403–409.

Nordquist, G. (1970). The synthesis of DNA and RNA in normal human endometrium in short-term incubation *in vitro* and its response to oestradiol and progesterone. *J. Endocrinol.* **48**, 17–28.

Northcutt, R. C., Island, D. P., and Liddle, G. W. (1969). An explanation for the target organ unresponsiveness to testosterone in the testicular feminization syndrome. *J. Clin. Endocrinol. Metab.* **29**, 422–425.

Noteboom, W. D., and Gorski, J. (1965). Stereospecific binding of estrogens in the rat uterus. *Arch. Biochem. Biophys.* **111**, 559–568.

Notides, A., and Gorski, J. (1966). Estrogen-induced synthesis of a specific uterine protein. *Proc. Natl. Acad. Sci. U.S.A.* **56**, 230–235.

Ohno, S. (1971). Simplicity of mammalian regulatory systems inferred by single gene determination of sex phenotypes. *Nature (London)* **234**, 134–137.

Oka, T., and Schimke, R. T. (1969). Interaction of estrogen and progesterone in chick oviduct development. II. Effects of estrogen and progesterone on tubular gland cell function. *J. Cell Biol.* **43**, 123–137.

O'Malley, B. W. (1971). The action of progestogens at the cellular level. *Res. Reprod.* **3**, No. 3, 3–4.

O'Malley, B. W., and McGuire, W. L. (1968). Studies on the mechanism of action of progesterone in regulation of the synthesis of specific protein. *J. Clin. Invest.* **47**, 654–664.

O'Malley, B. W., and Hardman, J. C., (eds.) (1975a). Steroid Hormones. *In* "Methods in Enzymology" vol. 36, Hormone action; Part A, Steroid hormones. Academic Press, New York.

O'Malley, B. W., and Hardman, J. C., (eds.) (1975b). Nuclear structure and function. *In* "Methods in Enzymology" vol. 40, Hormone action; Part E, Nuclear structure and function. Academic Press, New York.

O'Malley, B. W., and Means, A. R. (eds.) (1973). "Receptors for Reproductive Hormones." Plenum Press, New York.

O'Malley, B. W., and Means, A. R. (1974). Female steroid hormones and target cell nuclei. *Science* **183**, 610–620.

O'Malley, B. W., McGuire, W. L., Kohler, P. O., and Korenman, S. G. (1969). Studies on the mechanism of steroid hormone regulation of synthesis of specific proteins. *Recent Prog. Horm. Res.* **25**, 105–153.

O'Malley, B. W., Sherman, M. R., and Toft, D. O. (1970). Progesterone "receptors" in the cytoplasm and nucleus of chick oviduct target tissue. *Proc. Natl. Acad. Sci. U.S.A.* **67**, 501–508.

O'Malley, B. W., Toft, D. O., and Sherman, M. R. (1971). Progesterone-binding components of chick oviduct. II. Nuclear components. *J. Biol. Chem.* **246**, 1117–1122.

O'Malley, B. W., Rosenfeld, G. C., Comstock, J. P., and Means, A. R. (1972a). Steroid hormone induction of a specific translatable messenger RNA. *Nature (London) New Biol.* **240**, 45–48.

O'Malley, B. W., Spelsberg, T. C., Schrader, W. T., Chytil, F., and Steggles, A. W. (1972b). Mechanisms of interaction of a hormone-receptor complex with the genome of a eukaryotic target cell. *Nature (London)* **235**, 141–144.

Oxender, D. L., and Christensen, H. N. (1963). Distinct mediating systems for the transport of neutral amino acids by the Ehrlich cell. *J. Biol. Chem.* **238**, 3686–3699.

Palmiter, R. D. (1969a). Hormonal induction and regulation of lactose synthetase in mouse mammary gland. *Biochem. J.* **113**, 409–417.

Palmiter, R. D. (1969b). What regulates the lactose content in milk? *Nature (London)* **221**, 912–914.

Palmiter, R. D., Oka, T., and Schimke, R. T. (1971). Modulation of ovalbumin synthesis by estradiol-17β and actinomycin D as studied in explants of chick oviduct in culture. *J. Biol. Chem.* **246**, 724–737.

Parkes, A. S., and Deanesly, R. (1966). The ovarian hormones. *In* "Marshall's Physiology of Reproduction" (A. S. Parkes, ed.), 3rd ed., Vol. 3, pp. 570–828. Longmans Green, London and New York.

Pasquilini, J. R., and Scholler, R. (1975). Receptors and steroid hormone action. *J. Steroid Biochem.* **6**, 459–514.

Pasteels, J. L. (1970). The control of prolactin secretion. *Hypothal., Proc. Workshop Conf., 1969* pp. 385–399.

Paterson, J. Y. F., and Harrison, F. A. (1968). The specific activity of plasma cortisol in sheep after intravenous infusion of [1,2-^{3}H$_2$]cortisol and its relation to the distribution of cortisol. *J. Endocrinol.* **40**, 37–47.

Paterson, J. Y. F., and Harrison, F. A. (1972). Protein binding and the metabolism of cortisol in sheep. *In* "The Endocrinology of Pregnancy and Parturition" (C. G. Pierrepoint, ed.), pp. 94–102. Alpha Omega Alpha Publ., Cardiff, Wales.

Paterson, J. Y. F., and Hills, F. (1967). The binding of cortisol by ovine plasma proteins. *J. Endocrinol.* **37**, 261–268.

Pikler, G. M., Webster, R. A., and Spelsberg, T. C. (1976). Nuclear binding of progesterone in hen oviduct. Binding to multiple sites *in vitro. Biochem. J.* **156**, 399–408.

Plotz, E. J., and Davis, M. E. (1957). Distribution of radioactivity in human maternal and fetal tissues following administration of C^{14}-4-progesterone. *Proc. Soc. Exp. Biol. Med.* **95**, 92–96.

Pollard, I. (1970). Ultrastructural evidence for the stimulation of nuclear ribonucleic acid synthesis by oestradiol-17β in the vaginal epithelium of the ovariectomized mouse. *J. Endocrinol.* **47**, 143–148.

Prop, F. J. A. (1963). Hypophyseal hormones and mammary glands in organ cultures. *Acta Physiol. Pharmacol. Neerl.* **12**, 172–176.

Puca, G. A., and Bresciani, F. (1969). Association constant and specificity of oestradiol-receptor interaction. *Nature (London)* **223**, 745–747.

Puca, G. A., Nola, E., Sica V., and Bresciani, F. (1971). Estrogen-binding proteins of calf uterus. Partial purification and preliminary characterization of two cytoplasmic proteins. *Biochemistry* **10**, 3769–3780.

Rao, R. B., Wiest, W. G., and Allen, W. M. (1974). Progesterone "receptor" in human endometrium. *Endocrinology* **95**, 1275–1281.

Raynaud-Jammet C., and Baulieu, E.-E. (1969). Action de l'oestradiol *in vitro:* augmentation de la biosynthèse d'acide ribonucléique dans noyaux utérins. *C. R. Acad. Sci., Ser. D* **268**, 3211–3214.

Revel, M., Hiatt, H. H., and Revel, J. (1964). Actinomycin D: an effect on rat liver homogenates unrelated to its action on RNA synthesis. *Science* **146**, 1311–1313.

Riegel, B., Hartop, W. L., and Kittinger, G. W. (1950). Studies on the metabolism of radio-progesterone in mice and rats. *Endocrinology* **47**, 311–319.

Riggs, T. R. (1970). Hormones and transport across cell membranes. *In* "Biochemical Actions of Hormones" (G. Litwak, ed.), Vol. 1, pp. 157–208. Academic Press, New York.

Riggs, T. R., Pan, M. W., and Feng, H. W. (1968). Transport of amino acids into the estrogen-primed uterus. I. General characteristics of the uptake *in vitro. Biochim. Biophys. Acta* **150**, 92–103.

Robertson, D. M., Mešter, J., Beilby, J., Steele, S. J., and Kellie, A. E. (1971). The measurement of high-affinity oestradiol receptors in human uterine endometrium and myometrium. *Acta Endocrinol. (Copenhagen)* **68**, 534–542.

Robison, G. A., Butcher, F. W., and Sutherland, E. W. (1971). "Cyclic AMP." Academic Press, New York.

Rochefort, H., and Baulieu, E. E. (1968). Récepteurs hormonaux: relations entre les "récepteurs" uterins de l'oestradiol, "8S" cytoplasmique, et "4S" cytoplasmique et nucléaire. *C. R. Acad. Sci., Ser. D* **267**, 662–665.

Romanoff, L. P., Grace, M. P., Barter, M. N., and Pincus, G. (1966). Metabolism of pregnenolone-7α-^{3}H and progesterone-4-^{14}C in young and elderly men. *J. Clin. Endocrinol. Metab.* **26**, 1023–1031.

Rosenfeld, M. G., Kissel, J. H., and O'Malley, B. W. (1970). Progesterone effects on adenylcyclase activity and cyclic adenosine 3′,5′-monophosphate levels in the chick oviduct. *Cytobios* **2**, 33–40.

Rosenthal, H. E., Slaunwhite, W. R., and Sandberg, A. A. (1969). Transcortin: a corticosteroid-binding protein of plasma. XI. Effects of estrogens or pregnancy in guinea pigs. *Endocrinology* **85**, 825–830.

Rosenthal, H. E., Paul, M. A., and Sandberg, A. A. (1974). Transcortin: a corticosteroid-binding protein of plasma. XII. Immunologic studies on transcortin in guinea-pig tissues. *J. Steroid Biochem.* **5**, 219–225.

Roskoski, R., Jr., and Steiner, D. F. (1967a). Cycloheximide and actinomycin D inhibition of estrogen-stimulated sugar and amino acid transport in rat uterus. *Biochim. Biophys. Acta* **135**, 347–349.

Roskoski, R., Jr., and Steiner, D. F. (1967b). The effect of estrogen on amino acid transport in rat uterus. *Biochim. Biophys. Acta* **135**, 727–731.

Rowlands, I. W., and Heap, R. B. (1964). Histological observations on the ovary and progesterone levels in the coypu, *Myocastor coypus. Symp. Zool. Soc. London* **15**, 335–352.

Sandberg, A. A., and Slaunwhite, W. R. (1958). The metabolic fate of progesterone-C^{14} in human subjects. *J. Clin. Endocrinol. Metab.* **18**, 253–265.

Sandberg, A. A., Rosenthal, H. H., Schneider, S. L., and Slaunwhite, W. R. (1966). Protein-steroid interactions and their role in the transport and metabolism of steroids. *In* "Steroid Dynamics" (G. Pincus, T. Nakao, and J. F. Tait, eds.), pp. 1–59. Academic Press, New York.

Schimke, R. T., McKnight, G. S., Shapiro, D. J., Sullivan, D., and Palacois, R. (1975). Hormonal regulation of ovalbumin synthesis in the chick oviduct. *Recent Prog. Horm. Res.* **31**, 175–212.

Schlosnagle, D. C., Bazer, F. W., Tsibris, J. C. M., and Roberts, R. M. (1974). An iron-containing phosphatase induced by progesterone in the uterine fluid of pigs. *J. Biol. Chem.* **249**, 7574–7579.

Schomberg, D. W., Jones, P. H., Featherstone, W. R., and Erb, R. E. (1966). Identification of metabolites of progesterone-4-^{14}C in domestic sow urine. *Steroids* **8**, 277–288.

Schrader, W. T., and O'Malley, B. W. (1972). Progesterone-binding components of chick oviduct. *J. Biol. Chem.* **247**, 51–59.

Schrader, W. T., Buller, R. E., Kuhn, R. W., and O'Malley, B. W. (1974). Molecular mechanisms of steroid hormone action. *J. Steroid Biochem.* **5**, 989–999.

Schrader, W. T., Heuer, S. S., and O'Malley, B. W. (1975). Progesterone receptors of chick oviduct: identification of 6S receptor dimers. *Biol. Reprod.* **12**, 134–142.

Seal, U. S., and Doe, R. P. (1966). Corticosteroid-binding globulin: biochemistry, physiology and phylogeny. *In* "Steroid Dynamics" (G. Pincus, T. Nakao, and J. F. Tait, eds.), pp. 63–88. Academic Press, New York.

Segal, S. J., and Scher, W. (1967). Estrogens, nucleic acids, and protein synthesis in uterine metabolism. *In* "Cellular Biology of the Uterus" (R. M. Wynn, ed.), pp. 114–150. North-Holland Publ., Amsterdam.

Sekeris, C. E. (1967). Effect of hormones on cell nuclei. *Colloq. Ges. Physiol. Chem.* **5**, 126–151.

Sherman, M. R., Carvol, P. L., and O'Malley, B. W. (1970). Progesterone-binding components of chick oviduct. I. Preliminary characterization of cytoplasmic components. *J. Biol. Chem.* **245**, 6085–6096.

Sherman, M. R., Atienza, S. B. P., Shansky, J. R., and Hoffman, L. M. (1974). Progesterone receptors of chick oviduct. Steroid-binding "subunit" formed with divalent cations. *J. Biol. Chem.* **249**, 5351–5363.

Shirley, I. M., and Cooke, B. A. (1968). Ring A and side-chain reduction of [4-^{14}C] progesterone by rat liver *in vitro. J. Endocrinol.* **40**, 477–483.

Shoppee, C. W. (1958). "Chemistry of the Steroids," 1st ed. Butterworth, London.

Short, R. V. (1961). Progesterone. *In* "Hormones in Blood" (C. H. Gray and A. L. Bacharach, eds.), pp. 379–437. Academic Press, New York.

Slotin, C. A., Heap, R. B., Christiansen, J. M., and Linzell, J. L. (1970). Synthesis of progesterone by the mammary gland of the goat. *Nature (London)* **225**, 385–386.

Slotin, C. A., Harrison, F. A., and Heap, R. B. (1971). Kinetics of progesterone metabolism in the pregnant sheep. *J. Endocrinol.* **49**, xxx–xxxii.

Sluyser, M. (1971). Interaction of steroid hormones and histones. *Biochem. Soc. Symp.* **32**, 31–47.

Smith, D. E., and Gorski, J. (1968). Estrogen control of uterine glucose metabolism. An analysis based on the transport and phosphorylation of 2-deoxyglucose. *J. Biol. Chem.* **243**, 4169–4174.

Smith, J. A., Martin, L., and King, R. J. B. (1969). Effects of oestradiol-17β and progesterone on nuclear-protein synthesis in the mouse uterus. *Biochem. J.* **114**, 59P.

Smith, J. A., Martin, L., King, R. J. B., and Vertes, M. (1970). Effects of oestradiol-17β and progesterone on total nuclear-protein synthesis in epithelial and stromal tissues of the mouse uterus, and of progesterone on the ability of these tissues to bind oestradiol-17β. *Biochem. J.* **119**, 773–784.

Smith, H. E., Smith, R. G., Toft, D. O., Neergaard, J. R., Burrows, E. P., and O'Malley, B. W. (1974). Binding of steroids to progesterone receptor proteins in chick oviduct and human uterus. *J. Biol. Chem.* **249**, 5924–5932.

Snart, R. S., and Wilson, M. J. (1967). Uptake of steroid hormones into artificial phospholipid/cholesterol membrane. *Nature (London)* **215**, 964.

Spaziani, E., and Suddick, R. P. (1967). Hexose transport and blood flow rate in the uterus: effects of estradiol, puromycin and actinomycin D. *Endocrinology* **81**, 205–212.

Spaziani, E., and Szego, C. M. (1959). Early effects of estradiol and cortisol on water and electrolyte shifts in the uterus of the immature rat. *Am. J. Physiol.* **197**, 355–359.

Spelsberg, T. C., Steggles, A. W., and O'Malley, B. W. (1971). Progesterone-binding components of chick oviduct. III. Chromatin acceptor sites. *J. Biol. Chem.* **246**, 4188–4197.

Spelsberg, T. (1967). Nuclear binding of progesterone in chick oviduct. Multiple binding sites *in vivo* and transcriptional response. *Biochem. J.* **156**, 391–398.

Spelsberg, T. C., Steggles, A. W., Chytil, F., and O'Malley, B. W. (1972). Progesterone-binding components of chick oviduct. V. Exchange of progesterone-binding capacity from target to non target tissue chromatins. *J. Biol. Chem.* **247**, 1368–1374.

Sperry, P. J., Woo, S. L. C., Means, A. R., and O'Malley, B. W. (1976). Partial purification of a progesterone-inducible messenger RNA (avidin) from hen oviduct. *Endocrinology* **99**, 315–325.

Stedman, E., and Stedman, E. (1950). Cell specificity of histones. *Nature (London)* **166**, 780–781.

Steggles, A. W., Marx, J., and King, R. J. B. (1971). The role of endogenous hormones in the synthesis of oestradiol receptor proteins. *Proc. Int. Congr. Horm. Steroids, 3rd, 1970.* Excerpta Med. Found. Int. Congr. Ser. No. 210, p. 148, Abstract No. 303.

Stone, G. M. (1963). The uptake of tritiated oestrogens by various organs of the ovariectomized mouse following subcutaneous administration. *J. Endocrinol.* **27**, 281–288.

Stone, G. M., and Baggett, B. (1965). The uptake of some tritiated estrogenic and non-estrogenic steroids by the mouse uterus and vagina in vivo and in vitro. *Steroids* **6**, 277–299.

Stone, G. M., and Martin, L. (1964). The uptake of tritiated estradiol and estrone by the uterus of the ovariectomized mouse following local application. *Steroids* **3**, 699–706.

Stone, G. M., Baggett, B., and Donnelly, R. B. (1963). The uptake of tritiated oestrogens by various organs of the ovariectomized mouse following intravenous administration. *J. Endocrinol.* **27**, 271–280.

Stumpf, W. E. (1968). Subcellular distribution of ^{3}H-estradiol in rat uterus by quantitative autoradiography—a comparison between ^{3}H-estradiol and ^{3}H-norethynodrel. *Endocrinology* **83**, 777–782.

Stupnicki, R., and Williams, K. I. H. (1968). Urinary metabolites of 4-^{14}C-progesterone in the ewe. *Steroids* **12**, 581–587.

Stupnicki, R., McCracken, J. A., and Williams, K. I. H. (1969). Progesterone metabolism in the ewe. *J. Endocrinol.* **45**, 67–74.

Sunshine, G. H., Williams, D. J., and Rabin, B. R. (1971). Role for steroid hormones in the interaction of ribosomes with the endoplasmic membranes of rat liver. *Nature (London), New Biol.,* **230**, 133–136.

Sutherland, E. W., Hardman, J. G., Butcher, R. W., and Broadus, A. E. (1969). The biological role of cyclic AMP (some areas of contrast with cyclic GMP). *Prog. Endocrinol., Proc. Int. Congr. Endocrinol., 3rd, 1968* pp. 26–32.

Szego, C. M. (1965). Role of histamine in mediation of hormone action. *Fed. Proc., Fed. Am. Soc. Exp. Biol.* **24**, 1343–1352.

Szego, C. M., and Davis, J. S. (1967). Adenosine 3′,5′-monophosphate in rat uterus: acute elevation by estrogen. *Proc. Natl. Acad. Sci. U.S.A.* **58**, 1711–1718.

Tait, J. F., and Burstein, S. (1964). *In vivo* studies of steroid dynamics in man. *Hormones* **5**, 441–557.

Talalay, P. (1957). Molecular specificity in the enzymic oxidation of steroids. *Rec. Chem. Prog.* **18**, 31–49.

Talalay, P., and Wang, V. S. (1955). Enzymic isomerisation of Δ^5-3-ketosteroids. *Biochim. Biophys. Acta* **18**, 300–301.

Talalay, P., and Williams-Ashman, H. G. (1958). Activation of hydrogen transfer between pyridine nucleotides by steroid hormones. *Proc. Natl. Acad. Sci. U.S.A.* **44**, 15–26.

Talalay, P., and Williams-Ashman, H. G. (1960). Participation of steroid hormones in the enzymatic transfer of hydrogen. *Recent Prog. Horm. Res.* **16**, 1–47.

Talwar, G. P., and Segal, S. J. (1964). Prevention of hormone action by local application of actinomycin D. *Proc. Natl. Acad. Sci. U.S.A.* **50**, 226–230.

Talwar, G. P., Sopori, M. L., Biswas, D. K., and Segal, S. J. (1968). Nature and characteristics of the binding of oestradiol-17β to a uterine macromolecular fraction. *Biochem. J.* **107**, 765–774.

Tata, J. R. (1968). Hormonal regulation of growth and protein synthesis. *Nature (London)* **219**, 331–337.

Tata, J. R. (1970). Regulation of protein synthesis by growth and developmental hormones. *In* "Biochemical Actions of Hormones" (G. Litwack, ed.), Vol. 1, pp. 89–133. Academic Press, New York.

Taylor, J. L., and Haydon, D. A. (1965). The interaction of progesterone with lipid films at the air-water interface. *Biochim. Biophys. Acta* **94**, 488–493.

Taylor, N. E., Hodgkin, D. C., and Rollett, J. S. (1960). The X-ray crystallographic determination of the structure of bromomiroestrol. *J. Chem. Soc.* **3**, 3685–3695.

Taylor, W. (1954). The metabolism of progesterone by animal tissues *in vitro*. I. Factors influencing the metabolism of progesterone by rat liver and the investigation of the products of metabolism. *Biochem. J.* **56**, 463–470.

Taylor, W. (1965). Biliary excretion of neutral steroid hormone metabolites in animals. *In* "The Biliary System." *Proc. NATO Adv. Study Inst.*, pp. 399–416.

Taylor, W., and Scratcherd, T. (1961). The metabolism of [4-¹⁴C]progesterone in the cat: biliary and urinary excretion of conjugated metabolites. *Biochem. J.* **81**, 398–405.

Taylor, W., and Scratcherd, T. (1967). Steroid metabolism in the rabbit. Biliary and urinary excretion of metabolites of [4-¹⁴C]testosterone. *Biochem. J.* **104**, 250–253.

Telfer, M. A. (1953). Influence of estradiol on nucleic acids, respiratory enzymes, and the distribution of nitrogen in the rat uterus. *Arch. Biochem. Biophys.* **44**, 111–119.

Telfer, M. A., and Hisaw, F. L. (1957). Biochemical responses of the rabbit endometrium and myometrium to oestradiol and progesterone. *Acta Endocrinol. (Copenhagen)* **25**, 390–404.

Teng, C.-S., and Hamilton, T. H. (1968). The role of chromatin in estrogen action in the uterus. I. The control of template capacity and chemical composition and the binding of ³H-estradiol-17β. *Proc. Natl. Acad. Sci. U.S.A.* **60**, 1410–1417.

Teng, C.-S., and Hamilton, T. H. (1969). Role of chromatin in estrogen action in the uterus. II. Hormone-induced synthesis of nonhistone acidic proteins which restore histone-inhibited DNA-dependent RNA synthesis. *Proc. Natl. Acad. Sci. U.S.A.* **63**, 465–472.

Teng, C.-S., and Hamilton, T. H. (1970). Regulation by estrogen of organ-specific synthesis of a nuclear acidic protein. *Biochem. Biophys. Res. Commun.* **40**, 1231–1238.

Teng, C. S., and Teng, C. T. (1976). Studies on sex-organ development. Oestrogen-receptor translocation in the developing chick Müllerian duct. *Biochem. J.* **154**, 1–9.

Tepperman, J., and Tepperman, H. M. (1960). Some effects of hormones on cells and cell constituents. *Pharmacol. Rev.* **12**, 301–354.

Terenius, L. (1966). Specific uptake of oestrogens by the mouse uterus *in vitro*. *Acta Endocrinol. (Copenhagen)* **53**, 611–618.

Terenius, L. (1968). Structural characteristics of oestrogen binding in the mouse uterus: inhibition of 17β-oestradiol binding *in vitro* by a plant oestrogen, Miroestrol. *Acta Pharmacol. Toxicol.* **26**, 15–21.

Terenius, L. (1969). Oestrogen binding sites in the chicken oviduct similar to those of the mouse uterus. *Acta Endocrinol. (Copenhagen)* **60**, 79–90.

Thijssen, J. H. H., and Zander, J. (1966). Progesterone-4-¹⁴C and its metabolites in the blood after intravenous injection into women. *Acta Endocrinol. (Copenhagen)* **51**, 563–577.

Thomas, P. J. (1973). Steroid hormones and their receptors. *J. Endocrinol.* **57**, 333–359.

Toft, D. O. (1972). The interaction of uterine estrogen receptors with DNA. *J. Steroid Biochem.* **3**, 515–529.

Toft, D. and Gorski, J. (1966). A receptor molecule for estrogens: isolation from the rat uterus and preliminary characterization. *Proc. Natl. Acad. Sci. U.S.A.* **55,** 1574–1581.

Toft, D. Shyamala, G., and Gorski, J. (1967). A receptor molecule for estrogens: studies using a cell-free system. *Proc. Natl. Acad. Sci. U.S.A.* **57,** 1740–1743.

Tomkins, G. M., and Ames, B. N. (1967). The operon concept in bacteria and higher organisms. *Natl. Cancer Inst., Monogr.* **27,** 211–234.

Tomkins, G. M., and Thompson, E. B. (1967). Hormonal control of protein synthesis at the translational level. *Colloq. Ges. Physiol. Chem.* **18,** 107–120.

Tomkins, G. M., Gelehrter, T. D., Granner, D., Martin, D., Samuels, H. H., and Thompson, E. B. (1969). Control of specific gene expression in higher organisms. *Science* **166,** 1474–1480.

Trachewsky, D., and Segal, S. J. (1968). Differential synthesis of ribonucleic acid in uterine nuclei: evidence for selective gene transcription induced by estrogens. *Eur. J. Biochem.* **4,** 279–285.

Trams, G., Engel, B., Lehmann, F., and Maass, H. (1973). Specific binding of oestradiol in human uterine tissue. *Acta Endocrinol. (Copenhagen)* **72,** 351–360.

Tsai, M.-J., Towle, H. C., Harris, S. E., and O'Malley, B. W. (1976). Effect of estrogen on gene expression in the chick oviduct. Comparative aspects of RNA chain initiation in chromatin using homologous *versus Escherichia coli* RNA polymerase. *J. Biol. Chem.* **251,** 1960–1968.

Turkington, R. W., and Hill, R. L. (1969). Lactose synthetase: progesterone inhibition of the induction of α-lactalbumin. *Science* **163,** 1458–1460.

Turner, C. W. (1939). "The Comparative Anatomy of the Mammary Glands." University Cooperative Store, Columbia, Missouri.

Ui, H., and Mueller, G. C. (1963). The role of RNA synthesis in early estrogen action. *Proc. Natl. Acad. Sci. U.S.A.* **50,** 256–260.

Ungar, F., Dorfman, R. I., Stecher, R. M., and Vignos, P. J. (1951). Metabolism of the steroid hormones, metabolism of progesterone and related steroids. *Endocrinology* **49,** 440–448.

van der Molen, H. J., and Aakvaag, A. (1967). Progesterone. *In* "Hormones in Blood" (C. H. Gray and A. L. Bacharach, eds.), 2nd rev. ed., Vol. 2, pp. 221–303. Academic Press, New York.

Villee, C. A., and Hagerman, D. D. (1953). Effects of estradiol on the metabolism of human placenta *in vitro. J. Biol. Chem.* **205,** 873–882.

Villee, C. A., and Hagerman, D. D. (1958). On the identity of the estrogen-sensitive enzyme of human placenta. *J. Biol. Chem.* **233,** 43–48.

Villee, C. A., Hagerman, D. D., and Joel, P. B. (1960). An enzymatic basis for the physiologic functions of estrogens. *Recent Prog. Horm. Res.* **16,** 46–69.

Vonderhaar, B., and Mueller, G. C. (1969). Binding of estrogen receptor to estradiol immobilized on insoluble resins. *Biochim. Biophys. Acta* **176,** 626–631.

Vonderhaar, B. K., Kim, U. H., and Mueller, G. C. (1970a). The heterogeneity of soluble estrogen receptors from rat uteri and their modification by temperature, imidazole compounds and estradiol. *Biochim. Biophys. Acta* **208,** 517–527.

Vonderhaar, B. K., Kim, U. H., and Mueller, G. C. (1970b). The subunit character of soluble estrogen receptors from rat uteri and their modification *in vitro. Biochim. Biophys. Acta* **215,** 125–133.

Walaas, O. (1952). Effect of oestrogens on the glycogen content of the rat uterus. *Acta Endocrinol. (Copenhagen)* **10,** 175–192.

Webster, R. A., Pikler, G. M., and Spelsberg, T. C. (1976). Nuclear binding of progesterone in hen oviduct. Role of acidic chromatin proteins in high-affinity binding. *Biochem. J.* **156,** 409–418.

Weissmann, G. (1965). Studies of lysosomes. VI. The effect of neutral steroids and bile acids on lysosomes *in vitro. Biochem. Pharmacol.* **14,** 525–535.

Weissmann, G., Sessa, G., and Weissmann, S. (1966). The action of steroids and triton X-100 upon phospholipid/cholesterol structures. *Biochem. Pharmacol.* **15,** 1537–1551.

Westphal, U. (1958). Diffusion of progesterone and desoxycorticosterone into nitrocellulose (Lusteroid): a potential source of error in ultracentrifugation. *J. Lab. Clin. Med.* **51,** 473–478.

Westphal, U. (1971). "Steroid–Protein Interactions," Monogr. Endocrinol., Vol. 4. Springer-Verlag, Berlin and New York.

Westphal, U., and Ashley, B. D. (1959). Steroid-protein interactions. VI. Stereochemical aspects of interaction between Δ^4-3-ketosteroids and human serum albumin. *J. Biol. Chem.* **234,** 2847–2851.

Wichmann, K. (1967). On the metabolism and subcellular distribution of progesterone in the myometrium of the pregnant rat. *Acta Endocrinol. (Copenhagen), Suppl.* **116,** 3–98.

Wichmann, K., Luukkainen, T., and Adlercreutz, H. (1967). The distribution of injected C^{14}-progesterone in the subcellular fractions of rat myometrium before and after parturition. *Ann. Med. Exp. Biol. Fenn.* **45,** 393–398.

Wiest, W. G. (1963a). *In vitro* metabolism of progesterone and 20α-hydroxypregn-4-en-3-one by tissues of the female rat. *Endocrinology* **73,** 310–316.

Wiest, W. G. (1963b). Extrahepatic metabolism of progesterone in pseudopregnant rats. Identification of reduction products. *J. Biol. Chem.* **238,** 94–99.

Wiest, W. G. (1969). Progesterone interactions in the rat uterus. *Ciba Found. Study Group* **34,** 56–72.

Williams, D. J., and Rabin, B. R. (1969). The effects of aflatoxin B and steroid hormones on polysome binding to microsomal membranes as measured by the activity of an enzyme catalysing disulphide interchange. *FEBS Lett.,* **4,** 103–107.

Williams-Ashman, H. G., and Liao, S. (1964). Sex hormones and hydrogen transport by isolated enzyme systems. *In* "Actions of Hormones on Molecular Processes" (G. Litwack and D. Kritchevsky, eds.), pp. 482–508. Wiley, New York.

Williams-Ashman, H. G., Liao, S., Hancock, R. L., Jurkowitz, L., and Silverman, D. A. (1964). Testicular hormones and the synthesis of ribonucleic acids and proteins in the prostate gland. *Recent Prog. Horm. Res.* **20,** 247–292.

Willmer, E. N. (1961). Steroids and cell surface. *Biol. Rev. Cambridge Philos. Soc.* **36,** 368–398.

Wilson, E. W., and King, R. J. B. (1969). *In-vitro* stimulation of RNA synthesis by oestradiol-17β. *J. Endocrinol.* **43,** xl-xli.

Wilson, J. D. (1963). The nature of the RNA response to estradiol administration by the uterus of the rat. *Proc. Natl. Acad. Sci. U.S.A.* **50,** 93–100.

Wilson, J. D., and Walker, J. D. (1969). The conversion of testosterone to 5α-androstan-17β-ol-3-one (dihydrotestosterone) by skin slices of man. *J. Clin. Invest.* **48,** 371–379.

Woo, S. L. C., Rosen, J. M., Liarakos, C. D., Choi, Y. C., Busch, H., Means, A. R., O'Malley, B. W., and Robberson, D. L. (1975). Physical and chemical characterization of purified ovalbumin messenger RNA. *J. Biol. Chem.* **250,** 7027–7039.

Woo, S. L. C., Smith, R. G., Means, A. R., and O'Malley, B. W. (1976). The ovalbumin gene. Partial purification of the coding strand. *J. Biol. Chem.* **251,** 3868–3874.

Wyss, R. H., Karsznia, R., Heinricks, W. Le R., and Herrmann, W. L. (1968). Inhibition of uterine receptor binding of estradiol by anti-oestrogens (clomiphene and CL-868). *J. Clin. Endocrinol. Metab.* **28,** 1824–1828.

Yokoyama, A., Shinde, Y., and Ôta, K. (1969). Endocrine control of changes in lactose content of the mammary gland in rats shortly before and after parturition. *In* "Lactogenesis: the Initiation of Milk Secretion at Parturition" (M. Reynolds and S. J. Folley, eds.), pp. 65–71. Univ. of Pennsylvania Press, Philadelphia.

Yu, J.Y.-L., Campbell, L. D., and Marquardt, R. R. (1971). Sex hormone control mechanisms. I. Effect of estrogen and progesterone on major cellular components in chicken (*Gallus domesticus*) oviducts. *Can. J. Biochem.* **49,** 348–356.

Zander, J. (1959). Gestagens in human pregnancy. *In* "Recent Progress in the Endocrinology of Reproduction" (C. W. Lloyd, ed.), pp. 255–277. Academic Press, New York.

Zarrow, M. X. (1961). Gestation. *In* "Sex and Internal Secretions" (W. C. Young, ed.), 3rd ed., pp. 958–1031. Williams & Wilkins, Baltimore, Maryland.

Zarrow, M. X. (1965). The biological profile of progesterone and a consideration of the bioassay of progestogens. *Horm. Steroids; Biochem., Pharmacol., Ther., Proc. Int. Congr., 1st, 1962* Vol. 2, pp. 239–252.

Zarrow, M. X., Shoger, R. L., and Lazo-Wasem, E. A. (1954). The rate of disappearance of exogenous progesterone from the blood. *J. Clin. Endocrinol. Metab.* **14,** 645–652.

Zimbelman, R. G., Lauderdale, J. W., Sokolowski, J. H., and Schalk, T. G. (1970). Safety and pharmacologic evaluations of melengestrol acetate in cattle and other animals: a review. *J. Am. Vet. Med. Assoc.* **157,** 1528–1536.

Zimmering, P. E., Lieberman, S., and Erlanger, B. F. (1967). Binding of steroids to steroid-specific antibodies. *Biochemistry* **6,** 154–164.

3

The Physiological Effects of Estrogens and Progesterone

C. A. Finn and Janet E. Booth

I. INTRODUCTION

Ovarian hormones have widespread effects on the tissues of the body. For convenience these can be divided into effects on the peripheral target organs, where the final stages of reproduction occur, and effects on the central target organs, especially the hypothalamic–pituitary system. The influence of the ovarian hormones on the latter principally concerns the synchronization of ovarian activity and the behavioral responses of the animal. One of the biggest problems of research in this field is the distinction between the pharmacological effects of large amounts of hormones and the physiological effects of smaller, and perhaps normal, amounts (see Boling, 1969).

II. EFFECTS ON PERIPHERAL TARGET ORGANS

A. Introduction

Differentiation of the Müllerian duct system in the embryo into the adult female duct system, comprising the uterine tubes, uterus, and vagina (see Volume II, Chapter 2), appears to be largely independent of the ovarian hormones (Burns, 1961). Postnatally, however, the ovary influences the growth and function of the reproductive tract, and removal of the ovaries results in its atrophy. Atrophy also occurs after hypophysectomy or after damage to the hypothalamus. The atrophic state is reversible; growth and function can be reinstated by the administration of the ovarian hormones, estrogens and progesterone. Through them the ovary, and indirectly the pituitary and brain, initiate and synchronize the complex changes in the reproductive tract which take place during reproductive life.

In this section, the effects of estrogens and progesterone on the uterine tubes, uterus, and vagina of the adult mammal are considered. An attempt is made also to show how the hormones control the functions of these organs in such a way that the reproductive processes are synchronized.

B. Uterine Tubes

In the adult mammal the epithelial cells lining the uterine tubes (also known as the oviducts or Fallopian tubes) are tall and columnar with three basic cell types: ciliated cells, secretory cells, and peg cells (Hadek, 1955; Nilsson and Reinius, 1969). The latter are thought to be depleted secretory cells. A connective tissue stroma lies between the epithelium and the outermost muscle layer. Within the Fallopian tubes spermatozoa are capacitated, fertilization of ova occurs, and the initial growth and development of the zygotes takes place. During their passage along the tube, the eggs are dividing and growing, probably using nutrients from the tubal secretions (Hartman, 1939). The time taken for the eggs to reach the uterus is very critical, premature subjection of the eggs to the uterine environment being detrimental to their survival (McLaren and Michie, 1956; Noyes and Dickmann, 1960).

Ovarian hormones influence the uterine tubes in several ways. They regulate the differentiation and growth of cilia (Flerkó, 1954), the secretory activity of the epithelium (Hadek, 1955), and the motor activity of the muscles (Corner, 1923; Boling, 1969). After ovariectomy, the epithelial cells become flattened, cilia are lost, and other signs of secretory activity, such as the Golgi apparatus and endoplasmic reticulum, are reduced (Nilsson and

Reinius, 1969). Estrogen causes an increase in the rate of secretion (Bishop, 1956; Mastroianni *et al.,* 1961; Greenwald, 1969; McDonald and Bellvé, 1969) and alters its composition (Holmdahl and Mastroianni, 1965). The importance of estrogen in motor activity is doubtful. Many workers have shown that there is an increased activity of the musculature after estrogen treatment. Boling (1969), however, believes that the doses used were, in most cases, far higher than the assumed normal physiological level and he has demonstrated that, if low physiological doses of estrogen are given, the musculature remains quiescent and only becomes active after the cessation of hormone treatment. He has shown also that administration of progesterone after estrogen initiates motility. This would explain the great activity of the tubal musculature which occurs after the peak of estrogen secretion at the time of ovulation and the fact that ova reach the uterus even if ovariectomy is performed soon after mating (Alden, 1942; Adams, 1958; Wu *et al.,* 1971).

The effect of ovarian hormones on the transport of ova has been studied in several species of laboratory mammals. High doses of estrogens have usually been used, and should probably be considered above the normal physiological level. The widespread use of the ovarian hormones as contraceptive drugs makes the results of such experiments of considerable importance.

To summarize the results from many laboratories it appears that, depending upon the dose and the time of administration, exogenous estrogen given to mated intact females affects egg transport, either by accelerating the passage of the ova into the uterus or by retaining the ova in the part of the oviduct between the ampulla and isthmus—the so-called tube-locking effect (Noyes *et al.,* 1959; Greenwald, 1961; Deanesly, 1963; Chang, 1966; Humphrey and Martin, 1968; Blandau, 1969). Greenwald (1967) showed that, in guinea pigs, rabbits, and hamsters, low doses of estrogen accelerated tubal transport while high doses caused tube locking. In the rat, tube locking never occurred, and in the mouse the minimum dose to produce both effects was similar. Humphrey (1968), working on the mouse, has shown that a dose of estradiol which has a tube-locking effect when given on day 1 of pregnancy will cause acceleration of tubal transport when given on days 2 or 3.

The influence of progesterone is uncertain. In the rabbit, Greenwald (1961) was unable to demonstrate a clear-cut effect but the results suggested an accelerating effect on egg transport. Harper (1966) and Chang (1966) concluded that progestins slow egg transport in the rabbit oviduct. There is little effect in the mouse (Humphrey, 1968), and large doses of progesterone given during the period of ovum transport have no effect on subsequent implantation (Martin, 1965; Finn and Martin, 1971). Some recent work by

Wu *et al.* (1971) has shown that progesterone on day 3 is necessary for normal egg transport in mice hypophysectomized on day 1. Hypophysectomy on day 2 does not interfere with normal egg transport and, thus, the pituitary appears to cause progesterone secretion at about day 3. At present, it is difficult to reconcile this observation with the finding (noted above) that ovariectomy does not interfere with egg transport.

C. Uterus

The functions of the endometrium and myometrium are regulated by the ovarian hormones. The activity of the ovary during pregnancy and the extent to which its endocrine functions can be effected by the placenta are discussed in Chapter 8 of Volume II. This chapter focuses on the cellular changes in the uterus in response to estrogens and progesterone.

1. Endometrium

The endometrium is primarily involved in the attachment and nutrition of the developing embryo and in the formation of the maternal placenta. The luminal surface is lined by a single layer of columnar epithelial cells from which glands, similarly lined by epithelial cells, pass into the stroma, which is separated from the epithelium by a basement membrane. The stromal cells are predominantly fibroblasts, together with variable numbers of white blood cells. In some species the fibroblasts can differentiate into a specialized cell, the decidual cell, which plays an important part in implantation (De Feo, 1967; Finn and Porter, 1975).

During implantation, the blastocyst attaches to the endometrium, attachment occurring at a well-defined time for each species. At this time, the cells of the endometrium can respond to the presence of the blastocyst (McLaren, 1970), but they remain in a fully receptive state for only a short period before becoming insensitive (De Feo, 1963; Psychoyos, 1963; Yochim and De Feo, 1963; Meyers, 1970; Finn and Martin, 1974). The endometrium then regresses and has to go through another cycle of change before it becomes sensitive again. For successful pregnancy, uterine sensitivity must coincide with the arrival in the uterus of the zygote which has reached at least the blastocyst stage (Adams, 1970).

Cyclic development of endometrial sensitivity is synchronized with ovulation and ovum transport through the changing levels of ovarian hormones during the estrous cycle. In most mammals the uterus goes through a phase of sensitivity during every estrous cycle regardless of the presence of a male. In others (e.g., mouse and rat), the cycle is abbreviated in the absence of copulation and full uterine preparation occurs only after mating; thus, the animal may become pregnant or, if the mating is infertile, pseudopregnant. The pseudopregnant cycle in the latter class of animals is comparable with the estrous cycle of others and the menstrual cycle of primates. A further modification of this cyclic pattern occurs in a few animals (e.g., stoat, lactating rat) in which the uterus remains partly prepared and the blastocyst does not attach and remains dormant but viable (Greenwald, 1958; Enders, 1967). This state of delayed implantation in the lactating rat may be terminated by ovarian hormones which induce full sensitivity (Krehbiel, 1941; Weichert, 1942). Termination of suckling normally provides the cue for implantation in rats and mice, but no environmental stimuli have yet been identified to explain cessation of the implantation delay in species like the stoat and badger.

Estimates of the rate and amount of secretion of the ovarian hormones during the estrous cycle and during pregnancy (Brown, 1955; Shaikh and Abraham, 1969; Somerville, 1971; Cox *et al.*, 1971; Hotchkiss *et al.*, 1971; Henricks *et al.*, 1972) and indirect biological estimates (Barnea *et al.*, 1968; Finn and Martin, 1969) show that, just before ovulation, relatively high levels of estrogen are secreted and reach a maximum before ovulation. After this estrogen peak, there appears to be, at least in some species, some secretion of progesterone (Lindner and Zmigrod, 1967; Hashimoto *et al.*, 1968), which may be necessary for behavioral estrus. At the time of estrus and for a day or two afterwards, neither hormone is secreted in significant quantities. The corpus luteum then forms and progesterone is secreted in larger amounts. Small quantities of estrogen are also secreted, and, at least in the rat and mouse, these two hormones together, acting on an endometrium primed by estrogen secreted before ovulation (Finn and Martin, 1969, 1974), bring about the changes necessary for implantation. In some species both hormones appear to be necessary for the maintenance of pregnancy. The amounts secreted increase until the end of gestation when the relationship between the two hormons again becomes critical in the control of parturition (see Finn and Porter, 1975; also Volume II, Chapter 8).

a. The Effect of Estrogens. Most studies of the effects of the sex hormones have been on rodents. These provide the framework for the following account, although, when possible, reference is made to work on other species, especially primates.

After ovariectomy the uterus atrophies and the blood supply is much reduced. The epithelium is thin with a smooth, deeply eosinophilic surface. Where the cells are ciliated (e.g., rabbits) the cilia tend to disappear (Larsen, 1962). The cytoplasm stains only lightly and there is usually some subnuclear vacuolation. In material fixed in osmic acid, these vesicles are seen to contain lipid. The small nuclei do not contain prominent granules, and are regularly aligned, resulting in prominent bands of supra- and subnu-clear cytoplasm.

Changes which are characteristic of estrus take place in the uterus after the administration of estrogen. Growth of the endometrium occurs a few hours after injection. The initial increase in weight of the uterus is due to intercellular edema in the stroma (Carroll, 1945). In some rodents fluid also collects in the uterine lumen. The response occurs within 4 to 6 hours and forms the basis of a rapid assay for estrogens (Astwood, 1938; Cole, 1950). The fluid which collects in the lumen differs from ascites fluid in having a lower protein and higher mucopolysaccharide content. It is considered, therefore, to be a secretion of the epithelial cells (Homberger *et al.*, 1963). The edema is thought to be due, at least in part, to an increase in permeability of the blood vessels of the endometrium (Hechter *et al.*, 1941; Kalman, 1955). Using the electron microscope, it has been shown that pores appear in the walls of the capillaries, soon after estrogen administration, through which pass plasma proteins and fluid (Ham *et al.*, 1970). Estrogens also cause changes in the connective tissue of the stroma and large quantities of acid sulfomucopolysaccharides are produced (Joseph *et al.*, 1954; Zachariae, 1958). It has been suggested that there is a disaggregation of a mucoprotein matrix with uptake of water (Moses and Catchpole, 1955). Similar changes have been shown to be responsible for the swelling of the sexual skin of some primates (Moses and Catchpole, 1955).

Vascular changes are a very important aspect of sex-hormone action. Considerable variation in blood flow occurs in the uterus (Krichesky, 1943; Kalman, 1958; Dickson *et al.*, 1969; Bindon, 1969; Greiss and Anderson, 1969), mostly due to the growth and destruction of blood vessels. Variation in blood flow is especially pronounced during the menstrual cycle of primates (Daron, 1936; Prill and Gotz, 1961).

Coincidental with the vascular effects, changes take place in the epithelial cells of the uterus. These have been described comprehensively from light microscope studies (Burrows, 1949) and more recently from electron microscope pictures. The epithelial cells lining the lumen and glands increase in size, especially in height (Nilsson, 1958a,b,c). This increase can be demonstrated as early as 4 hours after an estrogen injection. The initial increase is due to uptake of water by the cells, probably for hydration of the cytoplasmic proteins, a necessary preliminary to cell growth in interphase (Alfert and Bern, 1951; Schultz *et al.*, 1969). The nuclei increase in size and the cytoplasm stains darkly due to greater numbers of ribosomes. The lipid droplets disappear (Gillman, 1940; Alden, 1947), and the mitochondria, endoplasmic reticulum, and Golgi apparatus increase in size (Horning, 1942–1944; Elftman, 1958, 1963). The microvilli on the luminal surface increase in number and length and are covered with a substance which Nilsson (1959) refers to as "uterine substance." This is probably analogous to the fuzzy coat found on intestinal cells. It is considered to be part of the glycocalyx (Ito, 1969) and is formed in the Golgi apparatus whence it passes to the luminal surface (Bennet and Leblond, 1970). The hormonal control of this cell coating is very interesting in view of the importance of the epithelial cell surface for attachment of the blastocyst.

The oxygen consumption of the uterus rises significantly in response to estrogens (Khayyal and Scott, 1931) and many enzymes become active (Barker and Warren, 1966; Dugan et al., 1968). Alkaline phosphatase was one of the first of these enzymes to be investigated (Atkinson and Elftman, 1947; Atkinson and Engle, 1947; Arzac and Blanchet, 1948; Sani and Hanau, 1952; Moss et al., 1954; Colville, 1968). There is very little of the enzyme in the uterus of the ovariectomized animal, but after estrogen treatment it occurs in considerable quantities on the epithelial luminal surface and in the glands (Figs. 1 and 2), perhaps associated with the "fuzz" described earlier. Other uterine enzymes which have been shown to be controlled by the ovarian hormones include carbonic anhydrase (Lutwak-Mann, 1955; Bialy and Pincus, 1962; Ogawa and Pincus, 1962), β-glucuronidase (Harris and Cohen, 1951; Fishman and Farmelant, 1953), peptidase (Albers et al., 1961), and many enzymes involved in carbohydrate metabolism (Barker and Warren, 1966; Murdoch and White, 1968; Wilson, 1969).

After estrogen treatment or during the follicular stage of the cycle, the epithelial cells undergo proliferation (Allen et al., 1937; Schmidt, 1943; Bensley, 1951; Martin and Finn, 1968). Using colchicine to arrest mitosis, maximum division of the luminal and glandular epithelial cells is found 24 hours and 72 hours respectively after estrogen injection (Finn and Martin, 1971). DNA replication in epithelial cells, indicated by uptake of [^{3}H]thymidine, takes place before cell division at about 14 hours after estrogen administration (Beato et al., 1968).

The duration of estrogen secretion before ovulation varies considerably in different species, ranging from about 2 days in rodents to about 5 days in man. The changes induced in the uterus by estrogens secreted at this time are important in preparing the organ for the subsequent response to progesterone. It has been known for some time that the typical glandular response of the rabbit uterus to progesterone is not produced unless the animals are primed with estrogen before progesterone administration. More recently it has been demonstrated that priming by estrogen secreted before ovulation plays a part in inducing maximum sensitivity to a decidual stimulus about 100 hours after mating in the mouse (Finn and Martin, 1970).

Uterine changes at estrus are probably also necessary for the successful transport of spermatozoa to the oviduct, their capacitation (Chang, 1970), and removal of excess spermatozoa after fertilization.

b. The Effect of Progesterone. As discussed above, progesterone is usually secreted following a period of estrogen secretion, and although progesterone may act simultaneously with estrogen, it is difficult to demonstrate an endometrial response to progesterone alone.

The main response of the endometrium to progesterone, which forms the basis for a common assay for the hormone, is the development of uterine glands in the rabbit (Corner, 1928; Clauberg, 1930; McPhail, 1934) which gives the characteristic "lace like" appearance to the endometrium first described by Bouin and Ancel (1910). This reaction does not occur unless the animals have been primed with estrogen. In man and other primates, proliferation of the gland cells seems to be brought about by estrogens, and the glands are then converted to a secretory condition by progesterone or progesterone and estrogen acting together (Engle and Smith, 1938; Everett, 1962; Kohorn and Tchao, 1969). In the mouse, proliferation of the cells of the glands occurs during the period following estrogen stimulation, that is, on day 3 of pregnancy or 3 days after priming, and the administration of progesterone during this time will prevent mitosis in the glandular epithelium. Progesterone treatment after mitosis has occurred stimulates the glands to secrete, although the secretion is very limited unless a small quantity of estrogen is also given (Finn and Martin, 1971). The chemical nature of the secretion at this time has not been fully determined because of the difficulty of collection. It is PAS-positive and contains mucopolysaccharide (Austad and Garm, 1959; Heap and Lamming, 1962, 1963). It appears to differ from the uterine secretion at the time of estrus (Heap, 1962); there is significantly less potassium and a higher sodium to potassium ratio (Clemetson *et al.,* 1970). In the human endometrium during the menstrual cycle, the glandular secretion during the follicular phase consists mainly of sulfomucins, whereas during the luteal phase carboxymucins predominate (Hester *et al.,* 1970). The secretion, sometimes called uterine milk, is thought in many species to be a nutritional substrate for the blastocyst (Amoroso, 1952; Beier, 1968a,b).

In addition to stimulating secretion of the uterine glands, progesterone is a factor in the differentiation of the endometrium in preparation for implantation of the ovum. There are considerable differences between species in the extent to which the endometrium, after being primed by estrogen, can be prepared for the reception of the blastocyst by progesterone alone (Marcus and Shelesnyak, 1970; Finn and Porter, 1975). In the rat, mouse, and lemming, additional estrogen is necessary during the period of progesterone activity for full uterine sensitivity to the blastocysts

(Mayer, 1959; Psychoyos, 1967; Aldeen and Finn, 1970), whereas in the rabbit (Marcus *et al.*, 1964; Rennie and Davies, 1965), guinea pig (Deanesly, 1960), and hamster (Orsini and Meyer, 1959), estrogen does not appear to be essential for nidation. The situation in man is unknown, although estimates of estrogen in the urine during the menstrual cycle show that, in addition to the high levels of estrogen secreted during the follicular phase, there is a secondary peak in the luteal phase (Brown, 1955). Similarly Cox *et al.* (1971) have shown that a secondary rise in estrogen secretion 3 to 4 days after ovulation occurs in sheep, and this they suggest may be involved in the development of the endometrium.

During the period between ovulation and implantation characteristic changes take place in the uterus. These have been studied extensively in the mouse, in which the normal changes have been reproduced artificially by appropriate hormone treatment. After treatment with estrogen the lumen becomes dilated and very irregular in outline (Fig. 1a). After 3 or 4 days of treatment with progesterone, it becomes slit shaped, with the slit orientated in a mesometrial–antimesometrial direction (Fig. 1b). The epithelium is thick and has a smooth, deeply eosinophilic surface. The cytoplasm stains palely and there is extensive subnuclear vacuolation, caused by lipid accumulation. The nuclei are small, without enlarged nucleoli and granules. Later, the surface of the epithelium becomes raised into wave-like projections (Martin *et al.*, 1970). A similar change occurs in pregnant mice on days 5 and 6 of pregnancy, and has been termed "corrugation" (Finn and McLaren, 1967). Fixation of uteri in glutaraldehyde showed that the lumen is completely closed at this time (Fig. 2), and the corrugation seen in Bouin-fixed sections is presumably due to shrinkage and separation of the apposed uterine surfaces. Nilsson (1966a,b) has suggested that closure of the lumen is due to increased adhesiveness of the epithelial surface and that it is important for attachment of the blastocyst.

The closure reaction has been studied in detail only in the mouse, although it has been demonstrated with the light microscope in the rat and guinea pig (Marcus, 1974). Under the electron microscope, two stages of closure are apparent. In the first stage, the microvilli from apposing surfaces, which under the influence of progesterone have lost their "fuzz" of glycocalyx (Fig. 3), become closely interlocked (Fig. 4a). In the second stage, the epithelial surfaces alter and cell membranes from opposite cells become closely apposed, obliterating the uterine lumen (Fig. 4b). It is not possible to differentiate between the first and second stage of closure under the light microscope, and progesterone alone cannot bring about the fully differentiated second stage. Progesterone is sufficient for differentiation to the first stage, which can be maintained for some time as during delay of implantation (Enders, 1967), but estrogen is required for full differentiation of the uterine epithelial cells in preparation for attachment of the blastocyst.

The responsiveness of the endometrium can be tested by the ability of the uterus to implant a blastocyst or to undergo a decidual reaction to an artificial stimulus such as an intraluminal injection of oil (Finn and Keen, 1962) or trauma (Loeb, 1908). Using either technique, it has been demonstrated that in rats and mice there is a peak in uterine sensitivity, after progesterone and estrogen treatment. This is maintained for a short time, but if an implantation stimulus is not applied during peak sensitivity the uterus becomes refractory and does not become sensitive again until after another hormonal cycle (Psychoyos, 1963; Finn, 1966). Little is known about the onset and decline of sensitivity in species which do not appear to require

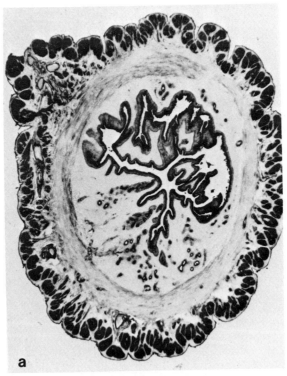

a

Fig. 1.   Cross section of mouse uterus (a) at estrus showing the open, irregular lumen and the presence of alkaline phosphatase in the epithelial cells of the lumen and glands (reproduced by permission from Finn, 1970) and (b) on day 5 of pseudopregnancy showing the slit-shaped lumen and the disappearance of alkaline phosphatase from the luminal cells.

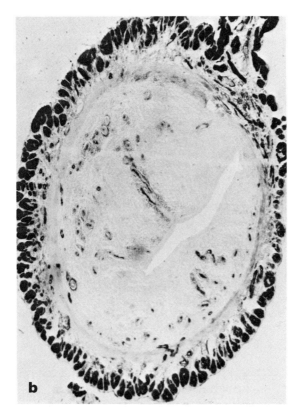

Fig. 1. (Continued)

estrogen during the luteal phase. Possibly estrogen secreted before ovulation plays a determining role in the timing of sensitivity and implantation. Even in the mouse, priming of the uterus by estrogen affects the degree of sensitivity obtained (Finn and Martin, 1970), although the time of onset of sensitivity is determined by estrogen secreted during the luteal phase.

In a few species (rabbit: Larsen, 1962; ferret: Buchanan, 1966), the epithelium shows a further modification at the time of implantation and multinucleated cells (rabbit) or symplasmic masses of cells (ferret) are formed. These are present in areas of the uterus not in contact with a blastocyst, and their formation may be hormonally controlled, although symplasma formation in some other species requires the stimulus of a blastocyst.

Progesterone also inhibits the mitotic response of the uterine epithelial cells to estrogen. This effect may be apparent after only 1 day but is most noticeable after 3 days treatment with progesterone (Martin and Finn,

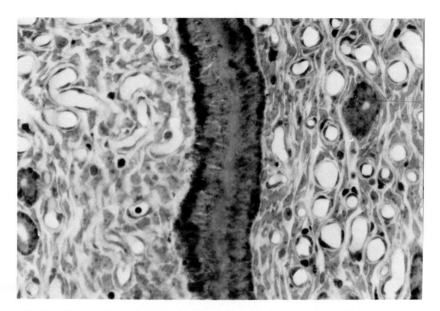

Fig. 2.   Section of a mouse uterus, fixed by perfusion with glutaraldehyde, showing closure of the lumen after treatment with progesterone and estrogen. (Reproduced by permission from Finn, 1970.)

1969). It appears that estrogen can stimulate the epithelial cells to divide or to differentiate in preparation for attachment of the blastocyst. In the absence of progesterone, division of the epithelial cells invariably occurs. After progesterone treatment, small doses of estrogen initiate differentiation, shown morphologically by the onset of the second stage of closure. Larger doses of estrogen stimulate division, in spite of progesterone treatment, and in consequence produce insensitivity to a decidual stimulus (Martin and Finn, 1968).

The effect of progesterone on the blood vessels of the uterus has been studied intensively in primates, in which it has been shown that although estrogen alone stimulates development of the coiled arteries, full development only follows progesterone treatment (Phelps, 1946). Full development of the blood vessels of the endometrium in the rat (Williams, 1948) and rabbit (Gillman, 1941) also depends on progesterone as well as estrogen. Menstruation in primates occurs after differentiation of the endometrium in response to the ovarian hormones. It appears to be initiated by the fall in blood levels of these hormones at the end of the cycle (Krohn, 1949). The exact cause of the bleeding is not known. The first stage is regression of the progestational development in the endometrium, especially of the connective tissue elements in the stroma, with sloughing of tissue and bleeding from the denuded arteries (Hisaw and Hisaw, 1961). At one time it was thought that ischemia of the spiral arteries played an important part in the onset of menstruation. However, New World monkeys do not possess spiral arteries but do menstruate. Menstrual bleeding can be induced experimentally in primates by the administration of progesterone or estrogen for several days and then stopping treatment. Bleeding follows more quickly and is more intense following the cessation of progesterone (Hisaw, 1942; Eckstein, 1949). The administration of estrogen and progesterone produces a well-developed progestational endometrium in about 10 days, but if this treatment is continued for a longer period, the endometrium becomes atrophic (Hisaw et al., 1937); the luminal epithelium is lost and the glands

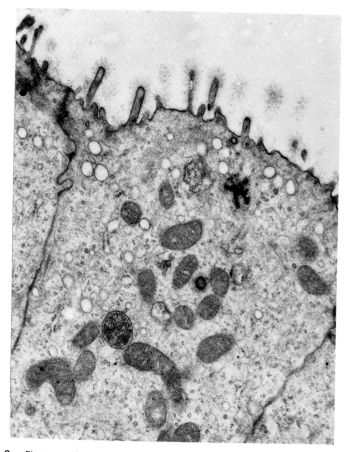

Fig. 3. Electron micrograph of the surface of a luminal epithelial cell showing the "fuzz" on the microvilli.

and coiled arteries disappear. However, when progesterone treatment is discontinued, bleeding takes place from the uterus in spite of the atrophic condition and the loss of the coiled arteries. Withdrawal of estrogen while progesterone treatment continues does not precipitate bleeding (Hisaw, 1950).

Progesterone has a pronounced effect on the cells of the endometrial stroma (Hooker, 1945). This is reflected by the initial change in the nucleus and was utilized by Hooker and Forbes (1947) for a biological assay of progesterone. The uterine stromal cells of ovariectomized mice have fusiform, dense nuclei with clumped chromatin and no visible nucleoli. After the administration of progesterone, the nuclei become plump and oval shaped with conspicuous nucleoli and fine, evenly dispersed chromatin.

Cell division in the stroma in response to ovarian hormones has been studied mainly in the ovariectomized mouse. The administration of progesterone alone or estrogen alone results in few mitotic figures appearing in the stroma, even after treatment with colchicine. However, if a small

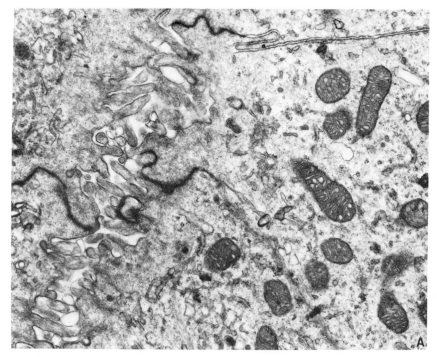

Fig. 4.   Electron micrographs of mouse uterus: (A) in the first stage of luminal closure; (B) in the second stage of closure. (Pictures by Mrs. Jennifer Downie.)

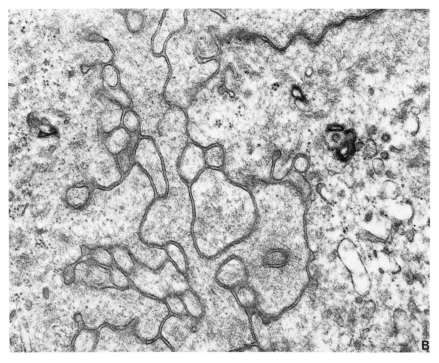

Fig. 4. (Continued)

quantity of estrogen is given after 3 or 4 days treatment with progesterone, large numbers of stromal cells undergo division with very little division of the epithelial cells. This pattern of cell division is very different from that in the endometrium of an ovariectomized animal not treated with progesterone but given the same dose of estradiol. In these conditions mitosis is confined to the epithelium. Thus, progesterone, although by itself unable to stimulate significant stromal cell division, modifies the response of the endometrium to a subsequent dose of estrogen, inhibiting an epithelial response but predisposing to a stromal response (Martin and Finn, 1968). A further factor which affects stromal cell division is priming by estrogen before the start of progesterone treatments. Thus, the administration of a single dose of progesterone or of progesterone plus estradiol to unprimed ovariectomized mice does not stimulate stromal cell division, whereas a single dose given within 6 days of estrogen priming elicits stromal mitosis, with maximum response on the fourth or fifth days after the end of priming. By contrast, in the unprimed ovariectomized animal the stromal cells do not start dividing unless progesterone has been administered for at least 3 days, followed by a low dose of estrogen.

In the pregnant animal there is an abrupt change in endometrial cell division, from being almost entirely epithelial on day 3 to mostly stromal on day 4 (Finn and Martin, 1967). It is likely that the high levels of estrogen secreted during proestrus together with the progesterone and estrogen secreted during the luteal phase following mating all play a part in eliciting this pattern of cell division in the endometrium. The relevance of stromal cell division to the physiological events at implantation is not known. The increase in mitosis the day before implantation could be necessary for the subsequent transformation of stromal fibroblasts into decidual cells at implantation. It may be necessary for the stromal cells to undergo mitosis before they can differentiate, as has been suggested in the mammary gland (Stockdale and Topper, 1966), or the increased rate of mitosis may provide a bank of cells for further differentiation.

Following the burst of cell division in the stroma in response to estrogen in the progesterone-treated mouse, the stroma becomes refractory to a further dose of estrogen for at least 48 hours. In these circumstances it is the epithelial cells lining the lumen that divide (Finn *et al.,* 1969). This refractoriness of the stroma may be related to the insensitivity to a decidual stimulus discussed above (p. 160). After implantation in those species which exhibit a decidual reaction, progesterone is necessary for the growth and maintenance of the decidua (Yochim and De Feo, 1962). Even in the presence of continuous and adequate supplies of progesterone, artificially induced decidual cells have a limited life span of only about 11 days in rats (Atkinson, 1944) and 7 days in mice (Finn and Porter, 1975). This is different from the situation in some primates in which continuous progesterone and estrogen treatment causes the development and maintenance of a large decidual reaction in the stroma (Hisaw and Hisaw, 1961).

The control of the cellular activity of the endometrium is clearly a very complex process, with estrogen and progesterone interacting together in a temporal pattern to produce integrated changes in the various tissues for the successful attachment of the blastocyst and formation of the placenta (Finn and Martin, 1974).

2. The Myometrium

Csapo (1969) postulates four primary factors in the control of myometrial function: uterine volume, oxytocin, estrogen, and progesterone. Of these, the gonadal hormones appear to have a modulating influence on the stimulating effect of the others. Like the cells of the endometrium, the muscle cells of the uterus are to a large extent controlled by the ovarian hormones. The myometrial cells vary in size according to the physiological state of the animal. They are small in the castrated female and large during late pregnancy (Stieve, 1929; Mark, 1956; Dessouky, 1968, 1969), and there are significant differences between animals in diestrus and estrus (Ross and Klebanoff, 1967). The principal physiological event in which the myometrial changes are implicated is parturition, but myometrial activity is also involved in sperm transport and in menstruation.

Estrogen promotes synthesis of actomyosin (Csapo, 1950; Csapo and Corner, 1953; Michael and Schofield, 1969), of glycogen (Boettiger, 1946; Fitch, 1963), and of glycogen synthetase (Bo *et al.*, 1967). In the estrogen-treated myometrial cell, there are more glycogen particles, free ribosomes, endoplasmic reticulum, and a more extensive Golgi complex than in that of an ovariectomized control animal (Ross and Klebanoff, 1967). Treatment with progesterone causes the myometrial cells to enlarge and the myofilaments to become more abundant and prominent. The number of mitochondria is increased by both estrogen and progesterone. When both hormones are given simultaneously they act synergistically rather than antagonistically (Bo *et al.*, 1969).

The contractility of the myometrium is influenced by estrogen and progesterone (Corner, 1923; Frank *et al.*, 1925; Finn and Porter, 1975). Estrogen is required for normal contractility in response to changes in volume of luminal contents or to oxytocin, and estrogen deficiency results in failure of excitability of the uterus. Progesterone has an inhibitory effect on uterine contractility. According to the progesterone-block theory of Csapo (1956), progesterone prevents response of the uterus to oxytocin or volume changes by blocking the propagation of the impulse along the muscle (Kuriyama, 1961).

The response of the uterus to sympathetic nervous stimulation is also affected by sex hormones. In an animal under the influence of estrogen, hypogastric nerve excitation is stimulatory and this is evidently due to the dominance of the α-receptors. In the progesterone-treated animal, the β-receptors are dominant and hypogastric nerve excitation is inhibitory (Marshall, 1969).

3. Utero-ovarian Relationships

There appears to be a functional interrelationship between the uterus and ovaries, related to the transient life of the corpus luteum. This was first shown by Loeb (1923) although the significance of his experiments remained unrealized until recently (Wiltbank and Casida, 1956; Rowlands, 1961; Perry and Rowlands, 1961). Many recent reviews have considered the luteolytic activities of the uterus (Bland and Donovan, 1966; Melampy and Anderson, 1968; Anderson *et al.*, 1969; Caldwell *et al.*, 1969; Caldwell,

1970; Donovan, 1971). The way progesterone is supplied during the period a conceptus is in the uterus varies from species to species. Some obtain sufficient progesterone from extraovarian sources, usually the placenta (see Amoroso and Finn, 1962; also Volume II, Chapter 8), whilst in others the corpus luteum of the cycle is maintained and may be increased in size or activity during gestation. The uterus is known to participate in the control of the development and regression of the corpus luteum in the pig, sheep, cow, guinea pig, and hamster (see Melampy and Anderson, 1968).

Surgical removal of the uterus in the cyclic sow, cow, guinea pig, and ewe, or the pseudopregnant rat or hamster, leads to the prolongation of the corpus luteum for periods which are well beyond the normal duration of metestrus, and may be for as long as normal gestation. Destruction of the endometrium, for example, by the instillation of irritant or corrosive substances into the uterine lumen, has a similar effect (Butcher *et al.,* 1962). Experiments involving unilateral or partial hysterectomy have demonstrated that the effect of the uterus on the ovary is, at least to some extent, a local one, for when one horn is removed the corpus luteum on the ipsilateral side is maintained while that in the contralateral ovary regresses (Bland and Donovan, 1966; Moor and Rowson, 1966a). Anderson *et al.* (1961) have shown that in the pig only a quarter of the uterus is required for the corpora lutea to regress normally at the end of the ovarian cycle. The inference from these experiments is that the uterus secretes a substance which has a luteolytic effect (Anderson, 1968; Mazer and Wright, 1968). Confirmation that a chemical substance is involved has been obtained by experiments involving homografts of uterine tissue in hysterectomized animals. When the grafts survived with a functional endometrium, luteal regression occurred and normal estrous cycles recommenced, although in many cases the length of metestrus was extended. The chemical substance concerned is thought to be prostaglandin $F_{2\alpha}$ (Donovan, 1971; Pharriss, 1971), although its mode of action is at present unknown. There is also evidence that estrogens may be involved in the production or release of the luteolytic substance by the uterus (Bland and Donovan, 1968, 1970).

It, thus, seems that the intact uterus plays an important part in the cyclic regression of the corpus luteum in the pig, ewe, cow, and guinea pig and in the regression of the corpus luteum of pseudopregnancy in the hamster and rat, but not in the mouse (Dewar, 1973). It would, therefore, be necessary for the luteolytic activity of the uterus to be suppressed in normal pregnancies. This is presumably effected by the young embryos which counteract the luteolytic effect of the uterus. In a series of ingenious experiments involving transfer of ova into sheep at various stages of the estrous cycle, Moor and Rowson (1966b) have shown that the presence of the embryo is essential on the twelfth to thirteenth day after estrus for maintenance of the corpora lutea. Intrauterine infusions of frozen and thawed homogenates of 14-day-old sheep embryos, maintained sheep corpora lutea, whereas similar of 25-day-old sheep embryos, 14-day-old pig embryos, or serum and white blood cells were ineffective, as were 14-day-old sheep embryo homogenates when infused into extrauterine sites. These experiments demonstrate that the early sheep embryo has a specific action on the luteolytic mechanism of the uterus, and that this is critical in maintaining a functional corpus luteum until the placenta can take over the production of progesterone (see Volume II, Chapter 8).

D. Vagina

Cornification of the surface layer of epithelial cells in response to estrogens (Morau, 1889; Stockard and Papanicolaou, 1917) and an increase in the glycogen content of the cells (Biggers, 1953) are the most obvious effects of ovarian hormones on the vagina. The response to exogenous hormones occurs with either systemic or local administration. Cornified cells are easily demonstrated in smears taken from the vagina of treated rodents, and the response, although quantal in nature, is the basis of a very simple and convenient assay for estrogens (Emmens, 1950).

Cornification of the vaginal cells is the last of a chain of reactions to estrogens, and it begins about 48 hours after injection of the hormone. Soon after the administration of estrogen there is an increase in the RNA content of the cells. Correlated with this is a change in the structure of the nucleolus (Pollard *et al.*, 1966) that can be seen as early as 1 hour after intravaginal application of the hormone. The nucleoli of the stratum germinativum increase in size and develop large, deeply staining intranucleolar bodies. The number of mitochondria then increases (Pollard *et al.*, 1966) as does the metabolic rate of the cells. The latter can be demonstrated manometrically or indirectly by the use of 2-3-5-triphenyltetrazolium chloride (Martin, 1960). Martin (1960) has shown that estrogen increases metabolic rate in a dose-dependent manner and has adapted the reaction as a very sensitive assay for the hormone. The tetrazolium reaction occurs mainly in the basal layers of the epithelium. The DNA content of the vaginal cells, as shown by uptake of [^{3}H]thymidine and quantitative Feulgen cytophotometria, increases approximately 7 hours after the administration of estrogen (Beato *et al.*, 1968). This is in preparation for cell division, which is at a maximum at 24 hours (Allen *et al.*, 1937; Biggers and Claringbold, 1955; Martin and Claringbold, 1960). Cornification is probably a result of increased cell division, rather than a direct effect of estrogen. Biggers and Claringbold (1955) have made a detailed histological and histochemical study of the mouse vagina after estrogen administration. In the resting condition, there are three layers of epithelial cells. About 18 hours after the injection of estrogen, mitosis starts in the basal layer and several new layers of cells are produced which push the original outer layer further from the germinal layer. Keratinization then takes place in the outer of the new layers, leaving the original outer layer uncornified on the surface (Long and Evans, 1922). This layer is mucified and stains positively with PAS. Eventually all the new growth is shed. The fact that the original outer layer does not cornify indicates that cornification is not simply a result of distance of the cells from the germinal layer but of changes in the cell itself. The time sequence for the changes in the mouse vagina in response to estrogens is shown in Fig. 5.

The number of cells responding to estrogen by mitosis is dose-dependent, and this response, like the tetrazolium reaction, provides the basis for a better assay for estrogen than does cornification, especially for testing the effect of antiestrogens. For example, progesterone, vitamin A, and testosterone prevent cornification of the vagina but do not block mitosis. It is likely, therefore, that they are not true antagonists of estrogen but rather have a modifying effect on the secondary response to estrogen. Progesterone and testosterone cause mucification of the vagina (Klein, 1937) and increased production of sialic acids when given at the same time as estrogen (Carlborg, 1966), thus producing the vaginal picture typical of pregnancy.

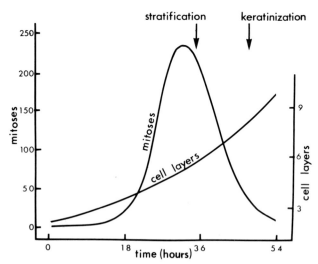

Fig. 5.   Graph showing the variation in the number of cells in mitosis and cell layers in the vagina at various times after the intravaginal injection of estrone. The beginning of stratification and keratinization is indicated. (After Biggers and Claringbold, 1955.)

In species such as the guinea pig, the vagina is closed in the absence of hormonal stimulation (Jagiello, 1965). Deanesly (1966) has shown that the vagina of the ovariectomized guinea pig will open if estrogen is given for 3 to 4 days in sufficient quantity. Earlier workers had claimed that a small amount of progesterone was necessary to obtain vaginal opening, but Deanesly's results refute this and show that progesterone when given in combination with estrogen causes mucification. Estrogen also causes opening of the vagina at puberty in other rodents, and premature opening can be brought about by injections of estrogen.

Another recently reported action of estrogen on the vagina is the stimulation of a male-attracting pheromone in the rhesus monkey. Michael and Keverne (1970) have demonstrated that ovariectomized monkeys given estrogen secrete a substance in the vagina which stimulates the mounting reaction in males.

III. CENTRAL EFFECTS

A. Introduction

As well as acting on their target organs in the reproductive tract and the mammary glands, ovarian hormones can also influence the function of the anterior pituitary gland (adenohypophysis). It is also held that estrogen and progesterone act on the hypothalamus, the part of the brain to which the pituitary gland is attached, and that the hypothalamus is an integrating and controlling center for pituitary function (Everett, 1961, 1964, 1972; Bogdanove, 1964; Sawyer, 1964; Flerkó, 1966; Szentágothai et al., 1972; Martini et al., 1968; Davidson, 1969). The hypothalamus is a central part of a complex central nervous system which includes the midbrain, the limbic forebrain, the amygdala, hippocampus, and septum, and it is widely held that it controls the anterior pituitary via the hypophysial portal blood system. The general belief is that these vessels carry chemotransmitters or releasing factors from the hypothalamus to the anterior pituitary, and that these chemical substances control the secretion of the pituitary tropic hormones. Besides affecting hypothalamic activity, estrogen and progesterone have been reported to affect the activities of the limbic system and midbrain (Komisaruk, 1971), the appearance of the ependymal cells lining the third ventricle (Knowles, 1971), and possibly also the activity of the pineal gland (Wurtman et al., 1968). This review, however, will be concerned only with the way in which it is supposed that estrogen and progesterone act on the hypothalamus and anterior pituitary to control gonadotropin secretion.

(The correlation of ovarian function with the activities of the hypothalamus and pituitary is discussed in Volume III, Chapter 4; and the influence of ovarian steroids on behavior is discussed in Volume II, Chapter 9.)

The first section reviews studies of the structure and the synthetic and secretory activities of the pituitary; and the second section deals with the evidence that estrogen and progesterone act on the hypothalamus.

B. Studies on the Anterior Pituitary

It is a well-established fact that the ovaries and adenohypophysis interact (Moore and Price, 1932) and that the adenohypophysis secretes two gonadotropins, luteinizing hormone (LH), and follicle-stimulating hormone (FSH), as well as prolactin (Greep, 1973). The interaction of ovarian hormones with the adenohypophysis has been investigated by studying pituitary cytology and by measuring pituitary and circulating LH and FSH in animals under different hormonal conditions. It is of interest to know whether estrogen and progesterone act on more than one type of cell in the anterior pituitary, that is, whether LH and FSH are produced by the same or different cells under all conditions. The balance of evidence today is that both LH and FSH are secreted by the same cells, and that ovarian steroids do have a direct action on the secretory elements of the adenohypophysis and so directly influence gonadotropin secretion.

In several species, including the rat (Purves and Griesbach, 1951), bat (Herlant, 1956), dog (Purves and Griesbach, 1957), and rabbit (Kanematsu and Sawyer, 1963b), two types of basophilic gonadotropin-secreting cells, or gonadotrops, have been distinguished by differential staining techniques from other basophilic cells in the anterior pituitary (Reese *et al.*, 1943; Purves, 1961). These cells may correspond to the respective sources of LH and FSH. Purves and Griesbach (1954) found that when female rats are treated with 250 μg testosterone daily, the centrally located gonadotrops become fewer, while the peripheral gonadotrops remain unaffected. Greep and Chester Jones (1950) had shown that testosterone treatment produces a pituitary gland rich in FSH but lacking in LH. Purves and Griesbach accordingly concluded that the peripheral gonadotrops are concerned with FSH and the more central ones with LH secretion. Further studies (see Herlant, 1964, for review) support this conclusion, and it is now clear that alterations in the levels of ovarian steroids cause differential changes in the LH- and FSH-producing cells of the anterior pituitary. These changes can be seen most strikingly in bats from temperate regions (Herlant, 1956). After mating, the female hibernates and ovulation and fertilization are delayed. During this period "LH-type" basophils are almost completely absent from the anterior pituitary, while the "FSH-type" basophils are numerous and show clear evidence of secretory activity.

Recently, however, Nakane (1970, 1971) has questioned the one cell–one hormone idea. Using peroxidase-labeled antibodies he has shown that, in the rat anterior pituitary, gonadotropin-producing cells may contain both LH and FSH. Most gonadotrops in the center of the gland contained FSH, the remainder containing LH, whereas the peripheral cells contained both LH and FSH. It has also been noted that gonadotropin-secreting cells are intermingled with prolactin-secreting cells, and this fact too may be of physiological significance. The balance of evidence from recent cytological studies therefore suggests that LH and FSH can be secreted by the same cells. Moreover there is evidence that changes in circulating gonadal steroids can lead to discrete and reproducible alterations in the histological appearance of the adenohypophysis.

It has long been known that ovariectomy, which implies the removal of the influence of ovarian steroids, results in the development of "castration cells" in the adenohypophysis, associated with increased gonadotropin secretion, and that the systemic administration of estrogen restores the normal picture (Bogdanove, 1963a). Thus, Bogdanove (1963b) was able to prevent the development of castration cells in rats ovariectomized 2 to $2\frac{1}{2}$ weeks before by the implantation of minute pellets approximately 0.5 mm in diameter and 0.5 mm long containing estradiol dipropionate and choles-terol in a ratio of less than one part estrogen to 2000 parts cholesterol. Pellets containing one part estradiol dipropionate to 1000 parts cholesterol were ineffective when implanted into the hypothalamus. Similar, but subcutaneous implants of diluted estradiol diproprionate also had no effect on pituitary histology, although some vaginal stimulation was noted with higher estradiol:cholesterol ratios. Lisk (1963) also reported inhibition of castration cells in ovariectomized rats with intrapituitary estrogen implants. In both this study and that of Bogdanove (1963b), the inhibition was limited to a small area around the implant. These results indicate a direct effect of estrogen on pituitary structure and might also be presumed to indicate altered function, although no assays for gonadotropins were carried out. Gersten and Baker (1970) have used the peroxidase-labeled antibody tech-nique to examine the effect of implanted estradiol on different anterior pituitary cell types in the rat. Prolactin and gonadotropin cells in the immediate vicinity of the implant were affected; hypertrophy of prolactin cells occurred and LH cells were reduced in size whereas corticotropin cells were unaffected. The authors concluded that this provided evidence for a direct feedback of estradiol on the anterior pituitary, stimulating prolactin and inhibiting LH secretion. This confirms the increased prolactin secretion measured in rabbits and rats with estrogen implants in the anterior pituitary (Kanematsu and Sawyer, 1963a; Ramirez and McCann, 1964).

Kanematsu and Sawyer (1963b) apparently failed to prevent the development of castration cells in the rabbit by implanting stainless steel tubes containing less than 0.1 mg estradiol benzoate into the anterior pituitary at the time of ovariectomy. This failure could be explained by the small area of contact between the estrogen and the pituitary tissue, and the consequent poor distribution of the hormone throughout the gland (Bogdanove, 1963b).

It is clear that the development of castration cells is associated with an increase in the gonadotropic potency of the anterior pituitary gland and of circulating gonadotropins (Greep and Chester Jones, 1950; Chester Jones and Ball, 1962). Specific quantitative and sensitive assay methods have revealed differences between species in the pituitary response to ovariectomy and differences between the nature and resulting level of gonadotropins.

Bogdanove *et al.* (1971), working in three different laboratories, compared radioimmunoassay methods and five variations of the ovarian ascorbic acid depletion bioassay for LH. Pituitary and serum LH levels in gonadectomized rats were measured by bioassay and immunoassay methods. Good agreement between the immuno- and bioassays was found for pituitary LH, but the bioasssay tended to overestimate serum levels of this gonadotropin. Since even the high serum LH and FSH levels found in the gonadectomized rat are at the lower limit of sensitivity of the bioassay method, data on blood values of gonadotropins based on bioassays are not as reliable as those based on immunoassays.

Using quantitative bioassay methods, Parlow (1964b,c) found a tenfold increase over preoperative levels of LH and FSH in the pituitaries of female rats ovariectomized 50 days before. Similarly, in the sow, Parlow *et al.* (1964) found increases in pituitary LH and FSH 25 to 90 days after ovariectomy. Bioassay studies demonstrated an increase in pituitary FSH after ovariectomy in the ewe (McDonald and Clegg, 1966; Arendarcik and Maracek, 1968). Specific radioimmunoassays have shown that pituitary LH also increases following ovariectomy in this species (Roche *et al.*, 1970). Although pituitary FSH increases greatly in ovariectomized mice, pituitary LH does not change (Parlow, 1964a). In the guinea pig, a slight post-ovariectomy rise in pituitary FSH is not accompanied by any change in LH (Parlow and Hendrich, 1970). Swanson *et al.* (1971) reported that, 60 days after heifers were ovariectomized, pituitary FSH had doubled, but that pituitary LH had declined. Larger amounts of FSH than LH were found in the pituitaries of postmenopausal women (Swerdloff and Odell, 1968). This confirms earlier bioassay data on gonadotropin levels in urine (Hamburger and Johnson, 1957; Keller, 1966).

Assays of pituitary extracts, therefore, confirm that ovariectomy leads to an increase in pituitary gonadotropin content. In some of the species studied both LH and FSH may be increased, whereas in others only FSH increases and pituitary LH may even decline.

The degree of change in plasma levels of LH and FSH following ovariectomy also varies between species. While neither LH nor FSH could be detected in the plasma of the ovariectomized guinea pig (Parlow and Hendrich, 1970), in the mouse, plasma FSH increased following ovariectomy, but plasma LH remained undetectable by bioassay (Parlow, 1964a). With the introduction of sensitive radioimmunoassays, it has been possible to follow changes in blood levels of LH in individual animals after ovariectomy. In the ewe, serum levels of LH increase after ovariectomy (Roche *et al.,* 1970); in rats, Gay and Midgley (1969) and Blackwell and Amoss (1971) found that they increase slowly after removal of the ovaries and begin to "plateau" after 30 to 35 days. In the hamster, increases in serum LH can be detected within $3\frac{1}{2}$ hours after ovariectomy, but later the levels rarely exceed those found during the preovulatory LH surge at proestrus (Goldman and Porter, 1970). In the sow, serum LH is increased 24 to 52 days after ovariectomy, but the levels are not significantly higher than those found on the day of estrus (Rayford *et al.,* 1971). In the heifer, plasma LH increased within 3 days after ovariectomy, leveled off at 30 days, and remained at double the control level (Swanson *et al.,* 1971). In the rhesus monkey, Atkinson *et al.* (1970) found that plasma LH levels increase tenfold within 3 weeks of ovariectomy. Perhaps of more significance than these reported interspecific variations in changes of plasma LH levels after ovariectomy is the fact that in ovariectomized rats (Gay and Midgley, 1969), sheep (Roche *et al.,* 1970), monkeys (Atkinson *et al.,* 1970; Dierschke *et al.,* 1970), cattle (Swanson *et al.,* 1971), and women (Albert and Mendoza, 1966; Orr *et al.,* 1970), serum LH levels fluctuate considerably at all times. A regular rhythm of LH secretion with peaks every hour has been observed in the ovariectomized rhesus monkey (Dierschke *et al.,* 1970), while Lawton and Smith (1970) have reported circadian rhythms in circulating LH in rats which had been ovariectomized 5 months earlier. These observations indicate that changes can occur in the rate of release and/or synthesis of LH in the absence of ovarian steroids and that statements about interspecific differences in serum LH after ovariectomy need to be viewed with caution.

Measurements of pituitary and circulating LH and FSH in ovariectomized animals following estrogen and progesterone replacement therapy should help to reveal the part played by different steroids in controlling the synthesis, storage, and release of each gonadotropin. The early experimental data on this subject have been reviewed previously (Greep, 1961; Chester Jones and Ball, 1962; van Rees, 1964). More sensitive and specific assays have since been used (Swerdloff and Odell, 1968; Monroe *et al.*, 1968; Niswender *et al.*, 1969) and with these technical improvements it is becoming possible to understand the situations in which ovarian steroids will stimulate or inhibit gonadotropin synthesis and release.

It has been reported that the levels of circulating LH in adult ovariectomized rats, as measured by bioassay, fall within 24 hours of single injections of large doses (20 to 50 μg) of estradiol benzoate (McCann and Taleisnik, 1961; Caligaris *et al.*, 1971a), and within 3 days of smaller doses (0.1 μg or 0.2 μg/100 gm body weight) (McCann and Taleisnik, 1961; Barraclough and Haller, 1970). More sensitive radioimmunoassays have since shown that single small doses of estradiol benzoate (0.2 to 5.0 μg) reduce plasma LH and FSH within 24 hours in adult ovariectomized rats (Ajika *et al.*, 1972). The fall in circulating gonadotropins was associated with an increase in pituitary LH and FSH within 48 hours of the estrogen injection. This result suggests that these doses of estradiol benzoate reduced the release of gonadotropins more than it did their synthesis. Using bioassay methods, Parlow (1964c) found that the rise in circulating and pituitary LH was prevented by daily doses of 0.1 to 0.4 μg estradiol in ovariectomized immature rats, whereas the rise in FSH was prevented by not less than 2 μg estradiol daily. This indicates that LH is more sensitive than FSH to the inhibitory effects of estradiol on their respective synthesis and release.

Other experiments in which estrogen has been administered in a different regimen have yielded contrary results. Gonadotropin secretion was observed by Hohlweg (1934) to be stimulated in intact immature rats, and Johnson (1971) later found that a single injection of 1 μg estradiol benzoate given to such animals increased plasma and pituitary FSH within 24 hours. In the intact adult rat, a fall in pituitary LH is accompanied by a rise in plasma LH after 7 daily injections of estradiol, 0.06 μg/100 gm body weight (Callantine *et al.*, 1966). A single injection of estradiol benzoate can increase plasma LH and FSH in the intact anestrous ewe (Goding *et al.*, 1969; Beck *et al.*, 1973; Reeves *et al.*, 1974). In the intact rhesus monkey, LH and FSH surges, similar to those observed at spontaneous ovulation, can be induced during the follicular phase of the menstrual cycle by sustaining a raised level of plasma estrogen for 12 hours or more (Yamaji *et al.*, 1971; Dierschke *et al.*, 1973). Small amounts of estrogen can also stimulate the secretion of LH from explanted rat pituitaries *in vitro* (Piacsek and Meites, 1966; Schneider and McCann, 1970). Exposure of the pituitaries to larger amounts of estrogen inhibited LH release (Schneider and McCann, 1970).

Estrogen can also stimulate gonadotropin release in ovariectomized animals. Pelletier and Signoret (1969) and Radford *et al.* (1969) found that an injection of estradiol benzoate induced an increase in plasma LH within 12 to 36 hours in ovariectomized ewes which had been previously treated with progesterone. A single dose of estrogen can stimulate a surge of LH secretion in ovariectomized rats and rhesus monkeys in which the characteristically high levels of circulating gonadotropins have been suppressed by previous estrogen treatment (Swelheim, 1965; Caligaris *et al.*, 1971a; Karsch *et al.*, 1973a). These results show that a positive feedback effect of estrogen on LH and FSH release is superimposed on, and independent of, its negative feedback on gonadotropin secretion. They also support the hypothesis that the ovulatory surge of gonadotropins in the intact animal is partly triggered by increased estradiol levels. Increases in plasma estradiol have been shown to precede the initiation of the LH surge in the human menstrual cycle (Brown *et al.*, 1958; Burger *et al.*, 1968; Baird and Guevara, 1969; Corker *et al.*, 1969; Abraham *et al.*, 1972), the menstrual cycles of the baboon and rhesus monkey (V. C. Stevens *et al.*, 1970; Hotchkiss *et al.*, 1971), and the estrous cycles of the sow (Raeside, 1963), ewe (Norman *et al.*, 1968; Cox *et al.*, 1971), cow (Mellin and Erb, 1966), and rat (Hori *et al.*, 1968; Yoshinaga *et al.*, 1969; Brown-Grant *et al.*, 1970). Ferin *et al.* (1969, 1974) inhibited ovulation in rats and rhesus monkeys by using antibodies raised to estradiol, coupled to bovine serum albumin, and Shirley *et al.* (1968) and Labhsetwar (1970a,b) have shown similar effects with synthetic antiestrogens.

Progesterone is a relatively poor inhibitor of LH release in the ovariectomized animal (Rothchild, 1965; Caligaris *et al.*, 1971b). A single injection of 25 mg progesterone caused only a slight fall in plasma LH measured by bioassay in the ovariectomized rat (McCann, 1962), but using the radioimmunoassay technique, Kalra *et al.* (1973a) found that a single injection of only 5 or 10 mg reduced circulating LH, and 10 mg also reduced FSH. Daily treatment of ovariectomized rats with 2 mg progesterone decreased pituitary FSH after 5 days (Martini *et al.*, 1970). Yamaji *et al.* (1972) reported that progesterone had no effect on plasma LH in ovariectomized rhesus monkeys, although it can inhibit an estrogen-induced LH and FSH surge in the intact animal (Dierschke *et al.*, 1973). In the intact and ovariectomized sheep, McDonald and Clegg (1967) found that daily treatment with 5 to 40 mg progesterone did not significantly depress blood levels of LH, but 4 or 5 days after treatment had ceased, plasma LH was increased. Daily progesterone treatment can prevent ovulation in the rat, but LH release and ovulation follow its withdrawal (Hoffmann and Schwartz, 1965). A combination of estrogen and progesterone provides more effective inhibition of LH and FSH secretion than either steroid alone in both the ovariectomized rat (Blake *et al.,* 1972) and rhesus monkey (Karsch *et al.,* 1973b). Unfortunately, the physiological significance of these findings remains obscure because they cannot be explained in terms of the hormonal changes known to take place during the cycle (Knobil, 1974).

Like estrogen, progesterone can stimulate LH secretion in the ovariectomized rat (Caligaris *et al.,* 1971b) and can advance ovulation when given at certain times of the estrous cycle (Everett, 1948; Nallar *et al.,* 1966; Zeilmaker, 1966; McDonald and Gilmore, 1971a). This effect is distinct from the rise in circulating LH which follows progesterone withdrawal, and depends on prior exposure to estrogen (Brown-Grant, 1969; Caligaris *et al.,* 1971b). In most species the evidence is against an obligatory role for progesterone in the natural induction of ovulation. There is no preovulatory increase in circulating progesterone in the estrous cycles of either the ewe, the cow, or the sow (Moore *et al.,* 1969; Kazama and Hansel, 1970; Henricks *et al.,* 1972) or in the human or rhesus monkey menstrual cycles (Johansson and Wide, 1969; Krey *et al.,* 1973). In the rat, progesterone secretion by the ovaries does not increase until after the ovulatory LH surge has started (Piacsek *et al.,* 1971), and ovulation is not prevented by an injection of antiserum to progesterone on the morning of proestrus. Progesterone of adrenal origin is similarly not essential for ovulation to occur in the rat (Brown-Grant, 1972). Nevertheless, it still remains possible that preovulatory secretion of progesterone or of other ovarian progestins such as 20α-hydroxyprogesterone in the rabbit and rat, and 17α-hydroxyprogesterone in primates may sustain and reinforce the LH surge once it has started (Hilliard *et al.,* 1967; Brown-Grant, 1972; Swerdloff *et al.,* 1972). Progesterone can also stimulate FSH synthesis and release in the rat. Daily progesterone injections can increase pituitary FSH in intact female rats (van Rees, 1959). This response may not depend on the presence of estrogen, for a single injection of 25 mg progesterone can increase pituitary and circulating FSH in ovariectomized rats (Blake *et al.,* 1972; Kalra *et al.,* 1973a). However, it has been reported that progesterone treatment during the late luteal phase and early follicular phase of the menstrual cycle does not alter FSH secretion in intact rhesus monkeys (Resko *et al.,* 1974).

In summary, the levels of LH and FSH secretion are affected by estrogen and progesterone. Removal of the influence of these steroids by ovariectomy leads to increased gonadotropin secretion and release, but species differ in the magnitude of this response and in the relative increases in LH and FSH. In the absence of ovarian steroids, gonadotropin secretion appears not to be at a constant high level, but to fluctuate somewhat. After replacement therapy with large doses of estrogen, secretion patterns are more constant, and the levels of pituitary and circulating LH and FSH are reduced. The responses vary in accordance with the intensity and duration of the estrogenic stimulus. Suitably spaced small doses of estrogen, on the other hand, may stimulate LH and FSH secretion, and this facilitatory effect of estrogen on gonadotropin secretion is independent of the reduction of gonadotropin levels. Progesterone by itself is a poor inhibitor of LH secretion in the ovariectomized animal, but estrogen and progesterone combined inhibit gonadotropin secretion more effectively than either steroid alone. Prolonged progesterone treatment can inhibit gonadotropin secretion and ovulation in the intact animal, but a single injection can stimulate LH release in the estrogen-primed ovariectomized rat and FSH synthesis and release in the absence of estrogen. Some of these actions of estrogen and progesterone can be correlated with the pattern of steroid and gonadotropin secretion during the ovulatory cycle.

C. Studies on the Hypothalamus

The concept that the ovaries and adenohypophysis interact is qualified by the knowledge that the central nervous system is included in the feedback circuit. Interest now focuses on the question of whether all the changes in pituitary gonadotropin synthesis and secretion described in the previous section are directly due to the action of gonadal steroids on the gland or partly indirectly through their actions on specific brain cells. Thus, gonadal steroids could act on pituitary gonadotrops to alter gonadotropin release either directly by altering their secretory activity or by causing changes in their sensitivity to hypothalamic gonadotropin-releasing factors. They could also act on the brain to cause changes in the discharge of hypothalamic gonadotropin-releasing factors. This might be reflected by changes in their concentration in the hypothalamic nuclei concerned and by changes in the electrical activity of the latter. The evidence seems to suggest that steroids act both at the hypothalamus and pituitary. Such a dual system might allow for more subtle control than either process alone.

The possible effects of estrogen and progesterone on hypothalamic nuclei have been investigated in a number of ways: by implanting the solid hormones in selected areas of the brain and pituitary; by measuring the content of gonadotropin-releasing activity in the hypothalamus and the response of the anterior pituitary to extracts of hypothalamus, to synthetic releasing hormone, or to electrical stimulation of the hypothalamus; by identifying the brain cells which bind estrogen and progesterone; and by recording the activities of one or several hypothalamic neurons. This section considers the adequacy of these techniques and the results they have yielded.

1. Local Application of Hormones

Experiments in which estrogen and progesterone have been applied directly to the hypothalamus or anterior pituitary are difficult to interpret, and it is impossible to draw definite conclusions from the results.

First, the hormones have to be inserted into the brain as pellets, or implanted in stainless steel cannulae, either as pure steroids or diluted with cocoa butter or cholesterol. Precise localization in specific nuclei is all but impossible, and other neural tissue is inevitably damaged. The effects observed may therefore be nonspecific. Alterations in gonadotropin secretion reported after the hypothalamus has been damaged might also merely reflect the results of altering blood flow and causing hypothalamic ischemia or of changed nutritional status of the animal following damage to nerve pathways controlling food intake. A further difficulty relates to the question of controls. Empty cannulae, or cannulae or pellets containing an inert substance can be inserted, but such experiments usually involve different animals, and it is extremely difficult to replicate the precise anatomical conditions which affect minute areas of the brain in different animals. The positions of implants may differ sufficiently to account for variable or inconsistent results.

Second, the tissue surrounding the hormone implant is probably exposed to unphysiologically high levels of the steroid, often for long periods, and information about the diffusion of the steroid from the implantation site is usually lacking. To some extent this difficulty can be overcome by testing the effects of steroid implants elsewhere in the brain. The issue of local diffusion is particularly relevant to experiments in which steroids are implanted in the median eminence. Such implants might have a direct effect; or they might diffuse into the hypophysial portal vessels and so be carried to the anterior pituitary on which they might also exert a direct effect (Bogdanove, 1964); or they might diffuse into systemic vessels. A steroid substance implanted in the median eminence could in short be distributed throughout the anterior pituitary more effectively than the same substance implanted in the gland itself. Thus, when implants in the median eminence, but not the anterior pituitary, are reported to alter gonadotropin secretion, the obvious conclusion, that the steroid is having an action on the median eminence, has to be viewed with caution.

An experiment by Palka et al. (1966) illustrates the impossibility of deciding whether the changes in gonadotropin secretion which follow the local application of estrogen to the median eminence or the anterior pituitary are due to direct actions of the steroid on the release of LH from the gonadotrops, to changes in gonadotrop sensitivity to hypothalamic gonadotropin-releasing factors, or to changes in the secretion of gonadotropin-releasing factors. Stainless steel tubes containing 1–2 μg of radioactively labeled estradiol were implanted into one side of the median eminence or anterior pituitary of intact female rats. In rats with estradiol implanted in the median eminence, plasma LH was found to be elevated 4 days later but not in those with implants in the anterior pituitary. Since a significant amount of radioactivity was present in the anterior pituitary after both procedures, the effect of the median eminence implants on LH secretion could have been due to any of the three causes listed above, while the lack of effect of intrapituitary implants could have been due to poor distribution of the steroid. When labeled estradiol was left in the median eminence or anterior pituitary for 18 days, plasma LH was undetectable in both groups of rats, all of which had significant levels of radioactivity in their pituitaries. Once again, there is no way of telling by which mechanism the hormone brought about its effects.

The inhibitory action of hypothalamic estrogen implants on gonadotropin secretion has been assessed by the reduction of ovarian and uterine weights in intact animals, by the prevention of ovarian compensatory hypertrophy in unilaterally ovariectomized animals, and by measurement of pituitary and plasma LH in intact and ovariectomized animals. For the reasons discussed above, it is never logically possible to decide whether the inhibition that is observed is due to a reduction in pituitary sensitivity to gonadotropin-releasing factors or to the direct inhibition of gonadotropin synthesis and release. Flerkó and Szentágothai (1957) showed that ovarian tissue grafted in rats, close to the paraventricular nuclei of the hypothalamus, caused a decrease in uterine weight, while grafts in the mammillary area and anterior pituitary failed to have this effect. This result does not indicate precisely where the ovarian hormones might have been acting, because substances diffusing into the third ventricle could have easily reached the pituitary by way of the median eminence. Many other workers have reported that the local application of estradiol, but not control substances, to the median eminence–arcuate region of intact rats, rabbits, and sheep caused reductions in ovarian and uterine weights (Lisk, 1960; Davidson and Sawyer, 1961; Kanematsu and Sawyer, 1963a,b,c, 1964; Chowers and McCann, 1967; Przékop and Domański, 1970). Similarly, in the unilaterally ovariectomized rat, compensatory hypertrophy of the remaining ovary was prevented by estrogen implants in the anterior hypothalamus or mammillary complex (Littlejohn and de Groot, 1963) as well as in the median eminence (Fendler and Endroczi, 1965–1966), whereas compensatory hypertrophy still occurred when the implants were made in the preoptic area or the lateral hypothalamus (Littlejohn and de Groot, 1963). In intact rabbits, autopsied 8 weeks after implantation of tubes containing estrogen into the median eminence, pituitary LH was lower than in control animals implanted with empty tubes; estrogen implants elsewhere in the brain or in the anterior pituitary were not found to affect pituitary LH levels (Kanematsu and Sawyer, 1963c). The reduction in pituitary LH by estrogen implants in the median eminence probably reflects reduced LH release into the circulation, for Kawakami *et al.* (1969) implanted estrogen into the arcuate nucleus of rabbits and 3 weeks later found that ovarian steroid biosynthesis was markedly reduced. In both rabbits and rats, implantation of estradiol into the median eminence on the day of ovariectomy is said to inhibit the rise in plasma and pituitary LH which normally follows this operation (Kanematsu and Sawyer, 1964; Ramirez *et al.*, 1964), whereas estradiol implants elsewhere in the brain failed to block this response. None of these experiments, however, excludes the possibility of an action of estrogen on the pituitary. Equally, direct implantation of estrogen into the anterior pituitary resulted in different changes in pituitary and plasma LH in rabbits and rats. This may have been due to differences in the extent to which the anterior pituitary was exposed to estrogen. Ramirez *et al.* (1964) implanted a 27-gauge (0.4-mm diameter) stainless steel tube containing estradiol diluted with 13 parts of cholesterol on each side of the anterior pituitary in rats which were ovariectomized on the same day. Two weeks later, the postovariectomy rise in plasma LH was inhibited, but not the rise in pituitary LH. In similar experiments in rabbits, a single 24-gauge (0.56-mm diameter) tube containing undiluted estradiol benzoate was implanted into the anterior pituitary, and the animals killed 8 weeks later (Kanematsu and Sawyer, 1964). The postovariectomy increase of pituitary LH was inhibited, but not that of plasma LH. It is clearly impossible to draw any firm conclusions from such experiments about the mode of action of estrogen implants in the median eminence of ovariectomized animals.

In the experiments described above, concerning the inhibitory action of estrogen on gonadotropin secretion, relatively large amounts of estrogen had been implanted into the hypothalamus or anterior pituitary for long periods, often for weeks at a time. To study the facilitatory action of estrogen on gonadotropin secretion, Weick and Davidson (1970) developed a modified implantation technique by which the pituitary could be exposed to steroids for only a few hours. This "double cannula" technique has made it possible to show that exposure of the anterior pituitary to estrogen for only 4 hours can stimulate gonadotropin release, but unfortunately again the technique does not make it possible to say what causes the response. The method involves the prior implantation, under stereotaxic guidance, of a stainless steel outer cannula of 1.22-mm diameter, closed with a cap, into the anterior pituitaries of female rats. After the rats had undergone three consecutive 5-day estrous cycles and had become used to manipulation of the cap of the outer cannula, the cap was removed on the second day of diestrus and, without anesthesia, an inner cannula (0.3-mm diameter) containing pure estradiol benzoate or cholesterol was inserted. Exposure of the pituitary to the steroid or cholesterol could be terminated at any time by removing the inner cannula and then replacing the cap. Ten out of twelve rats had ovulated 48 hours after double-cannula implantation of estradiol benzoate into the anterior pituitary. That is, ovulation took place 24 hours earlier than expected. None of seven controls implanted with cholesterol had ovulated within this period. Weick *et al.* (1971) later showed that in 19 out of 26 rats which are said to have had at least two consecutive 5-day estrous cycles previously, the inner cannula containing estrogen had to be in the anterior pituitary for only 4 hours in order for ovulation to occur within 48 hours. None of 9 rats with similar estrogen implants elsewhere in the brain had ovulated in this space of time. The significance of these two experiments (Weick and Davidson, 1970; Weick *et al.*, 1971) depends upon the credence attached to the belief that the rats were undergoing completely regular 5-day estrous cycles. The authors give no indication of the variance of the cycle lengths in their colony, and the experiments have not been repeated in other laboratories.

The same limitations apply to similar experiments carried out on rats described as undergoing regular 4-day estrous cycles. In these animals, progesterone injected subcutaneously on the first day of diestrus is said to delay ovulation by 24 hours so that the cycle is 5 days long (Everett, 1948). If, the day after the progesterone injection, estrogen was implanted into the anterior pituitary, ovulation was induced in 8 out of 9 rats at the expected time, 48 hours later, restoring the cycle to 4 days duration (Weick *et al.*, 1971). Only 2 out of 7 cholesterol-implanted controls ovulated by this time. This difference is statistically significant ($P < 0.025$) and is in agreement with the results of systemic estrogen administration (Everett, 1948). Normal ovulation can be blocked in cyclic rats by implanting a compound with antiestrogenic properties into the anterior pituitary (Bainbridge and Labhsetwar, 1971; Billard and McDonald, 1973), and precocious ovulation can be induced in prepubertal rats by implanting solid estrogen into the anterior pituitary (Döcke and Dörner, 1965). These results all indicate that estrogen can stimulate gonadotropin release and cause ovulation by acting directly on the anterior pituitary. By themselves, however, they again give no indication of the mechanism by which the response is brought about. The estrogen could either stimulate gonadotropin secretion by acting directly on the gonadotrops, or increase their sensitivity to hypothalamic gonadotropin-releasing factors. A change in the secretion of releasing factors from the hypothalamus does not seem to be involved.

Progesterone has also been applied directly to the median eminence in attempts to study the mechanism of gonadotropin release. No studies have been reported in which its diffusion from the site of implantation has been followed, but it seems reasonable to assume that, like estrogen, molecules of progesterone implanted in the median eminence would pass into the anterior pituitary. Smith *et al.* (1969) reported that the implantation of large pellets of progesterone, weighing 400 μg, into the median eminence of rats significantly lowered ovarian and uterine weights after 28 days and interrupted the estrous cycle. Similar implants of cholesterol are said not to have had these effects, neither did implants of progesterone in the midbrain, anterior hypothalamus, or anterior pituitary. For the reasons discussed above, these results may not indicate an action of progesterone on gonadotropin control at the median eminence level. For the same reasons, the report of Döcke and Dörner (1969), that implantation of 10 μg pellets of progesterone into either side of the median eminence, but not the preoptic area or the anterior pituitary, stimulated ovulation within 3 days, does not definitely indicate that the median eminence is a site where progesterone influences gonadotropin release. Weick and Davidson (1970) in any event could not repeat the results of Döcke and Dörner. They used a double cannula technique and a lower dose of progesterone than did Döcke and Dörner (1969).

While they do not lend themselves to precise generalizations, the results of experiments in which estrogen or progesterone has been implanted into the hypothalamus or anterior pituitary may be summarized as follows. In the short term, exposure of the median eminence and/or anterior pituitary to these hormones for a few hours or days may stimulate sufficient gonadotropin release to cause ovulation. More prolonged exposure results in reduced gonadotropin synthesis and release. These effects can be achieved by doses too low to have any effect when given systemically. However, the nature of the experiments is not such as to reveal the mechanisms by which estrogen and progesterone bring about changes in gonadotropin secretion. To achieve this end, steroid implantation must be combined with other techniques.

2. Effects of Gonadal Steroids on Releasing Factors

a. At the Hypothalamic Level. Attempts have been made to determine how gonadal steroids affect the hypothalamus by measuring "hypothalamic gonadotropin-releasing activity" at different stages of the estrous cycle, after ovariectomy and after steroid replacement therapy. In addition to the releasing factors for anterior pituitary hormones, the hypothalamus contains many different biologically active substances, including neurotransmitters and the peptide hormones, oxytocin and antidiuretic hormone (ADH). Substances with gonadotropin-releasing activity must therefore be separated from other active substances which could interfere with the assays. The level of gonadotropin-releasing activity in the hypothalamus is, however, so low that it is difficult to measure reliably, and in experiments with small animals, large numbers of hypothalami must be pooled to obtain measurable amounts of gonadotropin-releasing factor. It is possible therefore that the range of the responses of individual animals may be wide enough to obscure any general trends, and it is not surprising that the reported results of such experiments conflict greatly, especially when ovariectomy and steroid replacement are involved. The balance of evidence, however, tends to suggest that in the rat estrogen causes a fall in hypothalamic gonadotropin-releasing activity and, in both the rat and the sheep, that a drop in hypothalamic LH-releasing activity occurs at the time of the preovulatory LH surge. In the rat, this may correspond to an increase in LH-releasing factor (LH-RF) in the hypophysial portal blood.

Early work on changes in hypothalamic gonadotropin-releasing activity depended on the bioassay of gonadotropins released in rats given various hypothalamic extracts. For example, Chowers and McCann (1965) and Ramirez and Sawyer (1965) measured LH-releasing activity by injecting acid extracts of pooled rat hypothalami into gonadotropin-treated immature rats and taking the percentage ovarian ascorbic acid depletion (OAAD) as an index of the LH released. They took what precautions they could against any contamination of the hypothalamic extract with LH and checked the specificity of the assay for LH. They also took precautions against possible variations in the response of the test rats and in every assay compared the OAAD of an experimental group with that from a control group. More recently the LH released by hypothalamic extracts in test rats has been measured radioimmunologically (Smith and Davidson, 1974).

An alternative method for the assay of gonadotropin-releasing activity is to incubate the extract with anterior pituitary glands cultured under standard conditions *in vitro*. The amount of gonadotropin released into the incubating medium can then be assayed either biologically (Piacsek and Meites, 1966; Crighton *et al.*, 1973) or radioimmunologically (Jackson *et al.*, 1971; Kalra *et al.*, 1973b). This avoids the problem of the sensitivity of assay rats to the hypothalamic extract but still requires that samples of extract from rats should be pooled before assay. Hypothalami from larger animals such as sheep can be assayed separately (Crighton *et al.*, 1973). But obviously, assays for gonadotropin-releasing activity are subject to the same limitations of sensitivity and specificity as are assays for pituitary and plasma gonadotropins. They also ultimately depend on the thoroughness of the extraction procedure to which the hypothalamic tissue is subjected. If one group of workers extracts more gonadotropin-releasing activity than another, the reported results of the two groups are bound to be different.

In the rat, LH-releasing activity from pooled hypothalami was reported to decline during proestrus, to remain low on the day of estrus, and to reach a maximum in late diestrus (Chowers and McCann, 1965; Ramirez and Sawyer, 1965). These changes closely match alterations in pituitary LH content during the rat estrous cycle (Schwartz and Bartosik 1962). Kalra *et al.* (1973b), using different assay techniques, confirmed that hypothalamic LH-releasing activity is reduced at proestrus. This reduction occurs during the preovulatory LH surge and is apparently preceded by a slight rise in hypothalamic LH-releasing activity. Hypothalamic FSH-releasing activity has also been reported to fall during proestrus, to remain low at estrus, and to reach a maximum in late diestrus (Negro-Vilar *et al.*, 1970; Kalra *et al.*, 1973b), and these changes are paralleled by those in pituitary FSH (Caligaris *et al.*, 1967; Goldman and Mahesh, 1968; McClintock and Schwartz, 1968). Similarly, in the ewe, hypothalamic LH-releasing activity is said to be low after ovulation, reach a peak at diestrus, and decline again at the onset of behavioral estrus (Jackson *et al.*, 1971; Crighton *et al.*, 1973). Pituitary LH declined within 12 hours, and plasma LH reached a peak 16 hours after the onset of estrous behavior (Crighton *et al.*, 1973).

In general, the reports of experiments on rat and sheep tissue suggest that hypothalamic LH-releasing activity rises to a peak before ovulation and then falls dramatically at the time of the preovulatory surge of LH secretion. Given that the fall in hypothalamic LH-releasing activity reflects secretion of the releasing factor responsible for the preovulatory increase in circulating LH and that hypothalamic releasing factors are carried to the pituitary in the hypophysial portal vessels, increased amounts of LH-RF should be present in the portal vessels before or during the LH surge. Not surprisingly, attempts have been made to correlate variations in the LH-releasing activity in hypophysial portal blood with the preovulatory LH surge. It need hardly be said, however, that the technical difficulties involved in the collection of hypophysial portal blood are enormous. Some dilution of the samples by cerebrospinal fluid (CSF) and blood from systemic vessels at the base of the brain is all but impossible to prevent, and it is difficult to see how contamination by LH-containing blood leaking back from the pituitary sinusoids can be avoided. In experiments carried out by Fink and Harris (1970), LH is nonetheless said to have been removed from pooled samples of presumed portal blood collected from the cut pituitary stalk. Yet it was estimated that the portal blood was about 50% diluted by blood and fluid from sources other than the hypophysial vessels. In spite of this, it is said that LH-releasing activity (as measured by OAAD) was lower in hypophysial portal plasma after ovulation, on the day of estrus, than at proestrus, an observation which accords with the reported reduction in hypothalamic LH-releasing activity at estrus (Chowers and McCann, 1965; Ramirez and Sawyer, 1965). On the other hand, no peak of LH-releasing activity in hypophysial portal blood was apparent during proestrus, corresponding to the fall in hypothalamic gonadotropin-releasing activity which Chowers and McCann, Ramirez and Sawyer, and Kalra *et al.* (1973b) write also occurs at this time.

It has recently become possible to measure circulating levels of LH-RF with antibodies raised to synthetic LH-RF (Kerdelhué *et al.*, 1973; Nett *et al.*, 1973). This technique has the advantage of making it possible to assay LH-RF in individual animals, and it has been used to measure LH-RF in blood collected from the cut pituitary stalks of male and female rats (Fink and Jamieson, 1974a,b). Subsequently, Fink and Jamieson (1976) reported that using this method they found higher levels of LH-RF in blood collected from the cut pituitary stalks of individual female rats at proestrus than at diestrus or estrus. This increased level of LH-RF secretion corresponds to the fall in hypothalamic LH-releasing activity at proestrus (Chowers and McCann, 1965; Ramirez and Sawyer, 1965; Kalra *et al.*, 1973b). On the other hand the LH-RF measured might not all have come from portal blood. In combination with the increased sensitivity of the anterior pituitary to LH-RF in the hormonal environment of proestrus (see Section III,C,2b), increased LH-RF secretion could account for the preovulatory LH surge of the rat. So far, the rat is the only animal in which increased secretion of LH-RF has been demonstrated at the time of ovulation.

There have been two conflicting reports concerning the changes in peripheral plasma LH-RF during the sheep estrous cycle. Kerdelhué *et al.* (1973) reported an increase in immunologically assayable LH-RF in the systemic plasma of one cyclic ewe on the day of estrus 1 hour before the start of the LH peak to 8 hours after its end. However, this was not confirmed by Nett *et al.* (1974).

Obviously, measurements of such changes as may occur in hypothalamic and plasma gonadotropin-releasing activity during the estrous cycle do not provide any direct information about the influence of gonadal steroids on the hypothalamus. Nevertheless, one possibility is that the preovulatory rise in plasma estrogen is associated with the reported fall in hypothalamic LH-releasing activity at proestrus. On this hypothesis, changes in hypothalamic gonadotropin-releasing activity following ovariectomy and steroid replacement should give more direct information about the influence of gonadal hormones on the hypothalamus. On the other hand, any changes which might be recorded in hypothalamic gonadotropin-releasing activity following ovariectomy could reflect changes due both to the removal of the influence of gonadal steroids on some unspecified hypothalamic nuclei and to the direct action on the hypothalamus of increased circulating gonadotropins. For example, an increase in hypothalamic gonadotropin-releasing activity following ovariectomy might be interpreted as being due to a decline in steroid negative feedback. A fall might be caused by the action of the increased plasma gonadotropins on the gonadotropin-releasing factor-producing cells in the hypothalamus. Such an interpretation would be supported by the presence of measurable quantities of LH-releasing activity in the peripheral plasma of hypophysectomized rats, as reported by Nallar and McCann (1965).

Reports conflict, however, about the effects of ovariectomy on hypothalamic gonadotropin-releasing activity in rats. The differences in the results of such experiments may be due to the differing assay techniques used by the various authors, to different anatomical procedures, to differing responses of individual rats, or to differences in the postoperative interval. Early reports implied that ovariectomy did not alter hypothalamic LH-releasing activity (Chowers and McCann, 1965) but that hypothalamic FSH-releasing activity was increased (David et al., 1965; Mittler and Meites, 1966). Other reports said that a fall occurred in hypothalamic LH-releasing activity (Moszkowska and Kordon, 1965; Piacsek and Meites, 1966) and FSH-releasing activity (McCann et al., 1968). Such reductions in hypothalamic gonadotropin-releasing activity do not parallel the increases in pituitary FSH and LH which occur after ovariectomy in the rat, but Burger et al. (1972) state that the peripheral plasma of ovariectomized rats, besides containing large amounts of LH, also contains measurable amounts of LH-releasing activity which is absent from intact animals. This finding may be a reflection of increased secretion of LH-RF from the hypothalamus following removal of gonadal steroid inhibition by ovariectomy. It also suggests that the increased LH secretion does not suppress hypothalamic LH-RF production. At the time of puberty in the rat, the rise of plasma gonadotropins and gonadal steroids is said to be accompanied by reductions in hypothalamic LH-releasing activity (Ramirez and Sawyer, 1966) and FSH-releasing activity (Corbin and Daniels, 1967). These changes occur at about the same time as the levels of pituitary LH and FSH fall. Based on the reported changes in hypothalamic gonadotropin-releasing activity during the estrous cycle and at puberty in the female rat, a plausible working hypothesis might be that gonadal steroids reduce hypothalamic gonadotropin-releasing activity.

The results of gonadal steroid treatment of ovariectomized rats provide support for such a view. Piacsek and Meites (1966) found that daily treatment of ovariectomized rats with 0.8 μg estradiol benzoate prevented the postovariectomy rise in pituitary LH and reduced the level of hypothalamic

LH-releasing activity below that of ovariectomized controls. Ajika *et al.* (1972) state that a single injection of 0.2 to 5.0 μg estradiol benzoate caused a fall in plasma LH and FSH after 24 hours in ovariectomized rats and led to depressed hypothalamic LH and FSH-releasing activity after another 24 hours. Daily treatment with progesterone, or estrogen combined with progesterone, did not affect hypothalamic LH-releasing activity after ovariectomy (Piacsek and Meites, 1966; Martini *et al.*, 1970), but progesterone alone depressed hypothalamic FSH-releasing activity (Martini *et al.*, 1970). In contrast to the results of steroid treatment of ovariectomized rats, daily treatment of intact rats with 2 to 50 μg estradiol benzoate did not affect hypothalamic LH-releasing activity, although pituitary LH was reduced (Chowers and McCann, 1965). However, implantation of estrogen in the median eminence led to a significant reduction in hypothalamic LH-releasing activity compared to that of cholesterol-implanted controls. As pointed out in Section III,C,1 the effects on gonadotropin secretion of estrogen implants in the median eminence could be due to actions either on the anterior pituitary or the hypothalamus. The reduction in hypothalamic LH-releasing activity could therefore be secondary to a fall in circulating LH. However, in view of the results of Burger *et al.* (1972), it is possible that the levels of gonadotropin-releasing activity and circulating gonadotropins are not reciprocally related. The fall in hypothalamic LH-releasing activity reported by Chowers and McCann (1965) could therefore be due to a direct effect of the estrogen implant on the cells which produce LH-RF.

   b. At the Pituitary Level. There is clear evidence that ovarian steroids can alter the responsiveness of the anterior pituitary to hypothalamic gonadotropin-releasing factors, although the precise physiological significance of these findings in different species is by no means certain. Highly purified LH-RF of ovine or porcine origin, or synthetic decapeptide LH-RF (Matsuo *et al.*, 1971a,b) have been used to study the LH response of the pituitary in animals under different hormonal conditions. In intact animals, undergoing spontaneous estrous or menstrual cycles, the pituitary appears to be most sensitive to LH-RF during the immediate preovulatory phase, after the peak of follicular estrogen secretion. The evidence from some species suggests that estrogens are at least partly responsible for the increased pituitary response to LH-RF. On the other hand, the ovulation-inhibiting effects of luteal progesterone could be due to a reduction in pituitary sensitivity to LH-RF.

   The amount of LH released in response to a standard dose of LH-RF varies according to the stage of the reproductive cycle and is greatest after the preovulatory secretion of estrogen (rat: Aiyer *et al.*, 1974a; hamster: Arimura *et al.*, 1972; ewe: Reeves *et al.*, 1971b; rhesus monkey: Krey *et al.*,

1972, 1973; women: Yen et al., 1972). However, it is not at all certain that estrogens are entirely responsible for the preovulatory increase in pituitary sensitivity to LH-RF. In the ovariectomized rhesus monkey, pituitary sensitivity to LH-RF exceeds that observed in intact animals, and when plasma estradiol is raised to physiological levels, the LH response to LH-RF is greatly reduced (Krey et al., 1973). A similar situation may exist in women (Siler and Yen, 1973; Thompson et al., 1973; Yen et al., 1974). In these species, therefore, the pituitary may be most sensitive to LH-RF when gonadotropin secretion is highest, whether this occurs during the preovulatory LH surge or after ovariectomy. In the rat, however, ovariectomy does not enhance the LH response to purified hypothalamic extract above that seen at proestrus (Libertun et al., 1974a), and in this and other species, estradiol apparently sensitizes the pituitary to LH-RF. For example, anestrous ewes injected with estradiol benzoate showed a greater pituitary response to LH-RF than control ewes given oil (Reeves et al., 1971a), while estradiol-17β given to heifers before they were slaughtered increased the response to LH-RF in vitro (Hobson and Hansel, 1974). Furthermore, in the rat, Arimura and Schally (1971) found that 10 μg estradiol benzoate, which represents a large dose, administered on the first day of diestrus did not affect plasma LH, but when the same dose was injected 24 hours before an intracarotid injection of LH-RF, it increased the amount of LH released in response to LH-RF above that observed in control rats injected with oil. Estradiol benzoate has since been shown to have a biphasic effect on pituitary sensitivity to LH-RF in the ovariectomized rat (Libertun et al., 1974b). Rats which had received an intravenous injection of 100 ng estradiol benzoate in oil immediately before the beginning of an infusion of LH-RF lasting 6 hours showed lower plasma LH during the first 2 hours of the infusion than did rats which had received the oil alone. In another experiment, the LH-RF infusion was delayed for 6 hours after the injection of estradiol benzoate. Under these conditions, the LH response 2 hours after the start of the LH-RF infusion was much higher in the rats which had received the estradiol than in those that received oil. The maximum recorded response to LH-RF thus occurred 8 hours after the injection of estradiol benzoate. The latter finding is compatible with the timing of the preovulatory secretion of estrogens, and the initial increase in pituitary sensitivity to LH-RF which occurs before the LH surge in the proestrous rat (Brown-Grant et al., 1970; Aiyer et al., 1974a). Subsequently, Aiyer and Fink (1974) confirmed that this initial increase in the pituitary response to LH-RF in the proestrous rat depends on the preovulatory secretion of estrogen. They further suggested that a second phase of increased pituitary sensitivity to LH-RF which occurs during the LH surge is partly due to an action of progesterone on the estrogen-primed pituitary and partly due to a priming effect of LH-RF itself on the gland (Aiyer *et al.*, 1974b).

Like estrogen, progesterone can affect pituitary sensitivity to LH-RF. Many workers have investigated the ovulation-inhibiting mechanisms of luteal progesterone, exogenous progesterone, and synthetic progestagens. Theoretically, the mechanism could involve one or more of the following possibilities: reduction in hypothalamic gonadotropin-releasing factor content, as discussed in Section III,C,2,a; reduced gonadotropin-releasing factor secretion in response to ovulation-inducing stimuli; or reduction in pituitary sensitivity to LH-RF. Results of experiments on rats and rabbits indicate that progestagens inhibit ovulation by acting both on the hypothalamus and pituitary. The pituitary response to both LH-RF and endogenous ovarian hormones may be altered.

Precautions must be taken in investigations of inhibitory effects to avoid the use of supraphysiological stimuli. For example, if electrical stimulation of the hypothalamus is to be used to overcome inhibition of ovulation by progesterone, the threshold stimulus for inducing ovulation must be demonstrated in control animals. Failure to do this might lead to unjustifiable conclusions. For example, Exley *et al.* (1968) and Harris and Sherratt (1969) did not find that the synthetic progestagen, chlormadinone acetate, prevented ovulation in estrous rabbits whose basal hypothalamus was stimulated electrically, and they concluded that the inhibitory action of chlormadinone on mating-induced ovulation does not depend on an action on the pituitary. However, the intensity of stimulation was probably well above threshold for the induction of ovulation in this species and could have released sufficient gonadotropin-releasing factor to have overcome any pituitary inhibition. This view is supported by the results of McDonald and Gilmore (1971b), who showed that administration of norethindrone (17α-19-nortestosterone, another synthetic progestagen) to rats necessitated an increase in the threshold current for inducing ovulation when the preoptic area was stimulated. The increased preoptic threshold was not associated with any change in the sensitivity of the pituitary to ovulation induced by crude, unpurified bovine hypothalamic extract infused into the carotid artery. It may be concluded, therefore, that norethindrone prevents ovulation by affecting the secretion of gonadotropin-releasing factor from the hypothalamus.

In experiments where hypothalamic extracts, or synthetic LH-RF are used to overcome progestagen-induced inhibition of ovulation, it is important that the threshold dose of gonadotropin-releasing factor should always first be determined in untreated control animals. Without this, no valid conclusion can be drawn. Schally *et al.* (1968, 1970) showed that a variety of synthetic antifertility agents did not influence pituitary LH release, as measured by bioassay or radioimmunoassay, in rats injected with a relatively large dose of LH-RF. Although one hypothesis could be that these compounds prevent ovulation by acting on the brain, the possibility that antifertility agents act on the pituitary is not excluded by these experiments. Such an action was demonstrated by Arimura and Schally (1970) and Debeljuk et al. (1972) who showed that progesterone itself could reduce the amount of LH released in response to intravenous LH-RF in both intact and ovariectomized rats.

Hilliard and co-workers have investigated the effect of progesterone and synthetic progestagens on the capacity of an unpurified extract of rabbit hypothalamus to induce ovulation in rabbits and rats. Their results suggest that these steroids act on the pituitary. The hypothalamic extract was prepared from blocks of tissue which "included the basal hypothalamus between the optic chiasma and mammillary bodies as well as the tuber cinereum" (Hilliard et al., 1966). When the extract was infused directly into the anterior pituitary of estrous rabbits and proestrous rats, it stimulated ovulation (Hilliard et al., 1966; K. T. Stevens et al., 1970). Extracts of cerebral cortex made from comparable amounts of tissue are said to have proved ineffective (Hilliard et al., 1966). Since ovulation was subsequently stimulated by intrapituitary infusions of purified LH-RF, it is likely that this peptide was the active constituent of the crude hypothalamic extract (Hilliard et al., 1971). When the hypothalamic extract or LH-RF was infused into the pituitaries of animals pretreated with norethindrone, chlormadinone, or progesterone, ovulation did not occur (Hilliard et al., 1966, 1971; Spies et al., 1969; K. T. Stevens et al., 1970). The block could be overcome by increasing the dose of hypothalamic extract or LH-RF, and Hilliard et al. (1971) suggested that the steroids had reduced the sensitivity of the pituitary to LH-RF. But since progesterone also reduced the number of animals ovulating in response to threshold doses of LH (Spies et al., 1969; K. T. Stevens et al., 1970), the apparent reduction in pituitary sensitivity to LH-RF may be due to the ovaries' inability to respond to a normal surge of LH. Part of the response of the rabbit ovary to LH is to secrete 20α-hydroxyprogesterone. This steroid is carried to the pituitary where it stimulates further LH release (Hilliard et al., 1967). Progestagens do not interfere with the secretion of 20α-hydroxyprogesterone in response to LH, but appear to counteract its facilitatory effect on the pituitary (Spies et al., 1969). Progesterone appears to affect the response to LH by a similar mechanism in the rat (K. T. Stevens et al., 1970). The inhibitory effect of exogenous progestagens on ovulation, in at least these two species, therefore, appears to depend partly on a reduction in the ability of the pituitary to maintain sufficient LH secretion to induce ovulation.

It is difficult to assess the extent to which endogenous progesterone affects the response of the pituitary to LH-RF. In the pseudopregnant or pregnant rabbit, as opposed to the nonpregnant animal, intrapituitary infusion of hypothalamic extract or purified LH-RF does not stimulate ovulation (Hilliard *et al.*, 1971). However, the lower levels of progesterone reached during the luteal phase of the cycle in other species do not appear to have such an inhibitory effect on ovulation. The pituitary is no less responsive to LH-RF during the luteal phase than during the follicular phase of the cycle in the rhesus monkey (Krey *et al.*, 1973), women (Nillius and Wide, 1972), or ewes (Reeves *et al.*, 1971b). Experiments in which circulating progesterone has been raised to physiological levels in anestrous ewes have yielded conflicting results. Cumming *et al.* (1972) found that 24-hour intravenous infusions of progesterone did not affect the increase in plasma LH during intracarotid infusions of LH-RF lasting 3 hours. On the other hand, Pant and Ward (1973) reported that daily subcutaneous injections of progesterone significantly reduced the response to a single intravenous injection of LH-RF.

Combinations of estrogen and progesterone have been shown in some experiments to alter the sensitivity of the pituitary to LH-RF, and in others to have no effect. In the intact anestrous ewe and diestrous rat, a decrease was demonstrated (Debeljuk *et al.*, 1972), but in the ovariectomized rat, single large doses of estrogen combined with progesterone apparently sensitized the pituitary to LH-RF (Libertun *et al.*, 1974a).

The general conclusions that can be drawn from these experiments are not very clear-cut. It appears that the administration of gonadal and synthetic steroids alters the response of the pituitary to LH-RF. In intact animals, the pituitary is most sensitive to LH-RF during the preovulatory phase of the cycle, after the peak of follicular estrogen secretion. The levels of circulating progesterone reached during pseudopregnancy or pregnancy apparently diminish the pituitary's capacity to respond to LH-RF, but the levels reached during the luteal phase of the cycle do not seem to affect the amount of LH released in response to LH-RF. Exogenous progestagens seem to prevent ovulation by various means including (in the rabbit and rat at least) a reduction in the ability of the pituitary to maintain an ovulatory surge of LH secretion.

3. Localization of Ovarian Steroids in Brain and Pituitary

The sites of cells which can selectively accumulate steroids have been investigated by assessing the radioactivity in homogenates or sections of tissues following an injection of labeled hormone. These techniques give more precise information about possible sites of hormone feedback in the brain and pituitary gland than studies in which steroids have been implanted. However, at the moment neither method is suitable for demonstrating possible sites of action of progesterone, perhaps because this hormone is metabolized too quickly.

a. Uptake of Radioactive Hormones in Tissue Homogenates. Numerous studies have shown that the hypothalamus and anterior pituitary selectively accumulate and retain radioactive estradiol (Eisenfeld and Axelrod, 1966; Kato and Villee, 1967; McGuire and Lisk, 1969; Whalen and Maurer, 1969; Kato, 1970a,b). Over 80% of the injected [3H]estradiol retained in the anterior hypothalamus and anterior pituitary is in the form of estradiol, but in the basal hypothalamus significant amounts of the metabolite estrone are found (Luttge and Whalen, 1970).

It is assumed that the estradiol retained in the hypothalamus and anterior pituitary is bound by specific "receptors," macromolecular proteins present in the cells (Thomas, 1973). The nature of the receptors has been studied by determining which other compounds can interfere with the uptake of radioactive estradiol. Injections of estrone or estriol 30 seconds before administration of [3H]estradiol-17β reduced uptake in homogenates of hypothalamus and anterior pituitary (Eisenfeld and Axelrod, 1966). Radioactive estradiol uptake was also depressed when labeled and unlabeled steroid were given simultaneously, but not when the unlabeled steroid was injected 30 minutes after the [3H]estradiol (Kato and Villee, 1967). On the other hand, unlabeled estrogens did not affect the uptake of [3H]estradiol-17β in the cerebral cortex or cerebellum (Eisenfeld and Axelrod, 1966; Kato and Villee, 1967), indicating the absence of specific estrogen receptors in these tissues. These findings in tissue homogenates were confirmed by Anderson and Greenwald (1969) using autoradiography. Administration of 2 μg estradiol-17β 1 hour before the injection of [3H]estradiol-17β decreased the accumulation of labeled estradiol in cells from the preoptic area to the basal hypothalamus and anterior pituitary. Cells accumulating estradiol were not found in the olfactory tubercle, caudate nucleus and putamen, or cerebral cortex. Synthetic estrogens and estrogen antagonists have a similar effect to that of estradiol on the accumulation of [3H]estradiol-17β by homogenates of hypothalamus and anterior pituitary (Eisenfeld and Axelrod, 1967). On the other hand, pretreatment with progestagens or estradiol-17α did not reduce uptake (Eisenfeld and Axelrod, 1967; Kato and Villee, 1967). These studies suggest that estradiol accumulates in both the hypothalamus and anterior pituitary and that a specific receptor is involved in the process.

Some comparisons may be made of the properties of the estradiol receptor system present in the hypothalamus and anterior pituitary and that present in the uterus (Eisenfeld, 1970). Both systems appear to be selective in concentrating and retaining estradiol and are stereospecific (Kato and Villee, 1967). The receptors do not appear to bind nonestrogenic hormones and can be saturated by estradiol, that is, their estradiol-binding capacity is limited (King *et al.*, 1965; Kato and Villee, 1967; Eisenfeld, 1970).

In view of this limited binding capacity, attempts have been made to estimate the extent of uptake of endogenous estradiol by injecting labeled hormone and measuring the radioactivity retained in the hypothalamus and pituitary at different stages of the reproductive cycle. Since binding capacity is limited, the uptake of exogenous hormone should be inversely related to the amount of endogenous hormone previously bound to the receptors. The affinity of the receptors for estradiol is such that a high degree of binding will take place in the presence of physiological concentrations of estradiol (Thomas, 1973). As expected, Kato *et al.* (1969) found changes in uptake in the anterior hypothalamus during the estrous cycle of the rat, with a lower uptake during proestrus than at diestrus. No cyclic changes were detected in pooled samples of median eminence resected from blocks of hypothalamic tissue or in individual samples of anterior pituitary, thigh muscle, or blood plasma.

It is also possible to study estradiol uptake and receptors *in vitro*. Kato (1970a) confirmed that the uptake of [^{3}H]estradiol by rat anterior hypothalamus *in vitro* was lower at proestrus than at diestrus. Ginsburg *et al.* (1972) measured the available binding capacity for estradiol in hypothalami and anterior pituitaries of rats at different stages of the estrous cycle by studying the formation of complexes between different quantities of [^{3}H]estradiol and the tissue receptors *in vitro*. Unbound [^{3}H]estradiol was separated from the [^{3}H]estradiol–receptor complexes and discarded. Measurement of the radioactivity of the complex made it possible to calculate the amount of [^{3}H]estradiol bound to the receptors and so to estimate the number of available binding sites. As before, during the proestrous peak of circulating estradiol, the number of available estrogen receptor sites in the hypothalamus and anterior pituitary fell by 70% and 50%, respectively, compared with values found at diestrus.

Estrogens may not have to remain bound to their receptors in the hypothalamus for their effects on sexual behavior to be observed. Michael (1965) reported that the hypothalamus retained radioactivity for 11 to 20 hours following injection of [^{3}H]hexestrol in ovariectomized cats—much longer than did other parts of the brain (e.g., thalamus, midbrain, and cerebral cortex). However, there is an interval of a few days between an injection of hexestrol and the display of estrous behavior, suggesting that the action of estrogen in stimulating sexual behavior is to trigger some neural activity in the hypothalamus.

Studies of progesterone uptake have not proved conclusive: it has not been possible to determine binding sites for progesterone or even to demonstrate that progesterone is bound (see Volume III, Chapter 2). The hypothalamus and pituitary of the rat can take up [³H]progesterone rapidly, maximal activity occurring 2 minutes after injection. This is followed by an equally rapid decline in activity, background levels being reached in 60 minutes (Laumas and Farooq, 1966). During the estrous cycle of the rat, no significant differences in radioactivity were seen between anterior pituitary, hypothalamus, and other brain regions after short-term exposure to [³H]progesterone (Seiki *et al.,* 1968, 1969). However, exposure of ovariectomized rats to larger amounts of the labeled steroid for 1 to 12 hours suggested that a selective uptake of progesterone occurred in the pituitary gland, anterior and posterior hypothalamus, and the median eminence (Seiki and Hattori, 1971). Thus, short-term exposure of rats to radioactive progesterone reveals that the steroid is rapidly accumulated by the hypothalamus and anterior pituitary but is not retained for any length of time. The results of Seiki and Hattori (1971), revealing a selective uptake of progesterone by the hypothalamus and pituitary, may nevertheless indicate the presence of specific progesterone-binding receptors, similar to those which exist for estradiol. For a fuller discussion of this topic, the reader is referred to Chapter 2 of this volume and Thomas (1973).

   b. **Autoradiographic Localization of Estradiol.** Attempts to locate steroid receptor sites autoradiographically are beset with many difficulties. Much confusion has arisen as a result of poorly controlled experimental conditions and ill-chosen techniques. These disadvantages are discussed by Stumpf (1969) who warns against acceptance of such pictorial data "without sufficient concern for their validity". Most of the problems have arisen because of diffusion of the steroid from its site of uptake during tissue processing, especially during liquid fixation. Different methods give reproducible but different results, so that disagreements arise over which cells contain receptors and which areas of the cell bind the hormone. Only by avoiding all fluid treatments during the preparation of tissues and autoradiograms is agreement obtained between biochemical and autoradiographic data for steroid hormone uptake (Stumpf, 1969). Thus, Stumpf uses freeze-drying for tissue fixation, and a "dry mount" method for preparing the autoradiograms (Stumpf, 1968). When interpreting autoradiographic studies of the central nervous system, it must also be remembered that not all authors make the same distinctions between brain nuclei, so that care must be taken in making comparisons.

   Using dry mount autoradiography, Stumpf (1968) investigated the distribution of [³H]estradiol in the hypothalamus of the ovariectomized female rat. Evidence of estradiol uptake was present in the nuclei of neurons; glial and ependymal cells were devoid of activity (see also Attramadal, 1970a).

The distribution of radioactively-marked neurons was well-defined, some regions being heavily labeled while adjacent areas were unlabeled. Labeled cells were found in the arcuate nucleus, lateral ventromedial hypothalamic nucleus, periventricular nucleus, medial preoptic nucleus, suprachiasmatic nucleus, and border cells of the nucleus of the stria terminalis. These nuclei are involved in many different functions besides steroid feedback. It is interesting, therefore, that no labeling was found in the magnocellular parts of the supraoptic and paraventricular nuclei, premamillary and mamillary nuclei, and lateral, dorsal, ventral, and posterior hypothalamic nuclei which are not thought to be involved in the control of gonadotropin release. The distribution of labeled neurons closely resembled the areas innervated by the terminals of the stria terminalis, a fiber tract which arises in the amygdala. The labeled cells were also in areas which are known to influence anterior pituitary function. The amygdala can influence hypothalamic and pituitary function (see Volume II, Chapter 9 and Volume III, Chapter 4) and Stumpf and Sar (1971) have demonstrated selective uptake of radioactive estradiol by certain of the amygdaloid nuclei. Stumpf (1970) proposed the existence of hormone–neuron systems as an alternative to the idea of discrete "sex-centers." These systems involve groups of neurons and nerve fiber tracts which have been shown to accumulate labeled estradiol and are thought to be concerned in the control of anterior pituitary function. The regions involved include the periventricular brain, stria terminalis, ventral amygdalohypothalamic tract, and parts of the fornix.

Autoradiographic studies on the anterior pituitary have shown a widespread distribution of labeled estradiol (Stumpf, 1968; Attramadal, 1970b). Labeling was present in all three cell types, acidophils, basophils, and chromophobes, and, although most of the radioactivity was confined to the cell nucleus, significant amounts were also present in the cytoplasm. This has been interpreted as a direct effect of estradiol on prolactin-producing cells (acidophils) and gonadotropin cells (basophils). The uptake by chromophobes can be explained on the basis that these cells are probably degranulated acidophils or basophils.

In summary, it appears that the nervous system and anterior pituitary, like the uterus, vagina, and mammary glands, contain cells which can take up and retain estradiol. This mechanism is specific for estradiol and in general is not influenced by other nonestrogenic hormones. Estradiol distribution within the CNS is well-defined and the steroid is absent from certain specific nuclei, whereas in the anterior pituitary the distribution is more widespread. Radioactive uptake studies and autoradiography provide indirect evidence for the localization of the receptors, at least for estradiol; evidence for the existence of specific progesterone receptors is still equivocal. However, the fact that a particular cell or neuron can take up and retain a hormone does not necessarily mean that the particular cell is the site of action of the hormone. To show this beyond doubt, it would be necessary to demonstrate that alterations in the activity of the cell accompany or follow the uptake of the hormone and result in alterations in pituitary function. The effects of ovarian hormones on changes in the activity of cells in the hypothalamus have been studied (see below), but it has not been possible to demonstrate that the responsive cells are those which accumulate ovarian steroids.

4. Effects on Neural Activity

The influence of ovarian hormones on neural activity has been studied by recording cortical and subcortical EEG patterns and the firing rates of single cells (unit activity) or small groups of cells (multiple unit activity) in different parts of the brain. Monitoring EEG patterns in conscious, freely moving animals has yielded abundant data, but several problems inherent in this method make interpretation of the results difficult (Beyer and Sawyer, 1969). Multiple- and single-unit recording is more suitable for monitoring activity changes in response to hormone treatment. Since these methods record only the activity of single or small groups of cells within a short distance of the tip of the electrode, the location and identification of the responding neurons is much easier. However, changes in EEG patterns must also be monitored at the same time to check whether a hormone is affecting a specific area of the brain or having a generalized effect on neural activity. Experiments do indeed show that estrogens and progesterone have both general effects on EEG patterns and specific effects on the activity of single neurons.

Much of the work on recording neural activity has been carried out on anesthetized animals. Therefore the data obtained must be interpreted with caution as many of the responses are modified by anesthetics. Furthermore, some studies have used pharmacological doses of hormones, so that the physiological significance of the results is questionable. For example, Kawakami and Sawyer (1959) investigated the effects of large doses of estradiol benzoate and progesterone on arousal thresholds in the brain-stem reticular formation and on EEG afterreaction thresholds inside and outside the hypothalamus. In the estrous or estrogen-primed rabbit, an injection of progesterone lowered both thresholds within 4 to 6 hours. After 24 hours, both thresholds were raised and the animals became anestrous. Kawakami and Sawyer therefore concluded that ovarian steroids may affect sexual behavior and pituitary function by altering brain thresholds. This proposition needs, however, to be viewed with caution. Although the dose of progesterone used in these studies (2 mg) was probably within the physiological range (Hilliard *et al.*, 1968), the dose of estradiol benzoate which they administered was several hundred times that required to induce sexual receptivity (McDonald *et al.*, 1970). Nevertheless, changes in arousal thresholds in the limbic system and reticular system have since been shown to occur following the administration of low doses of estrogen and progesterone in the ovariectomized rat and during the estrous cycle (Terasawa and Timiras, 1968, 1969). This suggests that physiological doses of ovarian hormones can influence neural activity in areas other than the hypothalamus. Such changes may or may not subsequently modify the activity of hypothalamic neurons (Feldman and Dafny, 1970; see also Volume II, Chapter 9).

A technique originally developed to investigate the control of posterior pituitary function (Cross and Green, 1959) has been used to investigate the influence of ovarian hormones on the activity of single hypothalamic cells of rats under urethane anesthesia (Barraclough and Cross, 1963; Cross and Silver, 1965; Lincoln, 1967; Lincoln and Cross, 1967). The responses to a variety of sensory stimuli were compared in animals at different stages of the estrous cycle and in ovariectomized animals given ovarian steroids. Some of the stimuli were nonspecific (pain, cold, or noise), but mechanical stimulation of the uterine cervix is a specific reproductive stimulus (Shelesnyak, 1931). The numbers of neurons responding to different sensory stimuli varied during the estrous cycle, and the number responding to cervical probing was lowest during pseudopregnancy (Barraclough and Cross, 1963; Cross and Silver, 1965). If the ovaries of pseudopregnant rats were removed immediately before the recording was made, more units responded, suggesting that progesterone selectively depresses excitation of hypothalamic neurons by cervical stimulation. This is supported by the decreased response of lateral hypothalamic neurons to cervical probing following administration of 400 μg progesterone (Cross and Silver, 1965).

This postulated selective action of progesterone on the response of hypothalamic units to stimuli arising from the genital tract is disputed by Ramirez *et al.* (1967) and Beyer *et al.* (1967), who argue that since progesterone induces generalized inhibitory changes in brain activity, the responses of individual neurons or groups of neurons simply reflect this nonspecific depression (Beyer *et al.*, 1967). The depression of spontaneous firing rates of hypothalamic and thalamic neurons induced by even low doses of progesterone is invariably accompanied by a change in the cortical EEG from a desynchronized (aroused) pattern to a synchronized (sleeplike) pattern (Ramirez *et al.*, 1967). The synchronized EEG pattern and reduced hypothalamic multiple unit activity caused by an injection of progesterone in intact rats under urethane anesthesia was associated with a sharp rise in systemic blood pressure (Beyer *et al.*, 1967). Beyer and his colleagues suggested that the hormonally induced changes in blood pressure (perhaps caused by direct actions on the baroceptors) caused alterations in the activity of the brain-stem reticular formation which governs the arousal state of the animal. This hypothesis is supported by the observations of Komisaruk et al. (1967) who simultaneously recorded systemic blood pressure, cortical EEG and single- and multiple-unit activity in the hypothalamus, cortex, or thalamus of intact female rats under urethane anesthesia. They found that intravenous progesterone caused a fall in the activity recorded from single cells or groups of cells. This fall was always correlated with a sleeplike EEG pattern. Stimuli such as vaginal or rectal probing, tail pinching, noise, heat or cold, all of which normally induced EEG arousal, and an increase in single- and multiple-unit activity were less effective after the administration of progesterone. It was concluded that, under the experimental conditions employed, any specific or differential effects of progesterone were exerted through a nonspecific depression of arousal. Thus, under the influence of progesterone, the effect of a stimulus on neural activity depends upon the ability of the stimulus to induce arousal. Endroczi (1968), who used different electrophysiological measurements, has also reported on the widespread inhibitory effects of progesterone.

Terasawa and Sawyer (1970) dispute this general conclusion, and argue that progesterone has a specific effect on neural activity. In their experiments, the activity of groups of cells in the arcuate nucleus of the basal hypothalamus and the cortical EEG pattern were recorded in ovariectomized rats. The animals were primed with estrogen and the recording started before a subcutaneous injection of 2.5 mg progesterone. The progesterone caused synchronization of the cortical EEG within 60 minutes, regardless of the time when the injection was given, but a diurnal variation was observed in the response of the hypothalamus. When progesterone was given during the morning, hypothalamic multiple-unit activity increased briefly. It was then depressed for a period before finally increasing again for several hours. This activity was not related to changes in the cortical EEG but was related to the time of onset of sexual receptivity. The changes in multiple-unit activity in the arcuate nucleus may therefore represent a specific effect of progesterone on neural activity in the hypothalamus.

The influence of estrogens on unit activity and unit responses to sensory stimulation in the anesthetized rat has been studied by Lincoln (1967) and Lincoln and Cross (1967). Spontaneous activity in the anterior hypothalamus and preoptic and septal regions was lower in ovariectomized, estradiol-treated rats and in rats with high levels of endogenous estrogen due to exposure to constant light than in untreated ovariectomized animals. In the lateral hypothalamus, estradiol appeared to increase spontaneous activity (Lincoln, 1967). Stimuli such as pain, cold, and vaginal probing excited units in the lateral hypothalamus and inhibited units in the septum. In the preoptic area and anterior hypothalamus, some units were excited while others were inhibited (Lincoln and Cross, 1967). The time course of the responses closely corresponded to activation of the cortical EEG, suggesting nonspecific arousal effects. In the preoptic, lateral and anterior hypothalamic areas, estrogens increased the proportion of units whose activity was depressed by vaginal probing. In the lateral and anterior hypothalamic areas, a similar increase was observed when activity was depressed by cold or pain. In the septum, estrogen reduced the proportion of units depressed by these stimuli. Thus, under the influence of estrogens, certain stimuli which induce EEG arousal appear to strengthen inhibitory responses in the anterior and lateral hypothalamus and the preoptic area. The inhibitory action of estrogens on unit activity may be related to the inhibition of ovulation resulting from chronic exposure to estrogens.

To overcome some of the difficulties encountered in recording unit activity in animals with intact brains, Cross and Kitay (1967) and Cross and Dyer (1970, 1972) have introduced the "diencephalic island" preparation. This is in effect a type of decerebration which permits the recording of spontaneous neural activity within the hypothalamus of unanesthetized rats in the absence of all afferent inputs. Significant differences in the spontaneous firing rates of anterior hypothalamic units were found during the estrous cycle, the highest rate occurring during proestrus and the lowest during estrus (Cross and Dyer, 1970, 1972). This increased activity depended on the presence of the ovaries, and there was some indication that it was a response to declining estrogen levels (Cross and Dyer, 1972). The effects of ovarian hormones on hypothalamic activity are therefore still exerted in the absence of influences from outside the diencephalon. The increased hypothalamic activity is not an artifact caused by brain damage as Dyer *et al.* (1972) demonstrated the presence of increased spontaneous activity in the ventral part of the anterior hypothalamus but not in the thalamus of anesthetized proestrous rats with intact brains.

The experimental data therefore demonstrate that both estrogens and progestagens have general effects on EEG patterns and activation thresholds in different parts of the brain and also on the arousal state of the animal. Effects of progesterone on hypothalamic and extrahypothalamic activity, especially the responses of neurons to sensory stimulation, may depend on the animals' arousal and the arousal-inducing capacity of the stimulus. Progesterone can also induce changes in hypothalamic activity which are independent of EEG changes and may be related to the induction of estrous behavior. Estrogens, on the other hand, appear to inhibit hypothalamic activity specifically and to strengthen inhibitory responses to certain arousal-inducing stimuli. Declining levels of circulating estrogen at proestrus may account for the increased activity present in the anterior hypothalami of rats with intact brains and rats with no afferent connections to the diencephalon.

REFERENCES

Abraham, G. E., Odell, W. D., Swerdloff, R. S., and Hopper, K. (1972). Simultaneous radioimmunoassay of plasma FSH, LH, progesterone, 17-hydroxyprogesterone and estradiol-17β during the menstrual cycle. *J. Clin. Endocrinol. Metab.* **34,** 312–318.

Adams, C. E. (1958). Egg development in the rabbit: the influence of postcoital ligation of the uterine tubes and of ovariectomy. *J. Endocrinol.* **16,** 283–293.

Adams, C. E. (1970). Egg-uterus interrelationship. *In* "Mechanisms Involved in Conception" (G. Raspé, ed.), pp. 149–162. Pergamon, Oxford.

Aiyer, M. S., and Fink, G. (1974). The role of sex steroid hormones in modulating the responsiveness of the anterior pituitary gland to luteinizing hormone releasing factor in the female rat. *J. Endocrinol.* **62,** 553–572.

Aiyer, M. S., Fink, G., and Greig, F. (1974a). Changes in the sensitivity of the pituitary gland to luteinizing hormone releasing factor during the oestrous cycle of the rat. *J. Endocrinol.* **60,** 47–64.

Aiyer, M. S., Chiappa, S. A., and Fink, G. (1974b). A priming effect of luteinizing hormone releasing factor on the anterior pituitary gland in the female rat. *J. Endocrinol.* **62,** 573–588.

Ajika, K., Krulich, L., Fawcett, C. P., and McCann, S. M. (1972). Effects of estrogen on plasma and pituitary gonadotropins and prolactin and on hypothalamic releasing and inhibiting factors. *Neuroendocrinology* **9,** 304–315.

Albers, H. J., Bedford, J. M., and Chang, M. C. (1961). Uterine peptidase activity in the rat and rabbit during pseudopregnancy. *Am. J. Physiol.* **201,** 554–556.

Albert, A., and Mendoza, D. (1966). Daily fluctuations in excretion of FSH and LH by a postmenopausal woman. *J. Clin. Endocrinol. Metab.* **26,** 371–373.

Aldeen, K. M., and Finn, C. A. (1970). The implantation of blastocysts in the Russian Steppe Lemming (*Lagurus lagurus*). *J. Exp. Zool.* **173,** 63–78.

Alden, R. H. (1942). Aspects of the egg-ovary-oviduct relationship in the albino rat. 1. Egg passage and development following ovariectomy. *J. Exp. Zool.* **90,** 159–179.

Alden, R. H. (1947). Implantation of the rat egg. II. Alteration in osmophilic epithelial lipids of the rat uterus under normal and experimental conditions. *Anat. Rec.* **97,** 1–13.

Alfert, M., and Bern, H. (1951). Hormonal influence on nuclear synthesis. 1. Estrogen and uterine gland nuclei. *Proc. Natl. Acad. Sci. U.S.A.* **37,** 202–205.

Allen, E., Smith, G. M., and Gardner, W. U. (1937). Accentuation of the growth effect of theelin on genital tissues of the ovariectomised mouse by arrest of mitosis with colchicine. *Am. J. Anat.* **61,** 321–341.

Amoroso, E. C. (1952). Placentation. *In* "Marshall's Physiology of Reproduction" (A. S. Parkes, ed.), 3rd. ed. Vol. 2, pp. 127–311. Longmans Green, London and New York.

Amoroso, E. C., and Finn, C. A. (1962). Ovarian activity during gestation, ovum transport and implantation. *In* "The Ovary" (S. Zuckerman, ed.), 1st ed., Vol. 1, pp. 451–537. Academic Press, New York.

Anderson, C. H., and Greenwald, G. S. (1969). Autoradiographic analysis of estradiol uptake in the brain and pituitary of the female rat. *Endocrinology* **85**, 1160–1165.

Anderson, L. L., Butcher, R. L., and Melampy, R. M. (1961). Subtotal hysterectomy and ovarian function in gilts. *Endocrinology* **69**, 571–580.

Anderson, L. L., Bland, K. P., and Melampy, R. M. (1969). Comparative aspects of uterine-luteal relationships. *Recent Prog. Horm. Res.* **25**, 57–99.

Anderson, R. R. (1968). Uterine luteolysis in the rat: evidence for blood-borne and local actions. *J. Reprod. Fertil.* **16**, 423–431.

Arendarcik, J., and Maracek, I. (1968). FSH and LH level in serum following bilateral ovariectomy in sheep. *Endocrinol. Exp.* **2**, 193–197.

Arimura, A., and Schally, A. V. (1970). Progesterone suppression of LH releasing hormone-induced stimulation of LH release in rats. *Endocrinology* **87**, 653–657.

Arimura, A., and Schally, A. V. (1971). Augmentation of pituitary responsiveness to LH-releasing hormone (LH-RH) by estrogen. *Proc. Soc. Exp. Biol. Med.* **136**, 290–293.

Arimura, A., Debeljuk, L., and Schally, A. V. (1972). LH release by LH-releasing hormone in golden hamsters at various stages of the estrous cycle. *Proc. Soc. Exp. Biol. Med.* **140**, 609–612.

Arzac, J. P., and Blanchet, E. (1948). Alkaline phosphatase and glycogen in human endometrium. *J. Clin. Endocrinol. Metab.* **8**, 315–324.

Astwood, E. B. (1938). A six hour assay for the quantitative determination of estrogen. *Endocrinology* **23**, 25–31.

Atkinson, L. E., Bhattacharya, A. N., Monroe, S. E., Dierschke, D. J., and Knobil, E. (1970). Effects of gonadectomy on plasma LH concentration in the rhesus monkey. *Endocrinology* **87**, 847–849.

Atkinson, W. B. (1944). The persistence of deciduomata in the mouse. *Anat. Rec.* **88**, 271–280.

Atkinson, W. B., and Elftman, H. (1947). Mobilization of alkaline phosphatase in the uterus of the mouse by estrogen. *Endocrinology* **40**, 30–36.

Atkinson, W. B., and Engle, E. T. (1947). Studies on endometrial alkaline phosphatase during human menstrual cycle and in hormone treated monkey. *Endocrinology* **40**, 327–333.

Attramadal, A. (1970a). Cellular localization of ³H-oestradiol in the hypothalamus. *Z. Zellforsch. Mikrosk. Anat.* **104**, 572–581.

Attramadal, A. (1970b). Cellular localization of ³H-oestradiol in the hypophysis. *Z. Zellforsch. Mikrosk. Anat.* **104**, 597–614.

Austad, R., and Garm, O. (1959). Periodic acid-Schiff-positive material and alkaline phosphatase in the uterine wall of the pig during the sexual cycle. *Nature (London)* **184**, 999–1000.

Bainbridge, J. G., and Labhsetwar, A. P. (1971). The role of oestrogens in spontaneous ovulation; location of site of action of positive feedback of oestrogen by intracranial implantation of the anti-oestrogen ICI 46474. *J. Endocrinol.* **50**, 321–327.

Baird, D. T., and Guevara, A. (1969). Concentration of unconjugated estrone and estradiol in peripheral plasma in non-pregnant women throughout the menstrual cycle in castrate and post-menopausal women and in men. *J. Clin. Endocrinol. Metab.* **29,** 149 155.

Barker, K. L., and Warren, J. C. (1966). Estrogen control of carbohydrate metabolism in the rat uterus: pathways of glucose metabolism. *Endocrinology* **78,** 1205 1212.

Barnea, A., Gershonowitz, T., and Shelesnyak, M. C. (1968). Assessment of ovarian secretion of oestrogens during the oestrous cycle by the use of biological criteria. *J. Endocrinol.* **41,** 281 288.

Barraclough, C. A., and Cross, B. A. (1963). Unit activity in the hypothalamus of the cyclic female rat: effect of genital stimuli and progesterone. *J. Endocrinol.* **26,** 339 359.

Barraclough, C. A., and Haller, E. W. (1970). Positive and negative feedback effects of estrogen on pituitary LH synthesis and release in normal and androgen sterilized female rats. *Endocrinology* **86,** 542 551.

Beato, M., Lederer, B., Boquoi, E., and Sandritter, W. (1968). Effect of estrogens and gestagens on the initiation of DNA synthesis in the genital tract of ovariectomized mice. *Exp. Cell Res.* **52,** 173 179.

Beck, T. W., Nett, T. M., and Reeves, J. J. (1973). Serum FSH and LH in anestrous ewes treated with 17β-estradiol. *J. Anim. Sci.* **37,** 300.

Beier, H. M. (1968a). Investigations from the point of view of biochemistry and development physiology on the protein milieu for the development of the blastocyst of the rabbit (*Oryctolagus cuniculus*). *Zool. Jahrb.* **85,** 72 190.

Beier, H. M. (1968b). Uteroglobin: a hormone-sensitive endometrial protein involved in blastocyst development. *Biochim. Biophys. Acta* **160,** 289 291.

Bennet, G., and Leblond, C. P. (1970). Formation of cell coat material for the whole surface of columnar cells in the rat small intestine, as visualized by radioautography with L-fructose[3]H. *J. Cell Biol.* **46,** 409 416.

Bensley, C. M. (1951). Cyclic fluctuations in the rate of epithelial mitosis in the endometrium of the rhesus monkey. *Contrib. Embryol. Carnegie Inst.* **34,** 87 98.

Beyer, C., and Sawyer, C. H. (1969). Hypothalamic unit activity related to control of the pituitary gland. *In* "Frontiers in Neuroendocrinology" (W. F. Ganong and L. Martini, eds.), pp. 255 287. Oxford Univ. Press, London and New York.

Beyer, C., Ramirez, V. D., Whitmoyer, D. I., and Sawyer, C. H. (1967). Effects of hormones on the electrical activity of the brain in the rat and rabbit. *Exp. Neurol.* **18,** 313 326.

Bialy, G., and Pincus, G. (1962). Effects of estrogen and progesterone on the uterine carbonic anhydrase of immature rats. *Endocrinology* **70,** 781 785.

Biggers, J. D. (1953). The carbohydrate components of the vagina of the normal and ovariectomised mouse during oestrogenic stimulation. *J. Anat.* **87,** 327 336.

Biggers, J. D., and Claringbold, P. J. (1955). Mitotic activity in the vaginal epithelium of the mouse following local oestrogenic stimulation. *J. Anat.* **89,** 124 131.

Billard, R., and McDonald, P. G. (1973). Inhibition of ovulation in the rat by intrahypothalamic implants of an anti-oestrogen. *J. Endocrinol.* **56,** 585 590.

Bindon, B. M. (1969). Blood flow in the reproductive organs of the mouse after hypophysectomy, after gonadotrophin treatment, during the oestrous cycle and during early pregnancy. *J. Endocrinol.* **44,** 523 536.

Bishop, D. W. (1956). Active secretion in the rabbit oviduct. *Am. J. Physiol.* **187,** 347–352.

Blackwell, R. E., and Amoss, M. S. (1971). A sex difference in the rate of rise of plasma LH in rats following gonadectomy. *Proc. Soc. Exp. Biol. Med.* **136,** 11–14.

Blake, C. A., Norman, R. L., and Sawyer, C. H. (1972). Effects of estrogen and/or progesterone on serum and pituitary gonadotropin levels in ovariectomized rats. *Proc. Soc. Exp. Biol. Med.* **141,** 1100–1113.

Bland, K. P., and Donovan, B. T. (1966). The uterus and the control of ovarian function. *Adv. Reprod. Physiol.* **1,** 179–216.

Bland, K. P., and Donovan, B. T. (1968). Gonadal hormones and the regression of the corpora lutea in the guinea-pig. *Nature (London)* **220,** 179–180.

Bland, K. P., and Donovan, B. T. (1970). Oestrogen and progesterone and the function of the corpora lutea in the guinea-pig. *J. Endocrinol.* **47,** 225–230.

Blandau, R. (1969). Gamete transport—comparative aspects. *In* "The Mammalian Oviduct" (E. S. E. Hafez and R. Blandau, eds.), pp. 129–162. Univ. of Chicago Press, Chicago, Illinois.

Bo, W. J., Maraspin, L., and Smith, M. M. (1967). Glycogen synthetase activity in the rat uterus. *J. Endocrinol.* **38,** 33–37.

Bo, W. J., Odor, D. L., and Rothrock, M. L. (1969). Ultrastructure of uterine smooth muscle following progesterone or progesterone-estrogen treatment. *Anat. Rec.* **163,** 121–132.

Boettiger, E. G. (1946). Changes in the glycogen and water content of the rat uterus. *J. Cell. Comp. Physiol.* **27,** 9–14.

Bogdanove, E. M. (1963a). Failure of anterior hypothalamic lesions to prevent either pituitary reactions to castration or the inhibition of such reactions by estrogen treatment. *Endocrinology* **72,** 638–642.

Bogdanove, E. M. (1963b). Direct gonad-pituitary feedback: an analysis of effects of intracranial estrogenic depots on gonadotropin secretion. *Endocrinology* **73,** 696–712.

Bogdanove, E. M. (1964). The role of the brain in the regulation of pituitary gonadotropin secretion. *Vitam. Horm. (N.Y.)* **22,** 205–260.

Bogdanove, E. M., Schwartz, N. B., Reichert, L. R., and Midgley, A. R. (1971). Comparisons of pituitary: serum luteinizing hormone (LH) ratios in the castrated rat by radioimmunoassay and OAAD bioassay. *Endocrinology* **88,** 644–652.

Boling, J. L. (1969). Endocrinology of oviductal musculature. *In* "The Mammalian Oviduct" (E. S. E. Hafez and R. J. Blandau, eds.), pp. 163–181. Univ. of Chicago Press, Chicago, Illinois.

Bouin, P., and Ancel, P. (1910). Recherches sur les fonctions du corps jaune gestatif. I. Sur le déterminisme de la préparation de l'uterus à la fixation de l'oeuf. *J. Physiol. Pathol. Gen.* **12,** 1–16.

Brown, J. B. (1955). Urinary excretion of oestrogens during the menstrual cycle. *Lancet* **1,** 320–323.

Brown, J. B., Klopper, A., and Loraine, J. A. (1958). The urinary excretion of oestrogens, pregnanediol and gonadotrophins during the menstrual cycle. *J. Endocrinol.* **17,** 401–410.

Brown-Grant, K. (1969). The induction of ovulation by ovarian steroids in the adult rat. *J. Endocrinol.* **43,** 553–562.

Brown-Grant, K. (1972). The role of steroid hormones in the control of gonadotropin secretion in female mammals. In "Steroid Hormones and Brain Function" (C. H. Sawyer and R. A. Gorski, eds.), pp. 269–294. Univ. of California Press, Berkeley, California.

Brown-Grant, K., Exley, D., and Naftolin, F. (1970). Peripheral plasma oestradiol and luteinizing hormone concentrations during the oestrous cycle of the rat. J. Endocrinol. 48, 295–296.

Buchanan, G. D. (1966). Reproduction in the ferret (Mustela furo). I. Uterine histology and histochemistry during pregnancy and pseudopregnancy. Am. J. Anat. 118, 195–216.

Burger, H. G., Catt, K. J., and Brown, J. B. (1968). Relationship between plasma luteinizing hormone and urinary estrogen excretion during the menstrual cycle. J. Clin. Endocrinol. Metab. 28, 1508–1512.

Burger, H. G., Fink, G., and Lee, V. W. K. (1972). Luteinizing hormone releasing factor in ultrafiltrates of blood collected from the pituitary stalk of ovariectomized rats, and of rats subjected to electrical stimulation of the preoptic area. J. Endocrinol. 54, 227–237.

Burns, R. K. (1961). Role of hormones in the differentiation of sex. In "Sex and Internal Secretions" (W. C. Young, ed.), 3rd ed., Vol. 1, pp. 76–160. Williams & Wilkins, Baltimore, Maryland.

Burrows, H. (1949). "Biological Actions of Sex Hormones." Cambridge Univ. Press, London and New York.

Butcher, R. L., Chu, K. Y., and Melampy, R. M. (1962). Utero-ovarian relationships in the guinea-pig. Endocrinology 71, 810–815.

Caldwell, B. V. (1970). Uterine factors influencing corpus luteum function. Adv. Biosci. 4, 399–415.

Caldwell, B. V., Rowson, L. E. A., Moor, R. M., and Hay, M. F. (1969). The utero-ovarian relationship and its possible role in infertility. J. Reprod. Fertil., Suppl. 8, 59–76.

Caligaris, L., Astrada, J. J., and Taleisnik, S. (1967). Pituitary FSH concentrations in the rat during the estrous cycle. Endocrinology 81, 1261–1266.

Caligaris, L., Astrada, J. J., and Taleisnik, S. (1971a). Release of luteinizing hormone induced by estrogen injection into ovariectomized rats. Endocrinology 88, 810–815.

Caligaris, L., Astrada, J. J., and Taleisnik, S. (1971b). Biphasic effect of progesterone on the release of gonadotropin in rats. Endocrinology 89, 331–338.

Callantine, M. R., Humphrey, R. R., and Nesset, B. L. (1966). LH release by 17β-estradiol in the rat. Endocrinology 79, 455–456.

Carlborg, L. G. (1966). Quantitative determination of sialic acids in the mouse vagina. Endocrinology 78, 1093–1099.

Carroll, W. R. (1945). Variations in the water content of the rat uterus during continuous estrogenic treatment. Endocrinology 36, 266–271.

Chang, M. C. (1966). Effect of oral administration of medroxyprogesterone acetate and ethinyl estradiol on the transportation and development of rabbit eggs. Endocrinology 79, 939–948.

Chang, M. C. (1970). Hormonal regulation of sperm capacitation. In "Mechanisms Involved in Conception" (G. Raspé, ed.), pp. 13–24. Pergamon, Oxford.

Chester Jones, I., and Ball, J. N. (1962). Ovarian–pituitary relationships. *In* "The Ovary" (S. Zuckerman, ed.), 1st ed., Vol. 1, pp. 361–434. Academic Press, New York.

Chowers, I., and McCann, S. M. (1965). Content of luteinizing hormone–releasing factor and luteinizing hormone during the estrous cycle and after changes in gonadal steroid titers. *Endocrinology* **76**, 700–708.

Chowers, I., and McCann, S. M. (1967). Comparison of the effect of hypothalamic and pituitary implants of estrogen and testosterone on reproductive system and adrenal of female rats. *Proc. Soc. Exp. Biol. Med.* **124**, 260–266.

Clauberg, C. (1930). Sexual Hormones im Besanderen des Hormones des Corpus Luteum. *Zentralbl. Gynaekol.* **54**, 2757–2770.

Clemetson, C. A. B., Mallikarjuneswara, V. R., Moshfeghi, M. M., Carr, J. J., and Wilds, J. H. (1970). The effects of oestrogen and progesterone on the sodium and potassium concentration of rat uterine fluids. *J. Endocrinol.* **47**, 309–319.

Cole, D. F. (1950). The effects of oestradiol on the rat uterus. *J. Endocrinol.* **7**, 12–23.

Colville, E. A. (1968). The ultrastructure of the human endometrium. *J. Obstet. Gynaecol. Br. Commonw.* **75**, 342–350.

Corbin, A., and Daniels, E. L. (1967). Changes in concentration of female rat pituitary FSH and stalk-median eminence follicle stimulating hormone releasing factor with age. *Neuroendocrinology* **2**, 304–314.

Corker, C. S., Naftolin, F., and Exley, D. (1969). Interrelationships between plasma luteinizing hormone and oestradiol in the human menstrual cycle. *Nature (London)* **222**, 1063.

Corner, G. W. (1923). Cyclic variation in uterine and tubal contraction waves. *Am. J. Anat.* **32**, 345–351.

Corner, G. W. (1928). Physiology of the corpus luteum. I. The effect of very early ablation of the corpus luteum upon embryos and uterus. *Am. J. Physiol.* **86**, 74–81.

Cox, R. I., Mattner, P. E., and Thorburn, G. D. (1971). Changes in ovarian secretion of oestradiol-17β around oestrus in the sheep. *J. Endocrinol.* **49**, 345–346.

Crighton, D. B., Hartley, B. M., and Lamming, G. E. (1973). Changes in the luteinizing hormone releasing activity of the hypothalamus, and in pituitary gland and plasma luteinizing hormone during the oestrous cycle of the sheep. *J. Endocrinol.* **58**, 377–385.

Cross, B. A., and Dyer, R. G. (1970). Characterization of unit activity in hypothalamic islands with special reference to hormone effects. *In* "The Hypothalamus" (L. Martini, M. Motta, and F. Fraschini, eds.), pp. 115–122. Academic Press, New York.

Cross, B. A., and Dyer, R. G. (1972). Ovarian modulation of unit activity in the anterior hypothalamus of the cyclic rat. *J. Physiol. (London)* **222**, 25P.

Cross, B. A., and Green, J. D. (1959). Activity of single neurones in the hypothalamus: effect of osmotic and other stimuli. *J. Physiol. (London)* **148**, 544–569.

Cross, B. A., and Kitay, J. I. (1967). Unit activity in diencephalic islands. *Exp. Neurol.* **19**, 316–330.

Cross, B. A., and Silver, I. A. (1965). Effect of luteal hormones on the behaviour of hypothalamic neurons in pseudopregnant rats. *J. Endocrinol.* **31**, 251–263.

Csapo, A. (1950). Actomyosin formation by estrogen action. *Am. J. Physiol.* **162**, 406–410.

Csapo, A. (1956). Progesterone 'block.' *Am. J. Anat.* **98**, 273–291.

Csapo, A. (1969). The four direct regulatory factors of myometrial function. *Ciba Found. Study Group* **34**, 13–42.

Csapo, A., and Corner, G. W. (1953). The effect of estrogen on the isometric tension of rabbit uterine strips. *Science* **117**, 162–164.

Cumming, I. A., Buckmaster, J. M., Cerini, J. C., Cerini, M. E., Chamley, W. A., Findlay, J. K., and Goding, J. R. (1972). Effect of progesterone on the release of luteinizing hormone induced by a synthetic gonadotrophin-releasing factor in the ewe. *Neuroendocrinology* **10**, 338–348.

Daron, G. H. (1936). The arterial pattern of the tunica mucosa of the uterus in macacus rhesus. *Am. J. Anat.* **58**, 349–419.

David, M. A., Fraschini, F., and Martini, L. (1965). Parallelisme entre le contenu hypophysaire en FSH et le contenu hypothalamique on FSH–RF (FSH–releasing factor). *C. R. Acad. Sci., (Paris) Ser. D* **261**, 2249–2251.

Davidson, J. M. (1969). Feedback control of gonadotropin secretion. *In* "Frontiers in Neuroendocrinology" (W. F. Ganong and L. Martini, eds.), pp. 343–388. Oxford Univ. Press, London and New York.

Davidson, J. M., and Sawyer, C. H. (1961). Effects of localized intracerebral implantation of oestrogen on reproductive function in the female rabbit. *Acta Endocrinol. (Copenhagen)* **37**, 385–393.

Deanesly, R. (1960). Implantation and early pregnancy in ovariectomized guinea-pigs. *J. Reprod. Fertil.* **1**, 242–248.

Deanesly, R. (1963). Further observations on the effects of oestradiol on tubal eggs and implantation in the guinea-pig. *J. Reprod. Fertil.* **5**, 49–57.

Deanesly, R. (1966). Pro-oestrus in the guinea-pig: hormonal stimulation of the vaginal epithelium. *J. Reprod. Fertil.* **12**, 205–212.

Debeljuk, L., Arimura, A., and Schally, A. V. (1972). Effect of estradiol and progesterone on the LH release induced by LH-releasing hormone (LH-RH) in intact diestrous rats and anestrous ewes. *Proc. Soc. Exp. Biol. Med.* **139**, 774–777.

De Feo, V. J. (1963). Temporal aspects of uterine sensitivity in the pseudo-pregnant or pregnant rat. *Endocrinology* **72**, 305–316.

De Feo, V. J. (1967). Decidualization. *In* "Cellular Biology of the Uterus" (R. M. Wynn, ed.), pp. 191–290. North-Holland Publ., Amsterdam.

Dessouky, D. A. (1968). Electron microscopic studies of the myometrium of the guinea-pig. *Am. J. Obstet. Gynecol.* **100**, 30–41.

Dessouky, D. A. (1969). Fine structural changes of the uterine smooth muscle boundary during gestation. *Am. J. Obstet. Gynecol.* **103**, 1117–1124.

Dewar, A. D. (1973). Effects of hysterectomy on corpus luteum activity in the cyclic, pseudopregnant and pregnant mouse. *J. Reprod. Fertil.* **33**, 77–89.

Dickson, W. M., Bosc, M. J., and Locatelli, A. (1969). Effect of estrogen and progesterone on uterine blood flow of castrate sows. *Am. J. Physiol.* **217**, 1431–1434.

Dierschke, D. J., Bhattacharya, A. N., Atkinson, L. E., and Knobil, E. (1970). Circhoral oscillations of plasma LH levels in the ovariectomized rhesus monkey. *Endocrinology* **87**, 850–853.

Dierschke, D. J., Yamaji, T., Karsch, F. J., Weick, R. F. Weiss, G., and Knobil, E. (1973). Blockade by progesterone of estrogen-induced LH and FSH release in the rhesus monkey. *Endocrinology* **92**, 1496–1501.

Döcke, F., and Dörner, G. (1965). The mechanism of the induction of ovulation by oestrogens. *J. Endocrinol.* **33**, 491–499.

Döcke, F., and Dörner, G. (1969). A possible mechanism by which progesterone facilitates ovulation in the rat. *Neuroendocrinology* **4**, 139–149.

Donovan, B. T. (1971). The control of ovarian function. *Acta Endocrinol. (Copenhagen)* **66**, 1–15.

Dugan, F. A., Radhakrishnamurthy, B., Rudman, R. A., and Berenson, G. S. (1968). Stimulation of synthesis of glycoprotein enzymes in the rat uterus by oestradiol. *J. Endocrinol.* **42**, 261–266.

Dyer, R. G., Pritchett, C. J., and Cross, B. A. (1972). Unit activity in the diencephalon of female rats during the oestrous cycle. *J. Endocrinol.* **53**, 151–160.

Eckstein, P. (1949). The induction of progesterone withdrawal bleeding in spayed rhesus monkeys. *J. Endocrinol.* **6**, 405–411.

Eisenfeld, A. J. (1970). ³H-Estradiol: *in vitro* binding to macromolecules from the rat hypothalamus, anterior pituitary and uterus. *Endocrinology* **86**, 1313–1326.

Eisenfeld, A. J., and Axelrod, J. (1966). Effect of steroid hormones, ovariectomy, estrogen pretreatment, sex and immaturity on the distribution of ³H-estradiol. *Endocrinology* **79**, 38–42.

Eisenfeld, A. J., and Axelrod, J. (1967). Evidence for estradiol binding sites in the hypothalamus—effect of drugs. *Biochem. Pharmacol.* **16**, 1781–1785.

Elftman, H. (1958). Estrogen control of the phospholipids of the uterus. *Endocrinology* **62**, 410–415.

Elftman, H. (1963). Estrogen-induced changes in the Golgi apparatus and lipid of the uterine epithelium of the rat in the normal cycle. *Anat. Rec.* **146**, 139–144.

Emmens, C. W. (1950). Estrogens. *In* "Hormone Assay" (C. W. Emmens, ed.), pp. 391–417. Academic Press, New York.

Enders, A. C. (1967). The uterus in delayed implantation. *In* "Cellular Biology of the Uterus" (R. M. Wynn, ed.), pp. 151–190. North-Holland Publ., Amsterdam.

Endroczi, E. (1968). Effects of hormones on brainstem and diencephalic structures in controlling behavioral reactions. *In* "Progress in Endocrinology" (C. Gual, ed.), pp. 317–320. Excerpta Med. Found., Amsterdam.

Engle, E. T. and Smith, P. E. (1938). The endometrium of the monkey and oestrone–progesterone balance. *Am. J. Anat.* **63**, 349–365.

Everett, J. W. (1948). Progesterone and oestrogen in the experimental control of ovulation time and other features of the estrous cycle in the rat. *Endocrinology* **43**, 389–405.

Everett, J. W. (1961). The mammalian female reproductive cycle and its controlling mechanisms. *In* "Sex and Internal Secretions" (W. C. Young, ed.), 3rd ed., Vol. 1, pp. 497–555. Williams & Wilkins, Baltimore, Maryland.

Everett, J. W. (1962). The influence of oestriol and progesterone on the endometrium of the guinea-pig *in vitro*. *J. Endocrinol.* **24**, 491–496.

Everett, J. W. (1964). Central neural control of reproductive function of the adenohypophysis. *Physiol. Rev.* **44**, 373–431.

Everett, J. W. (1972). Brain, pituitary gland and the ovarian cycle. *Biol. Reprod.* **6**, 3–12.

Exley, D., Gellert, R. J., Harris, G. W., and Nadler, R. D. (1968). The site of action of chlormadinone acetate (6-chloro-⁶Δ-dehydro-17α-acetoxyprogesterone) in blocking ovulation in the mated rabbit. *J. Physiol. (London)* **195**, 697–714.

Feldman, S., and Dafny, I. V. (1970). Effects of extra-hypothalamic structures on sensory projections to the hypothalamus. *In* "The Hypothalamus" (L. Martini, M. Motta, and F. Fraschini, eds.), pp. 103–114. Academic Press, New York.

Fendler, K., and Endroczi, E. (1965–1966). Effects of hypothalamic steroid implants on compensatory ovarian hypertrophy of rats. *Neuroendocrinology* **1**, 129–137.

Ferin, M., Tempone, A., Zimmering, P. E., and Vande Wiele, R. L. (1969). Effects of antibodies to 17β-estradiol and progesterone on the estrous cycle of the rat. *Endocrinology* **85**, 1070–1078.

Ferin, M., Dyrenfurth, I., Cowchock, S., Warren, M., and Vande Wiele, R. L. (1974). Active immunization to 17β-estradiol and its effects upon the reproductive cycle of the rhesus monkey. *Endocrinology* **94**, 765–776.

Fink, G., and Harris, G. W. (1970). The luteinizing hormone releasing activity of extracts of blood from the hypophysial portal blood vessels of rats. *J. Physiol. (London)* **208**, 221–241.

Fink, G., and Jamieson, M. G. (1974a). Effect of electrical stimulation of the preoptic area on luteinizing hormone releasing factor in pituitary stalk blood. *J. Physiol. (London)* **237**, 37P–38P.

Fink, G., and Jamieson, M. G. (1974b). Effect of preoptic stimulation on luteinizing hormone releasing factor in pituitary stalk blood during the oestrous cycle of the rat. *J. Physiol. (London)* **241**, 128P–130P.

Fink, G., and Jamieson, M. G. (1976). Immunoreactive luteinizing hormone releasing factor in rat pituitary stalk blood: effects of electrical stimulation of the medial preoptic area. *J. Endocrinol.* **68**, 71–87.

Finn, C. A. (1966). Endocrine control of endometrial sensitivity during the induction of the decidual cell reaction in the mouse. *J. Endocrinol.* **36**, 239–248.

Finn, C. A. (1970). The ageing uterus and its influence on reproductive capacity. *J. Reprod. Fertil., Suppl.* **12**, 31–38.

Finn, C. A., and Keen, P. M. (1962). Studies on deciduomata formation in the rat. *J. Reprod. Fertil.* **4**, 215–216.

Finn, C. A., and McLaren, A. (1967). A study of the early stages of implantation in mice. *J. Reprod. Fertil.* **13**, 259–267.

Finn, C. A., and Martin, L. (1967). Patterns of cell division in the mouse uterus during early pregnancy. *J. Endocrinol.* **39**, 593–597.

Finn, C. A., and Martin, L. (1969). Hormone secretion during early pregnancy in the mouse. *J. Endocrinol.* **45**, 57–65.

Finn, C. A., and Martin, L. (1970). The role of the oestrogen secreted before oestrus in the preparation of the uterus for implantation in the mouse. *J. Endocrinol.* **47**, 431–438.

Finn, C. A., and Martin, L. (1971). Endocrine control of the proliferation and secretion of uterine glands in the mouse. *Acta Endocrinol. (Copenhagen), Suppl.* **155**, 139.

Finn, C. A., and Martin, L. (1974). The control of implantation. *J. Reprod. Fertil.* **39**, 195–206.

Finn, C. A., and Porter, D. G. (1975). "The Uterus". Elek Science, London.

Finn, C. A., Martin, L., and Carter, J. (1969). A refractory period following oestrogenic stimulation of cell division in the mouse uterus. *J. Endocrinol.* **44**, 121–126.

Fishman, W. H., and Farmelant, M. H. (1953). Effect of androgens and estrogens on β-glucuronidase in inbred mice. *Endocrinology* **52**, 536–545.

Fitch, K. L. (1963). A study of uterine glycogen during the oestrous cycle of the dog. *J. Morphol.* **113**, 331–343.

Flerkó, B. (1954). Die Epithelien des Eileiters und ihre hormonalen Reaktionen. *Z. Mikrosk.-Anat. Forsch.* **61**, 99–118.

Flerkó, B. (1966). Control of gonadotrophin secretion in the female. *In* "Neuroendocrinology" (L. Martini and W. F. Ganong, eds.), Vol. 1, pp. 613–668. Academic Press, New York.

Flerkó, B., and Szentágothai, J. (1957). Oestrogen sensitive nervous structures in the hypothalamus. *Acta Endocrinol. (Copenhagen)* **26**, 121–127.

Frank, R. T., Bonham, C., and Gustavson, R. E. (1925). A new method of assaying the potency of the female sex hormone based upon its effect on the spontaneous contraction of the uterus of the white rat. *Am. J. Physiol.* **74**, 395–399.

Gay, V. L., and Midgley, A. R. (1969). Response of the adult rat to orchidectomy and ovariectomy as determined by LH radioimmunoassay. *Endocrinology* **84**, 1359–1364.

Gersten, B. E., and Baker, B. L. (1970). Local action of intrahypophyseal implants of estrogen as revealed by staining with peroxidase-labeled antibody. *Am. J. Anat.* **128**, 1–20.

Gillman, J. (1940). Fat: an index of oestrogen and progesterone activity in the human endometrium. *Nature (London)* **146**, 402.

Gillman, J. (1941). Profound vascular changes induced in the uterus of the castrated rabbit by combinations of oestradiol benzoate and progesterone. *Endocrinology* **29**, 336–342.

Ginsburg, M., MacLusky, N. J., Morris, I. D., and Thomas, P. J. (1972). Cyclical fluctuation of oestradiol receptors in hypothalamus and pituitary. *J. Physiol. (London)* **224**, 72P–74P.

Goding, J. R., Catt, K. J., Brown, J. M. Kaltenbach, C. C., Cumming, I. A., and Mole, B. J. (1969). Radioimmunoassay for ovine luteinizing hormone. Secretion of luteinizing hormone during estrus and following estrogen administration in the sheep. *Endocrinology* **85**, 133–142.

Goldman, B. D., and Mahesh, V. B. (1968). Fluctuations in pituitary FSH during the ovulatory cycle in the rat and a possible role of FSH in the induction of ovulation. *Endocrinology* **83**, 97–106.

Goldman, B. D., and Porter, J. C. (1970). Serum LH levels in intact and castrated golden hamsters. *Endocrinology* **87**, 676–679.

Greenwald, G. S. (1958). Formation of deciduomata in the lactating mouse. *J. Endocrinol.* **17**, 24–28.

Greenwald, G. S. (1961). A study of the transport of ova through the rabbit oviduct. *Fertil. Steril.* **12**, 80–95.

Greenwald, G. S. (1967). Species differences in egg transport in response to exogenous estrogen. *Anat. Rec.* **157**, 163–172.

Greenwald, G. S. (1969). Endocrinology of oviductal secretions. *In* "The Mammalian Oviduct" (E. S. E. Hafez and R. J. Blandau, eds.), pp. 183–201. Univ. of Chicago Press, Chicago, Illinois.

Greep, R. O. (1961). Physiology of the anterior hypophysis in relation to reproduction. *In* "Sex and Internal Secretions" (W. C. Young, ed.), 3rd ed., Vol. 1, pp. 240–301. Williams & Wilkins, Baltimore, Maryland.

Greep, R. O. (1973). A vista of research on mammalian gonadotropins. *Biol. Reprod.* **8**, 2–10.

Greep, R. O., and Chester Jones, I. (1950). Steroid control of pituitary function. *Recent Prog. Horm. Res.* **5**, 197–261.

Greiss, F. C., and Anderson, S. G. (1969). Uterine vascular changes during the ovarian cycle. *Am. J. Obstet. Gynecol.* **103**, 629–640.

Hadek, R. (1955). The secretory processes in the sheep's oviduct. *Anat. Rec.* **121**, 187–201.

Ham, K. N., Hurley, J. V., Lopata, A., and Ryan, G. B. (1970). A combined isotopic and electron microscopic study of the response of the rat uterus to exogenous oestradiol. *J. Endocrinol.* **46**, 71–81.

Hamburger, C., and Johnson, S. G. (1957). Studies on urinary gonadotrophins. III. Qualitative differences between urinary gonadotrophins of young males and of post menopausal women. *Acta Endocrinol. (Copenhagen)* **26**, 1–29.

Harper, M. J. K. (1966). Hormonal control of transport of eggs in cumulus through the ampulla of the rabbit oviduct. *Endocrinology* **78**, 568–574.

Harris, G. W., and Sherratt, R. M. (1969). The action of chlormadinone acetate (6-chloro-Δ^6-dehydro-17α-acetoxyprogesterone) upon experimentally induced ovulation in the rabbit. *J. Physiol. (London)* **203**, 59–66.

Harris, R. S., and Cohen, S. L. (1951). The influence of ovarian hormones on the enzymic activity of tissues. *Endocrinology* **48**, 264–272.

Hartman, C. G. (1939). Ovulation, fertilization and the transport and viability of eggs and spermatozoa. *In* "Sex and Internal Secretions" (E. Allen, ed.), 2nd ed., pp. 674–733. Williams & Wilkins, Baltimore, Maryland.

Hashimoto, I., Henricks, D. M., Anderson, L. L., and Melampy, R. M. (1968). Progesterone and pregn-4-en-20α-ol-one in ovarian venous blood during various reproductive states in the rat. *Endocrinology* **82**, 333–341.

Heap, R. B. (1962). Some chemical constituents of uterine washings: a method of analysis with results from various species. *J. Endocrinol.* **24**, 367–378.

Heap, R. B., and Lamming, G. E. (1962). The influence of ovarian hormones on some chemical constituents of the uterine washings of the rat and rabbit. *J. Endocrinol.* **25**, 57–68.

Heap, R. B., and Lamming, G. E. (1963). An acid-soluble component of uterine washings. *J. Endocrinol.* **27**, 265–266.

Hechter, O., Krohn, L., and Harris, J. (1941). The effect of estrogen on the permeability of the uterine capillaries. *Endocrinology* **29**, 386–392.

Henricks, O. M., Guthrie, H. D., and Handlin, D. L. (1972). Plasma estrogen, progesterone and luteinizing hormone levels during the estrous cycle in pigs. *Biol. Reprod.* **6**, 210–218.

Herlant, M. (1956). Correlations hypophysogenitales chez la femelle de la chauve-souris, *Myotis myotis* (Borkhausen). *Arch. Biol.* **67**, 89–180.

Herlant, M. (1964). The cells of the adenohypophysis and their functional significance. *Int. Rev. Cytol.* **17**, 299–382.

Hester, L. L., Kellett, W. W., Spicer, S. S., Williamson, H. O., and Pratt-Thomas, H. R. (1970). Effects of the intrauterine contraceptive device on endometrial enzyme and carbohydrate histochemistry. *Am. J. Obstet. Gynecol.* **106**, 1144–1151.

Hilliard, J., Croxatto, H. B., Hayward, J. N., and Sawyer, C. H. (1966). Norethindrone blockade of LH release to intrapituitary infusion of hypothalamic extract. *Endocrinology* **79**, 411–419.

Hilliard, J., Penardi, R., and Sawyer, C. H. (1967). A functional role for 20α-hydroxy-pregn-4-en-3-one in the rabbit. *Endocrinology* **80**, 901–909.

Hilliard, J., Spies, H. G., and Sawyer, C. H. (1968). Cholesterol storage and progestin secretion during pregnancy and pseudopregnancy in the rabbit. *Endocrinology* **82,** 157–165.

Hilliard, J., Schally, A. V., and Sawyer, C. H. (1971). Progesterone blockade of the ovulatory response to intrapituitary infusion of LH-RH in rabbits. *Endocrinology* **88,** 730–736.

Hisaw, F. L. (1942). The interaction of the ovarian hormones in experimental menstruation. *Endocrinology* **30,** 301–308.

Hisaw, F. L. (1950). Factors influencing endometrial growth in monkeys (*Macaca mulatta*). *In* "A Symposium on Steroid Hormones" (E. S. Gordon, ed.), pp. 259–276. Univ. of Wisconsin Press, Madison.

Hisaw, F. L., and Hisaw, F. L., Jr. (1961). Action of estrogen and progesterone on the reproductive tract of lower primates. *In* "Sex and Internal Secretions" (W. C. Young, ed.), 3rd ed., Vol. 1, pp. 556–589. Williams & Wilkins, Baltimore, Maryland.

Hisaw, F. L., Greep, R. O., and Fevold, H. L. (1937). The effect of oestrin-progestin combinations on the endometrium, vagina and sexual skin of monkeys. *Am. J. Anat.* **61,** 483–494.

Hobson, W., and Hansel, W. (1974). Increased *in vitro* pituitary response to LH-RH after *in vivo* estrogen treatment. *Proc. Soc. Exp. Biol. Med.* **146,** 470–474.

Hoffmann, J. C., and Schwartz, N. B. (1965). Timing of ovulation following progesterone withdrawal in the rat. *Endocrinology* **76,** 626–631.

Hohlweg, W. (1934). Veranderungen des Hypophysenvarderlappens und des Ovariums nach Behandlungen mit grossen Dosen von Follikelhormon. *Klin. Wochenschr.* **13,** 92–95.

Holmdahl, T. H., and Mastroianni, L. (1965). Continuous collection of rabbit oviduct secretions at low temperature. *Fertil. Steril.* **16,** 587–595.

Homberger, F., Bernfeld, P., Tregier, A., Grossman, M. S., and Harpel, P. (1963). Endometrial secretions. *Ann. N.Y. Acad. Sci.* **106,** 683–691.

Hooker, C. W. (1945). A criterion of luteal activity in the mouse. *Anat. Rec.* **93,** 333–347.

Hooker, C. W., and Forbes, T. R. (1947). A bio-assay for minute amounts of progesterone. *Endocrinology* **41,** 158–169.

Hori, T., Ide, M., and Miyake, T. (1968). Ovarian estrogen secretion during the estrous cycle and under the influence of exogenous gonadotropins in rats. *Endocrinol. Jpn.* **15,** 215–222.

Horning, E. S. (1942–1944). The reaction of the uterine epithelium of the rat to oestrogenic stimulation. *J. Endocrinol.* **3,** 260–267.

Hotchkiss, J., Atkinson, L. E., and Knobil, E. (1971). Time course of serum estrogen and luteinizing hormone concentrations during the menstrual cycle of the rhesus monkey. *Endocrinology* **89,** 177–183.

Humphrey, K. W. (1968). The effects of oestradiol-3,17β on tubal transport in the laboratory mouse. *J. Endocrinol.* **42,** 17–26.

Humphrey, K. W., and Martin, L. (1968). The effect of oestrogen and anti-oestrogens on ovum transport in mice. *J. Reprod. Fertil.* **15,** 191–197.

Ito, S. (1969). Structure and function of the glycocalyx. *Fed. Proc., Fed. Am. Soc. Exp. Biol.* **28,** 12–25.

Jackson, G. L., Roche, J. F., Foster, D. L., and Dziuk, P. J. (1971). LH-releasing activity in the hypothalamus of anestrous and cyclic ewes. *Biol. Reprod.* **5,** 5–12.

Jagiello, G. (1965). Effects of selected hormones on the closed vaginal membrane of the ovariectomized guinea-pig. *Proc. Soc. Exp. Biol. Med.* **118**, 412–414.

Johansson, E. D. B., and Wide, L. (1969). Periovulatory levels of plasma progesterone and luteinizing hormone in women. *Acta Endocrinol. (Copenhagen)* **62**, 82–88.

Johnson, D. C. (1971). Temporal aspects of oestrogen stimulated increases in plasma and pituitary follicle stimulating hormone. *Acta Endocrinol. (Copenhagen)* **66**, 89–98.

Joseph, N. R., Angel, M. B., and Catchpole, H. R. (1954). Homeostasis of connective tissue. *Arch. Pathol.* **58**, 40–58.

Kalman, S. M. (1955). Some studies on estrogens and uterine permeability. *J. Pharmacol. Exp. Ther.* **115**, 442–448.

Kalman, S. M. (1958). The effect of oestrogens on uterine blood flow in the rat. *J. Pharmacol. Exp. Ther.* **124**, 179–181.

Kalra, P. S., Fawcett, C. P., Krulich, L., and McCann, S. M. (1973a). The effects of gonadal steroids on plasma gonadotropins and prolactin in the rat. *Endocrinology* **92**, 1256–1268.

Kalra, S. P., Krulich, L., and McCann, S. M. (1973b). Changes in gonadotropin releasing factor content in the rat hypothalamus following electrochemical stimulation of anterior hypothalamic area and during the estrous cycle. *Neuroendocrinology* **12**, 321–333.

Kanematsu, S., and Sawyer, C. H. (1963a). Effects of intrahypothalamic and intrahypophysial estrogen on pituitary prolactin and lactation in the rabbit. *Endocrinology* **72**, 243–251.

Kanematsu, S., and Sawyer, C. H. (1963b). Effects of hypothalamic and hypophysial estrogen implants on pituitary gonadotrophic cells in ovariectomized rabbits. *Endocrinology* **73**, 687–695.

Kanematsu, S., and Sawyer, C. H. (1963c). Effects of hypothalamic estrogen implants on pituitary LH and prolactin in rabbits. *Am. J. Physiol.* **205**, 1073–1076.

Kanematsu, S., and Sawyer, C. H. (1964). Effects of hypothalamic and hypophysial estrogen implants on pituitary and plasma LH in ovariectomized rabbits. *Endocrinology* **75**, 579–585.

Karsch, F. J., Dierschke, D. J., Weick, R. F., Yamaji, T., Hotchkiss, J., and Knobil, E. (1973a). Positive and negative feedback control by estrogen of luteinizing hormone secretion in the rhesus monkey. *Endocrinology* **92**, 799–804.

Karsch, F. J., Weick, R. F., Hotchkiss, J., Dierschke, D. J., and Knobil, E. (1973b). An analysis of the negative feedback control of gonadotropin secretion utilizing chronic implantation of ovarian steroids in ovariectomized rhesus monkeys. *Endocrinology* **93**, 478–486.

Kato, J. (1970a). *In vitro* uptake of tritiated oestradiol by the rat anterior hypothalamus during the oestrous cycle. *Acta Endocrinol. (Copenhagen)* **63**, 577–584.

Kato, J. (1970b). *In vitro* uptake of tritiated oestradiol by the anterior hypothalamus and hypophysis of the rat. *Acta Endocrinol. (Copenhagen)* **64**, 687–695.

Kato, J., and Villee, C. A. (1967). Factors affecting uptake of estradiol-6,7-^{3}H by the hypophysis and hypothalamus. *Endocrinology* **80**, 1133–1138.

Kato, J., Inaba, M., and Kobayashi, T. (1969). Variable uptake of tritiated oestradiol by the anterior hypothalamus in the postpubertal female rat. *Acta Endocrinol. (Copenhagen)* **61**, 585–591.

Kawakami, M., and Sawyer, C. H. (1959). Neuroendocrine correlates of changes in brain activity thresholds by sex steroids and pituitary hormones. *Endocrinology* **65,** 652–668.

Kawakami, M., Seto, K., Yoshida, K., and Miyamoto, T. (1969). Biosynthesis of ovarian steroids in the rabbit. Influence of progesterone or estradiol implantation into the hypothalamus and limbic structures. *Neuroendocrinology* **5,** 303–321.

Kazama, N., and Hansel, W. (1970). Preovulatory changes in the progesterone level of bovine peripheral blood plasma. *Endocrinology* **86,** 1252–1256.

Keller, P. J. (1966). Studies on pituitary gonadotropins in human plasma. II. FSH and LH in male and post menopausal plasma. *Acta Endocrinol. (Copenhagen)* **52,** 348–356.

Kerdelhué, B., Jutisz, M., Gillessen, D., and Studer, R. O. (1973). Obtention d'immunoserums spécifiques du facteur hypothalamique (LH, FSH-RH) qui stimule la libération des gonadotropines hypophysaires et développement du dosage radioimmunologique de ce peptide. *C. R. Acad. Sci., (Paris) Ser. D* **276,** 593–597.

Khayyal, M. A., and Scott, C. M. (1931). The oxygen consumption of the isolated uterus of the rat and mouse. *J. Physiol. (London)* **72,** 13–14.

King, R. J., Gordon, J., and Inman, D. R. (1965). The intracellular localization of oestrogen in rat tissues. *J. Endocrinol.* **32,** 9–15.

Klein, M. (1937). The mucification of the vaginal epithelium in rodents. *Proc. R. Soc. London, Ser. B* **124,** 23–29.

Knobil, E. (1974). On the control of gonadotropin secretion in the rhesus monkey. *Recent Prog. Horm. Res.* **30,** 1–46.

Knowles, F. G. W. (1971). Secretory cells in the ependyma. *In* "Subcellular Organization and Function in Endocrine Tissues" (H. Heller and K. Lederis, eds.), pp. 875–881. Cambridge Univ. Press, London and New York.

Kohorn, E. I., and Tchao, R. (1969). Conversion of proliferative endometrium to secretory epithelium by progesterone in organ culture. *J. Endocrinol.* **45,** 401–405.

Komisaruk, B. R. (1971). Strategies in neuroendocrine neurophysiology. *Am. Zool.* **11,** 741–754.

Komisaruk, B. R., McDonald, P. G., Whitmoyer, D. I., and Sawyer, C. H. (1967). Effects of progesterone and sensory stimulation on EEG and neuronal activity in the rat. *Exp. Neurol.* **19,** 494–507.

Krehbiel, R. H. (1941). The effects of theelin on delayed implantation in the pregnant lactating rat. *Anat. Rec.* **81,** 381–392.

Krey, L. C., Yamaji, T., Butler, W. R., and Knobil, E. (1972). Phases of the menstrual cycle and responsiveness to synthetic LH-RH in the rhesus monkey. *Fed. Proc., Fed. Am. Soc. Exp. Biol.* **31,** 276.

Krey, L. C., Butler, W. R., Weiss, G., Weick, R. F., Dierschke, D. J., and Knobil, E. (1973). Influences of endogenous and exogenous gonadal steroids on the actions of synthetic LRF in the rhesus monkey. *In* "Hypothalamic Hypophysiotropic Hormones" (C. Gual and E. Rosenberg, eds.), Int. Congr. Ser. No. 263, pp. 39–47. Excerpta Med. Found., Amsterdam.

Krichesky, B. (1943). Vascular changes in the rabbit uterus and in intraocular endometrial transplants during pregnancy. *Anat. Rec.* **87,** 221–233.

Krohn, P. L. (1949). The induction of uterine bleeding in normal monkeys by means of progesterone. *J. Endocrinol.* **6,** xxxii.

Kuriyama, H. (1961). The effect of progesterone and oxytocin on the mouse myometrium. *J. Physiol. (London)* **159**, 26–39.

Labhsetwar, A. P. (1970a). Role of estrogens in ovulation: a study using the estrogen-antagonist I.C.I. 46,474. *Endocrinology* **87**, 542–551.

Labhsetwar, A. P. (1970b). The role of oestrogens in spontaneous ovulation; evidence for positive oestrogen feedback in the 4-day oestrous cycle. *J. Endocrinol.* **47**, 481–493.

Larsen, J. F. (1962). Electron microscopy of the uterine epithelium in the rabbit. *J. Cell Biol.* **14**, 49–62.

Laumas, K. R., and Farooq, A. (1966). The uptake *in vivo* of [1,2-³H]progesterone by the brain and genital tract of the rat. *J. Endocrinol.* **36**, 95–96.

Lawton, I. E., and Smith, S. W. (1970). LH secretory patterns in intact and gonadectomized male and female rats. *Am. J. Physiol.* **219**, 1019–1022.

Libertun, C., Cooper, K. J., Fawcett, C. P., and McCann, S. M. (1974a). Effects of ovariectomy and steroid treatment of hypophyseal sensitivity to purified LH-releasing factor (LRF). *Endocrinology* **94**, 518–525.

Libertun, C., Orias, R., and McCann, S. M. (1974b). Biphasic effect of estrogen on the sensitivity of the pituitary to luteinizing hormone-releasing factor (LRF). *Endocrinology* **94**, 1094–1100.

Lincoln, D. W. (1967). Unit activity in the hypothalamus, septum and preoptic area of the rat: characteristics of spontaneous activity and the effect of oestrogen. *J. Endocrinol.* **37**, 177–189.

Lincoln, D. W., and Cross, B. A. (1967). Effect of oestrogen on the responsiveness of neurones in the hypothalamus, septum and preoptic area of rats with light induced persistent oestrus. *J. Endocrinol.* **37**, 191–203.

Lindner, H. R., and Zmigrod, A. (1967). Microdetermination of progestins in rat ovaries: progesterone and 20α-hydroxy-pregn-4-en-3-one content during prooestrus, oestrus and pseudopregnancy. *Acta Endocrinol. (Copenhagen)* **56**, 16–26.

Lisk, R. D. (1960). Estrogen-sensitive centers in the hypothalamus of the rat. *J. Exp. Zool.* **145**, 197–205.

Lisk, R. D. (1963). Maintenance of normal pituitary weight and cytology in the spayed rat following estradiol implants in the arcuate nucleus. *Anat. Rec.* **146**, 281–291.

Littlejohn, B. M., and de Groot, J. (1963). Estrogen sensitive areas in the rat brain. *Fed. Proc., Fed. Am. Soc. Exp. Biol.* **22**, 571.

Loeb, L. (1908). The experimental production of the maternal placenta and the formation of the corpus luteum. *J. Am. Med. Assoc.* **50**, 1897–1901.

Loeb, L. (1923). The effect of extirpation of the uterus on the life and function of the corpus luteum in the guinea-pig. *Proc. Soc. Exp. Biol. Med.* **20**, 441–443.

Long, J. A., and Evans, H. M. (1922). The oestrous cycle in the rat and its associated phenomena. *Mem. Univ. Calif.* **6**, 1–148.

Luttge, W. G., and Whalen, R. E. (1970). Regional localization of estrogenic metabolities in the brain of male and female rats. *Steroids* **15**, 605–612.

Lutwak-Mann, C. (1955). Carbonic anhydrase in the female reproductive tract. *J. Endocrinol.* **13**, 26–38.

McCann, S. M. (1962). Effect of progesterone on plasma luteinizing hormone activity. *Am. J. Physiol.* **202**, 601–604.

McCann, S. M., and Taleisnik, S. (1961). The effect of estrogen on plasma luteinizing hormone (LH) activity in the rat. *Endocrinology* **69**, 909–914.

McCann, S. M., Watanabe, S., Crighton, D. B., Beddow, D., and Dhariwal, A. P. S. (1968). The physiology and biochemistry of luteinizing hormone releasing factor and follicle stimulating hormone releasing factor. *In* "Pharmacology of Hormonal Polypeptides and Proteins" (J. Meites, ed.), pp. 112–122. Plenum, New York.

McClintock, J. A., and Schwartz, N. B. (1968). Changes in pituitary and plasma follicle stimulating hormone concentrations during the rat estrous cycle. *Endocrinology* **83**, 433–441.

McDonald, M. F., and Bellvé, A. R. (1969). Influence of oestrogens and progesterone on flow of fluid from the Fallopian tube in the ovariectomized ewe. *J. Reprod. Fertil.* **20**, 51–61.

McDonald, P. G., and Clegg, M. T. (1966). Some factors affecting gonadotropin levels in sheep. *Proc. Soc. Exp. Biol. Med.* **121**, 482–485.

McDonald, P. G., and Clegg, M. T. (1967). The effect of progesterone on serum luteinizing hormone concentrations in the ewe. *J. Reprod. Fertil.* **13**, 75–82.

McDonald, P. G., and Gilmore, D. P. (1971a). The effect of ovarian steroids on hypothalamic thresholds for ovulation in the female rat. *J. Endocrinol.* **49**, 421–429.

McDonald, P. G., and Gilmore, D. P. (1971b). The effect of norethindrone on hypothalamic and pituitary thresholds for the induction of ovulation in the rat. *Neuroendocrinology* **7**, 46–53.

McDonald, P. G., Vidal, N., and Beyer, C. (1970). Sexual behaviour in the ovariectomized rabbit after treatment with different amounts of gonadal hormones. *Horm. Behav.* **1**, 161–172.

McGuire, J. L., and Lisk, R. D. (1969). Localization of oestrogen receptors in the rat hypothalamus. *Neuroendocrinology* **4**, 289–295.

McLaren, A. (1970). Early embryo-endometrial relationships. *In* "Ovo-implantation, Human Gonadotrophins and Prolactin" (J. F. Hubinon *et al.*, eds.), pp. 18–37. Karger, Basel.

McLaren, A., and Michie, D. (1956). Studies on the transfer of fertilized mouse eggs to uterine foster mothers. I. Factors affecting the implantation and survival of native and transferred eggs. *J. Exp. Biol.* **33**, 394–416.

McPhail, M. K. (1934). The assay of progestin. *J. Physiol. (London)* **83**, 145–156.

Marcus, G. J. (1974). Hormonal control of proliferation in the guinea-pig uterus. *J. Endocrinol.* **63**, 89–97.

Marcus, G. J., and Shelesnyak, M. C. (1970). Steroids in nidation. *Adv. Steroid Biochem. Pharmacol.* **2**, 373–440.

Marcus, G. J., Shelesnyak, M. C., and Kraicer, P. F. (1964). Studies on the mechanism of nidation. X. The oestrogen-surge, histamine release and decidual function in the rat. *Acta Endocrinol. (Copenhagen)* **47**, 255–264.

Mark, J. S. T. (1956). An electron microscope study of uterine smooth muscle. *Anat. Rec.* **125**, 473–493.

Marshall, J. M. (1969). The effect of ovarian hormones on the uterine response to adrenergic nerve stimulation and to adrenergic amines. *Ciba Found. Study Group* **34**, 89–101.

Martin, L. (1960). The use of 2-3-5-triphenyltetrazolium chloride in the biological assay of oestrogens. *J. Endocrinol.* **20**, 187–197.

Martin, L. (1965). Interactions between oestradiol and progestogens in the uterus of the mouse. *J. Endocrinol.* **26**, 31–39.

Martin, L., and Claringbold, P. J. (1960). The mitogenic action of oestrogens in the vaginal epithelium of the ovariectomized mouse. *J. Endocrinol.* **20**, 173–186.

Martin, L., and Finn, C. A. (1968). Hormonal regulation of cell division in epithelial and connective tissues of the mouse uterus. *J. Endocrinol.* **41**, 363–371.

Martin, L., and Finn, C. A. (1969). Duration of progesterone treatment required for a stromal response to oestradiol-17β in the uterus of the mouse. *J. Endocrinol.* **44**, 279–280.

Martin, L., Finn, C. A., and Carter, J. (1970). Effects of progesterone and oestradiol-17β on the luminal epithelium of the mouse uterus. *J. Reprod. Fertil.* **21**, 461–469.

Martini, L., Fraschini, F., and Motta, M. (1968). Neural control of anterior pituitary functions. *Recent Prog. Horm. Res.* **24**, 439–485.

Martini, L., Piva, F., and Motta, M. (1970). The hypothalamic-pituitary axis and ovulation. *In* "Ovo-implantation, Human Gonadotrophins and Prolactin" (J. F. Hubinon *et al.,* eds.), pp. 170–180. Karger, Basel.

Mastroianni, L., Beer, F., Shah, U., and Clewe, T. H. (1961). Endocrine regulation of oviduct secretions in the rabbit. *Endocrinology* **68**, 92–100.

Matsuo, H., Baba, Y., Nair, R. M. G., and Schally, A. V. (1971a). Structure of the porcine LH- and FSH-releasing hormone. I. The proposed amino acid sequence. *Biochem. Biophys. Res. Commun.* **43**, 1334–1339.

Matsuo, H., Baba, Y., Nair, R. M. G., Arimura, A., and Schally, A. V. (1971b). Synthesis of the porcine LH- and FSH-releasing hormone by the solid phase method. *Biochem. Biophys. Res. Commun.* **45**, 822–827.

Mayer, G. (1959). Recent studies on hormonal control of delayed implantation and superimplantation in the rat. *Mem. Soc. Endocrinol.* **6**, 76–83.

Mazer, R. S., and Wright, P. A. (1968). A hamster uterine luteolytic extract. *Endocrinology* **83**, 1065–1070.

Melampy, R. M., and Anderson, L. L. (1968). Role of the uterus in corpus luteum function. *J. Anim. Sci.* **27**, Suppl. 1, 77–96.

Mellin, T. N., and Erb, R. E. (1966). Estrogen metabolism and excretion during the bovine estrous cycle. *Steroids* **7**, 589–606.

Meyers, K. P. (1970). Hormonal requirements for the maintenance of oestradiol-induced inhibition of uterine sensitivity in the ovariectomized rat. *J. Endocrinol.* **46**, 341–346.

Michael, C. A., and Schofield, B. M. (1969). The influence of the ovarian hormones on the actomyosin content and the development of tension in uterine muscle. *J. Endocrinol.* **44**, 501–511.

Michael, R. P. (1965). Oestrogens in the central nervous system. *Br. Med. Bull.* **21**, 87–91.

Michael, R. P., and Keverne, E. B. (1970). A male sex-attractant pheromone in rhesus monkey vaginal secretions. *J. Endocrinol.* **46**, xx–xxi.

Mittler, J. C., and Meites, J. (1966). Effects of hypothalamic extract and androgen on pituitary FSH release *in vitro. Endocrinology* **78**, 500–504.

Monroe, B. E., Parlow, A. F., and Midgley, A. R. (1968). Radioimmunoassay for rat luteinizing hormone. *Endocrinology* **83**, 1004–1012.

Moor, R. M., and Rowson, L. E. A. (1966a). Local uterine mechanisms affecting luteal function in the sheep. *J. Reprod. Fertil.* **11**, 307–310.

Moor, R. M., and Rowson, L. E. A. (1966b). The corpus luteum of the sheep: functional relationship between the embryo and the corpus luteum. *J. Endocrinol.* **34**, 233–239.

Moore, C. R., and Price, D. (1932). Gonad hormone functions and the reciprocal influence between gonads and hypophysis with its bearing on the problem of sex hormone antagonism. *Am. J. Anat.* **50,** 13–71.

Moore, N. E., Barrett, S., Brown, J. B., Schindler, I., Smith, M. A., and Smyth, B. (1969). Oestrogen and progesterone content of ovarian vein blood of the ewe during the oestrous cycle. *J. Endocrinol.* **44,** 55–62.

Morau, H. (1889). Des transformations épithéliales de la muqueuse du vagin de quelques ronguers. *J. Anat. (Paris)* **25,** 275–297.

Moses, L., and Catchpole, H. R. (1955). Determination of water soluble mucoprotein fractions in normal and estrogen stimulated tissues. *Fed. Proc. Fed. Am. Soc. Exp. Biol.* **14,** 104.

Moss, S., Wrenn, T. R., and Sykes, J. F. (1954). Alkaline phosphatase, glycogen and periodic acid-Schiff positive substances in the bovine uterus during the estrous cycle. *Endocrinology* **55,** 261–273.

Moszkowska, A., and Kordon, C. (1965). Contrôle hypothalamique de la fonction gonadotrope et variation du taux des GRF chez le rat. *Gen. Comp. Endocrinol.* **5,** 596–613.

Murdoch, R. N., and White, I. G. (1968). The effect of oestradiol and progesterone on the activity of enzymes in the endometrium and caruncles of the ovariectomized ewe. *J. Endocrinol.* **42,** 187–192.

Nakane, P. K. (1970). Classification of anterior pituitary cell types with immunoenzyme histochemistry. *J. Histochem. Cytochem.* **18,** 9–20.

Nakane, P. K. (1971). Application of peroxidase labelled antibodies to the intracellular localisation of hormones. *Acta Endocrinol. (Copenhagen), Suppl.* **153,** 190–202.

Nallar, R., and McCann, S. M. (1965). Luteinizing hormone-releasing activity in plasma of hypophysectomized rats. *Endocrinology* **76,** 272–275.

Nallar, R., Antunes-Rodrigues, J., and McCann, S. M. (1966). Effect of progesterone on the level of plasma luteinizing hormone (LH) in normal female rats. *Endocrinology* **79,** 907–920.

Negro-Vilar, A., Sar, M., and Meites, J. (1970). Changes in hypothalamic FSH-RF and pituitary FSH during the estrous cycle of rats. *Endocrinology* **87,** 1091–1093.

Nett, T. M., Akbar, A. M., Niswender, G. D., Hedlund, M. T., and White, W. F. (1973). A radioimmunoassay for gonadotropin-releasing hormone (Gn-RH) in serum. *J. Clin. Endocrinol. Metab.* **36,** 880–885.

Nett, T. M., Akbar, A. M., and Niswender, G. D. (1974). Serum levels of luteinizing hormone and gonadotropin releasing hormone in cycling, castrated and anestrous ewes. *Endocrinology* **94,** 713–718.

Nillius, S. J., and Wide, L. (1972). Variation in LH and FSH response to LH-releasing hormone during the menstrual cycle. *J. Obstet. Gynaecol. Br. Commonw.* **79,** 865–873.

Nilsson, O. (1958a). Ultrastructure of mouse uterine surface epithelium under different estrogenic influences. 1. Spayed animals and estrous animals. *J. Ultrastruct. Res.* **1,** 375–396.

Nilsson, O. (1958b). Ultrastructure of mouse uterine surface epithelium under different estrogenic influences. 2. Early effect of estrogen administered to spayed animals. *J. Ultrastruct. Res.* **2,** 73–95.

Nilsson, O. (1958c). Ultrastructure of mouse uterine surface epithelium under different estrogenic influences. 3. Late effect of estrogen administered to spayed animals. *J. Ultrastruct. Res.* **2,** 185–199.

Nilsson, O. (1959). Ultrastructure of mouse uterine surface epithelium under different estrogenic influences. 4. Uterine secretion. *J. Ultrastruct. Res.* **2,** 331–341.

Nilsson, O. (1966a). Structural differentiation of luminal membrane in rat uterus during normal and experimental implantations. *Z. Anat. Entwicklungsgesch.* **125,** 152–159.

Nilsson, O. (1966b). Estrogen-induced increase of adhesiveness in uterine epithelium of mouse and rat. *Exp. Cell Res.* **43,** 239–241.

Nilsson, O., and Reinius, S. (1969). Light and electron microscopic structure of the oviduct. *In* "The Mammalian Oviduct" (E. S. E. Hafez and R. J. Blandau, eds.), pp. 57–83. Univ. of Chicago Press, Chicago, Illinois.

Niswender, G. D., Reichert, L. E., Midgley, A. R., and Nalbandov, A. V. (1969). Radioimmunoassay for bovine and ovine luteinizing hormone. *Endocrinology* **84,** 1166–1173.

Norman, R. L., Eleftheriou, B. E., Spies, H. G., and Hoppe, P. (1968). Free plasma estrogens in the ewe during the estrous cycle. *Steroids* **11,** 667–671.

Noyes, R. W., and Dickmann, Z. (1960). Relationship of ovular age to endometrial development. *J. Reprod. Fertil.* **1,** 186–196.

Noyes, R. W., Adams, C. E., and Walton, A. (1959). The transport of ova in relation to the dosage of oestrogen in ovariectomized rabbits. *J. Endocrinol.* **18,** 108–117.

Ogawa, Y. and Pincus, G. (1962). Estrogen effects on the carbonic anhydrase content of mouse uteri. *Endocrinology* **70,** 359–364.

Orr, A. H., Ward, A. P., and Bagshawe, K. D. (1970). Radioimmunoassay measurement of luteinizing hormone excretion rates following bilateral oophorectomy. *J. Reprod. Fertil.* **21,** 207–213.

Orsini, M. W., and Meyer, R. K. (1959). Implantation of the castrate hamster in the absence of exogenous estrogen. *Anat. Rec.* **134,** 619–620.

Palka, Y. S., Ramirez, V. D., and Sawyer, C. H. (1966). Distribution and biological effects of tritiated estradiol implanted in the hypothalamo-hypophyseal region of female rats. *Endocrinology* **78,** 487–499.

Pant, H. C., and Ward, W. R. (1973). Effect of progesterone on pituitary responsiveness to luteinizing hormone releasing hormone (LH-RH) in intact anoestrous ewes. *J. Physiol. (London)* **232,** 45P–47P.

Parlow, A. F. (1964a). Effect of ovariectomy on pituitary and serum gonadotrophins in the mouse. *Endocrinology* **74,** 102–107.

Parlow, A. F. (1964b). Comparison of pituitary and serum gonadotrophins of the rat. *Endocrinology* **74,** 489–492.

Parlow, A. F. (1964c). Differential action of small doses of estradiol on gonadotrophins in the rat. *Endocrinology* **75,** 1–8.

Parlow, A. F., and Hendrich, C. E. (1970). Peculiar gonadotropin profile of the adult guinea-pig pituitary and its response to gonadectomy. *Endocrinology* **87,** 444–448.

Parlow, A. F., Anderson, L. L., and Melampy, R. M. (1964). Pituitary follicle stimulating hormone and luteinizing hormone concentrations in relation to reproductive stages of the pig. *Endocrinology* **75,** 365–376.

Pelletier, J., and Signoret, J-P. (1969). Contrôle de la décharge de LH dans le sang par la progesterone et le benzoate d'oestradiol chez la brebis castrée. *C. R. Acad. Sci., (Paris) Ser. D* **269**, 2595–2598.

Perry, J. S., and Rowlands, I. W. (1961). Effect of hysterectomy on the ovarian cycle of the rat. *J. Reprod. Fertil.* **2**, 332–340.

Pharriss, B. B. (1971). Prostaglandin induction of luteolysis. *Ann. N.Y. Acad. Sci.* **180**, 436–444.

Phelps, D. (1946). Endometrial vascular reactions and the mechanism of nidation. *Am. J. Anat.* **79**, 167–197.

Piacsek, B. E., and Meites, J. (1966). Effects of castration and gonadal hormones on hypothalamic content of luteinizing hormone releasing factor (LRF). *Endocrinology* **79**, 432–439.

Piacsek, B. E., Schneider, T. C., and Gay, V. L. (1971). Sequential study of luteinizing hormone (LH) and "progestin" secretion on the afternoon of proestrus in the rat. *Endocrinology* **89**, 39–45.

Pollard, I., Martin, L., and Shorey, C. D. (1966). The effects of intravaginal oestradiol on the cell structure of the vaginal epithelium of the ovariectomized mouse. *Steroids* **8**, 805–823.

Prill, H. G., and Gotz, F. (1961). Blood flow in the myometrium and endometrium of the uterus. *Am. J. Obstet. Gynecol.* **82**, 102–108.

Przékop, F., and Domański, E. (1970). Hypothalamic centers involved in the control of gonadotrophin secretion and ovulation in the sheep. *Acta Physiol. Pol.* **11**, 34–49.

Psychoyos, A. (1963). Précisions sur l'état de "non-receptivité" de l'uterus. *C. R. Acad. Sci., (Paris) Ser. D* **257**, 1153–1156.

Psychoyos, A. (1967). The hormonal interplay controlling egg-implantation in the rat. *Adv. Reprod. Physiol.* **2**, 257–278.

Purves, H. D. (1961). Morphology of the hypophysis related to its function. *In* "Sex and Internal Secretions" (W. C. Young, ed.), 3rd ed., Vol. 1, pp. 161–239. Williams & Wilkins, Baltimore, Maryland.

Purves, H. D., and Griesbach, W. E. (1951). The significance of the Gomori staining of the basophils of the rat pituitary. *Endocrinology* **49**, 652–662.

Purves, H. D., and Griesbach, W. E. (1954). The site of follicle stimulating and luteinizing hormone production in the rat pituitary. *Endocrinology* **55**, 785–793.

Purves, H. D., and Griesbach, W. E. (1957). A study on the cytology of the adenohypophysis of the dog. *J. Endocrinol.* **14**, 361–370.

Radford, H. M., Wheatley, I. S., and Wallace, A. L. C. (1969). The effects of oestradiol benzoate and progesterone on secretion of luteinizing hormone in the ovariectomized ewe. *J. Endocrinol.* **44**, 135–136.

Raeside, J. I. (1963). Urinary oestrogen excretion in the pig at oestrus and during the oestrous cycle. *J. Reprod. Fertil.* **6**, 421–426.

Ramirez, V. D., and McCann, S. M. (1964). Induction of prolactin secretion by implants of estrogen into the hypothalamo-hypophysial region of female rats. *Endocrinology* **75**, 206–214.

Ramirez, V. D., and Sawyer, C. H. (1965). Fluctuations in hypothalamic LH-RF (luteinizing hormone-releasing factor) during the rat estrous cycle. *Endocrinology* **76**, 282–289.

Ramirez, V. D., and Sawyer, C. H. (1966). Changes in hypothalamic luteinizing hormone-releasing factor (LH-RF) in the female rat during puberty. *Endocrinology* **78**, 958–964.

Ramirez, V. D., Abrams, R. M., and McCann, S. M. (1964). Effect of estradiol implants in the hypothalamo-hypophysial region of the rat on the secretion of luteinizing hormone. *Endocrinology* **75**, 243 248.

Ramirez, V. D., Komisaruk, B. R., Whitmoyer, D. I., and Sawyer, C. H. (1967). Effects of hormones and vaginal stimulation on the EEG. *Am. J. Physiol.* **212**, 1376 1384.

Rayford, P. L., Brinkley, H. J., and Young, E. P. (1971). Radioimmunoassay determination of LH concentration in the serum of female pigs. *Endocrinology* **88**, 707 713.

Reese, J. D., Koneff, A. A., and Wainman, P. (1943). Cytological differences between castration and thyroidectomy basophils in the rat hypophysis. *In* "Essays in Biology," pp. 471 485. Univ. of California Press, Berkeley.

Reeves, J. J., Arimura, A., and Schally, A. V. (1971a). Changes in pituitary responsiveness to luteinizing hormone-releasing hormone (LH-RH) in anestrous ewes pretreated with estradiol benzoate. *Biol. Reprod.* **4**, 88 93.

Reeves, J. J., Arimura, A., and Schally, A. V. (1971b). Pituitary responsiveness to purified luteinizing hormone-releasing hormone (LH-RH) at various stages of the estrous cycle in sheep. *J. Anim. Sci.* **32**, 123 127.

Reeves, J. J., Beck, T. W., and Nett, T. M. (1974). Serum FSH in anestrous ewes treated with 17β-estradiol. *J. Anim. Sci.* **38**, 374 377.

Rennie, P., and Davies, J. (1965). Implantation in the rabbit following administration of 20α-hydroxy-pregnen-3-one and 20β-hydroxy-pregnen-3-one. *Endocrinology* **76**, 535 537.

Resko, J. A., Norman, R. L., Niswender, G. D., and Spies, H. G. (1974). The relationship between progestins and gonadotropins during the late luteal phase of the menstrual cycle in rhesus monkeys. *Endocrinology* **94**, 128 135.

Roche, J. F., Foster, D. L., Karsch, F. J., and Dziuk, P. J. (1970). Effect of castration and infusion of melatonin on levels of luteinizing hormone in sera and pituitaries of ewes. *Endocrinology* **87**, 1205 1210.

Ross, R., and Klebanoff, S. J. (1967). Fine structural changes in uterine smooth muscle and fibroblasts in response to estrogen. *J. Cell Biol.* **32**, 155 167.

Rothchild, I. (1965). Interrelations between progesterone and the ovary, pituitary and central nervous system in the control of ovulation and the regulation of progesterone secretion. *Vitam. Horm.* (*N.Y.*) **23**, 209 327.

Rowlands, I. W. (1961). Effect of hysterectomy at different stages in the life cycle of the corpus luteum of the guinea-pig. *J. Reprod. Fertil.* **2**, 341 350.

Sani, G., and Hanau, R. (1952). Alkaline phosphatase in the genital organs of the female rat in relation to the sexual cycle and hormone administration. *Arch. Ital. Anat. Embriol.* **57**, 211 224.

Sawyer, C. H. (1964). Control of secretion of gonadotropins. *In* "Gonadotropins" (H. H. Cole, ed.), pp. 113 159. Freeman, San Francisco, California.

Schally, A. V., Carter, W. H., Saito, M., Arimura, A., and Bowers, C. Y. (1968). Studies on the site of action of oral contraceptive steroids. I. Effect of antifertility steroids on plasma LH levels and on the response to luteinizing hormone-releasing factor in rats. *J. Clin. Endocrinol. Metab.* **28**, 1747 1755.

Schally, A. V., Parlow, A. F., Carter, W. H., Saito, M., Bowers, C. Y., and Arimura, A. (1970). Studies on the site of action of oral contraceptive steroids. II. Plasma LH and FSH levels after administration of antifertility steroids and LH-releasing hormone (LH-RH). *Endocrinology* **86**, 530 541.

Schmidt, I. G. (1943). Proliferation in the genital tract of the normal mature guinea-pig treated with colchicine. *Am. J. Anat.* **73**, 59–80.

Schneider, H. P. G., and McCann, S. M. (1970). Estradiol and the neuroendocrine control of LH release *in vitro. Endocrinology* **87**, 330–338.

Schultz, R. H., Burcalow, H. B., Fahning, M. L., Graham, E. F., and Weber, A. F. (1969). A karyometric study of epithelial cells lining the glands of the bovine endometrium. *J. Reprod. Fertil.* **19**, 169–171.

Schwartz, N. B., and Bartosik, D. (1962). Changes in pituitary LH content during the rat estrous cycle. *Endocrinology* **71**, 756–762.

Seiki, K., and Hattori, M. (1971). A more extensive study on the uptake of labelled progesterone by the hypothalamus and pituitary gland of rats. *J. Endocrinol.* **51**, 793–794.

Seiki, K., Higashida, M., Imanishi, Y., Miyamoto, M., Kitagawa, T., and Kotani, M. (1968). Radioactivity in the rat hypothalamus and pituitary after injection of labelled progesterone. *J. Endocrinol.* **41**, 109–110.

Seiki, K., Miyamoto, M., Yamashita, A., and Kotani, M. (1969). Further studies on the uptake of labelled progesterone by the hypothalamus and pituitary of rats. *J. Endocrinol.* **43**, 129–130.

Shaikh, A. A., and Abraham, G. E. (1969). Measurement of estrogen surge during pseudopregnancy in rats by radioimmunoassay. *Biol. Reprod.* **1**, 378–380.

Shelesnyak, M. C. (1931). The induction of pseudopregnancy in the rat by means of electrical stimulation. *Anat. Rec.* **49**, 179–183.

Shirley, B., Wolinsky, J., and Schwartz, N. B. (1968). Effects of a single injection of an estrogen antagonist on the estrous cycle of the rat. *Endocrinology* **82**, 959–968.

Siler, T. M., and Yen, S. S. C. (1973). Augmented gonadotropin response to synthetic LRF in hypogonadal state. *J. Clin. Endocrinol. Metab.* **37**, 491–494.

Smith, E. R., and Davidson, J. M. (1974). Luteinizing hormone-releasing factor in rats exposed to constant light: effects of mating. *Neuroendocrinology* **14**, 129–138.

Smith, E. R., Weick, R. F., and Davidson, J. M. (1969). Influence of intracerebral progesterone on the reproductive system of female rats. *Endocrinology* **85**, 1129–1136.

Somerville, B. W. (1971). Daily variation in plasma levels of progesterone and estradiol throughout the menstrual cycle. *Am. J. Obstet. Gynecol.* **111**, 419–426.

Spies, H. G., Stevens, K. R., Hilliard, J., and Sawyer, C. H. (1969). The pituitary as a site of progesterone and chlormadinone blockade of ovulation in the rabbit. *Endocrinology* **84**, 277–284.

Stevens, K. T., Spies, H. G., Hilliard, J., and Sawyer, C. H. (1970). Site(s) of action of progesterone in blocking ovulation in the rat. *Endocrinology* **86**, 970–975.

Stevens, V. C., Sparks, S. J., and Powell, J. E. (1970). Levels of estrogens, progestogens and luteinizing hormone during the menstrual cycle of the baboon. *Endocrinology* **87**, 658–666.

Stieve, H. (1929). Das Mesenchym in der Wand der menschlichen Gebarmutter. *Zentralbl. Gynaekol.* **53**, 2706–2732.

Stockard, C. R., and Papanicolaou, G. N. (1917). The existence of a typical oestrous cycle in the guinea-pig—with a study of its histological and physiological changes. *Am. J. Anat.* **22**, 225–283.

Stockdale, F. E., and Topper, Y. J. (1966). The role of DNA synthesis and mitosis in hormone dependent differentiation. *Proc. Natl. Acad. Sci. U.S.A.* **56,** 1283–1289.

Stumpf, W. E. (1968). Estradiol-concentrating neurons: topography in the hypothalamus by dry-mount autoradiography. *Science* **162,** 1001–1003.

Stumpf, W. E. (1969). Too much noise in the autoradiogram. *Science* **163,** 958–959.

Stumpf, W. E. (1970). Estrogen-neurons and estrogen-neuron systems in the periventricular brain. *Am. J. Anat.* **129,** 207–217.

Stumpf, W. E., and Sar, M. (1971). Estradiol concentrating neurons in the amygdala. *Proc. Soc. Exp. Biol. Med.* **138,** 102–106.

Swanson, L. V., Kirton, K. T., Hackett, A. J., and Hafs, H. D. (1971). Pituitary and blood plasma levels of gonadotropin after ovariectomy of heifers. *J. Anim. Sci.* **32,** 678–681.

Swelheim, T. (1965). The influence of a single high dose of oestradiol benzoate on the ICSH-content in the serum of gonadectomized male and female rats. *Acta Endocrinol. (Copenhagen)* **49,** 231–238.

Swerdloff, R. S., and Odell, W. D. (1968). Gonadotropins: present concepts in the human. *Calif. Med.* **109,** 467–485.

Swerdloff, R. S., Jacobs, H. S., and Odell, W. D. (1972). Synergistic role of progestogens in estrogen induction of LH and FSH surge. *Endocrinology* **90,** 1529–1536.

Szentágothai, J., Flerkó, B., Mess, B., and Halász, B. (1972). "Hypothalamic Control of the Anterior Pituitary," 4th ed. Akadémiai Kiadó, Budapest.

Terasawa, E., and Sawyer, C. H. (1970). Diurnal variation in the effects of progesterone on multiple unit activity in the rat hypothalamus. *Exp. Neurol.* **27,** 359–374.

Terasawa, E., and Timiras, P. S. (1968). Electrical activity during the estrous cycle of the rat: cyclic changes in limbic structures. *Endocrinology* **83,** 207–216.

Terasawa, E., and Timiras, P. S. (1969). Cyclic changes in electrical activity of the rat midbrain reticular formation during the oestrous cycle. *Brain Res.* **14,** 189–198.

Thomas, P. J. (1973). Steroid hormones and their receptors. *J. Endocrinol.* **57,** 333–359.

Thompson, I. E., Arfania, J., and Taymor, M. L. (1973). Effects of estrogen and progesterone on pituitary response to stimulation by luteinizing hormone-releasing factor. *J. Clin. Endocrinol. Metab.* **37,** 152–155.

van Rees, G. P. (1959). The effect of progesterone on the ICSH and FSH content of anterior pituitary and blood. II. FSH; miscellaneous. *Acta Physiol. Pharmacol. Neerl.* **8,** 195–210.

van Rees, G. P. (1964). Interplay between steroid sex hormones and secretion of FSH and ICSH. *In* "Major Problems in Neuroendocrinology" (E. Bajusz and G. Jasmin, eds.), pp. 322–345. Karger, Basel.

Weichert, C. K. (1942). The experimental control of prolonged pregnancy in the lactating rat by means of oestrogen. *Anat. Rec.* **83,** 1–15.

Weick, R. F., and Davidson, J. M. (1970). Localization of the stimulatory feedback effect of estrogen on ovulation in the rat. *Endocrinology* **87,** 693–700.

Weick, R. F., Smith, E. R., Dominguez, R., Dhariwal, A. P. S., and Davidson, J. M. (1971). Mechanism of stimulatory feedback effect of estradiol benzoate on the pituitary. *Endocrinology* **88,** 293–301.

Whalen, R. E., and Maurer, R. A. (1969). Estrogen "receptors" in brain: an unsolved problem. *Proc. Natl. Acad. Sci. U.S.A.* **63,** 681–685.

Williams, M. F. (1948). The vascular architecture of the rat uterus as influenced by estrogen and progesterone. *Am. J. Anat.* **83,** 247–307.

Wilson, E. W. (1969). The effect of oestradiol-17β on enzymes concerned with metabolism of carbohydrate in human endometrium *in vitro. J. Endocrinol.* **44,** 63–68.

Wiltbank, J. N., and Casida, L. E. (1956). Alteration of ovarian activity by hysterectomy. *J. Anim. Sci.* **15,** 134.

Wu, J. T., Dickmann, Z., and Johnson, D. C. (1971). Effects of ovariectomy or hypophysectomy on day one of pregnancy on development and transport of fertilized rat eggs. *J. Endocrinol.* **49,** 507–513.

Wurtman, R. J., Axelrod, J., and Kelly, D. S. (1968). "The Pineal." Academic Press, New York.

Yamaji, T., Dierschke, D. J., Hotchkiss, J., Bhattacharya, A. N., Surve, A. H., and Knobil, E. (1971). Estrogen induction of LH release in the rhesus monkey. *Endocrinology* **89,** 1034–1041.

Yamaji, T., Dierschke, D. J., Bhattacharya, A. N., and Knobil, E. (1972). The negative feedback control by estradiol and progesterone of LH secretion in the ovariectomized rhesus monkey. *Endocrinology* **90,** 771–777.

Yen, S. S. C., VandenBerg, G., Rebar, R., and Ehara, Y. (1972). Variation in pituitary responsiveness to synthetic LRF during different stages of the menstrual cycle. *J. Clin. Endocrinol. Metab.* **35,** 931–934.

Yen, S. S. C., VandenBerg, G., and Siler, T. M. (1974). Modulation of pituitary responsiveness to LRF by estrogen. *J. Clin. Endocrinol. Metab.* **39,** 170–177.

Yochim, J. M., and De Feo, V. J. (1962). Control of decidual growth in the rat by steroid hormones of the ovary. *Endocrinology* **71,** 134–142.

Yochim, J. M., and De Feo, V. J. (1963). Hormonal control of the onset, magnitude and duration of uterine sensitivity in the rat by steroid hormones of the ovary. *Endocrinology* **72,** 317–326.

Yoshinaga, K., Hawkins, R. A., and Stocker, J. F. (1969). Estrogen secretion by the rat ovary *in vivo* during the estrous cycle and pregnancy. *Endocrinology* **85,** 103–112.

Zachariae, F. (1958). Autoradiographic (^{35}S) and histochemical studies of sulphomucopolysaccharides in the rabbit uterus, oviducts and vagina. Variations under hormonal stimulus. *Acta Endocrinol. (Copenhagen)* **29,** 118–134.

Zeilmaker, G. H. (1966). The biphasic effect of progesterone on ovulation in the rat. *Acta Endocrinol. (Copenhagen)* **51,** 461–468.

4

Hypothalamus–Pituitary Control of the Ovary

J. S. M. Hutchinson and P. J. Sharp

I. INTRODUCTION

This chapter provides a review of recent studies on mammals and, where possible, on other classes of vertebrates (Chester Jones and Ball, 1962, give references to the earlier literature).* In general, only stimulatory components of the hypothalamus–pituitary–ovarian regulatory system will be considered. The effects of ovarian secretions on this system and the effects of the external environment are dealt with elsewhere (Volume II, Chapter 10; Volume III, Chapter 3), as are atresia (Volume I, Chapter 6) and pregnancy (Volume II, Chapter 8).

The anterior lobe of the pituitary is essential for normal ovarian function in mammals and probably in all vertebrates. Clinical and experimental

* The literature survey for this chapter was completed in 1971.

work (Crowe *et al.*, 1910; Aschner, 1912; Smith, 1916) established a relationship between the anterior pituitary and the ovary. The establishment of a technique for hypophysectomy in the rat by Smith (1926a) encouraged study of gonadotropic hormones which were first detected in pituitary tissue by Smith (1926b) and Zondek and Aschheim (1927). Following these findings Fevold *et al.* (1931) described two active fractions prepared from the anterior pituitary, one of which brought about stimulation of follicle growth; the second caused the rupture of the follicle and its transformation into a corpus luteum. Accordingly, the two fractions were designated follicle-stimulating hormone (FSH) and luteinizing hormone (LH). Astwood and Fevold (1939) and Astwood (1941) later suggested that a third gonadotropin, luteotropin, activated the corpus luteum. The latter hormone, thought to be identical with prolactin, though possessing only lactogenic properties in many mammals, has luteotropic, corpus luteum-maintaining properties in others. However, the problem of the control of the secretory capacity and life span of the corpus luteum is very complex. It has become apparent that anterior pituitary and placental luteotropins and/or anterior pituitary or uterine luteolysins and possibly antiluteolysins are involved to various degrees in different mammals in the control of the corpus luteum (Rothchild, 1966; Anderson *et al.*, 1969; Moor, 1970).

That the central nervous system is somehow involved in the control of the ovary has long been recognized (Haighton, 1797). Clinical studies (Frölich, 1910) and subsequent experimental work (Harris, 1937; Haterius and Derbyshire, 1937) showed that the hypothalamus was implicated in the control of ovarian function. The lack of secretomotor innervation of the anterior pituitary and the discovery of the hypophysial portal system by Popa and Fielding (1930, 1933) led to the suggestion of a neurohumoral control of the anterior pituitary (Harris, 1937; Hinsey, 1937; Friedgood, 1970).

It is now widely held that the secretory functions of the anterior pituitary, including gonadotropin secretion, are controlled by neurohumors which are released into the primary plexus of the hypophysial portal system (Harris, 1955; McCann and Porter, 1969). However, the essentiality of the portal vessels and whatever humors they might convey for regulation of the anterior pituitary has been called into question by Thomson and Zuckerman (1953, 1954) and Zuckerman (1955).

II. GONADOTROPIC HORMONES

Three gonadotropins will be considered, FSH, LH, and prolactin (see Cole, 1969). Workers studying nonmammalian vertebrates are often dominated conceptually by the abundant information and enormous literature on mammals. For this reason, gonadotropic activities are described in terms of FSH and LH taken from the mammalian context.

The gonadotropins are protein in nature, and, with the exception of prolactin, also contain carbohydrate (Butt, 1967). Increasingly purified preparations are being prepared from a variety of species, and complete chemical analyses of such preparations are being made (Butt, 1967, 1969). The amino acid sequence of ovine prolactin has been determined (Li *et al.*, 1969).

The precise role of the gonadotropins has been difficult to ascertain. The lack of purity of preparations, species differences, problems of assay specificity, and the possibility of inherent FSH activity in LH and *vice versa* have been confusing factors (Section III,A,1). The use of nonhypophysectomized animals and of gonadotropins not of pituitary origin, such as pregnant mares' serum gonadotropin (PMSG) and human chorionic gonadotropin (HCG), has also complicated interpretations of results. Comparisons of the results of bioassays and immunoassays for measuring gonadotropic hormones have shown that the biologically and immunologically active groups on the molecules are not identical (Butt, 1969).

A. Cellular Origin of Gonadotropins

Adenohypophysial cells which form and secrete gonadotropins are generally basophilic in mammals (Purves, 1961; Allanson and Parkes, 1966) and in nonmammalian vertebrates (Benoit and Da Lage, 1963; van Oordt, 1968; Ball and Baker, 1969; Tixier-Vidal and Follett, 1973).

1. Fish

The adenohypophysis of the lamprey (Cyclostomata) is divided into three parts. The basophils of the midpart, the mesoadenohypophysis, or proximal pars distalis, show cyclic changes which correlate with seasonal changes in the ovary and may represent the gonadotrops (Evennett and Dodd, 1963b). In the dogfish (elasmobranchs), the ventral pituitary lobe of the sexually mature female becomes enlarged (Dodd *et al.*, 1960; Te Winkel, 1969) and contains basophils that differentiate as the fish becomes sexually mature (Mellinger, 1964). Similar correlations between seasonal ovarian changes and the number and staining properties of the basophils in the proximal pars distalis have been shown in teleost fish: goldfish and carp (Scruggs, 1951), trout and salmon (Robertson and Wexler, 1962a,b), plaice (Barr and Hobson, 1964), and black surfperch (Lagios, 1965). At spawning, basophils may be characterized by the appearance of large vacuoles as in *Mugilidea* spp. (Stahl *et al.*, 1960) and in goldfish (Yamamoto and Yamazaki, 1967). In other experiments, McBride and van Overbeeke (1969) removed the ovaries of sockeye salmon at various times in the breeding cycle and showed that the basophils of sexually mature fish underwent rapid degranulation and that the corresponding change did not occur in young adults.

2. Amphibians, Reptiles, and Birds

Up to three types of basophils have been described in these vertebrate classes (Mikami, 1958; Grignon and Grignon, 1965; St. Girons, 1967; van Oordt, 1968; Tixier-Vidal and Follett, 1973). Their appearance alters with changes in the ovarian condition. After ovariectomy, certain basophils become enlarged and vacuolated, a process that can be reversed by estrogen treatment as in the frog *Rana esculenta* (Rastogi and Chieffi, 1970a) and the lizard *Sceloporus cyanogenys* (Callard *et al.,* 1972a,b). Basophils, and particularly those presumed to be FSH cells, become strongly differentiated and/or degranulated at the onset of oogenesis and during vitellogenesis in the amphibians *Xenopus laevis* (Kerr, 1965), *Nectophrynoides occidentalis* (Zuber-Vogeli, 1968) and *Rana esculenta* (Smith, 1955; van Oordt, 1968; Rastogi and Chieffi, 1970b), in the tortoise (Herlant and Grignon, 1961), in the lizard (Eyeson, 1970), in the pigeon (Ljunggren, 1969), and the white-crowned sparrow (Matsuo *et al.,* 1969).

3. Mammals

The identification of separate FSH and LH cells remains controversial. For many years the cells of the anterior pituitary have been described as basophils, acidophils, and chromophobes, according to their reactivity towards the basic and acidic dyes. Many different staining methods were developed and it was deemed possible to identify some basophils as cells which produce FSH and LH, and various acidophil cells, one of which probably produces prolactin. With the introduction of histochemical methods, particularly the periodic acid–Schiff technique (McManus, 1946), electron microscopy, and immunological labeling (Midgley, 1969; LeLeux and Robyn, 1971) and the combination of these techniques with parallel hormone assays, data have been accumulated for a one cell–one hormone theory. However, there is some evidence from studies using immunological techniques that cells may contain both FSH and LH (Nakane, 1970).

B. Changes in Gonadotropins in Different Reproductive Stages

1. Fish

It has been known for many years that there are seasonal variations in the gonadotropic potency of teleost pituitaries (Gerbilsky, 1940; Fontaine and Chauvel, 1961; Barr and Hobson, 1964; Swift and Pickford, 1965; Singh, 1970). The gonadotropic content of the pituitary of *Pleuronectes platessa* has been measured at different times of the year using, as an assay, the spermiation response of *Xenopus laevis* (Barr and Hobson, 1964; Barr, 1965). Levels of gonadotropins were lowest after spawning but rose in the autumn at the time when vitellogenesis was beginning and remained high during the winter and spring as the ovaries continued to grow.

2. Birds

a. Season and Puberty. In seasonal breeders pituitary gonadotropin levels are highest when gonadal growth is stimulated by long daily photoperiods (Greeley and Meyer, 1953; Assenmacher *et al.*, 1962; King *et al.*, 1966). The way in which the length of day affects the release of gonadotropins has been studied in a number of birds (see Lofts *et al.*, 1970). In the Japanese quail there is a daily rhythm of photosensitivity; the photosensitive phase lasts for some $4\frac{1}{2}$ hours and starts about 11 hours after dawn (Follett and Sharp, 1969). Many seasonal breeders experience a postnuptial "refractory" period when the ovaries regress despite continuing long daily photoperiods. This may be due to a lack of gonadotropins, since pituitary gonadotropin levels are low in photorefractory birds (Greeley and Meyer, 1953; King *et al.*, 1966). During the rapid growth of the ovary, there is a steady increase in the gonadotropic potency of the pituitary reaching a peak just before the onset of breeding or laying (Breneman, 1955; Herrick *et al.*, 1962; King *et al.*, 1966). Once the ovary is mature, the gonadotropic content of the pituitary becomes stabilized at a lower level (Riley and Fraps, 1942; Saeki *et al.*, 1956; Nakajo and Imai, 1961).

b. Ovulatory Cycle. Bioassay techniques developed for mammalian gonadotropins have been used to measure FSH and LH activity during the ovulatory cycle of the domestic hen. The amount of LH in the pituitary decreases on two occasions and that of FSH on three occasions during a single ovulatory cycle (Nelson *et al.*, 1965; Tanaka and Yoshioka, 1967; Heald *et al.*, 1967; Imai and Nalbandov, 1971). The first fall in the gonadotropic potency, LH and FSH, of the pituitary occurs at the beginning of the cycle soon after the preceding ovulation. The second decrease, in LH, was found either 8 hours (Heald *et al.*, 1967; Tanaka and Yoshioka, 1967) or 14 hours (Nelson *et al.*, 1965) before the next ovulation. Determinations of plasma LH activity have shown peaks at both 8 and 14 hours before ovulation (Nelson *et al.*, 1965; Bullock and Nalbandov, 1967). The bioassay data must be accepted with caution since the assays involved are insensitive and, in the case of the LH bioassay, nonspecific (Jackson and Nalbandov, 1969a,b). The development of a radioimmunoassay for chicken LH has made it possible to study the ovulatory cycle in greater detail (Furr et al., 1973). An increase in plasma LH was observed 4 to 7 hours before ovulation.

3. Mammals

a. **Immaturity, Puberty, and Season.** The time of development of a definitive hypothalamus–hypophysial control system probably varies from species to species (Jost et al., 1970). In the rat, the full organization of the stalk–median eminence region along adult lines takes place during the second week of life (Florsheim and Rudko, 1968) when there is evidence of fairly high FSH and LH, though not prolactin, levels in the blood (Goldman et al., 1971). At this time the ovary may be only partly responsive to gonadotropins (Section III,D,3,b,i). There is a prepubertal rise of pituitary FSH and LH in the rat (see Labhsetwar, 1969; Fawke and Brown, 1970; Weisz and Ferin, 1970) and of LH and prolactin in the cow (Desjardins and Hafs, 1968; Sinha and Tucker, 1969). Puberty in the cow is associated with a decrease in pituitary LH. Pituitary FSH and LH levels fall just before puberty in the rat. Reports of the changes in gonadotropins taking place in humans at this time have varied, but they indicate increases in both FSH and LH secretion before and/or during puberty (see Lee et al., 1970; Rifkind et al., 1970). Such a rise in blood LH may not occur in the cow (Odell, 1970; Swanson et al., 1970), but there are high levels of prolactin before puberty (Swanson et al., 1970). In the rat, blood levels of FSH and LH (Kragt et al., 1970; Weisz and Ferin, 1970) reach a peak on day 14 or 15, that is, before the pituitary peak and at a time when there is active interstitial tissue in the ovaries and evidence of estrogen secretion (Cierciorowska and Russfield, 1968). Evidence from blood and pituitary values indicates that a second release of LH (Weisz and Ferin, 1970), probably accompanied by FSH (Goldman and Mahesh, 1968; Fawke and Brown, 1970), takes place to trigger the first ovulation at a time when the ovary has been shown to be maximally sensitive (Zarrow and Quinn, 1963). In addition, blood prolactin levels rise sharply at puberty (Voogt et al., 1970). The decreases in hypothalamic releasing factors which have been described as occurring at puberty (Section IV,B,2,b) are probably involved in the changes in gonadotropin levels.

The neural mechanisms controlling gonadotropin secretion (Section IV,B) and the factors affecting them may play a role at puberty. Among the various factors thought to be involved, and about which there has been speculation, are extrahypothalamic inhibitory influences (Critchlow and Bar-Sela, 1967), changes in the sensitivity of different positive and negative estrogen feedback areas in the hypothalamus and extrahypothalamic structures (Donovan and van der Werff ten Bosch, 1965; Davidson, 1969; Presl *et al.,* 1970; Swerdloff and Jacobs, 1971; and Volume III, Chapter 3), altered gonadal steroid patterns (Donovan, 1969), altered sensitivity of the ovary to gonadotropins perhaps produced by estrogen (Sections III,D,3,b,i, and III,D,3,b,ii; Ying and Greep, 1971), and changes in adrenal steroids (Gorski *et al.,* 1971). The stimulatory feedback effects of FSH and prolactin on the release of FSH (Ojeda and Ramirez, 1969) and FSH and LH (Clemens *et al.,* 1969) are also believed to be implicated (Section IV,B,4).

It has been speculated that at the onset of the breeding season in seasonal breeders a situation similar to that at puberty prevails. There is probably an inhibition of the release of gonadotropin during sexual quiescence which in some way is altered by light and/or other external environmental stimuli (Donovan, 1967a) resulting in altered levels or ratios of gonadotropins. There may be changes in ovarian sensitivity to gonadotropins and pituitary sensitivity to releasing factors. In the sheep, pituitary levels of FSH and LH during anestrus are similar to those of the luteal phase of the cycle (Robertson and Hutchinson, 1962; Roche *et al.,* 1970), but blood levels of LH remain low until late anestrus. The follicles present in the ovary during anestrus can be made to ovulate by treatment with exogenous gonadotropins (Hammond, 1961) and pituitary LH can be released by estrogen (Goding *et al.,* 1969). It has been said that hypothalamic extracts will not induce ovulation during midanestrus but will do so as the breeding season approaches (Domanski and Kochman, 1968).

b. Estrous (Menstrual) Cycle. In female mammals there is evidence during the estrous cycle for a low basal output of gonadotropins and for a surge of output associated with ovulation. Such periodic surges of gonadotropins are not evident in the male. It has been suggested that the neural factors concerned with gonadotropin control may be organized differently in the two sexes and a general concept of sexual differentiation of the hypothalamic regulation has been proposed (Harris, 1964, 1970b; Barraclough, 1966, 1967; Gorski, 1968). It is suggested that initially the preoptic area is undifferentiated, inherently female, and potentially cyclic. It is then supposed that during the critical differentiation period, early postnatal life in the rat, perhaps prenatally in other species, the preoptic area and/or higher nerve centers become sensitive to the organizing action of endogenous or exogenous androgen. Androgenic stimulation is then presumed to make the preoptic area acyclic, the male type; lack of stimulation results in this area developing a cyclicity which becomes established after puberty, the female type.

Changes of FSH and LH during the cycle have been studied in detail particularly in farm animals, rats, and man (Robertson, 1969; Gay *et al.,* 1970; Henzl and Segre, 1970). Results of studies of pituitary, blood, and urine gonadotropin levels indicate that there is a surge of LH from the pituitary just before ovulation in different species of mammals: rat (see Schwartz, 1969; Monroe *et al.*, 1969; Gay *et al.*, 1970; Daane and Parlow, 1971; Kalra *et al.*, 1971); mouse (Bingel and Schwartz, 1969; Desjardins *et al.*, 1970); hamster (Orsini and Schwartz, 1966; Goldman and Porter, 1970); cow (see Robertson, 1969; Henricks *et al.*, 1970; Odell, 1970; Snook *et al.*, 1971); sheep (see Robertson, 1969; Roche *et al.*, 1970; Scaramuzzi *et al.*, 1970); pig (see Robertson, 1969; Niswender *et al.*, 1970; Rayford *et al.*, 1971); rhesus monkey (Kirton *et al.*, 1970; Monroe *et al.*, 1970; Blakley, 1970); chimpanzee (McArthur and Perley, 1969); baboon and orangutan (Blakley, 1970); and man (see Henzl and Segre, 1970). The LH surge is accompanied by one of FSH in the rat (see Schwartz, 1969; Gay *et al.*, 1970; Daane and Parlow, 1971; Kalra *et al.*, 1971), hamster (Keever and Greenwald, 1967), cow, sheep, and pig (see Robertson, 1969), rhesus monkey (Simpson *et al.*, 1956), and man (Henzl and Segre, 1970) and by a surge of prolactin in the rat (Gay *et al.*, 1970; Neill, 1970; Neill and Reichert, 1971; Kalra *et al.*, 1971), cow (Swanson and Hafs, 1970), and sheep (Reeves *et al.*, 1970). These data accord with the hypothesis of a common release mechanism for FSH and LH and may indicate that both FSH and LH are required for ovulation (Section III,C,4). A detailed study of the ovulatory gonadotropin surge has been carried out only in the rat and man. In the rat, blood levels of FSH, LH, and prolactin all rise on the afternoon of proestrus (Gay *et al.*, 1970; Daane and Parlow, 1971; Kalra *et al.*, 1971). There is evidence that the LH surge is sharp whereas FSH is high on the afternoon of proestrus and the early morning of estrus and falls more slowly, perhaps because of its longer half-life (Gay *et al.*, 1970). Pentobarbital will block the LH surge and that part of the FSH surge which occurs on the afternoon of proestrus. The FSH circulating on the day of estrus may be involved in the stimulation of follicular growth, as many investigators have reported a raised FSH level in the early follicular phase of the longer human cycle (see Henzl and Segre, 1970). Frequent blood sampling in man (Thomas *et al.*, 1970) indicates that the ovulatory surge of LH is biphasic. Possible functions of these two peaks are not known but they may have specific roles in follicle rupture, luteinization, or steroid biosynthesis. Increase in blood progesterone levels may commence with the initiation of the second peak of LH.

   The large amount of data on pituitary levels of FSH and LH in different species gives little indication of their secretory rates during the nonovulatory stages of the cycle; indications so far on blood levels suggest that they are low. In man, there is evidence of a gradual rise in plasma LH during the follicular phase which may indicate an involvement of LH in follicular growth and estrogen secretion (Section III,C,2). Several workers have suggested that there are small rises in plasma LH during the luteal phase in the cow and again a few days before ovulation (see Snook *et al.*, 1971) and that these may be related to growth and atresia of follicles or to corpus luteum function (Karg and Schams, 1970). The indication may be that only low but constant levels of gonadotropins are required for maintaining follicular growth, corpora lutea, and steroidogenesis. A 24-hour periodicity is involved in the timing of ovulatory gonadotropin release in some rodents. Twenty-four hour rhythms of plasma FSH and LH have been shown to occur in human plasma during the follicular and luteal phases of the cycle by some workers (see Saxena *et al.*, 1969). Similar rhythms of plasma LH occur in the rat (Gay *et al.*, 1970) and cow (Madan and Johnson, 1971) and of pituitary prolactin in the rat (Clark and Baker, 1964). It has been suggested that these rhythms may be controlled by the pineal gland (Fraschini and Martini, 1970).

The above discussion has been confined to spontaneously ovulating animals. In rabbits, induced ovulators, there is a surge of LH following mating (Brown-Grant *et al.*, 1968; Bogdanove *et al.*, 1971) which is accompanied by a release of prolactin (Desjardins *et al.*, 1967).

For a discussion of the importance of steriod feedback mechanisms in the control of the changes in gonadotropins discussed in this section, see Volume III, Chapter 3.

c. Senescence. During senescence there is evidence that the ovary decreases in responsiveness to gonadotropins (Section III,D,3,b,i). Primary ovarian failure also develops in old age. There is an increase in FSH and LH in the pituitary of the aged rat (Labhsetwar, 1969, 1970a), and in the pituitary, blood, and urine of postmenopausal women (Albert, 1956; Saxena *et al.*, 1969; Loraine and Bell, 1971), possibly due to a fall in circulating steroids. However, an increase in gonadotropin levels may take place in late reproductive life in women (Papanicolaou *et al.*, 1969) and in rats still capable of responding to gonadotropin (Labhsetwar, 1969). As in the development from immaturity to maturity, ovarian sensitivity and hypothalamic changes may both be involved in the change from reproductive activity to senescence.

III. ACTIONS OF GONADOTROPINS

The ovary is involved in the production of ova and in the synthesis and secretion of steroids, and gonadotropins are involved in the control of both processes. It is not clear whether the primary effect is on the growth and development of cellular components of the ovary followed by steroid production in the tissues so formed or whether steroids are first produced to initiate growth and development. Probably both processes occur depending on the tissue, its stage of development, and the hormone.

In mammals, both FSH and LH are secreted throughout the estrous cycle and both apparently participate in some way in all phases of ovarian activity. Recently, more attention has been paid to the effects of combinations of gonadotropins as well as to the actions of a specific one. The combined action may be sequential and, without suitable preparation by one gonadotropin, the effect of the other may not be manifested. In general, FSH seems to be primarily a growth factor; LH seems to be basically a stimulator of secretion rather than of proliferation, requiring the prior action of FSH for its effect. The effects of LH on follicle growth, ovulation, and luteinization may be brought about by its steroidogenic action. Prolactin, in some species, seems primarily to be involved in corpus luteum function, with or without a combined action with FSH and LH. Clarification of these interrelationships is complicated by problems of purity of available preparations and of interpretation when gonadotropins of one species are tested in another (Section III,A,1).

Knowledge of the actions of gonadotropins in nonmammalian vertebrates is relatively scanty. Gonadotropin activities are often described in terms of FSH and LH taken from the mammalian context. Purified preparations are lacking and the situation has been further complicated by the frequent use of mammalian preparations.

A. Dependence of the Ovary on the Anterior Pituitary

Maturation of the ovary is controlled by the pituitary, that is, by the adenohypophysis or its equivalent. In Teleostei it is the proximal pars distalis and in the Elasmobranchii it is the ventral lobe which secretes gonadotropins (Dodd *et al.*, 1960). Representatives from each of the major Classes have been examined though these are but a few species among many (mammals: Rowlands and Parkes, 1966; birds: Parkes and Marshall, 1960; others: Dodd, 1960, 1972; Barr, 1968). In general, follicular maturation and ovulation cease after hypophysectomy. At the same time, all follicles exceeding a certain "critical size," which depends on the species, disappear. Follicles smaller than this persist and possibly grow despite hypophysial deficiency. Moreover, the maturation of the ovum can occur in the absence of gonadotropin in most vertebrates that have been investigated. The thecal cells of atretic follicles do not degenerate as completely as do the granulosa cells and ova, but in several species, such as the rat, they undergo a retrogressive process designated as "wheel cell formation." In mammals, preexistent corpora lutea usually follow the same pattern of regression after hypophysectomy as in the intact animal. However, in the rat such corpora lutea involute very slowly (Bunde and Greep, 1936).

Treatment with suitable combinations of gonadotropins can restore the normal structure of the ovary of the hypophysectomized animal.

1. Species Differences in Gonadotropins and Responses to Gonadotropins in Mammals

Information about the effectiveness in mammals of pituitary gonadotropins derived from species of other orders of vertebrates is scanty. Gonadotropins from birds are the most active and exhibit both FSH and LH activity (Greep, 1961). Pituitaries from fish and amphibians show very little activity in mammals (Witschi, 1955). Like the teleosts and amphibians, mammals also display examples of intraclass gonadotropic species specificity. For example, gonadotropins from different species give qualitatively different dose–response curves in bioassays which are not specific for FSH and LH (Albert, 1961; Rowlands and Parkes, 1966). In addition, Parlow and Reichert (1963) and Reichert (1967) have found that FSH preparations from different species give log dose–response curves with different slopes in the Steelman and Pohley (1953) assay. Similarly, Parlow (1963, 1968), Christiansen (1967), and Hutchinson *et al.* (1968) have found that LH preparations from different species give different slopes in the ventral prostate (VP) assay (Greep *et al.*, 1941), and several authors (Koed and Hamburger, 1967; Bell and Loraine, 1967) have found such differences in the ovarian ascorbic acid depletion (OAAD) assay (Parlow, 1961). Differences in relative responses between VP (chronic) and OAAD (acute) assays (Rosemberg *et al.*, 1964; Hutchinson *et al.*, 1968; Parlow, 1968) have been found for several LH preparations, and Bhalla *et al.* (1971) have found differences in the ability of LH from different species to cause estrogen secretion and to augment the action of FSH in increasing ovarian weight. These findings may reflect differences in half-lives of LH from different species (Parlow, 1968), but similar differences have been found by Reichert (1966) with the OAAD assay and the hyperemia assay (Ellis, 1961); in both assays responses are achieved in a few hours.

Species specificity is sometimes more pronounced. Simpson and van Wagenen (1958) found that ovulation could always be induced in the macaque monkey with monkey gonadotropins, whereas the use of sheep gonadotropins resulted in erratic or negative results. However, porcine FSH and HCG was found to be effective in this species (Knobil *et al.*, 1959). In human studies, human FSH and porcine FSH have proved to be equally effective (Dörner, 1967). Macdonald *et al.* (1971) found that rat prolactin maintains progesterone secretion from rat corpora lutea much longer than does sheep prolactin.

Many of the differences which have been discussed probably reflect a lack of constancy and purity of preparations and a structural difference of gonadotropins from species to species. There seem to exist in all gonadotropins two subunits, one of which is common to all glycoprotein hormones of a given species, while the other is structurally unique for the hormone in question (Reichert et al., 1969, 1970; Pierce et al., 1971). Geschwind (1966) has suggested, however, that all physiological differences in species response may not reflect chemical differences in the hormone itself, but rather the binding or lack of binding of the hormone to a unique nonhormonal protein. Some of the variation in response may be due to the fact that after chronic treatment with heterologous gonadotropins the ovary becomes refractory. This is probably due to the production of antibodies which neutralize subsequent injections of the hormone (Thompson, 1941; Rowlands and Parkes, 1966), although animals chronically treated with gonadotropins frequently display progonadotropic activity rather than antigonadotropic activity, i.e., their sera augment rather than inhibit the action of gonadotropins (Rowlands and Parkes, 1966; Cole, 1969). The cause of this effect is not clear, although it has been shown that antisera can markedly prolong the half-life of LH (Cole, 1969).

B. Nonmammalian Vertebrates

1. Follicular Growth and Ovulation

There are many reports about the effects on intact fish of fish pituitaries administered either as extracts or as implants (Pickford and Atz, 1957; Dodd, 1960; Ball, 1960; Ramaswami, 1962). In general, provided the appropriate experimental conditions are chosen, such treatment can produce ovarian growth, vitellogenesis, and premature spawning. Usually, immature ovaries did not go through all stages of oogenesis to complete maturation and ovulation, although this has been achieved in a few cases such as the eel (Fontaine et al., 1964). Studies on hypophysectomized fish show that administered gonadotropins act directly on the ovary and suggest that this is true also of intact fish and that stimulation does not occur in response to endogenous pituitary hormones. In early work, Vivien (1939, 1941) showed that injection of cod pituitary extracts into the goby, 15 days after hypophysectomy, restored vitellogenesis. Injection of homozooic pituitaries caused follicular development even in totally regressed ovaries of long-term hypophysectomized fish. Salmon pituitary gonadotropin has been found to induce vitellogenesis and ovulation in hypophysectomized goldfish (Yamazaki and Donaldson, 1968), guppies (Liley and Donaldson, 1969), and catfish (Sundararaj *et al.*, 1972). Pituitary extracts from species of other classes can induce ovarian growth and ovulation in fish (Pickford and Atz, 1957). However, the effective dose is usually high and the results are variable. In the lamprey, injections of a mixture of PMSG and HCG induce full ovarian maturation and ovulation (Evennett and Dodd, 1963a,b). Among the teleosts, differences in gonadotropic specificity can be as great between different species as between vertebrate classes. Thus, in some species, such as the stickleback, vitellogenesis can be stimulated with LH (Ahsan and Hoar, 1963), whereas LH has no effect in goldfish (Yamazaki, 1965). Similarly, Jalabert (1969) has shown that crude pituitary extracts from *Gambusia* maintained or restored vitellogenesis in hypophysectomized guppies, while extracts derived from spawning roach were without effect.

In Amphibia, principally *Bufo arenarum, Xenopus laevis,* and *Rana pipiens,* homologous pituitary extracts induce ovulation, but at least 3 hours are required before the ovary responds (Rugh, 1935; Houssay, 1947; Allende and Orias, 1955). Ramaswami and Lakshman (1960) showed that extracts of the pituitary of the skipper frog induced ovulation and that this property was destroyed by trypsin and pepsin but not by ptyalin. Since the first two enzymes are also known to destroy the activity of mammalian LH and ptyalin is known to destroy that of FSH, the latter authors concluded that the ovulatory hormone in the skipper frog has more in common with mammalian LH than with FSH. Amphibians are notably responsive to gonadotropins from widely differing sources. The ability of HCG to induce *Xenopus laevis* to extrude ova was the basis of a standard pregnancy test. However, though Anura are generally responsive to mammalian gonadotropins, they react poorly to elasmobranch and teleost pituitaries (Greep, 1961). Intraclass gonadotropic species specificity occurs in amphibians as in teleost fish. Thus, ovulation can be readily elicited in *Bufo* by extracts from pituitaries of this genus, but less so by those from *Heptodactylus ocellatus* and not at all by those from *Xenopus laevis* (Houssay, 1947).

Early work on reptiles showed that homologous pituitaries induced ovulation in *Xenodon merreni,* a South American snake (Houssay, 1931), and in *Anolis carolinensis* (Evans, 1935). More recently, Licht and Stockell-Hartree (1971) administered gonadotropins of mammalian, bird, and fish origin to *Anolis carolinensis*. The fish material had no effect, but the avian and mammalian preparations induced ovarian growth, ovulation, and steroidogenesis. The FSH-like preparations were more potent than LH-like fractions; a fact confirmed by other work on lizards (Jones, 1969a; Licht, 1970).

Injection of PMSG into various finches in the nonbreeding season establishes a normal graded series of follicles which then ovulate (Witschi and Riley, 1940; Vaugien, 1957; Bell and Hinde, 1963). In contrast, immature chickens treated with mammalian gonadotropins or chicken pituitary extracts show only a limited follicular response although steroidogenesis is stimulated and combs and oviducts develop (Taber, 1948; Das and Nalbandov, 1955; Taber *et al.*, 1958; Nalbandov, 1961; Fraps, 1961). Follicles in all stages of development are always present in the ovary of the laying hen. This range of size of follicle cannot be supported in either intact or hypophysectomized birds with mammalian FSH preparations, although extracts of chicken pituitaries will maintain them for a limited period (Opel and Nalbandov, 1961; Mitchell, 1967a). It seems that LH-like and FSH-like hormones are necessary for the maintenance of the follicular system. The FSH and LH obtained from chicken pituitaries (Stockell-Hartree and Cunningham, 1969) will together produce a follicular series in hypophysectomized laying hens (Mitchell, 1967a,b), although there may be an abnormally large number of smaller follicles. FSH alone was only able to delay ovarian regression after hypophysectomy and, in some cases, to support ovulation (Mitchell, 1970). Ferrando and Nalbandov (1969) have shown that the block to ovulation caused by systemic injection of dibenzyline can be overcome by systemic injections of LH, but local administration of LH in the wall of the largest follicle is not effective, FSH is also required. This adds force to the view that gonadotropins do not act in isolation but synergize with other hormones. Prolactin has been shown to reduce the weight of ovaries and oviducts in the domestic fowl (Bates *et al.*, 1935), pigeon (Bates *et al.*, 1937), and in *Zonotrichia leucophrys pugetensis* (Bailey, 1950). However, Juhn and Harris (1956) found that prolactin interrupted egg laying in the hen only temporarily and that it blocked the antiovarian influence of exogenously administered progesterone. Similarly, injections of prolactin interrupted, but did not halt, egg laying in the Californian quail (Jones, 1969b). Prolactin, when given to *Zonotrichia albicollis* simultaneously with FSH and LH, suppresses their stimulatory effects on the ovary (Meier, 1969). In a related species, *Zonotrichia leucophrys gambelii*, prolactin given with low doses of gonadotropins promotes ovarian growth (Meier and Farner, 1964). The ways in which prolactin and progesterone interact are not understood. Progesterone will induce ovulation, but given over a period of time invariably results in regression of the ovaries (Riddle *et al.*, 1935; Riddle and Lahr, 1944; Jones, 1969b). A review of the effects of progesterone and prolactin on the inhibition of vitellogenesis and on the maintenance of broody behavior during the breeding season is given by Riddle (1963).

2. Ovarian Secretions

Estrogens, androgens, and progestins have all been extracted from the ovaries of nonmammals. As in mammals, the general finding is that pituitary gonadotropins largely control ovarian steroidogenesis, but most of the evidence is indirect, depending on the examination of secondary sexual characteristics after gonadotropin treatment.

Secondary sexual characteristics fail to develop in lampreys after ovariectomy (Evennett and Dodd, 1963b) or hypophysectomy (Evennett and Dodd, 1963a; Larsen, 1969a,b). This implies a pituitary control of ovarian hormonal activity. Conversely, treatment of hypophysectomized lampreys with a mixture of PMSG and HCG induces the development of the female-type fins and cloacal labia (Evennett and Dodd, 1963b).

The relationship between the pituitary and ovarian secretion in elasmobranchs is not understood (Vivien, 1941; Dodd *et al.*, 1960), but hypophysectomy in teleost fish inhibits the development of, or causes the regression of, the ovary and those structures dependent on its secretions. For example, the oviductal epithelia and genital papilla regress in hypophysectomized *Gobius paganellus* (Ball, 1960). These can be restored by multiple implants of fish pituitaries (Vivien, 1941). Spawning color can be induced by gonadotropins in mature nonspawning female fish such as *Chloea sarchynnis* (Kinoshita, 1938) and the Pacific salmon, *Oncorhynchus nerka* (Palmer *et al.*, 1954). The reproductive behavior of female fish is also partly dependent on ovarian hormones (Liley, 1969). In the guppy it was abolished by hypophysectomy and was restored by administration of extracts of salmon pituitaries which first redeveloped the ovaries (Liley and Donaldson, 1969).

Ovarian atrophy occurs slowly in amphibians after hypophysectomy and the rate varies according to the season of the year (Gallien, 1940). The oviducts of hypophysectomized mature *Rana pipiens* took 8 months to regress to the prepubertal condition (Christensen, 1931), and in various urodeles, including *Ambystoma tigrinum,* the degeneration of the secondary sexual characteristics took from 1 to 3 years (Woronzowa and Blacher, 1930). The involuted oviducts seen in *Xenopus laevis* after hypophysectomy or starvation could be restored to normal by injections of mammalian pituitary extracts only in the presence of the ovaries. Gonadotropins stimulate the secretion of progestins and these induce ovulation (Burgers and Li, 1960; Wright, 1961), stimulate oocyte maturation (Schuetz, 1967, 1972; Subtelny *et al.*, 1968; Thornton and Evennett, 1969), and cause the release of oviductal jelly (Thornton and Evennett, 1969). Gonadotropins also stimulate the secretion of estrogen by the follicular cells and, consequently, growth of the oviducts and the biosynthesis of vitellogenin, the major yolk platlet precursor (Follett and Redshaw, 1973). The administration of pituitary extracts leads to the stimulation of the oviducts in *Anolis* (Evans, 1935), the horned lizard (Mellish, 1936), and the alligator (Forbes, 1937). FSH-like preparations, rather than LH-like, are the most effective in promoting oviductal growth in several species of lizards (Jones, 1969a; Licht, 1970; Licht and Stockell-Hartree, 1971). Callard *et al.* (1972a,b) observed that increased levels of plasma yolk proteins could be induced in hypophysectomized females of the blue spiny lizard, *Sceloporus cyanogenys,* by simultaneous injections of estrogen and mammalian growth hormone, but not by estrogen alone. This finding suggests that gonadotropins normally stimulate the production of ovarian estrogens which act synergistically with growth hormone to promote synthesis of yolk protein in the liver.

In birds, oviductal and comb growth can be induced in young pullets by treatment with either avian or mammalian pituitary extracts. In mature hens, these structures regress after hypophysectomy but can be maintained or stimulated to regrow by injections of gonadotropins (Opel and Nalbandov, 1961; Mitchell, 1967a,b).

C. Mammals

1. Follicular Growth and Oocyte Development

Experiments with hypophysectomized animals have shown that the early development of the follicle may not depend on gonadotropin stimulation (Hisaw, 1947; Young, 1961). The follicle becomes increasingly sensitive to gonadotropin once several layers of granulosa cells have been formed. While the proliferation and organization of the granulosa cells and the development of the thecal cells seem to be largely under FSH control, the conversion of the theca interna into an actively secreting tissue, the preovulatory enlargement, and the associated vascularization appear to be due to the action of LH (Greep, 1961). The effect of LH on the growth of the follicle may not be a direct one; it may act by stimulating estrogen production, which in turn stimulates cell division (Lostroh, 1959; Ryle, 1971). Other evidence that estrogen plays a role in the process is that, in the absence of gonadotropins, estrogen stimulates follicular growth (Williams, 1944), and that estrogens make the follicle more responsive to gonadotropic stimulation (Bradbury, 1961). There is evidence that a constant supply of gonadotropin is required for the continued growth of early follicles (Ryle, 1969) and the maintenance of follicles until a few hours before the ovulatory gonadotropin surge (Sasamoto and Kennan, 1971).

The oocyte reaches its full size before the appearance of the follicular antrum (Brambell, 1928), and its growth seems to be independent of gonadotropin action (Hisaw, 1947). The nucleus of the oocyte remains in the dictyate stage of the first meiotic division during the follicular changes.

At a set time before ovulation is due, the meiotic division is suddenly resumed, the first polar body is emitted and the egg becomes a secondary oocyte. It has been suggested that this process is under the control of gonadotropins, probably LH, which may act by stimulating glycolysis in granulosa cells (Ahrén *et al.*, 1969; Armstrong, 1970) or steroidogenesis, with estrogen, produced during follicular development, suppressing the completion of meiosis and progesterone, produced by the supporting granulosa cells (Channing, 1966) under LH stimulation, inducing it (Robertson and Baker, 1969).

2. Estrogen Secretion by the Follicle

There is evidence that the theca interna is the most active cellular element of the follicle in the synthesis of estrogens (Young, 1961), at least after the attainment of maturity. The role of FSH and LH in stimulating estrogen secretion has been debated over the years but no clear view has emerged. Greep *et al.* (1942) observed no signs of estrogen secretion in hypophysectomized immature rats when pig FSH alone was administered, but did when LH was also given. Although some studies (Li *et al.*, 1962; Eshkol and Lunenfeld, 1967; Donini *et al.*, 1968) with human urinary and sheep FSH preparations support the finding that FSH alone will not stimulate estrogen secretion, other studies do suggest that FSH free from LH is effective (Segaloff, 1966; Hori *et al.*, 1968, 1969; Cole, 1969; Petrusz *et al.*, 1970) if sufficiently high doses are given. In some studies the effect was probably due to LH contamination, but in others (Papkoff, 1965; Rosemberg and Joshi, 1968; Petrusz *et al.*, 1970) LH contamination was too slight to account for all the steroidogenic activity. Recent studies on the response of isolated rabbit ovarian follicles to gonadotropin *in vitro* (Mills *et al.*, 1971) indicated that only LH was capable of supporting and stimulating the biosynthesis of estrogen. FSH and prolactin were ineffective. The role of FSH in estrogen biosynthesis by the follicle will probably only be resolved when studies employing highly purified FSH and hypophysectomized animals or tissues of the same species are carried out.

3. Interstitial Tissue

The amount of interstitial tissue in the ovary varies from species to species (see Volume I, Chapter 4), and most of it probably arises from the theca interna, although in the rabbit the granulosa cells are also a significant source. The interstitial cells produce both androgens and estrogens (Young, 1961), but their control by gonadotropins is not well understood (Savard *et al.*, 1965). LH has been shown to repair this tissue in the rat after hypophysectomy (Simpson *et al.*,1942), and it has also been shown to cause cholesterol mobilization and steroidogenesis, leading to the secretion of progesterone and, more particularly, its reduction product, 20α-hydroxypregn-4-en-3-one (Armstrong et al., 1969; Hilliard et al., 1969). Prolactin is also involved in the control of the interstitial cells of the rabbit, probably maintaining their structure and steroid-producing capacity.

4. Ovulation

It is generally accepted that LH is involved in the process of ovulation (Greep, 1961; Harris and Campbell, 1966), and there is some evidence that FSH may also play a role. There is evidence for the common identity of follicle-stimulating hormone-releasing factor (FRF) and luteinizing hormone-releasing factor (LRF) (Section IV,B,2) and there is a surge of both FSH and LH during the estrous cycle in several species about the time of ovulation (Section II,B,3,b). Intrafollicular or intravenous injections of LH or FSH cause ovulation in the rabbit (Jones and Nalbandov, 1971), and relatively "pure" FSH has been shown to induce the ovulation of mature follicles in the rat (Carter et al., 1961; Lostroh and Johnson, 1966; Goldman and Mahesh, 1968; Harrington and Elton, 1969; Harrington et al., 1970). When given in combination, subthreshold doses of FSH and LH also cause ovulation in the rat and rabbit (Labhsetwar, 1970b; Jones and Nalbandov, 1971). An antiserum against FSH and LH blocks ovulation in hamsters, and the removal of much of the FSH antibody without loss of antiLH potency reduces the ovulation-blocking potential of the antiserum (Goldman and Mahesh, 1969).

The exact nature of the ovulatory stimulus is unknown, but it is associated with the stimulation of steroidogenesis by LH (Rondell, 1970; Lipner and Greep, 1971). Certain substances which inhibit enzymes at different sites in the steroid biosynthetic pathway also block ovulation (Lipner and Greep, 1971). Progesterone and LH have been said to cause an increase in distensibility of the follicular wall (Rondell, 1970), but the steroid involved may be neither progesterone nor estrogen (see Lipner and Greep, 1971), but 17α-hydroxyprogesterone, which is probably produced in the theca interna and which has been shown to cause ovulation in the rat (Lostroh, 1971). LH may stimulate the follicular tissue to secrete a steroid which has two effects in the ovulatory process. First, it may cause the release of a tissue enzyme (Espey and Lipner, 1965; Harvey and Rondell, 1970) that acts on the collagen framework of the follicle wall (Espey, 1967) to cause an increase in distensibility of the wall (Rondell, 1970) so that a continued influx of fluid into the follicle (Lipner and Smith, 1971) would take place, due to an increased permeability of the blood–follicle barrier (Zachariae, 1958). This process would eventually lead to follicular rupture.

The increase in fluid available may be related to the increased blood flow known to be caused by LH (Wurtman, 1964; McCracken and Baird, 1969). It has been suggested that histamine may be involved in this hyperemia (Szego and Gitin, 1964; Wurtman, 1964), but histamine does not seem to have a direct role in the induction of ovulation (Lipner, 1971).

5. Artificial Induction of Ovulation

The induction of ovulation and superovulation by exogenous gonadotropins has been extensively studied in various mammals (Hammond, 1961; Rowlands and Parkes, 1966; Hafez, 1969), including man (Gemzell et al., 1958; Gemzell, 1965; Crooke, 1970). Gonadotropins may be administered in order to produce an excess number of ova for transfer from genetically inferior to genetically superior farm animals. They may also be given in order to produce a near normal number of ovulations, to increase litter size, to induce ovulation during anestrus in farm animals or to overcome certain anovulatory conditions produced either by a lack or imbalance of gonadotropins in man, domestic animals, and even exotic species. Numerous factors influence these effects (Hafez, 1969; Sections III,A,2 and III,D), including the stage of reproduction (season, cycle, etc.), genetics (species, breed, strain), age, body weight, the nutritional condition, previous treatment with and dose of steroids, FSH:LH ratio, frequency of injection, and purity of gonadotropin.

6. Luteinization and Corpus Luteum Formation

Although, as shown above (Section III,C,4), ovulation can be induced by FSH, the corpora lutea which develop after ovulation do so poorly and do not secrete progesterone. Both these processes seem to require LH stimulation. Keyes (1969), Channing (1970), and Ellsworth and Armstrong (1971) have shown that granulosa cells need the stimulus of LH for luteinization to occur. *In vitro* studies (Channing, 1970) suggest that granulosa cells, depending on their size, probably require FSH and LH for luteinization, and that the stimulus of ovulation may also be important. That luteinization may be inhibited by the oocyte has been suggested by El-Fouly et al. (1970) who induced luteinization by removal of the oocyte. Luteinization without ovulation may occur as a result of an abnormal FSH–LH balance in the treatment schedule (Carter et al., 1961; Ponse, 1966; Lostroh and Johnson, 1966; Malven and Sawyer, 1966a; Mitchell and Yochim, 1968; Chateau, 1969). On the other hand accessory corpora lutea develop in many species without ovulation (see Volume I, Chapter 4). Luteinization probably requires lower levels of gonadotropin than does ovulation (Jones and Nalbandov, 1971; Ellsworth and Armstrong, 1971).

7. Corpus Luteum

Several studies of different aspects of the control of corpus luteum function have appeared in recent years (see, for example, Savard *et al.*, 1965; Rothchild, 1966; Short, 1967; Nalbandov and Cook, 1968; Greenwald and Rothchild, 1968; Denamur, 1968; Armstrong, 1968, 1970; Moor, 1968, 1970; Bartosik and Romanoff, 1969; Schomberg, 1969; Anderson *et al.*, 1969; Nalbandov, 1970; Caldwell, 1970; Rowson, 1970; Donovan, 1971) and the subject will therefore be covered only briefly in this section. The maintenance of the corpus luteum seems to be under dual control of a tropic system and a lytic system. In some species the lytic system may be dominant and act to "switch off" luteal function; in others the tropic system may dominate and the corpus luteum regress through failure of the tropic support.

The corpus luteum of most mammals is under the control of the pituitary. The term "luteotropin" has been applied to substances that maintain the corpus luteum and stimulate it to secrete progesterone and related progestational steroids (progestins). Lutein cells derived from granulosa cells are probably the major source of progesterone. In certain species, some luteal elements are derived from thecal cells (see Volume I, Chapter 4), and secrete estrogens and other steroids in addition to progestins (Savard *et al.*, 1965). There is evidence that LH is involved in estrogen production from the corpus luteum of the rat (Armstrong, 1968), but later data are not conclusive (Christiansen *et al.*, 1970).

Astwood (1941) used the term "luteotropin" for the pituitary hormone, prolactin, which maintains functional, progesterone-secreting corpora lutea in the rat, mouse (Dresel, 1935), and ferret (Donovan, 1967b), but not in other species. In contrast to prolactin (Armstrong, 1969), in several species LH has been shown to stimulate corpora lutea to produce progesterone *in vitro* (Nalbandov and Cook, 1968). However, such demonstrations of the ability of LH to stimulate progesterone production do not imply that LH is "the" luteotropin. In fact, it has become increasingly clear that no single known control mechanism can explain the maintenance of corpus luteum function. In some species, such as the pig, the corpus luteum of the cycle resulting from the action of gonadotropin is capable of functioning without continued hypophysial stimulation (du Mesnil du Buisson and Léglise, 1963; Anderson *et al.*, 1967). Other species, such as the guinea pig, seem to require a gonadotropic stimulus for a limited period, after which the corpus luteum is autonomous (Heap *et al.*, 1967). In yet other species, such as the sheep, the corpus luteum may be continuously dependent on pituitary hormone (Denamur *et al.*, 1966; Kaltenbach *et al.*, 1968). However, in all species the corpus luteum of early pregnancy seems to require luteotropin from the pituitary or from the placenta (Moor, 1970; and Volume II, Chapter 8).

The pituitary luteotropin varies from species to species. There is evidence that in many species there may not be a single luteotropin, but perhaps a combination of two or more of FSH, LH, and prolactin forming a luteotropic complex (Rothchild, 1966; Greenwald and Rothchild, 1968), as in the hamster and rat in which all three gonadotropins may be involved (see Greenwald, 1969; Spies and Niswender, 1971). The requirement for different pituitary luteotropic hormones may depend on the age and stage of development of the corpus luteum. In the pig, there seems to be a period of corpus luteum autonomy followed by a requirement for first LH and then prolactin (du Mesnil du Buisson and Denamur, 1969). Estrogens may act as luteotropic substances in some species and luteolytic substances in others (Nalbandov, 1970) and are presumably produced by the ovary in response to LH (and FSH) stimulation. In addition to uterine luteolytic factors, prolactin, though part of the luteotropic complex in the rat, paradoxically also seems to be involved in hastening the regression of the corpus luteum (Malven, 1969).

D. Factors Affecting the Actions of Gonadotropins

In this section, problems relating to the absorption of exogenous gonadotropins from an injection site into the blood and those related to the supply of gonadotropins reaching the ovary and its response to endogenous and exogenous gonadotropins will be discussed.

1. Absorption from Injection Sites

The response to an injection of gonadotropins will depend on the dose given and the mode of administration (Rowlands and Parkes, 1966). A single intravenous injection producing a rapid increase of hormone levels in the blood will usually give a maximal rapid response, whereas a maintained maximal response may be obtained by continuous infusion or the administration of divided doses over several days by subcutaneous or intraperitoneal injection. Such treatment will achieve a regular input and the maintenance of a steady level of hormone in the blood. A similar effect can be obtained by injecting a mixture of gonadotropins and various inert materials. An augmentation of response can be obtained by such procedures, presumably because the inert material delays the rate of absorption of the gonadotropin, which enables a steady blood level of gonadotropin to be maintained (Maxwell, 1934). Support for this idea has been extensive and it has been found that potentiation of gonadotropins with short half-lives in the blood (Parlow, 1968) is much greater than those with long half-lives (Deanesly, 1939; Armstrong and Greep, 1965; Hutchinson *et al.*, 1968). Some inert materials seem to exert their effects in the blood as well as at the injection site. Borth and Menzi (1969) have shown, for example, that gelatin and bovine serum albumin will potentiate the effects of HCG in short-term assays employing intravenous and intraperitoneal injection. Such materials may act by protecting the gonadotropin from inactivation or breakdown in the body. It has been suggested that the effect of human serum on assays of human pituitary and urinary gonadotropins in rats may be due to formation of a compound with the α- and β-globulins (Kaivola *et al.*, 1969). The effect of tannation of gonadotropins, perhaps involving altered absorption, delayed metabolism, and changed action, has been extensively studied (see Rosemberg, 1967); the effects of LH seem to be enhanced with chronic administration and decreased with acute administration.

2. Factors Affecting Gonadotropins in the Blood

Studies in the rat by Grosvenor (1967) and Parlow (1968) have shown that LH and prolactin from different species have different half-lives. Both homologous and heterologous FSH have a longer half-life in the rat than do homologous or heterologous LH (Bogdanove and Gay, 1968, 1969; Gay *et al.*, 1970). In man, FSH has a much longer half-life than does LH in blood (Kohler *et al.*, 1968; Coble *et al.*, 1969). The shapes of the decay curves indicate that both hormones are distributed in at least two compartments in the body. A similar distribution is suggested for rat prolactin (Kwa *et al.*, 1970). PMSG and HCG in the rat (Parlow, 1968) and HCG in man (Yen *et al.*, 1968) have been shown to have long half-lives. This may be of great importance as these preparations have been used frequently for the stimulation of follicular growth or the induction of ovulation in various species.

There is little conclusive evidence available as to the sites at which gonadotropic hormones are metabolized. Liver tissue has been shown to inactivate gonadotropins *in vitro* (Dasgupta *et al.*, 1964). Gay (1971) has indicated that the kidney is important in the metabolism of FSH and LH, and the ovary has also been implicated (Section III,D,3,a). Prolactin is probably metabolized by the mammary gland (Grosvenor, 1967), judging by the difference in half-life of prolactin in lactating and nonlactating rats.

The presence of binding proteins for gonadotropins in blood has been suggested (Geschwind, 1966) but not demonstrated. Various inhibitory factors have also been suggested, in pituitary extracts (Woods and Simpson, 1961), an anti-LH effect in human urine and an anti-FSH effect in monkey urine (Moudgal *et al.*, 1969). A factor which enhances gonadotropin action has been found (Hopkins, 1968) and may induce changes in the gonadotropin molecule. The physiological significance of the inhibitory and enhancing factors is not clear.

3. Factors Affecting Gonadotropins at the Ovarian Level

a. Ovarian Binding Sites; Metabolism of Gonadotropins by the Ovary. The use of radioactive and fluorescent labeling of FSH, LH, HCG, and prolactin has shown that specific binding sites for the different tropic hormones probably exist in the ovary (Sonenberg *et al.*, 1951; Eshkol and Lunenfeld, 1968; Fitko *et al.*, 1968; Espeland *et al.*, 1968; Rajaniemi *et al.*, 1970) and that the receptors may be in the cell membrane (Rajaniemi and Vanha-Perttula, 1971). Schwartz *et al.* (1971) have suggested that FSH may be bound to ovarian receptors for a long time or have a long duration of action. It has often been suggested that gonadotropins might be utilized or altered by the ovary while exerting their stimulatory action. Early experiments to test this hypothesis gave different results (see Naftolin *et al.*, 1968). Naftolin *et al.* (1968) and Llerena *et al.* (1969) have shown that, when measured by radioimmunoassay, the level of LH, but not FSH, in women during the follicular or luteal phases of the cycle, is less in ovarian veins than in peripheral blood, though this is not the case in women taking oral contraceptives. This perhaps implies that tissues responding to LH are capable of metabolizing it. Morell *et al.* (1968) and Rosemberg *et al.* (1968) found less FSH secreted by women whose ovaries responded to exogenous gonadotropin than in women with unresponsive ovaries, and Karsch *et al.* (1970) have suggested that follicles in the sheep may compete with corpora lutea for LH by metabolizing LH. However, Scaramuzzi *et al.* (1970) and Niswender and Cicmanec (1971) did not obtain evidence of ovarian metabolism of LH, FSH, or prolactin in the sheep when ovarian arterial–venous differences in blood levels of the hormones were measured. The half-lives of gonadotropins in the blood do not seem to be affected by the presence or absence of an active ovary (Bogdanove and Gay, 1969; Ross *et al.*, 1970), but a small ovarian contribution to metabolism may not be noticed in such studies.

b. Factors Affecting Ovarian Sensitivity to Gonadotropins. Changes in the ability of the ovarian tissue to respond to gonadotropins are a characteristic feature of all stages of ovarian development and may be brought about by gonadal maturity or atrophy, previous action upon the tissues by other gonadotropins or other hormones, or other stimulatory or inhibitory factors. The level of ovarian response depends, therefore, on age, stage of reproduction, and differences in genetic constitution.

i. Changes during reproductive life. There are many observations which indicate that the ovary from immaturity through to senescence is able to respond to increased gonadotropic stimulation, though ovarian sensitivity clearly varies throughout life (Price and Ortiz, 1944; Rowlands and Parkes, 1966; Butt *et al.*, 1970). In the majority of mammals studied, there is a postnatal period of variable length during which the ovary is relatively insensitive to gonadotropic stimulation. The interstitial tissue becomes competent to respond earlier than the follicular tissue (Ben-Or, 1963, 1970; Ponse, 1966). Once the capacity to respond has developed, there is a rapid increase in responsiveness, reaching a peak before puberty, at a time when large numbers of vesicular follicles are present in the ovary (Rowlands and Parkes, 1966; Hafez, 1969). The time of maximum sensitivity occurs at different ages depending on the species: about 4 weeks in the rat, mouse, and hamster (Ortiz, 1947; Zarrow and Wilson, 1961; Zarrow and Quinn, 1963) and about 16 weeks in the rabbit (Kennelly and Foote, 1965). There is evidence that in some species the ovaries are responsive before birth (e.g., giraffe: Amoroso, 1955; man: van Wagenen and Simpson, 1965). The ovary of the neonatal calf contains large numbers of vesicular follicles and is sensitive to gonadotropins (Marden, 1953). The ability to induce ovulation during the cycle, anestrus, and pseudopregnancy in different species depends on the presence of mature follicles (Rowlands and Parkes, 1966; van Rees *et al.*, 1968). LH-like gonadotropins will induce ovulation in many species during the preovulatory stage of the cycle (Thibault and Mauléon, 1964; Rowlands and Parkes, 1966). However, treatment too early in the cycle will lead to luteinization with trapped ova rather than ovulation (Section III,C,6). In old age the ovary tends to become unresponsive (Green, 1957; Burack and Wolfe, 1959), presumably due to its altered structure, although some ovarian response may still occur (Ortiz, 1955).

ii. Effect of the pituitary and other glands: synergism. Many studies of the immature rat (see Zarrow and Dinius, 1971) have shown that ovulation and superovulation can be caused by a single dose of PMSG. Ovulation and superovulation are probably not a direct result of the PMSG, but are brought about via the positive feedback of steroids, in response to the PMSG, on the hypothalamus–pituitary complex, producing a release of the gonadotropins necessary for ovulation. Such interactions are probably involved in other species when gonadotropins are employed to augment ovarian function (Thibault and Mauléon, 1964). In hypophysectomized animals various elements of the ovary atrophy and functional capacity decreases. The response to gonadotropin will vary with the time after hypophysectomy (Rowlands and Parkes, 1966). In the rat treated with FSH, the form of response will differ if even 1 day elapses between hypophysectomy and treatment (Fevold *et al.*, 1937). A 6 to 20 hour interval after hypophysectomy inhibits the response of the corpus luteum of the rat to the luteotropic action of prolactin (Astwood, 1941), and after 48 hours prolactin causes structural luteolysis (Malven and Sawyer, 1966b).

The concomitant involution of other glands of internal secretion after hypophysectomy may also affect the response of the ovary to gonadotropins. Adrenal–ovarian (Parkes and Deanesly, 1966) and thyroid–ovarian (Young, 1961; Kmentová and Schreiber, 1965) effects have been demonstrated and some of these may involve altered ovarian sensitivity to gonadotropins. The effect of ovarian steroids is probably mainly indirect on gonadotropin control mechanisms (see Volume III, Chapter 3), but an increase in ovarian sensitivity to gonadotropins caused by estrogen has been suggested (Young, 1961). Evidence for an opposite effect of progestagens is still debatable.

The problem of hormone synergism, in which the effect of two or more hormones exceeds that of each given independently, has been reported on several occasions, particularly for FSH and LH (Greep, 1961). The "need" of the ovary for stimulation from the pituitary at different stages of development is characterized by a sequential requirement for FSH, LH, and prolactin singly or in combination (Section III, C). The synergism between the gonadotropins and the relative requirement for them probably varies from time to time and from species to species. Perhaps FSH with LH dominates in the follicular phase, LH with FSH in ovulation and luteinization and a constant low level of one or more of FSH, LH, and prolactin during the luteal phase.

iii. Genetic effects. Differences between species and between different strains of the same species in response to gonadotropins and in the gonadotropin content of the pituitary have been reported (Mauléon and Pelletier, 1964). Correlations have been found between strains and the response of the ovaries to exogenous gonadotropins (Mauléon, 1969). Strain differences of rats and mice in gonadotropin bioassays are well recognized (Bell and Loraine, 1967; Loraine and Bell, 1971) with respect to differences in sensitivity and the maximal ovarian size attainable.

IV. FACTORS AFFECTING THE PITUITARY

Many lines of evidence indicate that the brain regulates gonadotropin secretion. Marshall (1936, 1942) emphasized the dependence of the breeding season in many species on environmental factors and suggested that these factors exerted their effects via a nervous control over the anterior pituitary gland. Gonadal abnormalities seen in humans with brain pathology have been duplicated in laboratory animals. The hypothalamus is the center that is believed at present to constitute the final control of the adenohypophysis (Harris, 1955, 1972). In addition to neural control, another factor which must be considered in the control of gonadotropin secretion is the part played by the secretions of the target organs, that is, the interrelationships of the brain, anterior pituitary, and ovaries comprising the so-called feedback mechanisms. Neural and feedback control of gonadotropin secretion will be considered in the following sections, and steroid feedback in mammals is discussed in Volume III, Chapter 3.

A. Nonmammalian Vertebrates

The hypothalamic control of the pituitary–ovarian axis in nonmammalian vertebrates has been discussed in a number of reviews (general: Dodd et al., 1971; birds: Farner et al., 1967; Nalbandov and Graber, 1969; amphibians: Jørgensen, 1970; fish and amphibians: Jørgensen, 1968, 1970; Jørgensen and Larsen, 1967; bony fish: Ball et al., 1972).

1. Fish

There is no clear evidence for central nervous system control of pituitary function in either cyclostomes or elasmobranchs (Dodd, 1972). Ectopically transplanted pituitaries in Lampetra fluviatilis supported normal ovarian development, but ovulation did not always occur (Larsen, 1965, 1969a,b, 1970). In selachians, there is no nervous or vascular connection between the hypothalamus and the ventral lobe of the pituitary (Mellinger, 1964; Meurling, 1967). In teleost fish, the pituitary is innervated by nerve fibers, many originating in the nucleus lateralis tuberis of the hypothalamus. Changes in the staining qualities of the neurosecretory material in this nucleus have been correlated with increased gonadal activity (Brehm, 1958; Billenstein, 1962; Honma and Tamura, 1965). In Ophiocephalus punctatus, and Carassius auratus, when the pituitary is transplanted to a distant site, and so removed from the hypothalamus, its ability to support gonadogenesis ceases (Roy, 1964; Johansen, 1967). In hypophysectomized mollies (Poecilia formosa, comprising gynogenetic female clones), a pituitary homotransplant to the caudal musculature failed to stimulate vitellogenesis after the birth of the current brood (Ball et al., 1965) and contained only inactive gonadotrops (Olivereau and Ball, 1966). Following mammalian work, Peter (1970) studied the effects of lesions on Carassius auratus. An inhibition of ovarian activity occurred after lesions in the posterior part of the nucleus lateralis tuberis or in the pituitary stalk but not after those placed in other hypothalamic regions. It had been supposed that, as in mammals (Section IV,B,3), the preopticohypophysial neurosecretory system is implicated in the control of gonadal activity (Honma and Tamura, 1965), but no effects were noted after destruction of this system (Peter, 1970) although the isotocin it contains may be involved in reproduction (Sawyer and Pickford, 1963) and oviposition (Wilhelmi *et al.,* 1955; Egami and Ishii, 1962). The ways in which estrogen secretions of the ovary may influence the hypothalamus–pituitary system are not known. Administered estrogens cause ovarian regression (see Dodd, 1960) and it is possible that the action is not direct but via the pituitary (Egami, 1954; Goswami and Sundararaj, 1968a,b) and the hypothalamus.

2. Amphibians and Reptiles

The ovaries of amphibians living in temperate latitudes grow rapidly in the autumn and are ready for spawning in the spring after hibernation. The hypothalamus has been shown to be involved in the stimulation of ovarian growth in *Rana temporaria* (Dierickx, 1964) and *Bufo bufo* (van Dongen *et al.,* 1966). In Dierickx's experiments, adenohypophysial tissue was auto-transplanted into the anterior chamber of the eye after the spring spawning when the ovaries were spent. When the frogs were killed in the autumn, the ovaries were still regressed and similar to those of the hypophysectomized controls (weight 0.2 to 0.5 gm), suggesting that the transplanted pituitaries were inactive. The ovaries of unoperated controls were large (weight 3.5 to 4.5 gm). In *Bufo bufo* the ovarian weights of toads bearing pituitary transplants were intermediate between those of hypophysectomized and normal controls (van Dongen *et al.,* 1966; Vijayakumar *et al.,* 1971), and it was concluded that in *Bufo* a limited secretion of gonadotropic hormone can occur without pituitary contact with the hypothalamus. Normal oogenesis and seasonal development of ovaries and oviducts can continue in *Rana temporaria* when the hypothalamus–pituitary complex is isolated from the rest of the brain (Dierickx, 1966, 1967b). The hypothalamic component excludes the well-known neurosecretory magnocellular preoptic nuclei and has many anatomical features in common with the hypophysiotropic area in mammals (Dierickx, 1965). Ovulation is dependent on the integrity of the preoptic area in both *Rana temporaria* (Dierickx, 1967a,b) and *Bufo* (Jørgensen, 1968, 1970). In *Bufo,* transection of the brain anterior to the optic chiasma suppresses normal ovarian activity (Jørgensen, 1968, 1970). Lisk (1967) found that testosterone or estradiol crystals implanted in the median eminence of the desert iguana prevented further ovarian development, whereas control implants were without effect. However, Callard *et al.* (1972a,b) found that implants of estradiol into the same region and in the anteromedial hypothalamus in the blue spiny lizard, *Sceloporus cyanogenys,* did not prevent ovarian growth but inhibited ovulation. There was also some evidence for a reduction in the production of ovarian steroids.

3. Birds

Among the earliest studies on hypothalamus–pituitary relationships in any species were those of Benoit and Assenmacher on the Pekin duck (see Benoit and Assenmacher, 1955, 1959; Assenmacher, 1958, 1970). The conclusion they drew from these experiments was that the hypothalamus controls the gonadotropic activity of the pituitary by way of the hypophysial portal vessels, whereby blood is drained from the median eminence and carried directly to the anterior pituitary. In a study on the significance of the hypophysial portal vessels in the hen, Shirley and Nalbandov (1956) showed that transection of these vessels, without consequent pituitary ischemia, results in the atrophy of the ovary, oviduct, and comb. In birds there is no evidence for a direct nervous control of the pituitary and an indirect control via the pars intermedia (see Section IV,B,1) can be discounted since this structure is not present in the avian pituitary. There is, however, anatomical evidence for a further neuroendocrine control system whereby changes in the composition of the cerebrospinal fluid could influence pituitary function. In the quail, Sharp (1972) has described a tanycyte–vascular system in which specialized ependymal cells (tanycytes) bordering the third ventricle are characterized by long processes extending to the blood vessels which cover the surface of the median eminence. This system is similar to that described in mammals (see Section IV,B,1).

Location of the precise hypothalamic areas concerned in the production and release of gonadotropin-releasing factors has proved difficult. The preopticohypophysial system seems not to have the importance once assigned to it (see Dodd *et al.,* 1971). Ralph and Fraps (1959a) found that lesions made anywhere in the median eminence diencephalon of the laying hen could result in cessation of oviposition for at least 2 weeks. Other workers have found that lesions in the preoptic and basal hypothalamus are particularly effective in blocking ovarian activity (Egge and Chiasson, 1963). There have been many attempts to show that the preoptic area in the hen is necessary for the maintenance of the ovary and for ovulation. Only lesions located in the preoptic area and made more than 6 hours before the expected ovulation of the first (C_1) egg of a clutch (Ralph, 1959) prevented ovulation. Also, insertion of electrodes, with or without current, into the preoptic area about 14 hours before the C_1 ovulation blocked ovulation in a significant number of experimental hens (Opel, 1963). Premature ovulation of the C_1 follicle by systemic injections of progesterone were most effectively blocked by lesions in the preoptic area or in its caudally directed fiber connections (Ralph and Fraps, 1959b). Progesterone implanted directly into the preoptic area or caudal neostriatum resulted in premature ovulation (Ralph and Fraps, 1960).

Antiadrenergic and anticholinergic drugs block normal or progesterone-induced ovulation (Zarrow and Bastian, 1953; van Tienhoven *et al.,* 1954). Phenobarbitone also blocks the action of progesterone (Fraps, 1955), whereas 5,5-diallybarbituric acid, pentobarbital sodium, and calcium ethylisopropyl barbiturate alone or with subovulatory doses of progesterone can cause premature ovulation (Fraps and Case, 1953). These drugs are generally considered to act via the central nervous system, but the effect on the ovaries might be direct (see Ferrando and Nalbandov, 1969).

In many birds, as shown by the domestic fowl, there are pauses or "rest periods" between the laying of clutches of eggs. To explain this, the central nervous system, probably the hypothalamus, may be assumed to exhibit a diurnal rhythm of sensitivity to ovarian hormones (see Fraps, 1970), which themselves vary according to the stage of follicular maturation. The release of the hormone inducing ovulation (OIH) would then be triggered when the sensitive phase of the neural threshold rhythm coincides with the appropriate levels of circulating ovarian hormones. Changes in the levels of estrogens and progesterone before ovulation and the lack of a peak of pituitary LH activity on the days preceding missed ovulations (Bullock and Nalbandov, 1967; Heald *et al.,* 1968) would support this hypothesis. Further, Opel (1964) was able to induce ovulation of follicles (other than the C_1 follicle) merely by inserting an electrode into various parts of the forebrain and diencephalon, a peak of sensitivity occurring some 7 hours before the expected ovulation.

The *in vitro* demonstration of LRF in the chicken hypothalamus (Jackson and Nalbandov, 1969a,b; Tanaka *et al.,* 1969), using the ovarian ascorbic acid depletion assay for LH, was not conclusive as the assay is sensitive to arginine vasotocin (Jackson and Nalbandov, 1969a; Ishii *et al.,* 1970), which is present in hypothalamic extracts and which has no LH or LRF activity. Follett (1970), using a total gonadotropin assay based on the uptake of ^{32}P by the testes of young chicks, showed that hypothalamic extracts from sexually mature quails were more effective in stimulating the release of gonadotropins from chicken pituitaries than were a wide range of pharmacologically active substances (e.g., oxytocin, vasopressin, noradrenaline, dopamine, and histamine) or extracts of the cerebral cortex, hindbrain, or liver. *In vivo* methods have also been used to demonstrate LRF in the avian hypothalamus. Infusion of avian hypothalamic extracts into the adenohypophysis of the hen has been shown to cause ovulation (Opel and Lepone, 1967; Clark and Fraps, 1967). It is interesting to note that, in birds, nerve terminals containing catecholamines are present in the external layer of the median eminence (see Sharp and Follett, 1970). Dopamine and noradrenaline by themselves cannot release gonadotropins from the chicken pituitary (Follett, 1970) and efforts to show that they affect gonadotropin release at the hypothalamic level have been unsuccessful (see Dodd et al., 1971).

B. Mammals

It is generally held that gonadotropin secretion from the anterior pituitary in mammals is under neural control and that the hypothalamus constitutes the main control center (Harris, 1955, 1972). A particular problem has been to elucidate the functional and structural connections between the hypothalamus and the anterior pituitary. This aspect and the importance of higher neural and feedback control mechanisms will be discussed in this section. Many of the data upon which conclusions have been based have been obtained from very few species, often only the rat and, hence, their general applicability should be treated with caution.

1. Hypothalamus-Anterior Pituitary Relationship

The possibility that the hypothalamic control of anterior pituitary function is by direct innervation was reviewed in detail by Harris (1955). He concluded that the anterior pituitary received very few if any nerve fibers from the hypothalamus. There is evidence, however, that the pars intermedia is supplied by nerve fibers (Vincent and Anand Kumar, 1969); hence an indirect hypothalamic nervous control of anterior pituitary function is possible (Zuckerman, 1970). The lack of secretomotor innervation of the anterior pituitary and the rediscovery of the hypophysial portal system by Popa and Fielding (1930, 1933) led to the proposal of a neurohumoral control of the anterior pituitary (Harris, 1937; Hinsey, 1937; Friedgood, 1970). It was suggested that this control is mediated via the hypophysial portal vessels which carry blood from the hypothalamus directly to the anterior pituitary (Harris, 1955).

Experiments involving pituitary stalk section (Harris, 1950) and pituitary transplantation (Greep, 1936; Desclin, 1950; Harris and Jacobson, 1952; Nikitovitch-Winer and Everett, 1958) showed that the anterior pituitary functions normally when vascularized by the hypophysial portal vessels but that tropic hormone production, including FSH and LH but not prolactin, falls markedly if it is supplied only by systemic vessels. Such findings support the proposition (Green and Harris, 1947) that the hypothalamic control of anterior pituitary secretion is exerted by means of various neurohumors which are liberated from nerve fibers ending on a network of capillaries in the median eminence region of the hypothalamus and that these substances are then transported via the portal vessels to regulate the secretion of anterior pituitary hormones. However, the arguments presented by Zuckermann (1954) and Holmes *et al.* (1959) that the impairment of anterior pituitary function may primarily be due to partial anemia followed by necrosis as a result of transecting the hypophysial stalk need to be borne in mind in evaluating the results of such experiments. Also, the existence of neurohumors is not dependent on the belief that they are transmitted to the gland by way of portal vessels (Zuckerman, 1970). Thomson and Zuckerman (1953, 1954) and Zuckerman (1955) reported that, in the ferret, complete interruption of the vessels of the pituitary stalk was compatible with an estrous response to extra illumination in winter. Löfgren (1960) suggested that the hypothalamus may influence anterior pituitary function by secreting substances into the cerebrospinal fluid which are subsequently transported to the anterior pituitary. The position is complicated further by the existence in the third ventricle in the region of the anterior hypothalamus of ependymal cells, the so-called tanycytes, which may play a role in the function of the hypothalamus–pituitary unit (Knowles and Anand Kumar, 1969; Knigge and Scott, 1970). These cells may secrete substances into the cerebrospinal fluid or absorb substances, perhaps releasing factors, from it and subsequently secrete at their terminals substances affecting anterior pituitary function.

2. Releasing Factors

The search for, isolation, purification, chemical identification, and synthesis of separate hypothalamic neurohumors, variously called hypophysiotropic hormones, releasing, and/or inhibiting factors or hormones, is being actively carried out for each of the anterior pituitary hormones (McCann and Porter, 1969; Schally and Kastin, 1970; Meites, 1970; Burgus and Guillemin, 1970). The neurohumors involved in reproductive function were designated LRF and FRF (see Section III,C,4) based on their biological activities. Whereas the hypothalamus is believed to stimulate the release of FSH and LH, it inhibits that of prolactin by a prolactin-inhibiting factor (PIF). In addition to LRF, FRF, and PIF there is evidence for the presence of gonadotropin-inhibiting factor(s) (Karavolas *et al.*, 1971) and a prolactin-releasing factor (Meites *et al.*, 1960; Mishkinsky *et al.*, 1968; Nicoll *et al.*, 1970). This suggests that there may be a dual system acting via releasing and inhibiting factors for the control of gonadotropin and prolactin secretion. The use of the generic term releasing factor is convenient for both types of factor (Gorski, 1968). There is a good case for using the term releasing hormone (Schally *et al.*, 1968) once the factor has been fully characterized.

Since the first demonstration of the activities of LRF (McCann *et al.*, 1960; Harris, 1961), FRF (Igarashi and McCann, 1964; Mittler and Meites, 1964), and PIF (Pasteels, 1961; Talwalker *et al.*, 1963), these releasing factors have been shown to be present in hypothalamic extracts from numerous mammalian species (McCann and Porter, 1969; Schally and Kastin, 1970), in which they act without apparent species specificity. The responses obtained *in vivo* and *in vitro* are related to the log dose of the factor administered (Meites, 1970; Jutisz *et al.,* 1970). It has been claimed that LRF, FRF, and PIF activities are present in rat hypophysial portal blood (Fink *et al.,* 1967; Kamberi *et al.,* 1970a,b).

The chemistry of the releasing factors, FRF, LRF, and PIF, has been extensively studied in recent years (McCann and Porter, 1969; Fawcett, 1970; Burgus and Guillemin, 1970; Schally and Kastin, 1970; Schally *et al.,* 1970b). Despite apparent evidence for the complete separation of FRF and LRF from several species (Burgus and Guillemin, 1970), evidence from studies in the pig and man indicated that LRF probably contains inherent FRF activity (Schally *et al.,* 1970a,b, 1971a; White, 1970). A decapeptide with the sequence (pyro) Glu-His-Trp-Ser-Tyr-Gly-Leu-Arg-Pro-Gly-NH$_2$ has been prepared from porcine stalk–median eminence tissue. The natural and synthetic decapeptide stimulates the release and synthesis of LH and FSH (Schally *et al.,* 1971a,b; Matsuo *et al.,* 1971; Baba *et al.,* 1971), suggesting that FRF and LRF are the same substance.

Such a common identity of FRF and LRF suggests several possibilities for control of gonadotropin output. First, that FSH and LH are always released together and that the specificity of effects are dependent on the state of the ovary and its gonadotropin-binding sites. Second, the specificity of FSH and LH release may be at the pituitary level and it may be dependent on the steroid environment or other factors. It has been shown, for example, that pituitary responsiveness to LRF is increased by estrogen (Arimura and Schally, 1971) and decreased by progesterone (Arimura and Schally, 1970; Hilliard *et al.,* 1971). Steroids may facilitate LRF/FRF action or compete with it at the pituitary receptor sites. Third, there may be other specific gonadotropin-releasing, -synthesizing or -inhibiting factors.

a. Location of Releasing Factors. It has been supposed that the releasing factors might originate in the median eminence or might be synthesized in regions apart from the median eminence and only stored and released there. The origin of the factors has been sought by several methods (Mess, 1969; Mess *et al.,* 1970; Harris, 1970a). Histological study of pituitary transplants in different regions of the hypothalamus has led certain workers (Halász *et al.,* 1962, 1965; Flament-Durand, 1965; Flament-Durand and Desclin, 1970) to plot a half moon-shaped area of the hypothalamus, the so-called hypophysiotropic area, which includes the arcuate nucleus and related parts of the periventricular system. Pituitary tissue transplanted to this area shows structural maintenance and functional activity, indicating the stimulation of synthesis and release of FSH and LH and the inhibition of prolactin. When placed outside this definite area of the hypothalamus, the transplants are reported to be composed of small chromophobe cells and show no functional activity. Both FRF and LRF activities are distributed in a medial basal zone extending from the suprachiasmatic–preoptic region to the median eminence (Quijada *et al.*, 1971). Such a finding is in keeping with the existence of a single gonadotropin-releasing factor. Previous findings that FRF and LRF activities had different hypothalamic distributions (Watanabe and McCann, 1968; Mess *et al.*, 1970) can only be explained if there are specific factors for the release of FSH and LH in addition to the decapeptide possessing both activities. PIF activity has been located in the dorsolateral part of the preoptic area and PRF activity in the median eminence and medial basal portion of the preoptic area (Krulich *et al.*, 1971).

b. Changes with Reproductive Stage. There is evidence of FRF and LRF activity in the hypothalamus prenatally and neonatally (Campbell and Gallardo, 1966; Corbin and Daniels, 1967). At puberty there is a fall in hypothalamic LRF and FRF activity in female rats (Ramirez and Sawyer, 1966; Corbin and Daniels, 1967; Watanabe and McCann, 1969). During the estrous cycle there are falls in the hypothalamic content of LRF (Chowers and McCann, 1965; Ramirez and Sawyer, 1965) and FRF (Négro-Vilar *et al.*, 1970) which appear to correspond to the increased secretion of LH and FSH that occurs before ovulation (Section II,B,3,b). Decreased hypothalamic PIF levels at proestrus and estrus in the rat (Sar and Meites, 1967) and increased hypothalamic LRF levels at proestrus, estrus, and later in the cow (Hackett and Hafs, 1969) are less easily related to changes in prolactin and LH at these times. Fink and Harris (1970) have found LRF activity in hypophysial portal blood at all stages of the estrous cycle except estrus in the rat. The sensitivity of the pituitary to LRF has been shown to vary with the stage of the cycle (Arimura and Schally, 1970; Reeves *et al.*, 1971) and with the season (Domanski and Kochmam, 1968). These changes are probably related to the interaction of releasing factor and steroids at the pituitary level in the regulation of gonadotropin release.

3. Levels of Neural Control over Gonadotropin Secretion

a. Hypophysiotropic Area. Although other possibilities exist (Section IV,B,1) the common belief is that the link between the hypothalamus and the anterior pituitary may be by neurohumors synthesized in a system of neurons located in an extended hypophysiotropic area. Their nerve endings terminate on the capillary loops of the portal system, and the neurohumors are then transferred to act on the anterior pituitary. Is the area producing the releasing factors merely a link in the final common path or does it exert a regulatory influence on the secretory function of the anterior pituitary?

Isolation of the hypophysiotropic area (Halász and Pupp, 1965; Halász and Gorski, 1967; Halász, 1969; Butler and Donovan, 1971) from the rest of the brain, by means of cuts with a special knife (Halász, 1969), leads to constant follicular activity and vaginal estrus in some cases or persistent corpora lutea and almost constant vaginal diestrus in others. There was no appreciable difference in the size of the hypothalamic island which could account for the two types of effect (Donovan, 1969). It is claimed that this area is capable of supporting basal gonadotropin secretion from the anterior pituitary without any neural inputs, to enable follicular growth and steroid biosynthesis to occur. However, this area cannot by itself support full anterior pituitary activity. The ovulatory surge of gonadotropins appears to be eliminated in rats with the hypophysiotropic area isolated from the rest of the brain. These experiments support the concept (Barraclough and Gorski, 1961; Flerkó, 1962) that the neural control of gonadotropin secretion can be divided into two functional levels. One level, represented by the hypophysiotropic area, is claimed to regulate the basal secretion of gonadotropins sufficient for ovarian development and steroidogenesis. A second higher level of control includes all other hypothalamic and extrahypothalamic structures that exert a stimulatory or inhibitory effect on the secretions of gonadotropins. Such control is probably exerted via the hypophysiotropic area (Réthelyi and Halász, 1970).

b. **Extrahypophysiotropic Hypothalamic Areas.** There is now little doubt that anterior connections of the hypothalamus with the rest of the brain are of the greatest importance in the control of ovarian function (Donovan, 1970). Ovulation takes place only when the hypophysiotropic area, after partial separation, remains connected to the rest of the brain anteriorly (Halász and Gorski, 1967). This finding confirms conclusions drawn from experiments involving stimulation and lesions (Everett, 1964, 1969; Flerkó, 1966; Tejasen and Everett, 1967; Gorski, 1968; Szentágothai et al., 1968; Halász, 1969; Arimura and Findlay, 1971) and supports the view that the impulse responsible for the cyclical surge of release of the gonadotropins (Section II,B,3,b) causing ovulation is mediated by a system of nerve fibers which arise in or pass through the preoptic hypothalamic region. Köves and Halász (1970) have presented evidence that the hypothalamus, when surgically isolated from the rest of the brain, can induce ovulation provided that the isolated area includes the preoptic area. In addition to the control of the ovulatory surge of gonadotropins, of which LH may be the most important, it has been suggested that control of FSH secretion from the hypophysiotropic area requires that afferent impulses reach it from or via the anterior hypothalamic area (Flerkó, 1966, 1970; Kalra et al., 1970). But recent experiments involving stimulation of the hypothalamus and plasma assays of FSH and LH indicate that, although LH may be controlled by the preoptic area and FSH by the anterior hypothalamic area, there is a large overlap in the functions of the two areas (Clemens *et al.,* 1971b; Kalra *et al.,* 1971). The differences observed may be quantitative and not qualitative differences in the response of the two areas or different responses of the pituitary to the releasing factor(s). This would be in keeping with the idea of a single releasing factor for FSH and LH (Section IV,B,2). The findings of Köves and Halász (1970) indicate that the preoptic area itself may be the origin of the neurogenic stimuli that cause the release of LRF and FRF in amounts necessary for ovulation. However, the number of ova seen in the oviducts was less than in control animals and the animals with preoptic isolation exhibited irregular vaginal cycles. These findings suggest that additional structures are involved in the control of an entirely normal, cyclic, ovulatory output of gonadotropin.

There is evidence for PIF and PRF in the preoptic area and median eminence (Section IV,B,2,a). Other areas of the hypothalamus, including the medial hypothalamus from the paraventricular nucleus to the ventral premamillary region, and a region in the thalamohypothalamic border (Pasteels, 1970), have been shown to be involved in the regulation of prolactin secretion, possibly in an indirect way.

c. Extrahypothalamic Areas. Several studies have shown that extrahypothalamic structures can influence pituitary–ovarian function. Neurons belonging to ovulation–LH control mechanisms are situated in the septum (Everett, 1964; Kalra *et al.,* 1971; Clemens *et al.,* 1971b). The amygdala and hippocampus have also been shown to be involved in the control of FSH and LH secretion (see Critchlow and Bar-Sela, 1967; Velasco and Taleisnik, 1969a,b; Lawton and Sawyer, 1970; Flerkó, 1970), possibly by reciprocal inhibitory and facilitatory mechanisms. The amygdala is probably also involved in the control of prolactin secretion (Tindal *et al.,* 1967). In addition, different regions of the midbrain appear to have stimulatory and inhibitory effects on LH secretion and ovulation (Carrer and Taleisnik, 1970) and may be important in the control of prolactin secretion (Flerkó and Bárdos, 1966). The pineal probably also participates in the control of gonadotropin secretion and may play a role in the control of circadian rhythms of FSH and LH secretion (Fraschini and Martini, 1970).

d. Method of Control of the Hypophysiotropic Area and Releasing-Factor Release. The data given above suggest that influences from other areas of the hypothalamus and from extrahypothalamic structures may be involved in the control of the secreting neurons of the hypophysiotropic area. However, the means whereby the brain may affect the rate of secretion of the various neurohumors is poorly understood. Neurons which contain noradrenaline, serotonin, and acetylcholinesterase occur in the hypothalamus and preoptic area and, furthermore, there is a tuberoinfundibular dopamine neuron system which ends near the primary plexus of the hypophysial portal vessels in the median eminence (Fuxe and Hökfelt, 1969, 1970; Shute, 1970). These systems may be involved at certain points in the transmission of information from the brain to the pituitary. The dopamine neurons may act to control the release of FRF, LRF, and PIF from the secretory neurons (see McCann, 1970; Fuxe and Hökfelt, 1969, 1970).

4. Gonadotropin Feedback Mechanisms

It is clear from the above discussion that hypothalamic and extra-hypothalamic mechanisms exert a major controlling function over the secretion of FSH, LH, and prolactin from the anterior pituitary. In many species these neural mechanisms are influenced by factors from the external environment (Volume II, Chapter 10). Inputs from the sense organs and from the reproductive tract may be involved in triggering or timing the events of the neural control system. An additional factor which must be considered in the control of gonadotropin secretion is the part played by the secretions of the target organs, that is, the interrelationships of the anterior pituitary and ovaries (Volume III, Chapter 3) which comprise the so-called feedback mechanisms. The gonadotropins act in a stimulatory (positive feedback) or inhibitory (negative feedback) manner to control their own secretion. The terms short, internal, or autofeedback have been used (Martini *et al.,* 1968; Motta *et al.,* 1969). Internal feedback may be applied to the whole process whereby pituitary hormones control their own secretion; short-loop feedback being specifically applied to influences mediated by the hypothalamus and autofeedback to direct effects on the anterior pituitary (Igarashi, 1969; Hirono *et al.,* 1970). The existence of internal feedback has been demonstrated by four techniques; the implantation of minute amounts of hormones into the brain or anterior pituitary, the systemic injection of hormone, the measurement of the effects of hypophysectomy, and investigation of the effects of hormone on electrical activity of the brain. Such experiments performed with LH suggest that it exerts a negative feedback effect. Implants of LH into the median eminence reduce pituitary and in some cases plasma LH levels in adult (Corbin, 1966; Corbin and Cohen, 1966; Dávid *et al.,* 1966) and prepubertal (Ojeda and Ramirez, 1969) rats. Chronic injection of PMSG reduces pituitary LH stores (Szontágh *et al.,* 1963). Hypophysectomy induces hypersecretion of LRF (Nallar and McCann, 1965; Pelletier, 1965). Electrophysiological studies have suggested that LH may have actions on the hypothalamus (Sawyer and Kawakami, 1961; Beyer and Sawyer, 1969; Sawyer, 1970) at the areas of basal or cyclic control of LH release. FSH displays a negative and a positive feedback effect depending on the age of the animal. In adults, implants of FSH into the median eminence reduce pituitary and plasma FSH levels (Corbin and Story, 1967; Arai and Gorski, 1968; Fraschini *et al.*, 1968; Hirono *et al.*, 1970). FSH implanted into the anterior pituitary decreased pituitary FSH but did not consistently decrease plasma FSH (Hirono *et al.*, 1970). Administration of FSH reduces FRF in plasma and the median eminence (Corbin *et al.*, 1970). Hypophysectomy leads to a rise in plasma FRF (Corbin *et al.*, 1970). In immature rats implants of FSH placed in the medial basal hypothalamus, a larger area than that implicated in a negative feedback system in the adult, stimulate FSH release (Ojeda and Ramirez, 1969). Implants of prolactin placed into the median eminence (see Voogt and Meites, 1971) or high levels of plasma prolactin (Meites, 1970; Pasteels, 1970) lead to a depression of pituitary prolactin content and release. Prolactin has also been shown to affect electrical activity in the hypothalamus (Clemens *et al.*, 1971a). Whereas FSH and LH seem to affect their own secretion without affecting the secretion of other anterior pituitary hormones, prolactin not only inhibits prolactin secretion but stimulates gonadotropin release (Voogt and Meites, 1971).

These results suggest that the gonadotropins control their own synthesis and release from the pituitary via a short-loop action at the hypothalamic level. Several studies show the possibility also of a direct autofeedback action at the pituitary level. The direct autofeedback actions of FSH seem to result in an inhibition of synthesis but not release of FSH. In addition to these internal feedbacks, an ultrashort-loop feedback in which FRF inhibits FRF release has been postulated (Hyyppä *et al.*, 1971). It is presumed that the short, ultrashort- and autofeedbacks could act via the hormones in the general circulation, and the short-loop feedback could act by a vascular pathway of the hypophysial portal system in which blood is carried from the anterior pituitary to the median eminence (see Szentágothai *et al.*, 1968). The possible role of internal feedback under physiological conditions is not known. The positive FSH feedback and the effects of prolactin on FSH and LH release may be relevant at puberty (Section II,B,3,a). The different feedback systems may be involved in some way in making the non-specific gonadotropin releasing mechanism specific at the pituitary level.

V. CONCLUSIONS

It seems that the pituitary glands of mammals and birds secrete two distinct gonadotropins, referred to as FSH and LH. However, in fish, amphibians, and possibly reptiles there may be only a single hormone. Despite chemical evidence, absolute purity of preparations remains a problem and data presented here (Sections II,III,IV) from studies of the cellular origin, hormone level, actions, and the neural control of the secretion of the gonadotropins indicate either an overlapping of actions or lack of identity. These difficulties may be resolved by chemical studies now in progress. It has been shown (Section III,A,1) that the gonadotropins consist of two nonidentical subunits. One of these, the "alpha" unit appears to be common to all glycoprotein hormones, FSH, LH, and TSH, of a given species, while the other, the "beta" unit, appears to be structurally unique for the hormone in question.

Immunological assays have added to our information about the rate of secretion and changes in blood levels of gonadotropins in different reproductive states (Section II,B). The association of ovulation with a peak level of gonadotropin in the blood has been confirmed for many species. There is evidence that the plasma level remains relatively constant at other times of the cycle. Repeated or continuous sampling should give information on possible episodic rhythms, pulsatile secretions, or other small changes taking place at other times.

It seems that the follicular phase of the estrous cycle, in which pituitary gonadotropins stimulate follicle growth, estrogen secretion, and ovulation, can be observed in all classes of vertebrates. The specific role of FSH and LH in these processes in mammals remains unresolved (Section III,C). The addition of a luteal phase in the estrous cycle is a mammalian phenomenon, and its control by gonadotropins differs in the various species (Section III,C,7) studied.

The importance of such factors as clearance rate and specific ovarian binding sites (Sections III,D,2; III,D,3) on the action of gonadotropins *in vivo* has been stressed. These phenomena and an understanding of the mode of action of gonadotropins on the ovary are areas of increasing interest. The functional state of the ovary affects its response (Section III,D,3,b) and is a factor of great importance for the success of the artificial induction of follicular growth and ovulation with gonadotropins (Section III,C,5).

The demonstration that the hypothalamus plays an important role in the regulation of the anterior pituitary and the possibility that the portal vessels provide a functional link between the hypothalamus and the anterior pituitary has stimulated much interest. Recent work (Section IV,B,2) has focussed on possible hormones produced by the hypothalamus which may stimulate gonadotropin secretion. The isolation and synthesis of a single hormone that stimulates the secretion of both FSH and LH agrees well with much existing knowledge on the release of FSH and LH from the anterior pituitary. However, the possibilities for other stimulatory or inhibitory factors, particularly a specific FRF, remain. The nature and importance of extrahypothalamic central nervous systems which control the hypothalamus and releasing-hormone release remain obscure. Gonadotropin feedback is important in the control of hypothalamic function (Section IV,B,4), but the relative importance and interaction of this mechanism and steroid feedback mechanisms (Volume III, Chapter 3) is unknown.

VI. ADDENDUM

No attempt will be made to cover the huge amount of information on the hypothalamus-pituitary-ovarian axis published since this chapter was written; only certain particular aspects will be briefly reviewed.

It is now firmly established that mammalian, avian, reptilian, and amphibian pituitaries synthesize FSH and LH as two distinct molecular entities (see Section II; mammals: Reichert and Ward, 1974; Sairam and Papkoff, 1974; Ross, 1975; Liu and Ward, 1975; birds: Farmer *et al.,* 1975; Godden and Scanes, 1975; reptiles: Licht and Papkoff, 1974a; amphibians: Licht and Papkoff, 1974b). FSH and LH are glycoproteins and have been shown in mammals to comprise two noncovalently linked dissimilar subunits which have been isolated and purified. The α-subunit is common to FSH, LH, and TSH while the β-subunit confers hormone specificity. The individual subunits show little or no biological activity. There is controversy over whether one or two gonadotropins are produced by the teleost pituitary. Only one gonadotropic fraction was found in the carp (Burzawa-Gerard, 1971) and Chinook salmon (Donaldson *et al.,* 1972), but improvements in purification techniques have led to the separation of two gonadotropin fractions in the chum salmon (Idler *et al.,* 1975).

The continued development of radioimmunoassays (Moudgal and Raj, 1974), the use of radioreceptor assays (Reichert *et al.,* 1973; Riechert and Bhalla, 1974), and the development of highly sensitive bioassays (Kramer *et al.,* 1974), have enabled changes in gonadotropins to be studied in a wide variety of species and in different reproductive states (see Section II,B). In addition to the seasonal and cyclical changes described previously, it has been shown that the pattern of blood gonadotropin concentrations is pulsatile (see Section V). Episodic spurts occur, of varying size, duration, and frequency, depending on reproductive and hormonal status (Ferin *et al.,* 1974; Naftolin, 1975). During the human menstrual cycle the magnitude and frequency of the LH pulses is greatest at the time of the preovulatory surge (see Section II,B,3,b) and lowest in the luteal phase (Yen *et al.,* 1975). In addition, it has been found that sexual maturity in man is associated with an increased magnitude of the LH pulses during sleep (Kapen *et al.,* 1974). Whether such diurnal changes continue into adulthood is controversial (Ferin *et al.,* 1974; Weitzman, 1974).

The presence of increased plasma FSH during the early follicular and late luteal phases of the menstrual cycle and the rise of FSH coincident with the preovulatory LH surge (see Section II,B,3,b), has stimulated interest in the precise role of FSH in ovulation, follicle development, oocyte maturation, luteinization, corpus luteum maintenance, and steroidogenesis (see Section III,C; Greenwald, 1974; Schwartz, 1974; Neal and Baker, 1975). FSH may not be involved in ovulation (see Section III,C,4), since the ovulating capacity of LH-free FSH is negligible (Lipner *et al.,* 1974) and antiserum against FSH does not block ovulation (Schwartz, 1974). Increased plasma FSH on the day of proestrus and the morning of estrus in short-cycle mammals, like the rat, mouse, and hamster, may regulate the development of follicles destined to ovulate in the next and successive cycles (Bast and Greenwald, 1974; Schwartz, 1974).

It has been confirmed that gonadotropins are not essential to support the early stages of follicle growth (see Section III,C,1; Nakano *et al.,* 1975). The later stages are controlled by an interplay of FSH, LH, and estrogens which stimulate the full development of the granulosa cells (Richards, 1975). The concentrations of gonadotropins in the follicular fluid play a critical role in follicular development (McNeilly, 1975). FSH enters the small antral follicles and stimulates the granulosa cells to secrete estradiol into the follicular fluid (McNatty *et al.,* 1975), possibly by the aromatization of testosterone (Moon *et al.,* 1975). The FSH and estradiol in the follicular fluid stimulate cell division (see Section III,C,1) and the development of FSH and LH receptors (Goldenberg *et al.,* 1972; Channing and Kammerman, 1974; Zeleznik *et al.,* 1974; Louvet and Vaitukaitis, 1975).

Entry of LH into the follicular fluid during the preovulatory period promotes progesterone secretion (McNatty *et al.,* 1975). Prolactin may be required to maintain an adequate pool of precursor for progesterone synthesis (Behrman *et al.,* 1970) and high concentrations of prolactin have been observed in early follicular-phase follicles in women (McNatty *et al.,* 1975). A direct effect of prolactin on the ovary is suggested by the observation that hyperprolactinemia is associated with hypogonadism and an impairment of the ovarian response to gonadotropins (Jacobs, 1975). If prolactin levels are suppressed with bromocriptine, normal gonadal function is resumed (Thorner *et al.,* 1975).

There has been interest in the mechanism of action of gonadotropins on the ovary and the interaction between gonadotropins, the adenyl cyclase system, prostaglandins, and steroids in the control of follicle growth, luteinization, oocyte maturation, corpus luteum maintenance, and steroidogenesis

(Channing, 1973; Lipner, 1973; Armstrong *et al.*, 1974; Hammerstein, 1974; Lindner *et al.*, 1974; Marsh and Le Maire, 1974; Rondell, 1974; Schuetz, 1974; Neal *et al.*, 1975). The initial event, in all gonadotropin action on the ovary, is the binding of the gonadotropin molecule to a receptor site on the cell membrane (Hutchinson, 1973; Channing and Kammerman, 1974; Dufau *et al.*, 1975). FSH binds almost exclusively to the granulosa cells in medium–large follicles while LH binds to the cells of the theca interna, granulosa, and corpus luteum (Channing and Kammerman, 1974; Amsterdam *et al.*, 1975; Nimrod *et al.*, 1976). Binding capacity varies with the functional states of the tissues, and a lack of ovarian response to gonadotropins (see Section III,D,3,b) may be due to the absence of specific binding sites (Hutchinson, 1973). The number of available gonadotropin receptor sites is regulated by the concentrations of gonadotropins, steroids, prostaglandins, and "ovarian inhibitor" to which the ovary is exposed (Channing and Kammerman, 1974; Hichens *et al.*, 1974; Channing, 1975; Louvet and Vaitukaitis, 1975; Yang *et al.*, 1976). The initial binding of gonadotropin results in the activation of membrane-bound adenyl cyclase and the intracellular formation of cyclic $3',5'$-adenosine monophosphate (cAMP). Cyclic AMP activates protein kinases which stimulate the metabolic changes leading to the appropriate ovarian response (Major and Kilpatrick, 1972).

It has been confirmed that synthetic luteinizing-hormone releasing hormone (LH-RH; see Section IV,B,2) stimulates the release of both FSH and LH in various mammalian species (Arimura and Schally, 1974; Flerkó, 1974). LH-RH is active in nonmammalian vertebrates including fish (Breton and Weil, 1973), amphibians (Thornton and Geschwind, 1974; Vellano *et al.*, 1974), and birds (van Tienhoven and Schally, 1972; Wilson and Sharp, 1975). There is no convincing evidence in mammals, for the existence of a specific FSH-releasing factor or more than one gonadotropin releasing factor (Schally *et al.*, 1976).

The total and relative amounts of FSH and LH released in response to LH-RH (see Section IV,B,2) vary with the steroid environment. Generally, in a low estrogen environment LH-RH tends to stimulate a greater release of FSH and in a high estrogen environment a greater release of LH (Yen *et al.*, 1975).

The development of radioimmunoassays (Arimura and Schally, 1975; Jeffcoate *et al.*, 1975) and immunohistochemical methods (Barry *et al.*, 1973; Sétáló *et al.*, 1975) has made it possible to measure LH-RH in tissues and body fluids and to localize LH-RH (see Section IV,B,2,a) within the hypothalamus. The immunohistochemical studies of Barry *et al.* (1973) and Naik (1975) show that in rodents the majority of LH-RH-containing cell bodies occur in the preoptic–septal area of the hypothalamus and in front of the hypophysiotropic area (see Section IV,B,3,a). This has been confirmed by showing that after separating the hypophysiotropic area from the rest of the brain, the amount of LH-RH in the median eminence is reduced (Weiner *et al.*, 1975; Brownstein *et al.*, 1976). This observation casts doubt on the view that the rodent preoptic–septal area is involved in triggering the preovulatory LH surge (see Section IV,B,3,b; Halász, 1972). The loss of cyclicity after deafferentation of the hypophysiotropic area in the rat could be due to a reduction in the number of LH-RH containing terminals in the median eminence. This may lead to insufficient LH-RH being available to stimulate a preovulatory LH surge. Complete separation of the hypophysiotropic area from the rest of the brain in the rhesus monkey does not prevent the continuation of menstrual cycles (Krey *et al.*, 1975). This seems significant because in primates a large proportion of the LH-RH containing cell bodies occur in the mediobasal hypothalamus (Barry and Carette, 1975).

The role of various neurotransmitters in the control of FSH and LH release has continued to be studied (see Section IV,B,3,d; Sawyer, 1975). Dopamine, noradrenaline, serotonin, acetylcholine, and γ-aminobutyric acid have been shown to be involved in the control of LH-RH release (McCann, 1974; Ondo, 1974; McCann and Moss, 1975).

There has been further support for a short-loop feedback by gonadotropins on the hypothalamus (see Section IV,B,4; Davies *et al.*, 1975). This feedback may act via gonadotropin stimulation of enzymes which inactivate LH-RH (Kuhl and Taubert, 1975). The sensitivity of this feedback system is low, since high blood concentrations of gonadotropins do not always depress their release from the pituitary (Gay, 1974).

LH-RH has been extensively used in clinical studies in man (Besser, 1974; Kelch and Clemens, 1975) and animals (Convey *et al.*, 1975). An orally active analog of LH-RH has been prepared (Gonzalez-Barcena *et al.*, 1975). It is anticipated that LH-RH therapy will be used to induce ovulation in women with amenorrhea, and some successful treatments have already been reported (Keller, 1973; Zanartu *et al.*, 1974).

REFERENCES

Ahrén, K., Hamberger, L., and Rubinstein, L. (1969). Acute *in vivo* and *in vitro* effects of gonadotrophins on the metabolism of the rat ovary. *In* "The Gonads" (K. W. McKerns, ed.), pp. 327–354. Appleton, New York.

Ahsan, S. N., and Hoar, W. S. (1963). Some effects of gonadotropic hormones on the threespine stickleback, *Gasterosteus aculeatus*. *Can. J. Zool.* **41,** 1045–1053.

Albert, A. (1956). Human urinary gonadotropin. *Recent Prog. Horm. Res.* **12,** 227–296.

Albert, A. (1961). Biologic fingerprinting of gonadotropins. *In* "Human Pituitary Gonadotropins" (A. Albert, ed.), pp. 114–121. Thomas, Springfield, Illinois.

Allanson, M., and Parkes, A. S. (1966). Cytological and functional reactions of the hypophysis to gonadal hormones. *In* "Marshall's Physiology of Reproduction" (A. S. Parkes, ed.), 3rd ed., Vol. 3, pp. 147–300. Longmans Green, London and New York.

Allende, de I. L. C., and Orias, O. (1955). Hypophysis and ovulation in the toad *Bufo arenarum* (Hensel). *Acta physiol. Lat. Am.* **5**, 57–81.

Amoroso, E. C. (1955). Endocrinology of pregnancy. *Br. Med. Bull.* **11**, 117–125.

Amsterdam, A., Koch, Y., Lieberman, M. E., and Lindner, H. R. (1975). Distribution of binding sites for human chorionic gonadotropin in the preovulatory follicle of the rat. *J. Cell Biol.* **67**, 894–900.

Anderson, L. L., Dyck, G. W., Mori, H., Henricks, D. M., and Melampy, R. M. (1967). Ovarian function in pigs following stalk transection or hypophysectomy. *Am. J. Physiol.* **212**, 1188–1194.

Anderson, L. L., Bland, K. P., and Melampy, R. M. (1969). Comparative aspects of uterine-luteal relationships. *Recent Prog. Horm. Res.* **25**, 57–99.

Arai, Y., and Gorski, R. A. (1968). Inhibition of ovarian compensatory hypertrophy by hypothalamic implantation of gonadotropin in androgen-sterilized rats: evidence for 'internal' feedback. *Endocrinology* **82**, 871–873.

Arimura, A., and Findlay, A. (1971). Hypothalamic map for the regulation of gonadotropin release. *Res. Reprod.* **3**, No. 1.

Arimura, A., and Schally, A. V. (1970). Progesterone suppression of LH-releasing hormone-induced stimulation of LH release in rats. *Endocrinology* **87**, 653–657.

Arimura, A., and Schally, A. V. (1971). Augmentation of pituitary responsiveness to LH-releasing hormone (LHRH) by estrogen. *Proc. Soc. Exp. Biol. Med.* **136**, 290–293.

Arimura, A., and Schally, A. V. (1974). Hypothalamic LH- and FSH-releasing hormone: reevaluation of the concept that one hypothalamic hormone controls the release of LH and FSH. *In* "Biological Rhythms in Neuroendocrine Activity" (M. Kawakami, ed.), pp. 73–90. Igaku Shoin, Tokyo.

Arimura, A., and Schally, A. V. (1975). Immunological studies on hypothalamic hormones with special reference to radioimmunoassay for TRH, LH-RH and GIF. *In* "Hypothalamic Hormones" (M. Motta, P. G. Crosignani, and L. Martini, eds.), pp. 27–42. Academic Press, New York.

Armstrong, D. T. (1968). Gonadotropins, ovarian metabolism, and steroid biosynthesis. *Recent Prog. Horm. Res.* **24**, 255–308.

Armstrong, D. T. (1969). Luteotropic roles of prolactin and luteinizing hormone in the rat. *In* "Progress in Endocrinology" (C. Gual, ed.), pp. 89–97. Excerpta Med. Found., Amsterdam.

Armstrong, D. T. (1970). Reproduction. *Annu. Rev. Physiol.* **32**, 439–470.

Armstrong, D. T., and Greep, R. O. (1965). Failure of deciduomal response to uterine trauma and effects of LH upon estrogen secretion in rats with ovaries luteinized by exogenous gonadotropins. *Endocrinology* **76**, 246–254.

Armstrong, D. T., Jackanicz, T. M., and Keyes, P. L. (1969). Regulation of steroidogenesis in the rabbit ovary. *In* "The Gonads" (K. W. McKerns, ed.), pp. 3–25. Appleton, New York.

Armstrong, D. T., Moon, Y. S., and Zamecnik, J. (1974). Evidence for a role of ovarian prostaglandins in ovulation. *In* "Gonadotropins and Gonadal Function" (N. R. Moudgal, ed.), pp. 345–356. Academic Press, New York.

Aschner, B. (1912). Uber die Funktion der Hypophyse. *Pflugers Arch. Gesamte Physiol. Menschen Tiere* **146**, 1–146.

Assenmacher, I. (1958). Recherches sur le contrôle hypothalamique de la fonction gonadotrope préhypophysaire chez le canard. *Arch. Anat. Microsc. Morphol. Exp.* **47**, 447–572.

Assenmacher, I. (1970). Importance de la liaison hypothalamo-hypophysaire dans le contrôle nerveux de la reproduction chez les oiseaux. *In* "La Photorégulation de la Reproduction chez les Oiseaux et les Mammifères" (J. Benoit and I. Assenmacher, eds.), pp. 167–185. CNRS, Paris.

Assenmacher, I., Tixier-Vidal, A., and Boisson, J. (1962). Contenu et hormones gonadotropes et en prolactine de l'hypophyse du Canard soumis à un traitement lumineux ou reserpinique. *C. R. Soc. Biol.* **156**, 1555–1560.

Astwood, E. B. (1941). The regulation of corpus-luteum function by hypophysial luteotrophin. *Endocrinology* **28**, 309–320.

Astwood, E. B., and Fevold, H. L. (1939). Action of progesterone on the gonadotrophic activity of the pituitary. *Am. J. Physiol.* **127**, 192–198.

Baba, Y., Matsuo, H., and Schally, A. V. (1971). Structure of porcine LH- and FSH-releasing hormone. II. Confirmation of the proposed structure by conventional sequential analyses. *Biochem. Biophys. Res. Commun.* **44**, 459–463.

Bailey, R. E. (1950). Inhibition with prolactin of light induced gonad increase in White-Crowned Sparrows. *Condor* **52**, 247–251.

Ball, J. N. (1960). Reproduction in female bony fishes. *Symp. Zool. Soc. London* **1**, 105–135.

Ball, J. N., and Baker, B. I. (1969). The pituitary gland: anatomy and physiology. *Fish Physiol.* **2**, 1–110.

Ball, J. N., Olivereau, M., Slicher, A. M., and Kallman, K. D. (1965). Functional capacity of ectopic pituitary transplants in the teleost *Poecilia formosa*, with a comparative discussion on the transplanted pituitary. *Philos. Trans. R. Soc. London, Ser. B* **249**, 69–99.

Ball, J. N., Baker, B. I., Olivereau, M., and Peter, R. E. (1972). Investigations on hypothalamic control of adenohypophysial functions in teleost fishes. *Gen. Comp. Endocrinol., Suppl.* **3**, 11–21.

Barr, W. A. (1965). The endocrine control of fishes. *Oceanogr. Mar. Biol.* **3**, 257–298.

Barr, W. A. (1968). Patterns of ovarian activity. *In* "Perspectives in Endocrinology" (E. J. W. Barrington and C. B. Jørgensen, eds.), pp. 163–237. Academic Press, New York.

Barr, W. A., and Hobson, B. M. (1964). Endocrine control of the sexual cycle in the plaice, *Pleuronectes platessa* L. IV. Gonadotropic activity of the pituitary gland. *Gen. Comp. Endocrinol.* **4**, 608–613.

Barraclough, C. A. (1966). Modifications in the CNS regulation of reproduction after exposure of prepubertal rats to steroid hormones. *Recent Prog. Horm. Res.* **22**, 503–529.

Barraclough, C. A. (1967). Modifications in reproductive function after exposure to hormones during the prenatal and early postnatal period. *In* "Neuroendocrinology" (L. Martini and W. F. Ganong, eds.), Vol. 2, pp. 61–99. Academic Press, New York.

Barraclough, C. A., and Gorski, R. A. (1961). Evidence that the hypothalamus is responsible for androgen-induced sterility in the female rat. *Endocrinology* **68**, 68–79.

Barry, J., and Carette, B. (1975). Immunofluorescence study of LRF neurons in primates. *Cell Tissue Res.* **164**, 163–178.

Barry, J., Dubois, M. P., and Poulain, P. (1973). LRF producing cells of the mammalian hypothalamus. *Z. Zellforsch. Mikrosk. Anat.* **146,** 351–366.

Bartosik, D. B., and Romanoff, E. B. (1969). The luteotrophic process. Effects of prolactin and LH on sterol and progesterone metabolism in bovine luteal ovaries perfused *in vitro. In* "The Gonads" (K. W. McKerns, ed.), pp. 211–243. Appleton, New York.

Bast, J. D., and Greenwald, G. S. (1974). Serum profiles of follicle stimulating hormone, luteinizing hormone and prolactin during the estrous cycle of the hamster. *Endocrinology* **94,** 1295–1299.

Bates, R. W., Lahr, E. L., and Riddle, O. (1935). The gross action of prolactin and follicle stimulating hormone on the mature ovary and sex accessories of fowl. *Am. J. Physiol.* **111,** 361–368.

Bates, R. W., Riddle, O., and Lahr, E. L. (1937). The mechanism of the anti-gonad action of prolactin in adult pigeons. *Am. J. Physiol.* **119,** 610–614.

Behrman, H. R., Orczyk, G. P., Macdonald, G. J., and Greep, R. O. (1970). Prolactin induction of enzyme controlling luteal cholesterol ester turnover. *Endocrinology* **87,** 1251–1256.

Bell, E. T., and Loraine, J. A., eds. (1967). "Recent Research in Gonadotrophic Hormones." Livingstone, Edinburgh.

Bell, R. Q., and Hinde, R. A. (1963). Brood patch sensitivity of female canaries brought into reproductive condition in winter. *Anim. Behav.* **11,** 561–565.

Benoit, J., and Assenmacher, I. (1955). Le contrôle hypothalamique de l'activité préhypophysaire gonadotrope. *J. Physiol. (Paris)* **47,** 427–567.

Benoit, J., and Assenmacher, I. (1959). The control by visible radiations of the gonadotrophic activity of the duck hypophysis. *Recent Prog. Horm. Res.* **15,** 143–164.

Benoit, J., and Da Lage, C., eds. (1963). "Cytologie de l'Adénohypophyse." CNRS, Paris.

Ben-Or, S. (1963). Morphological and functional development of the ovary of the mouse. 1. Morphology and histochemistry of the developing ovary in normal conditions and after FSH treatment. *J. Embryol. Exp. Morphol.* **11,** 1–11.

Ben-Or, S. (1970). Development of the ovary under different experimental conditions. *In* "Gonadotrophins and Ovarian Development" (W. R. Butt, A. C. Crooke, and M. Ryle, eds.), pp. 266–271. Livingstone, Edinburgh.

Besser, G. M. (1974). Clinical value of the gonadotrophin RH. *Acta Eur. Fertil.* **5,** 113–118.

Beyer, C., and Sawyer, C. H. (1969). Hypothalamic unit activity related to control of the pituitary gland. *In* "Frontiers in Neuroendocrinology" (W. F. Ganong and L. Martini, eds.), pp. 255–287. Oxford Univ. Press, London and New York.

Bhalla, R. C., Kovacic, N., and Parlow, A. F. (1971). Species differences in the capacity of luteinizing hormones to stimulate uterine weight and FSH-induced ovarian weight increase in immature rats. *Fed. Proc. Fed. Am. Soc. Exp. Biol.* **30,** 473.

Billenstein, D. E. (1962). The seasonal secretory cycle of the *nucleus lateralis tuberis* of the hypothalamus and its relation to reproduction in the eastern brook trout, *Salvelinus fontinalis. Gen. Comp. Endocrinol.* **2,** 111–112.

Bingel, A. S., and Schwartz, N. B. (1969). Pituitary LH content and reproductive tract changes during the mouse oestrous cycle. *J. Reprod. Fertil.* **19,** 215–222.

Blakley, G. A. (1970). Luteinizing hormone levels in nonhuman primates. *Folia Primatol.* **13**, 298–305.

Bogdanove, E. M., and Gay, V. L. (1968). Use of bioassays to study LH and FSH secretory kinetics in the rat. *In* "Gonadotropins 1968" (E. Rosemberg. ed.), pp. 131–137. Geron-X Inc., Los Altos, California.

Bogdanove, E. M., and Gay, V. L. (1969). Studies on the disappearance of LH and FSH in the rat; a quantitative approach to adenohypophysial secretory kinetics. *Endocrinology* **84**, 1118–1131.

Bogdanove, E. M., Hilliard, J., and Sawyer, C. H. (1971). Serum LH patterns in the female rabbit as determined by radioimmunoassay. *Program, 53rd Meet. U.S. Endocrine Soc.* p. 76.

Borth, R., and Menzi, A. (1969). Effect of freezing and thawing and of diluent on the potency of human chorionic gonadotrophin in three methods of bioassay. *Acta Endocrinol. (Copenhagen)* **61**, 89–95.

Bradbury, J. T. (1961). Direct action of estrogens on the ovary of the immature rat. *Endocrinology* **68**, 115–120.

Brambell, F. W. R. (1928). The development and morphology of the gonads of the mouse. Part III. The growth of the follicles. *Proc. R. Soc. London, Ser. B* **103**, 258–272.

Brehm, H. V. (1958). Uber Jahreszylische Veränderungen im *Nucleus Lateralis Tuberis* der Schleie (*Tinca vulgaris*). *Z. Zellforsch. Mikrosk. Anat.* **49**, 105–125.

Breneman, W. R. (1955). Reproduction in birds: the female. *In* "Comparative Physiology of Reproduction and the Effects of Sex Hormones in Vertebrates" (I. Chester Jones and P. Eckstein, eds.), pp. 94–113. Cambridge Univ. Press, London and New York.

Breton, B., and Weil, C. (1973). Effets du LH/FSH-RH synthétique et d'extraits hypothalamiques de Carpe sur la sécrétion d'hormone gonadotrope *in vivo* chez la Carpe (*Cyprinus carpio*, L.). *C. R. Acad. Sci., Ser. D* **277**, 2061–2064.

Brown-Grant, K., El Kabir, D. J., and Fink, G. (1968). The effect of mating on pituitary luteinizing hormone and thyrotrophic hormone content in the female rabbit. *J. Endocrinol.* **41**, 91–94.

Brownstein, M. J., Arimura, A., Schally, A. V., Palkovits, M., and Kizer, J. S. (1976). The effect of surgical isolation of the hypothalamus on its luteinizing hormone releasing hormone content. *Endocrinology* **98**, 662–666.

Bullock, D. W., and Nalbandov, A. V. (1967). Hormonal control of the hen's ovulation cycle. *J. Endocrinol.* **38**, 407–415.

Bunde, C. A., and Greep, R. O. (1936). Suppression of persisting corpora lutea in hypophysectomized rats. *Proc. Soc. Exp. Biol. Med.* **35**, 235–237.

Burack, E., and Wolfe, J. M. (1959). The effect of anterior hypophysial administration on the ovaries of old rats. *Endocrinology* **64**, 676–684.

Burgers, A. C. J., and Li, C. H. (1960). Amphibian ovulation *in vitro* induced by mammalian pituitary hormones and progesterone. *Endocrinology* **66**, 255–259.

Burgus, R., and Guillemin, R. (1970). Hypothalamic releasing factors. *Annu. Rev. Biochem.* **39**, 499–526.

Burzawa-Gerard, E. (1971). Purification d'une hormone gonadotrope hypophysaire de poisson téléostéen, la carpe (*Cyprinus carpio*, L.). *Biochimie* **53**, 545–552.

Butler, J. E. M., and Donovan, B. T. (1971). Effect of surgical isolation of the hypothalamus on reproductive function in the female rat. *J. Endocrinol.* **49**, 293–304.

Dávid, M. A., Fraschini, F., and Martini, L. (1966). Control of LH secretion: role of a 'short' feedback mechanism. *Endocrinology* **78**, 55–60.

Davidson, J. M. (1969). Feedback control of gonadotropin secretion. *In* "Frontiers in Neuroendocrinology" (W. F. Ganong and L. Martini, eds.), pp. 343–388. Oxford Univ. Press, London and New York.

Davies, A. G., Duncan, I. F., and Lynch, S. S. (1975). Autoradiographic localization of ^{125}I labelled FSH in the rat hypothalamus. *J. Endocrinol.* **66**, 301–302.

Deanesly, R. (1939). Modification of the effectiveness of gonadotrophic extracts. *J. Endocrinol.* **1**, 307–322.

Denamur, R. (1968). Formation and maintenance of corpora lutea in domestic animals. *J. Anim. Sci.* **27**, Suppl. 1, 163–180.

Denamur, R., Martinet, J., and Short, R. V. (1966). Sécrétion de la progestérone par les corps jaunes de la brebis après hypophysectomie, section de la tige pituitaire et hystérectomie. *Acta Endocrinol. (Copenhagen)* **52**, 72–90.

Desclin, L. (1950). A-propos du mécanisme d'action des oestrogènes sur le lobe antérieur de l'hypophyse chez le rat. *Ann. Endocrinol.* **11**, 656–659.

Desjardins, C., and Hafs, H. D. (1968). Levels of pituitary FSH and LH in heifers from birth through puberty. *J. Anim. Sci.* **27**, 472–477.

Desjardins, C., Kirton, K. T., and Hafs, H. D. (1967). Anterior pituitary levels of FSH, LH, ACTH and prolactin after mating in female rabbits. *Proc. Soc. Exp. Biol. Med.* **126**, 23–26.

Desjardins, C., Chapman, V. M., and Bronson, F. H. (1970). Hypophysial LH and FSH release and uterine nucleic acid changes during the mouse estrous cycle. *Anat. Rec.* **167**, 465–471.

Dierickx, K. (1964). The structure and activity of the hypophysis of *Rana temporaria* in normal and in experimental conditions. *Z. Zellforsch. Mikrosk. Anat.* **61**, 920–939.

Dierickx, K. (1965). The origin of the aldehyde-fuchsin-negative nerve fibres of the median eminence of the hypophysis: a gonadotropic centre. *Z. Zellforsch. Mikrosk. Anat.* **66**, 504–518.

Dierickx, K. (1966). Experimental identification of a hypothalamic gonadotropic centre. *Z. Zellforsch. Mikrosk. Anat.* **74**, 53–79.

Dierickx, K. (1967a). The gonadotropic centre of the tuber cinereum hypothalami and ovulation. *Z. Zellforsch. Mikrosk. Anat.* **77**, 188–203.

Dierickx, K. (1967b). The function of the hypophysis without preoptic neurosecrtory control. *Z. Zellforsch. Mikrosk. Anat.* **78**, 114–130.

Dodd, J. M. (1960). Gonadal and gonadotrophic hormones in lower vertebrates. *In* "Marshall's Physiology of Reproduction" (A. S. Parkes, ed.), 3rd ed., Vol. 1, Part 2, pp. 417–582. Longmans Green, London and New York.

Dodd, J. M. (1972). Ovarian control in cyclostomes and elasmobranchs. *Am. Zool.* **12**, 325–339.

Dodd, J. M., Evennett, P. J., and Goddard, C. K. (1960). Reproductive endocrinology in Cyclostomes and Elasmobranchs. *Symp. Zool. Soc. London* **1**, 77–103.

Dodd, J. M., Follett, B. K., and Sharp, P. J. (1971). Hypothalamic control of pituitary function in submammalian vertebrates. *Adv. Comp. Physiol. Biochem.* **4**, 113–223.

Domanski, E., and Kochman, K. (1968). Induction of ovulation in sheep by intrahypophysial infusion of hypothalamic extracts. *J. Endocrinol.* **42**, 383–389.

Donaldson, E. M., Yamazaki, F., Dye, H. M., and Philleo, W. W. (1972). Preparation of gonadotrophin from salmon (*Oncorhynchus tshwaytscha*) pituitary glands. *Gen. Comp. Endocrinol.* **18**, 469–481.

Donini, P., Puzzuoli, D., D'Alessio, I., and Donini, S. (1968). New approach to the biological determination of the LH. *Acta Endocrinol. (Copenhagen)* **58**, 463–472.

Donovan, B. T. (1967a). The effect of light upon reproductive mechanisms, as illustrated by the ferret. *Ciba Found. Study Group* **26**, 43–52.

Donovan, B. T. (1967b). The control of corpus luteum function in the ferret. *Arch. Anat. Microsc. Morphol. Exp.* **56**, Suppl., 315–325.

Donovan, B. T. (1969). Control of synthesis and release of anterior pituitary hormones *in vivo. Adv. Biosci.* **1** 187–200.

Donovan, B. T. (1970). "Mammalian Neuroendocrinology." McGraw-Hill, New York.

Donovan, B. T. (1971). The control of ovarian function. *Acta Endocrinol. (Copenhagen)* **66**, 1–15.

Donovan, B. T., and van der Werff ten Bosch, J. J. (1965). "Physiology of Puberty." Arnold, London.

Dörner, G. (1967). FSH:ICSH-relations of gonadotrophins and their therapeutic significance. *Acta Endocrinol. (Copenhagen)*, Suppl. **119**, 112.

Dresel, I. (1935). The effect of prolactin on the oestrous cycle of non-parous mice. *Science* **82**, 173.

Dufau, M. L., Podesta, E. J., and Catt, K. J. (1975). Physiological characteristics of the gonadotropin receptor-hormone complex formed *in vivo* and *in vitro. Proc. Natl. Acad. Sci. U.S.A.* **72**, 1272–1275.

du Mesnil du Buisson, F., and Denamur, R. (1969). Mécanismes de contrôle de la fonction lutéale chez la truie, la brebis et la vache. In "Progress in Endocrinology" (C. Gual, ed.), pp. 929–934. Excerpta Med. Found., Amsterdam.

du Mesnil du Buisson, F., and Léglise, P. C. (1963). Effet de l'hypophysectomie sur les corps jaunes de la truie. Résultats préliminaires. *C. R. Acad. Sci. (Paris), Ser. D* **257**, 261–263.

Egami, N. (1954). Inhibitory effect of estrone benzoate on ovarian growth in the loach. *J. Fac. Sci., Univ. Tokyo*, **7**, 113–119.

Egami, N., and Ishii, S. (1962). Hypophysial control of reproductive function in teleost fishes. *Gen. Comp. Endocrinol., Suppl.* **1**, 248–253.

Egge, A. S., and Chiasson, R. B. (1963). Endocrine effects of diencephalic lesions in the white leghorn hen. *Gen. Comp. Endocrinol.* **3**, 346–361.

El-Fouly, M. A., Cook, B., Nekola, M., and Nalbandov, A. V. (1970). Role of the ovum in follicular luteinization. *Endocrinology* **87**, 288–293.

Ellis, S. (1961). Bioassay of luteinizing hormone. *Endocrinology* **68**, 334–340.

Ellsworth, L. R., and Armstrong, D. T. (1971). Effect of LH on luteinization of ovarian follicles transplanted under the kidney capsule in rats. *Endocrinology* **88**, 755–762.

Eshkol, A., and Lunenfeld, B. (1967). Purification and separation of follicle stimulating hormone (FSH) and luteinizing hormone (LH) from human menopausal gonadotrophin (HMG). III Effects of a biologically apparently pure FSH preparation on ovaries and uteri of intact, immature mice. *Acta Endocrinol. (Copenhagen)* **54**, 91–95.

Eshkol, A., and Lunenfeld, B. (1968). Fate and localization of iodine labelled HCG in mice. In "Gonadotropins 1968" (E. Rosemberg, ed.), pp. 187–188. Geron-X Inc., Los Altos, California.

Espeland, D. H., Naftolin, F., and Paulsen, C. A. (1968). Metabolism of labelled

^{125}I-HCG by the rat ovary. *In* "Gonadotropins 1968" (E. Rosemberg, ed.), pp. 177–184. Geron-X Inc., Los Altos, California.

Espey, L. L. (1967). Ultrastructure of the apex of the rabbit graafian follicle during the ovulatory process. *Endocrinology* **81**, 267–276.

Espey, L. L., and Lipner, H. (1965). Enzyme-induced rupture of rabbit graafian follicle. *Am. J. Physiol.* **208**, 208–213.

Evans, L. T. (1935). The effects of pituitary implants and extracts on the genital system of the lizard *Anolis carolinensis. Science* **81**, 465.

Evennett, P. J., and Dodd, J. M. (1963a). Endocrinology of reproduction in the river lamprey. *Nature (London)* **197**, 715–716.

Evennett, P. J., and Dodd, J. M. (1963b). The pituitary gland and reproduction in the lamprey (*Lampetra fluviatilis*). *J. Endocrinol.* **26**, xiv–xv.

Everett, J. W. (1964). Preoptic stimulative lesions and ovulation in the rat: 'thresholds' and LH-release time in late diestrus and proestrus. *In* "Major Problems in Neuroendocrinology" (E. Bajusz and G. Jasmin, eds.), pp. 346–366. Karger, Basel.

Everett, J. W. (1969). Neuroendocrine aspects of mammalian reproduction. *Annu. Rev. Physiol.* **31**, 383–416.

Eyeson, K. N. (1970). Cell types in the distalis lobe of the pituitary of the West African rainbow lizard, *Agama agama* (L.). *Gen. Comp. Endocrinol.* **14**, 357–367.

Farmer, S. W., Papkoff, H., and Licht, P. (1975). Purification of turkey gonadotrophins. *Biol. Reprod.* **12**, 415–421.

Farner, D. S., Wilson, F. E. and Oksche, A. (1967). Neuroendocrine mechanisms in birds. *In* "Neuroendocrinology" (L. Martini and W. F. Ganong, eds.), Vol. 2, pp. 529–582. Academic Press, New York.

Fawcett, C. P. (1970). The present status of the chemistry of PIF, FRF and LRF. *In* "Hypophysiotropic Hormones of the Hypothalamus: Assay and Chemistry" (J. Meites, ed.), pp. 242–248. Williams & Wilkins, Baltimore, Maryland.

Fawke, L., and Brown, P. S. (1970). Pituitary content of follicle-stimulating hormone in the female rat. *J. Reprod. Fertil.* **21**, 303–312.

Ferin, M., Halberg, F., Richart, R. M., and Vande Wiele, R. L., eds. (1974). "Biorhythms and Human Reproduction." Wiley, New York.

Ferrando, G., and Nalbandov, A. V. (1969). Direct effect on the ovary of the adrenergic blocking drug dibenzyline. *Endocrinology* **85**, 38–42.

Fevold, H. L., Hisaw, F. L., and Greep, R. O. (1937). Comparative action of gonadstimulating hormones on the ovaries of rats. *Endocrinology* **21**, 343–345.

Fevold, H. L., Hisaw, F. L., and Leonard, S. L. (1931). The gonad-stimulating and the luteinizing hormones of the anterior lobe of the hypophysis. *Am. J. Physiol.* **97**, 291–301.

Fink, G., and Harris, G. W. (1970). The luteinizing hormone releasing activity of extracts of blood from the hypophysial portal vessels of rats. *J. Physiol. (London)* **208**, 221–241.

Fink, G., Nallar, R., and Worthington, W. C., Jr. (1967). Demonstration of luteinizing hormone releasing factor in hypophysial portal blood of pro-oestrous and hypophysectomized rats. *J. Physiol. (London)* **191**, 407–416.

Fitko, R., Chamski, J., and Sawicka, K. (1968). Lokalizacja gonadotropin przysadkowych w jajnikach owiec. *Endokrynol. Pol.* **19**, 657–669.

Flament-Durand, J. (1965). Observations on pituitary transplants into the hypothalamus of the rat. *Endocrinology* **77**, 446–454.

Flament-Durand, J., and Desclin, L. (1970). The hypophysiotropic area. *In* "The Hypothalamus" (L. Martini, M. Motta, and F. Fraschini eds.), pp. 245–257. Academic Press, New York.

Flerkó, B. (1962). Hypothalamic control of hypophyseal gonadotrophic function. *In* "Hypothalamic Control of the Anterior Pituitary" (J. Szentágothai *et al.*, eds.), 1st ed., pp. 192–264. Akadémiai Kiadó, Budapest.

Flerkó, B. (1966). Control of gonadotropin secretion in the female. *In* "Neuroendocrinology" (L. Martini and W. F. Ganong, eds.), Vol. 1, pp. 613–668. Academic Press, New York.

Flerkó, B. (1970). Control of follicle-stimulating hormone and luteinizing hormone secretion. *In* "The Hypothalamus" (L. Martini, M. Motta, and F. Fraschini, eds.), pp. 351–363. Academic Press, New York.

Flerkó, B. (1974). Hypothalamus mediation of neuroendocrine regulation of hypophysial gonadotrophic functions. *Physiol., Ser. One* **8**, 1–32.

Flerkó, B., and Bárdos, V. (1966). Prolongation of dioestrus in rats with diencephalic and mesencephalic lesions. *Exp. Brain Res.* **1**, 299–305.

Florsheim, W. H., and Rudko, P. (1968). The development of portal system function in the rat. *Neuroendocrinology* **3**, 89–98.

Follett, B. K. (1970). Gonadotrophin-releasing activity in the quail hypothalamus. *Gen. Comp. Endocrinol.* **15**, 165–179.

Follett, B. K., and Redshaw, M. R. (1973). The physiology of vitellogenesis. *In* "Physiology of the Amphibia" (B. Lofts, ed.), Vol. 2, pp. 219–308. Academic Press, New York.

Follett, B. K., and Sharp, P. J. (1969). Circadian rhythmicity in photoperiodically induced gonadotrophin release and gonadal growth in the quail. *Nature (London)* **223**, 968–971.

Fontaine, M., and Chauvel, M. (1961). Evaluation of gonadotropic activity of the pituitary gland of teleost fish and in particular of *Salmo salar L.* at different stages of their development and their migrations. *C. R. Soc. Biol.* **252**, 822–824.

Fontaine, M., Bertrand, E., Lopez, E., and Callamand, O. (1964). Sur la maturation des organes génitaux de l'Anguille femelle (*Anguilla anguilla* L.) et l'émission spontanée des oeufs en aquarium. *C. R. Acad. Sci. (Paris), Ser. D* **259**, 2907–2910.

Forbes, T. R. (1937). Studies on the reproductive system of the alligator. I. The effects of prolonged injection of pituitary whole gland extract in the immature alligator. *Anat. Rec.* **70**, 113–137.

Fraps, R. M. (1955). The varying effect of sex hormones in birds. *Mem. Soc. Endocrinol.* **4**, 205–219.

Fraps, R. M. (1961). Ovulation in the domestic fowl. *In* "Control of Ovulation" (C. A. Villee, ed.), pp. 133–162. Pergamon, Oxford.

Fraps, R. M. (1970). Photoregulation in the ovulation cycle of the domestic fowl. *In* "La Photorégulation de la Reproduction chez les Oiseaux et les Mammifères" (J. Benoit and I. Assenmacher, eds.), pp. 281–306. CNRS, Paris.

Fraps, R. M., and Case, J. F. (1953). Premature ovulation in the domestic fowl following administration of certain barbiturates. *Proc. Soc. Exp. Biol. Med.* **82**, 167–171.

Fraschini, F., and Martini, L. (1970). Rhythmic phenomena and pineal principles. *In* "The Hypothalamus" (L. Martini, M. Motta, and F. Fraschini, eds.), pp. 529–549. Academic Press, New York.

Fraschini, F., Motta, M., and Martini, L. (1968). A 'short' feedback mechanism controlling follicle stimulating hormone secretion. *Experientia* **24**, 270–271.

Friedgood, H. B. (1970). The nervous control of the anterior hypophysis. Harvard University Tercentenary Celebrations, 1936. *J. Reprod. Fertil., Suppl.* **10**, 3–14.

Fröhlich, A. (1910). Ein Fall von Tumor der Hypophysis cerebri ohne Akromegalie. *Wien. Klin. Rundsch.* **15**, 883–886.

Furr, B. J. A., Bonney, R. C., England, R. J., and Cunningham, F. J. (1973). Luteinizing hormone and progesterone in peripheral blood during the ovulatory cycle of the hen *Gallus domesticus. J. Endocrinol.* **57**, 159–169.

Fuxe, K., and Hökfelt, T. (1969). Catecholamines in the hypothalamus and the pituitary gland. *In* "Frontiers in Neuroendocrinology" (W. F. Ganong and L. Martini, eds.), pp. 47–96. Oxford Univ. Press, London and New York.

Fuxe, K., and Hökfelt, T. (1970). Central monoaminergic systems and hypothalamic function. *In* "The Hypothalamus" (L. Martini, M. Motta, and F. Fraschini, eds.), pp. 123–138. Academic Press, New York.

Gallien, L. (1940). Recherche sur la physiologie hypophysaire dans ses relations avec les gonades et le cycle sexuel chez la grenouille rousse (*Rana temporaria* L.). *Bull. Biol. Fr.* **74**, 1–42.

Gay, V. L. (1971). Effects of bilateral nephrectomy on serum concentrations and disappearance of LH and FSH in castrated rats. *Program, 53rd Meet. U.S. Endocrine Soc.* p. 79.

Gay, V. L. (1974). Decreased metabolism and increased serum concentrations of LH and FSH following nephrectomy of the rat: absence of short-loop regulatory mechanism. *Endocrinology* **95**, 1582–1588.

Gay, V. L., Midgley, A. R., Jr., and Niswender, G. D. (1970). Patterns of gonadotrophin secretion associated with ovulation. *Fed. Proc. Fed. Am. Soc. Exp. Biol.* **29**, 1880–1887.

Gemzell, C. A. (1965). Induction of ovulation with human gonadotropins. *Recent Prog. Horm. Res.* **21**, 179–198.

Gemzell, C. A., Diczfalusy, E., and Tillinger, G. (1958). Clinical effect of human pituitary follicle-stimulating hormone (FSH). *J. Clin. Endocrinol. Metab.* **18**, 1333–1348.

Gerbilsky, N. L. (1940). Seasonal changes of the gonadotropic potency of the pituitary gland in fishes. *C. R. Acad. Sci. USSR* **28**, 571–573.

Geschwind, I. I. (1966). Species specificity of anterior pituitary hormones. *In* "The Pituitary Gland" (G. W. Harris and B. T. Donovan, eds.), Vol. 2, pp. 589–612. Univ. of California Press, Berkeley.

Godden, P. M. M., and Scanes, C. G. (1975). Studies on the purification and properties of avian gonadotrophins. *Gen. Comp. Endocrinol.* **27**, 538–542.

Goding, J. R., Catt, K. J., Brown, J. M., Kaltenbach, C. C., Cumming, I. A., and Mole, B. J. (1969). Radioimmunoassay for ovine luteinizing hormone, secretion of luteinizing hormone during estrus and following estrogen administration in the sheep. *Endocrinology* **85**, 133–142.

Goldenberg, R. L., Vaitukaitis, J. L., and Ross, G. T. (1972). Estrogen and FSH interactions on follicle growth in rats. *Endocrinology* **90**, 1492–1498.

Goldman, B. D., and Mahesh, V. B. (1968). Fluctuations in pituitary FSH during the ovulatory cycle in the rat and a possible role of FSH in the induction of ovulation. *Endocrinology* **83**, 97–106.

Goldman, B. D., and Mahesh, V. B. (1969). A possible role of acute FSH release in ovulation in the hamster as demonstrated by utilization of antibodies to LH and FSH. *Endocrinology* **84**, 236–243.

Goldman, B. D., and Porter, J. C. (1970). Serum LH levels in intact and castrated golden hamsters. *Endocrinology* **87**, 676–679.

Goldman, B. D., Grazia, Y. R., Kamberi, I. A., and Porter, J. C. (1971). Serum gonadotropin concentrations in intact and castrated neonatal rats. *Endocrinology* **88**, 771–776.

Gonzalez-Barcena, D., Kastin, A. J., Miller, M. C., Schalch, D. S., Coy, D. H., Schally, A. V., and Escalante-Herrera, A. (1975). Stimulation of LH release after oral administration of an analogue of LHRH. *Lancet* **2**, 1126–1127.

Gorski, M., Adaniya, J., and Lawton, I. E. (1971). Adrenal involvement in determining time of onset of puberty in the rat. *Program, 53rd Meet. U.S. Endocrine Soc.* p. 201.

Gorski, R. A. (1968). The neural control of ovulation. *In* "Biology of Gestation" (N. S. Assali, ed.), Vol. 1, pp. 1–66. Academic Press, New York.

Goswami, S. V., and Sundararaj, B. I. (1968a). Effect of estradiol benzoate, human chorionic gonadotropin, and follicle stimulating hormone on unilateral ovariectomy-induced compensatory hypertrophy in catfish, *Heteropneustes fossilis* (Bloch). *Gen. Comp. Endocrinol.* **11**, 393–400.

Goswami, S. V., and Sundararaj, B. I. (1968b). Compensatory hypertrophy of the remaining ovary after unilateral ovariectomy at various phases of the reproductive cycle of Catfish, *Heteropneustes fossilis* (Bloch). *Gen. Comp. Endocrinol.* **11**, 401–413.

Greeley, F., and Meyer, R. K. (1953). Seasonal variation in testis stimulating activity of male pheasant pituitary glands. *Auk* **70**, 350–358.

Green, J. A. (1957). Some effects of advancing age on the histology and reactivity of the mouse ovary. *Anat. Rec.* **129**, 333–347.

Green, J. D., and Harris, G. W. (1947). The neurovascular link between the neurohypophysis and adenohypophysis. *J. Endocrinol.* **5**, 136–146.

Greenwald, G. S. (1969). Evidence for a luteotropic complex in the hamster and other species. *In* "Progress in Endocrinology" (C. Gual, ed.), pp. 921–926. Excerpta Med. Found., Amsterdam.

Greenwald, G. S. (1974). Role of follicle-stimulating hormone and luteinizing hormone in follicular development and ovulation. *In* "Handbook of Physiology" (Am. Physiol. Soc., J. Field, ed.), Sect. 7, Vol. IV, pp. 293–323. Williams & Wilkins, Baltimore, Maryland.

Greenwald, G. S., and Rothchild, I. (1968). Formation and maintenance of corpora lutea in laboratory animals. *J. Anim. Sci.* **27**, Suppl. **1**, 139–162.

Greep, R. O. (1936). Functional pituitary grafts in rats. *Proc. Soc. Exp. Biol. Med.* **34**, 754–755.

Greep, R. O. (1961). Physiology of the anterior hypophysis in relation to reproduction. *In* "Sex and Internal Secretions" (W. C. Young, ed.), 3rd ed., Vol. 1, pp. 240–301. Williams & Wilkins, Baltimore, Maryland.

Greep, R. O., van Dyke, H. B., and Chow, B. F. (1941). Use of anterior lobe of prostate gland in assay of metakentrin. *Proc. Soc. Exp. Biol. Med.* **46**, 644–649.

Greep, R. O., van Dyke, H. B., and Chow, B. F. (1942). Gonadotrophins of the swine pituitary. I. Various biological effects of purified thylakentrin (FSH) and pure metakentrin (ICSH). *Endocrinology* **30**, 635–649.

Grignon, G. and Grignon, M. (1965). Variations cycliques de l'activité des glandes endocrines chez les reptiles. *Proc. Int. Congr. Endocrinol., 2nd, 1964* Excerpta Med. Found. Int. Congr. Ser. No. 83, Part 1, pp. 106–113.

Grosvenor, C. E. (1967). Disappearance rate of exogenous prolactin from serum of female rats. *Endocrinology* **80**, 195–199.

Hackett, A. J., and Hafs, H. D. (1969). Pituitary and hypothalamic endocrine changes during the bovine estrous cycle. *J. Anim. Sci.* **28**, 531–536.

Hafez, E. S. E. (1969). Superovulation and preservation of mammalian eggs. *Acta Endocrinol. (Copenhagen)* **62**, Suppl. 140, 5–44.

Haighton, J. (1797). An experimental enquiry concerning animal impregnation. *Philos. Trans. R. Soc. London* **87**, 159–196.

Halász, B. (1969). The endocrine effects of isolation of the hypothalamus from the rest of the brain. *In* "Frontiers in Neuroendocrinology" (W. F. Ganong and L. Martini, eds.), pp. 307–342. Oxford Univ. Press, London and New York.

Halász, B. (1972). Hypothalamic mechanisms controlling pituitary function. *Prog. Brain Res.* **38**, 97–118.

Halász, B., and Gorski, R. A. (1967). Gonadotrophic hormone secretion in female rats after partial or total interruption of neural afferents to the medial basal hypothalamus. *Endocrinology* **80**, 608–622.

Halász, B., and Pupp, L. (1965). Hormone secretion of the anterior pituitary gland after physical interruption of all nervous pathways to the hypophysiotrophic area. *Endocrinology* **77**, 553–562.

Halász, B., Pupp, L., and Uhlarik, S. (1962). Hypophysiotrophic area in the hypothalamus. *J. Endocrinol.* **25**, 147–154.

Halász, B., Pupp, L., Uhlarik, S., and Tima, L. (1965). Further studies on the hormone secretion of the anterior pituitary transplanted into the hypophysiotrophic area of the hypothalamus. *Endocrinology* **77**, 343–355.

Hammerstein, J. (1974). Regulation of ovarian steroidogenesis: gonadotropins, enzymes, prostaglandins, cyclic-AMP, luteolysis. *Physiol. Ser. One* **8**, 279–311.

Hammond, J., Jr. (1961). Hormonal augmentation of fertility in sheep and cattle. *In* "Control of Ovulation" (C. A. Villee, ed.), pp. 163–176. Pergamon, Oxford.

Harrington, F. E., and Elton, R. L. (1969). Induction of ovulation in adult rats with follicle stimulating hormone. *Proc. Soc. Exp. Biol. Med.* **132**, 841–844.

Harrington, F. E., Bex, F. J., Elton, R. L., and Roach, J. B. (1970). The ovulatory effects of follicle stimulating hormone treated with chymotrypsin in chlorpromazine blocked rats. *Acta Endocrinol. (Copenhagen)* **65**, 222–228.

Harris, G. W. (1937). The induction of ovulation in the rabbit by electrical stimulation of the hypothalamo-hypophysial mechanism. *Proc. R. Soc. London, Ser. B* **122**, 374–394.

Harris, G. W. (1950). Oestrus rhythm, pseudopregnancy and the pituitary stalk in the rat. *J. Physiol. (London)* **111**, 347–360.

Harris, G. W. (1955). "Neural Control of the Pituitary Gland." Arnold, London.

Harris, G. W. (1961). The pituitary stalk and ovulation. *In* "Control of Ovulation" (C. A. Villee, ed.), pp. 56–74. Pergamon, Oxford.

Harris, G. W. (1964). Sex hormones, brain development and brain function. *Endocrinology* **75**, 627–648.

Harris, G. W. (1970a). Unsolved problems in the portal vessel-chemotransmitter hypothesis. *In* "Hypophysiotropic Hormones of the Hypothalamus: Assay and Chemistry" (J. Meites, ed.), pp. 1–14. Williams & Wilkins, Baltimore, Maryland.

Harris, G. W. (1970b). Hormonal differentiation of the developing central nervous system with respect to patterns of endocrine function. *Philos. Trans. R. Soc. London, Ser. B* **259**, 165–177.

Harris, G. W. (1972). Humours and hormones. *J. Endocrinol.* **53**, ii–xxiii.

Harris, G. W., and Campbell, H. J. (1966). The regulation of the secretion of luteinizing hormone and ovulation. *In* "The Pituitary Gland" (G. W. Harris and B. T. Donovan, eds.), Vol. 2, pp. 99–165. Univ. of California Press, Berkeley.

Harris, G. W., and Jacobson, D. (1952). Functional grafts of the anterior pituitary gland. *Proc. R. Soc. London, Ser. B,* **139**, 263–276.

Harvey, S., and Rondell, P. (1970). Effect of gonadotropins on ovarian follicular tissue *in vitro. Fed. Proc. Fed. Am. Soc. Exp. Biol.* **29**, 643.

Haterius, H. O., and Derbyshire, A. J., Jr. (1937). Ovulation in the rabbit following upon stimulation of the hypothalamus. *Am. J. Physiol.* **119**, 329–330.

Heald, P. J., Furnival, B. E., and Rookledge, K. A. (1967). Changes in the levels of luteinizing hormone in the pituitary of the domestic fowl during an ovulatory cycle. *J. Endocrinol.* **37**, 73–81.

Heald, P. J., Rookledge, K. A., Furnival, B. E., and Watts, G. D. (1968). Changes in luteinizing hormone content of the anterior pituitary of the domestic fowl during the interval between clutches. *J. Endocrinol.* **41**, 197–201.

Heap, R. B., Perry, J. S., and Rowlands, I. W. (1967). Corpus luteum function in the guinea-pig; arterial and luteal progesterone levels, and the effects of hysterectomy and hypophysectomy. *J. Reprod. Fertil.* **13**, 537–553.

Henricks, D. M., Dickey, J. F., and Niswender, G. D. (1970). Serum luteinizing hormone and plasma progesterone levels during the estrous cycle and early pregnancy in cows. *Biol. Reprod.* **2**, 346–351.

Henzl, M. R., and Segre, E. J. (1970). Physiology of human menstrual cycle and early pregnancy. A review of recent investigations. *Contraception* **1**, 315–338.

Herlant, M., and Grignon, G. (1961). Les modifications hypophysaires chez la tortue terrestre (*T. mauritanica* Durrer) au cours du cycle génital. *C. R. Acad. Sci. (Paris), Ser. D* **252**, 2303–2305.

Herrick, R. B., McGibbon, W. H., and McShan, W. H. (1962). Gonadotropic activity of chicken pituitary glands. *Endocrinology* **71**, 487–491.

Hichens, M., Grinwich, D. L., and Behrman, H. R. (1974). $PGF_{2\alpha}$ induced loss of corpus luteum gonadotrophin receptors. *Prostaglandins* **7**, 449–458.

Hilliard, J., Spies, H. G., and Sawyer, C. H. (1969). Hormonal factors regulating ovarian cholesterol mobilization and progestin secretion in intact and hypophysectomized rabbits. *In* "The Gonads" (K. W. McKerns, ed.), pp. 55–92. Appleton, New York.

Hilliard, J., Schally, A. V., and Sawyer, C. H. (1971). Progesterone blockade of the ovulatory response to intrapituitary infusion of LH-RH in rabbits. *Endocrinology* **88**, 730–736.

Hinsey, J. C. (1937). The relation of the nervous system to ovulation and other phenomena of the female reproductive tract. *Cold Spring Harbor Symp. Quant. Biol.* **5**, 269–279.

Hirono, M., Igarashi, M., and Matsumoto, S. (1970). Short- and auto-feedback control of pituitary FSH secretion. *Neuroendocrinology* **6**, 274–282.

Hisaw, F. L. (1947). Development of the graafian follicle and ovulation. *Physiol. Rev.* **27**, 95–119.

Holmes, R. L., Hughes, E. B., and Zuckerman, S. (1959). Section of the pituitary stalk in monkeys. *J. Endocrinol.* **18**, 305–318.

Honma, Y., and Tamura, E. (1965). Studies on the Japanese chars, the Iwana (genus *Salvelinus*). II. The hypothalamic neurosecretory system of the Nikko-Iwana, *Salvelinus leucomaenis pluvius* (Hilgendorf). *Bull. Jpn. Soc. Sci. Fish.* **31**, 878–887.

Hopkins, T. F. (1968). Enhancement of gonadotrophin activity by a rat cerebral tissue fraction. *J. Endocrinol.* **41**, 345–352.

Hori, T., Ide, M., and Miyake, T. (1968). Ovarian estrogen secretion during the estrous cycle and under the influence of exogenous gonadotropins in rats. *Endocrinol. Jpn.* **15**, 215–222.

Hori, T., Ide, M., and Miyake, T. (1969). Pituitary regulation of preovulatory estrogen secretion in the rat. *Endocrinol. Jpn.* **16**, 351–360.

Houssay, B. A. (1931). Action sexuelle de l'hypophyse sur les poissons et les reptiles. *C. R. Soc. Biol.* **106**, 377–378.

Houssay, B. A. (1947). La foncion sexual del sapo *Bufo arenarum* Hensel. *An. Acad. Nac. Cienc. Exactas, Fis. Nat. Buenos Aires* **12**, 103–124.

Hutchinson, J. S. M. (1973). Biochemical action of gonadotrophins on the mammalian ovary. *Bibl. Reprod.* **22**, 181–190, 363–372.

Hutchinson, J. S. M., Armstrong, D. T., and Greep, R. O. (1968). Comparison of luteinizing hormone from different species using the ventral prostate assay, with and without delay vehicle, and the ovarian ascorbic acid depletion assay. *J. Endocrinol.* **40**, 231–235.

Hyyppä, M., Motta, M., and Martini, L. (1971). 'Ultrashort' feedback control of follicle-stimulating hormone-releasing factor secretion. *Neuroendocrinology* **7**, 227–235.

Idler, D. R., Bazar, L. S., and Hwang, S. J. (1975). Fish gonadotrophin(s). III. Evidence for more than one gonadotrophin in chum salmon pituitary glands. *Endocr. Res. Commun.* **2**, 237–247.

Igarashi, M. (1969). Short and auto-feedback control of the adenohypophyseal function. *Endocrinol. Jpn., Suppl.* **1**, 63–68.

Igarashi, M., and McCann, S. M. (1964). A hypothalamic follicle stimulating hormone-releasing factor. *Endocrinology* **74**, 446–452.

Imai, K., and Nalbandov, A. V. (1971). Changes in FSH activity of anterior pituitary glands and of blood plasma during the laying cycle of the hen. *Endocrinology* **88**, 1465–1470.

Ishii, S., Sarkar, A. K., and Kobayashi, H. (1970). Ovarian ascorbic acid-depleting factor in pigeon median eminence extracts. *Gen. Comp. Endocrinol.* **14**, 461–466.

Jackson, G. L., and Nalbandov, A. V. (1969a). A substance resembling arginine vasotocin in the anterior pituitary gland of the cockerel. *Endocrinology* **84**, 1218–1223.

Jackson, G. L., and Nalbandov, A. V. (1969b). Luteinizing hormone releasing activity in the chicken hypothalamus. *Endocrinology* **84**, 1262–1265.

Jacobs, H. S. (1975). Female hypogonadism. *J. Endocrinol.* **66**, 12P–13P.

Jalabert, B. (1969). Response de l'ovaire de *Poecilia reticulata* (poisson teleostéen vivipare) normal ou hypophysectomise a dés injections d'extraits hypophysaires bruts de gardon et de gambusie. *Ann. Biol. Anim., Biochim. Biophys.* **9**, 315–329.

Jeffcoate, S. L., Holland, D. T., White, N., Fraser, H. M., Gunn, A., Crighton, D. B., Foster, J. P., Griffiths, E. C., Hooper, K. C., and Sharp, P. J. (1975). The radioimmunoassay of hypothalamic hormones (TRH and LHRH) and related peptides in biological fluids. *In* "The Hypothalamic Hormones" (M. Motta, P. G. Crosignani, and L. Martini, eds.), pp. 279–297. Academic Press, New York.

Johansen, P. N. (1967). The role of the pituitary in the resistance of the goldfish (*Carassius auratus* L.) to a high temperature. *Can. J. Zool.* **45**, 329–345.

Jones, E. E., and Nalbandov, A. V. (1971). Local induction of ovulation and luteinization in the rabbit. *Program, 53rd Meet. U.S. Endocrine Soc.* p. 117.

Jones, R. E. (1969a). Effects of mammalian gonadotropins on the ovaries and oviducts of the lizard, *Lygosoma laterale. J. Exp. Zool.* **171**, 217–222.

Jones, R. E. (1969b). Effect of prolactin and progesterone on gonads of breeding California quail. *Proc. Soc. Exp. Biol. Med.* **131**, 172–174.

Jørgensen, C. B. (1968). Central nervous control of adenohypophysial functions. *In* "Perspectives in Endocrinology" (E. J. W. Barrington and C. B. Jørgensen, eds.), pp. 469–542. Academic Press, New York.

Jørgensen, C. B. (1970). Hypothalamic control of hypophyseal function in anurans. *In* "The Hypothalamus" (L. Martini, M. Motta and F. Fraschini, eds.), pp. 649–661. Academic Press, New York.

Jørgensen, C. B., and Larsen, L. O. (1967). Neuroendocrine mechanisms in lower vertebrates. *In* "Neuroendocrinology" (L. Martini and W. F. Ganong, eds.), Vol. 2, pp. 485–528. Academic Press, New York.

Jost, A., Dupouy, J. P., and Geloso-Meyer, A. (1970). Hypothalamo-hypophyseal relationships in the fetus. *In* "The Hypothalamus" (L. Martini, M. Motta, and F. Fraschini, eds.), pp. 605–615. Academic Press, New York.

Juhn, M., and Harris, P. C. (1956). Responses in molt and lay of fowl to progesterone and gonadotrophins. *Proc. Soc. Exp. Biol. Med.* **92**, 709–711.

Jutisz, M., de La Llosa, M. P., Bérault, A., and Kerdelhué, B. (1970). Concerning the mechanisms of action of hypothalamic releasing factors on the adenohypophysis. *In* "The Hypothalamus" (L. Martini, M. Motta, and F. Fraschini, eds.), pp. 293–311. Academic Press, New York.

Kaivola, S., Seppala, M., and Seppala, I. J. T. (1969). Effect of human serum on the activity of human pituitary (HPG) and menopausal (HMG) gonadotrophins in bioassay. *Ann. Med. Exp. Biol. Fenn.* **47**, 43–47.

Kalra, S. P., Velasco, M. E., and Sawyer, C. H. (1970). Influences of hypothalamic deafferentation on pituitary FSH release and estrogen feedback in immature female parabiotic rats. *Neuroendocrinology* **6**, 228–235.

Kalra, S. P., Ajika, K., Krulich, L., Fawcett, G. P., Quijada, M., and McCann, S. M. (1971). Effects of hypothalamic and preoptic electrochemical stimulation on gonadotropin and prolactin release in proestrus rats. *Endocrinology* **88**, 1150–1158.

Kaltenbach, C. C., Graber, J. W., Niswender, G. D., and Nalbandov, A. V. (1968). Effect of hypophysectomy on the formation and maintenance of corpora lutea in the ewe. *Endocrinology* **82**, 753–759.

Kamberi, I. A., Mical, R. S., and Porter, J. C. (1970a). Follicle stimulating hormone releasing activity in hypophysial portal blood and elevation by dopamine. *Nature (London)* **227**, 714–715.

Kamberi, I. A., Mical, R. S., and Porter, J. C. (1970b). Prolactin-inhibiting activity in hypophysial stalk blood and elevation by dopamine. *Experientia* **26**, 1150–1151.

Kapen, S., Boyar, R. M., Finkelstein, J. W., Hellman, L., and Weitzman, E. D. (1974). Effect of sleep-wake cycle reversal on LH secretory pattern in puberty. *J. Clin. Endocrinol. Metab.* **39**, 293–299.

Karavolas, H. J., Meyer, R. K., Deighton, K. J., and Adrouny, S. (1971). Stimulation and inhibition of synthesis of ovulating hormone (OH) in organ culture by subcellular fractions of rat medial basal hypothalami. *Endocrinology* **88,** 969–975.

Karg, H., and Schams, D. (1970). Regulatory pattern of sexual functions. *In* "Mammalian Reproduction" (H. Gibian and E. J. Plotz, eds.), pp. 88–96. Springer-Verlag, Berlin and New York.

Karsch, F. J., Noveroske, J. W., Roche, J. F., Norton, H. W., and Nalbandov, A. V. (1970). Maintenance of ovine corpora lutea in the absence of ovarian follicles. *Endocrinology* **87,** 1228–1236.

Keever, J. E., and Greenwald, G. S. (1967). Effect of oestrogen and progesterone on pituitary gonadotrophic content of the cyclic hamster. *Acta Endocrinol. (Copenhagen)* **56,** 244–254.

Kelch, R. F., and Clemens, L. E. (1975). Clinical applications of hypothalamic hormones in man. *In* "Hypothalamic Hormones" (E. S. E. Hafez and J. R. Reel, eds.), pp. 129–150. Wiley, New York.

Keller, P. J. (1973). Treatment of anovulation with synthetic luteinizing hormone-releasing hormone. *Am. J. Obstet. Gynecol.* **116,** 689–705.

Kennelly, J. J., and Foote, R. H. (1965). Superovulatory response of pre- and postpubertal rabbits to commercially available gonadotrophins. *J. Reprod. Fertil.* **9,** 177–188.

Kerr, T. (1965). The development of the pituitary in *Xenopus laevis* Daudin. *Gen. Comp. Endocrinol.* **6,** 303–311.

Keyes, P. L. (1969). Luteinizing hormone: action on the graafian follicle *in vitro.* *Science* **164,** 846–847.

King, J. R., Follett, B. K., Farner, D. S., and Morton, M. L. (1966). Annual gonadal cycles and pituitary gonadotrophins in *Zonotrichia leucophrys gambelii. Condor* **68,** 476–487.

Kinoshita, Y. (1938). On the secondary sexual characters, with special remarks on the influence of hormone preparations upon the nuptial coloration in *Chloea sarchynnis* Jordan and Snyder. *J. Sci. Hiroshima Univ., Ser. B Div. 1* **6,** 5–22.

Kirton, K. T., Niswender, G. G., Midgley, A. R., Jr., Jaffe, R. B., and Forbes, A. D. (1970). Serum luteinizing hormone and progesterone concentration during the menstrual cycle of the rhesus monkey. *J. Clin. Endocrinol. Metab.* **30,** 105–110.

Kmentová, V., and Schreiber, V. (1965). Comparison of the sensitivity of the weight reaction of the ovaries to chorionic gonadotropin in adult intact, hypophysectomized, hypothyroid and immature rats. *Folia Biol. (Prague)* **11,** 305–310.

Knigge, K. M., and Scott, D. E. (1970). Structure and function of the median eminence. *Am. J. Anat.* **129,** 223–243.

Knobil, E., Kostyo, J. L., and Greep, R. O. (1959). Production of ovulation in the hypophysectomized rhesus monkey. *Endocrinology* **65,** 487–493.

Knowles, F., and Anand Kumar, T. C. (1969). Structural changes, related to reproduction, in the hypothalamus and in the pars tuberalis of the rhesus monkey. *Philos. Trans. R. Soc. London, Ser. B* **256,** 357–375.

Koed, H. J., and Hamburger, C. (1967). Ovarian ascorbic acid depletion test for luteinizing hormone. Comparison of LH from human and ovine origin. *Acta Endocrinol. (Copenhagen)* **56,** 619–625.

Kohler, P. O., Ross, G. T., and Odell, W. D. (1968). Metabolic clearance and production rates of human luteinizing hormone in pre- and post-menopausal women. *J. Clin. Invest.* **47**, 38–47.

Köves, K., and Halász, B. (1970). Location of the neural structures triggering ovulation in the rat. *Neuroendocrinology* **6**, 180–193.

Kragt, C. L., Bloch, G., and Cons, J. (1970). Radioimmunoassay (RIA) of sheep and rat FSH. *Fed. Proc. Fed. Am. Soc. Exp. Biol.* **29**, 439.

Kramer, R. M., Holdaway, I. M., Rees, L. H., McNeilly, A. S., and Chard, T. (1974). Technical aspects of the redox bioassay for luteinizing hormone. *Clin. Endocrinol. Oxf.* **3**, 375–382.

Krey, L. C., Butler, W. R., and Knobil, E. (1975). Surgical disconnection of the medial basal hypothalamus and pituitary function in the rhesus monkey. I. Gonadotropin secretion. *Endocrinology* **96**, 1073–1087.

Krulich, L., Quijada, M., and Illner, P. (1971). Localization of prolactin-inhibiting factor (PIF), P-releasing factor (PRF), growth hormone-RF (GRF) and GIF activities in the hypothalamus of the rat. *Program, 53rd Meet. U.S. Endocrine Soc.* p. 83.

Kuhl, H., and Taubert, H. D. (1975). Short-loop feedback mechanism of LH: LH stimulates hypothalamic L-cystine arylamidase to inactivate LH-RH in the rat hypothalamus. *Acta Endocrinol. (Copenhagen)* **78**, 649–663.

Kwa, H. G., Feltkamp, C. A., van der Gugten, A. A., and Verhofstad, F. (1970). Rate of elimination of prolactin as a determinant factor for plasma levels assayed in rats. *J. Endocrinol.* **48**, 299–300.

Labhsetwar, A. P. (1969). Age-dependent changes in the pituitary-gonadal relationship. 2. Study of pituitary FSH and LH content in the female rat. *J. Reprod. Fertil.* **20**, 21–28.

Labhsetwar, A. P. (1970a). Ageing changes in pituitary-ovarian relationships. *J. Reprod. Fertil., Suppl.* **12**, 99–117.

Labhsetwar, A. P. (1970b). Synergism between LH and FSH in the induction of ovulation. *J. Reprod. Fertil.* **23**, 517–519.

Lagios, M. D. (1965). Seasonal changes in the cytology of the adenohypophysis, testis and ovaries of the black surfperch, *Embiotoca jacksoni,* a viviparous percomorph fish. *Gen. Comp. Endocrinol.* **5**, 207–221.

Larsen, L. O. (1965). Effects of hypophysectomy in the cyclostome, *Lampetra fluviatilis. Gen. Comp. Endocrinol.* **5**, 16–30.

Larsen, L. O. (1969a). Effects of hypophysectomy before and during sexual maturation in the cyclostome *Lampetra fluviatilis. Gen. Comp. Endocrinol.* **12**, 200–208.

Larsen, L. O. (1969b). Hypophyseal functions in river lampreys. *Gen. Comp. Endocrinol., Suppl.* **2**, 522–527.

Larsen, L. O. (1970). The lamprey egg at ovulation (*Lampetra fluviatilis*). *Biol. Reprod.* **2**, 37–47.

Lawton, I. E., and Sawyer, C. H. (1970). Role of amygdala in regulating LH secretion in the adult female rat. *Am. J. Physiol.* **218**, 622–626.

Lee, P. A., Midgely, A. R., Jr., and Jaffe, R. B. (1970). Regulation of human gonadotropins. VI. Serum follicle stimulating and luteinizing hormone. *J. Clin. Endocrinol. Metab.* **31**, 248–253.

LeLeux, P. and Robyn, C. (1971). Immunohistochemistry of individual adenohypophysial cells. *Acta Endocrinol. (Copenhagen), Suppl.* **153**, 168–184.

Li, C. H., Moudgal, N. R., Trenkle, A., Bourdel, G., and Sadri, K. (1962). Some aspects of immunochemical methods for the characterisation of protein hormones. *Ciba Found. Colloq. Endocrinol. [Proc.]* **14**, 20–32.

Li, C. H., Dixon, J. S., Lo, T.-B., Pankov, Y. A., and Schmidt, K. D. (1969). Amino-acid sequence of ovine lactogenic hormone. *Nature (London)* **224**, 695–696.

Licht, P. (1970). Effects of mammalian gonadotropins (ovine FSH and LH) in female lizards. *Gen. Comp. Endocrinol.* **14**, 98–106.

Licht, P., and Stockell-Hartree, A. (1971). Actions of mammalian, avian and piscine gonadotrophins in the lizard. *J. Endocrinol.* **49**, 113–124.

Licht, P., and Papkoff, H. (1974a). Separation of two distinct gonadotrophins from the pituitary gland of the bull frog, *Rana catesbeiana. Endocrinology* **94**, 1587–1594.

Licht, P., and Papkoff, H. (1974b). Separation of two distinct gonadotrophins from the pituitary gland of the snapping turtle (*Chelydra serpentina*). *Gen. Comp. Endocrinol.* **22**, 218–237.

Liley, N. R. (1969). Hormones and reproductive behaviour in fishes. *Fish Physiol.* **3**, 73–115.

Liley, N. R., and Donaldson, E. M. (1969). The effects of salmon pituitary gonadotropin on the ovary and sexual behaviour of the female guppy, *Poecilia reticulata. Can. J. Zool.* **47**, 569–573.

Lindner, H. R., Tsafriri, A., Lieberman, M. E., Zor, M., Koch, Y., Bauminger, S., and Barnea, A. (1974). Gonadotropin action on cultures of Graafian follicles: induction of maturation division of the mammalian oocytes and differentiation of the luteal cell. *Recent Prog. Horm. Res.* **30**, 79–138.

Lipner, H. (1971). Ovulation from histamine depleted ovaries. *Proc. Soc. Exp. Biol. Med.* **136**, 111–114.

Lipner, H. (1973). Mechanism of mammalian ovulation. *In* "Handbook of Physiology" (Am. Physiol. Soc., J. Field, ed.), Sect. 7, Vol. II, Part 1, pp. 153–167. Williams & Wilkins, Baltimore, Maryland.

Lipner, H., and Greep, R. O. (1971). Inhibition of steroidogenesis at various sites of the biosynthetic pathway in relation to induced ovulation. *Endocrinology* **88**, 602–607.

Lipner, H., and Smith, M. S. (1971). A method for determining the distribution and source of protein in preovulatory rat ovaries. *J. Endocrinol.* **50**, 187–200.

Lipner, H., Hirsch, M. A., Moudgal, N. R., Macdonald, G. J., Ying, S.-Y., and Greep, R. O. (1974). Ovulation-inducing activity of FSH in the rat. *Endocrinology* **94**, 1351–1358.

Lisk, R. D. (1967). Neural control of gonad size by hormone feed-back in the desert iguana *Dipsosaurus dorsalis dorsalis. Gen. Comp. Endocrinol.* **8**, 258–266.

Liu, W.-K., and Ward, D. N. (1975). The purification and chemistry of pituitary glycoprotein hormones. *Pharmacol. Ther., B* **1**, 545–570.

Ljunggren, L. (1969). Seasonal studies of wood pigeon populations. II. Gonads, crop glands, adrenals and the hypothalamo-hypophysial system. *Viltrevy* **6**, 41–126.

Llerena, L. A., Guevara, A., Lobotsky, J., Lloyd, C. W., Weisz, J., Pupkin, M., Zanartu, J., and Puga, J. (1969). Concentration of luteinizing hormone and follicle-stimulating hormone in peripheral and ovarian venous plasma. *J. Clin. Endocrinol. Metab.* **29**, 1083–1089.

Löfgren, R. (1960). The infundibular recess, a component in the hypothalamo-adenohypophyseal system. *Acta Morph. Neerl. Scand.* **3**, 55–78.

Lofts, B., Follett, B. K., and Murton, R. K. (1970). Temporal changes in the pituitary-gonadal axis. *Mem. Soc. Endocrinol.* **18,** 545–575.

Loraine, J. A., and Bell, E. T. (1971). "Hormone Assays and Their Clinical Application," 3rd ed. Livingstone, Edinburgh.

Lostroh, A. J. (1959). The response of ovarian explants from postnatal mice to gonadotropins. *Endocrinology* **65,** 124–132.

Lostroh, A. J. (1971). Induction of ovulation with 20α-OH-progesterone (20α-OH-P) in the hypophysectomized rat. *Fed. Proc. Fed. Am. Soc. Exp. Biol.* **30,** 595.

Lostroh, A. J., and Johnson, R. E. (1966). Amounts of interstitial cell-stimulating hormone and follicle-stimulating hormone required for follicular development, uterine growth and ovulation in the hypophysectomized rat. *Endocrinology* **79,** 991–996.

Louvet, J. P., and Vaitukaitis, J. L. (1975). Induction of FSH receptors in rat ovaries by estrogen priming. *Program 57th Meet. Endocrine Soc.* p. 135.

McArthur, J. W., and Perley, R. (1969). Urinary gonadotropin excretion by infrahuman primates. *Endocrinology* **84,** 508–513.

McBride, J. R., and van Overbeeke, A. P. (1969). Cytological changes in the pituitary gland of the adult sockeye salmon (*Oncorhynchus nerka*) after gonadectomy. *J. Fish. Res. Board Can.* **26,** 1147–1156.

McCann, S. M. (1970). Neurohormonal correlates of ovulation. *Fed. Proc. Fed. Am. Soc. Exp. Biol.* **29,** 1888–1894.

McCann, S. M. (1974). Regulation of secretion of follicular stimulating hormone and luteinizing hormone. *In* "Handbook of Physiology" (Am. Physiol. Soc., J. Field, ed.). Sect. 7, Vol. IV, pp. 489–517. Williams & Wilkins, Baltimore, Maryland.

McCann, S. M., and Porter, J. C. (1969). Hypothalamic pituitary stimulating and inhibiting hormones. *Physiol. Rev.* **49,** 240–284.

McCann, S. M., and Moss, R. L. (1975). Putative neurotransmitter involved in discharging gonadotropin-releasing neuro-hormones and the action of LHRH on the CNS. *Life Sci.* **16,** 823–852.

McCann, S. M., Taleisnik, S., and Friedman, H. M. (1960). LH-releasing activity in hypothalamic extracts. *Proc. Soc. Exp. Biol. Med.* **104,** 432–434.

McCracken, J. A., and Baird, D. T. (1969). The study of ovarian function by means of transplantation of the ovary in the ewe. *In* "The Gonads" (K. W. McKerns, ed.), pp. 175–209. Appleton, New York.

Macdonald, G. J., Yoshinaga, K., and Greep, R. O. (1971). Maintenance of luteal function in rats by rat prolactin. *Proc. Soc. Exp. Biol. Med.* **136,** 687–688.

McManus, J. F. A. (1946). Histological demonstration of mucin after periodic acid. *Nature (London)* **158,** 202.

McNatty, K. P., Hunter, W. M., McNeilly, A. S., and Sawers, K. S. (1975). Changes in the concentration of pituitary and steroid hormones in the follicular fluid of human Graafian follicles throughout the menstrual cycle. *J. Endocrinol.* **64,** 555–571.

McNeilly, A. S. (1975). Intragonadal control of ovarian function. *J. Endocrinol.* **66,** 7P–9P.

Madan, M., and Johnson, H. D. (1971). Circulating plasma luteinizing hormone (LH) and its rhythmicity during bovine estrous cycle. *Program, 53rd Meet. U.S. Endocrine Soc.* p. 226.

Major, P., and Kilpatrick, R. (1972). Cyclic AMP and hormone action. Review. *J. Endocrinol.* **52,** 593–630.

Malven, P. V. (1969). Hypophysial regulation of luteolysis in the rat. *In* "The Gonads" (K. W. McKerns, ed.), pp. 367–382. Appleton, New York.

Malven, P. V., and Sawyer, C. H. (1966a). Formation of new corpora lutea in mature hypophysectomized rats. *Endocrinology* **78**, 1259–1263.

Malven, P. V., and Sawyer, C. H. (1966b). A luteolytic action of prolactin in hypophysectomized rats. *Endocrinology* **79**, 268–274.

Marden, W. G. R. (1953). The hormone control of ovulation in the calf. *J. Agric. Sci.* **43**, 381–406.

Marsh, J. M., and Le Maire, W. J. (1974). The role of cyclic AMP and prostaglandin in the action of luteinizing hormone. *In* "Gonadotropins and Gonadal Function" (N. R. Moudgal ed.), pp. 376–390. Academic Press, New York.

Marshall, F. H. A. (1936). Sexual periodicity and causes which determine it. *Philos. Trans. R. Soc. London, Ser. B* **226**, 423–456.

Marshall, F. H. A. (1942). Exteroceptive factors in sexual periodicity. *Biol. Rev. Cambridge Philos. Soc.* **17**, 68–90.

Martini, L., Fraschini, F., and Motta, M. (1968). Neural control of anterior pituitary functions. *Recent Prog. Horm. Res.* **24**, 439–485.

Matsuo, S., Vitums, A., King, J. R., and Farner, D. S. (1969). Light-microscope studies of the cytology of the adenohypophysis of the White-Crowned Sparrow, *Zonotrichia leucophrys gambelii*. *Z. Zellforsch. Mikrosk. Anat.* **95**, 143–176.

Matsuo, H., Baba, Y., Nair, R. M. G., Arimura, A., and Schally, A. V. (1971). Structure of the porcine LH and FSH releasing hormone. I. The proposed amino acid sequence. *Biochem. Biophys. Res. Commun.* **43**, 1334–1339.

Mauléon, P. (1969). Oogenesis and folliculogenesis. *In* "Reproduction in Domestic Animals" (H. H. Cole and P. T. Cupps, eds.), 2nd ed., pp. 187–215. Academic Press, New York.

Mauléon, P., and Pelletier, J. (1964). Variations génétiques du fonctionnement hypophysaire de trois souches de rattes immatures. Relations avec la fertilité. *Ann. Biol. Anim. Biochim. Biophys.* **4**, 105–112.

Maxwell, L. C. (1934). The quantitative and qualitative ovarian response to distributed dosage with gonadotrophic extracts. *Am. J. Physiol.* **110**, 458–463.

Meier, A. H. (1969). Antigonadal effects of prolactin in the white throated sparrow, *Zonotrichia albicollis*. *Gen. Comp. Endocrinol.* **13**, 222–225.

Meier, A. H., and Farner, D. S. (1964). A possible endocrine basis for premigatory fattening in the White-Crowned Sparrow *Zonotrichia leucophrys gambellii* (Nuttall). *Gen. Comp. Endocrinol.* **4**, 584–595.

Meites, J., ed. (1970). "Hypophysiotropic Hormones of the Hypothalamus: Assay and Chemistry." Williams & Wilkins, Baltimore, Maryland.

Meites, J., Talwalker, P. K., and Nicoll, C. S. (1960). Initiation of lactation in rats with hypothalamic or cerebral tissue. *Proc. Soc. Exp. Biol. Med.* **103**, 298–300.

Mellinger, J. (1964). Les relations neuro-vasculo-glandulaires dans l'appareil hypophysiare de la Rousette *Scyliorhinus caniculus* L. *Arch. Anat., Histol. Embryol.* **47**, 1–201.

Mellish, C. (1936). Effect of anterior pituitary extract and certain environmental conditions on the genital system of the horned lizard, *Phrynosoma cornutum*. *Anat. Rec.* **67**, 22–33.

Mess, B. (1969). Site and onset of production of releasing factors. *In* "Progress in Endocrinology" (C. Gual, ed.), pp. 564–570. Excerpta Med. Found., Amsterdam.

Mess, B., Zanisi, M., and Tima, L. (1970). Site of production of releasing and inhibiting factors. *In* "The Hypothalamus" (L. Martini, M. Motta, and F. Fraschini, eds.), pp. 259–276. Academic Press, New York.

Meurling, P. (1967). The vascularization of the pituitary in elasmobranchs. *Sarsia* **28,** 1–104.

Midgley, A. R., Jr. (1969). Immunological characterization of the gonadotropins. *In* "Reproduction in Domestic Animals" (H. H. Cole and P. T. Cupps, eds.), 2nd ed., pp. 47–66. Academic Press, New York.

Mikami, S. (1958). The cytological significance of regional patterns in the adenohypophysis of the fowl. *J. Fac. Agric., Iwate Univ.* **3,** 473–545.

Mills, T. M., Davies, P. G. A., and Savard, K. (1971). Stimulation of estrogen synthesis in rabbit follicles by luteinizing hormone. *Endocrinology* **88,** 857–862.

Mishkinsky, J., Khazen, K., and Sulman, F. G. (1968). Prolactin-releasing activity of the hypothalamus in post-partum rats. *Endocrinology* **82,** 611–613.

Mitchell, J. A., and Yochim, J. M. (1968). Relation of HCG-induced ovulation to the production of prolonged diestrus in the adult rat. *Endocrinology* **82,** 1142–1148.

Mitchell, M. E. (1967a). Stimulation of the ovary in hypophysectomized hens by an avian pituitary preparation. *J. Reprod. Fertil.* **14,** 249–256.

Mitchell, M. E. (1967b). The effects of avian gonadotrophin precipitate on pituitary-deficient hens. *J. Reprod. Fertil.* **14,** 257–263.

Mitchell, M. E. (1970). Treatment of hypophysectomized hens with partially purified avian FSH. *J. Reprod. Fertil.* **22,** 233–241.

Mittler, J. C., and Meites, J. (1964). *In vitro* stimulation of pituitary follicle stimulating hormone release by hypothalamic extract. *Proc. Soc. Exp. Biol. Med.* **117,** 309–313.

Monroe, S. E., Rebar, R. W., Gay, V. L., and Midgley, A. R., Jr. (1969). Radioimmunoassay determination of luteinizing hormone during the estrous cycle of the rat. *Endocrinology* **85,** 720–724.

Monroe, S. E., Atkinson, L. E., and Knobil, E. (1970). Patterns of circulating luteinizing hormone and their relation to plasma progesterone levels during the menstrual cycle of the rhesus monkey. *Endocrinology* **87,** 453–455.

Moon, Y. S., Dorrinton, J. H., and Armstrong, D. T. (1975). Stimulatory actions of follicle stimulating hormone on estradiol-17β secretion by hypophysectomized rat ovaries in organ culture. *Endocrinology* **97,** 244–247.

Moor, R. M. (1968). Foetal homeostasis: conceptus–ovary endocrine balance. *Proc. R. Soc. Med.* **61,** 1217–1225.

Moor, R. M. (1970). The role of the conceptus in the control and maintenance of pregnancy. *In* "Mammalian Reproduction" (H. Gibian and E. J. Plotz, eds.), pp. 351–355. Springer-Verlag, Berlin and New York.

Morell, M., Crooke, A. C., and Butt, W. R. (1968). Excretion of follicle stimulating hormone by patients treated with gonadotrophins. *J. Endocrinol.* **41,** 571–575.

Motta, M., Fraschini, F., and Martini, L. (1969). 'Short' feedback mechanisms in the control of anterior pituitary function. *In* "Frontiers in Neuroendocrinology" (W. F. Ganong and L. Martini, eds.), pp. 211–253. Oxford Univ. Press, London and New York.

Moudgal, N. R., and Raj, H. G. M. (1974). Pituitary gonadotropin. *In* "Methods of Hormone Radioimmunoassay" (B. M. Jaffe and H. R. Behrman, eds.), pp. 57–86. Academic Press, New York.

Moudgal, N. R., Sairan, M. R., and Madhwa Raj, H. G. (1969). Gonadotropin inhibitory substances. *Gen. Comp. Endocrinol., Suppl.* **2,** 162–170.

Naftolin, F. (1975). Gonadotrophin rhythms. *J. Endocrinol.* **66**, 9P–10P.

Naftolin, F., Espeland, D. H., Tremann, J. A., Dillard, E. A., and Paulsen, C. A. (1968). Serum HLH levels in ovarian and systemic vein blood by radioimmunoassay. *In* "Gonadotropins 1968" (E. Rosemberg, ed.), pp. 373–379. Geron-X Inc., Los Altos, California.

Naik, D. V. (1975). Immunoreactive LH-RH neurons in the hypothalamus identified by light and fluorescent microscopy. *Cell Tissue Res.* **157**, 423–436.

Nakajo, S., and Imai, K. (1961). Gonadotropin content of the cephalic and caudal lobe of the anterior pituitary in laying, non-laying and broody hens. *Poult. Sci.* **40**, 739–744.

Nakane, P. K. (1970). Classifications of anterior pituitary cell types with immunoenzyme histochemistry. *J. Histochem. Cytochem.* **18**, 9–20.

Nakano, R., Mizumo, T., Katayama, K., and Tojo, S. (1975). Growth of ovarian follicles in rats in the absence of gonadotrophin. *J. Reprod. Fertil.* **45**, 545–546.

Nalbandov, A. V. (1961). Mechanisms controlling ovulation of avian and mammalian follicles. *In* "Control of Ovulation" (C. A. Villee, ed.), pp. 122–131. Pergamon, Oxford.

Nalbandov, A. V. (1970). Comparative aspects of corpus luteum function. *Biol. Reprod.* **2**, 7–13.

Nalbandov, A. V., and Cook, B. (1968). Reproduction. *Annu. Rev. Physiol.* **30**, 245–278.

Nalbandov, A. V., and Graber, J. W. (1969). Neural control of the anterior and the posterior pituitary gland in birds. *In* "The Hypothalamus" (W. Haymaker, E. Anderson, and W. J. H. Nauta, eds.), pp. 311–325. Thomas, Springfield, Illinois.

Nallar, R., and McCann, S. M. (1965). Luteinizing hormone-releasing activity in plasma of hypophysectomized rats. *Endocrinology* **76**, 272–275.

Neal, P., and Baker, T. G. (1975). Response of mouse Graafian follicles in organ culture to varying doses of follicle-stimulating hormone and luteinizing hormone. *J. Endocrinol.* **65**, 27–32.

Neal, P., Baker, T. G., McNatty, K. P., and Scaramuzzi, R. J. (1975). Influence of prostaglandins and human chorionic gonadotrophin on progesterone concentration and oocyte maturation in mouse ovarian follicles maintained in organ culture. *J. Endocrinol.* **65**, 19–25.

Négro-Vilar, A., Sar, M., and Meites, J. (1970). Changes in hypothalamic FSH-RF and pituitary FSH during the estrous cycle of rats. *Endocrinology* **87**, 1091–1093.

Neill, J. D. (1970). Effect of 'stress' on serum prolactin and luteinizing hormone levels during the estrous cycle of the rat. *Endocrinology* **87**, 1192–1197.

Neill, J. D., and Reichert, L. E., Jr. (1971). Development of a radioimmunoassay for rat prolactin and evaluation of the NIAMD rat prolactin radioimmunoassay. *Endocrinology* **88**, 548–555.

Nelson, D. M., Norton, H. W., and Nalbandov, A. V. (1965). Changes in hypophysial and plasma LH levels during the laying cycle of the hen. *Endocrinology* **77**, 889–896.

Nicoll, C. S., Fiorindo, R. P., McKennee, C. T., and Parsons, J. A. (1970). Assay of hypothalamic factors which regulate prolactin secretion. *In* "Hypophysiotropic Hormones of the Hypothalamus: Assay and Chemistry" (J. Meites, ed.), pp. 115–144. Williams & Wilkins, Baltimore, Maryland.

Nikitovitch-Winer, M., and Everett, J. W. (1958). Functional restitution of pituitary grafts retransplanted from kidney to median eminence. *Endocrinology* **63**, 916–930.

Nimrod, A., Erickson, G. F., and Ryan, K. J. (1976). A specific FSH receptor in rat granulosa cells: properties of binding *in vitro*. *Endocrinology* **98**, 56–64.

Niswender, G. D., and Cicmanec, J. L. (1971). Arterial-venous differences in concentrations of gonadotropins across the ovaries of cyclic and pregnant ewes. *Program, 53rd Meet. U.S. Endocrine Soc.* p. 116.

Niswender, G. D., Reichert, L. E., Jr., and Zimmerman, D. R. (1970). Radio-immunoassay of serum levels of luteinizing hormone throughout the estrous cycle in pigs. *Endocrinology* **87**, 576–580.

Odell, W. D. (1970). Studies of hypothalamic-pituitary gonadal interrelations in prepubertal cattle. *In* "Gonadotrophins and Ovarian Development" (W. R. Butt, A. C. Crooke, and M. Ryle, eds.), pp. 371–385. Livingstone, Edinburgh.

Ojeda, S. R., and Ramirez, V. D. (1969). Automatic control of LH and FSH secretion by short feedback circuits in immature rats. *Endocrinology* **84**, 786–797.

Olivereau, M., and Ball, J. N. (1966). Histological study of functional ectopic pituitary transplants in a teleost fish (*Poecilia formosa*). *Proc. R. Soc. London, Ser. B,* **164**, 106–129.

Ondo, J. G. (1974). Gamma-aminobutyric acid effects on pituitary gonadotropin secretion. *Science* **186**, 738–739.

Opel, H. (1963). Delay in ovulation in the hen following stimulation of the preoptic brain. *Proc. Soc. Exp. Biol. Med.* **113**, 488–492.

Opel, H. (1964). Premature oviposition following operative interference with the brain of the chicken. *Endocrinology* **74**, 193–200.

Opel, H., and Lepone, P. D. (1967). Ovulating hormone-releasing factor in the chicken hypothalamus. *Poult. Sci.* **46**, 1302.

Opel, H., and Nalbandov, A. V. (1961). Follicular growth and ovulation in hypophysectomized hens. *Endocrinology* **69**, 1016–1028.

Orsini, M. W., and Schwartz, N. B. (1966). Pituitary LH content during the estrous cycle in female hamsters: comparisons with males and acyclic females. *Endocrinology* **78**, 34–40.

Ortiz, E. (1947). The postnatal development of the reproductive system of the golden hamster (*Cricetus auratus*) and its reactivity to hormones. *Physiol. Zool.* **20**, 45–67.

Ortiz, E. (1955). The relationship of advancing age to the reactivity of the reproductive system in the female hamster. *Anat. Rec.* **122**, 517–537.

Palmer, D. D., Burrows, R. E., Robertson, O. H., and Newman, H. W. (1954). Further studies on the reactions of adult blueback salmon to injected salmon and mammalian gonadotrophins. *Prog. Fish Cult.* **16**, 99–107.

Papanicolaou, A. D., Loraine, J. A., Dove, R. A., and Loudon, N. B. (1969). Hormone excretion patterns in peri-menopausal women. *J. Obstet. Gynaecol. Br. Commonw.* **76**, 308–316.

Papkoff, H. (1965). Some biological properties of a potent follicle stimulating hormone preparation. *Acta Endocrinol. (Copenhagen)* **48**, 439–445.

Parkes, A. S., and Deanesly, R. (1966). Relation between the gonads and the adrenal glands. *In* "Marshall's Physiology of Reproduction" (A. S. Parkes, ed.), 3rd ed., Vol. 3, pp. 1064–1111. Longmans Green, London and New York.

Parkes, A. S., and Marshall, A. J. (1960). The reproductive hormones in birds. *In* "Marshall's Physiology of Reproduction" (A. S. Parkes, ed.), 3rd ed., Vol. 1, Part 2, pp. 583–706. Longmans Green, London and New York.

Parlow, A. F. (1961). Bioassay of pituitary luteinizing hormone by depletion of ovarian ascorbic acid. *In* "Human Pituitary Gonadotropins" (A. Albert, ed.), pp. 300–310. Thomas, Springfield, Illinois.

Parlow, A. F. (1963). Species differences in luteinizing hormone (LH, ICSH) as revealed by the slope in the prostate assay. *Endocrinology* **73**, 509–512.

Parlow, A. F. (1968). Comparative bioassay of luteinizing hormone by three methods. *In* "Gonadotropins 1968" (E. Rosemberg, ed.), pp. 59–65. Geron-X Inc., Los Altos, California.

Parlow, A. F., and Reichert, L. E., Jr. (1963). Species differences in follicle-stimulating hormone as revealed by the slope in the Steelman-Pohley assay. *Endocrinology* **73**, 740–743.

Pasteels, J. L. (1961). Premiers résultats de culture combinée *in vitro* d'hypophyse et d'hypothalamus, dans le but d'en apprécier la sécrétion de prolactine. *C. R. Acad. Sci. (Paris), Ser. D* **253**, 3074–3075.

Pasteels, J. L. (1970). Control of prolactin secretion. *In* "The Hypothalamus" (L. Martini, M. Motta, and F. Fraschini, eds.), pp. 385–399. Academic Press, New York.

Pelletier, J. (1965). Effet du plasma de brebis su la décharge de LH chez la ratte. *C. R Acad. Sci. (Paris), Ser. D* **260**, 5624–5626.

Peter, R. E. (1970). Hypothalamic control of thyroid gland activity and gonadal activity in the goldfish, *Carassius auratus. Gen. Comp. Endocrinol.* **14**, 334–356.

Petrusz, P., Robyn, C., and Diczfalusy, E. (1970). Biological effects of human urinary follicle stimulating hormone. *Acta Endocrinol. (Copenhagen)* **63**, 454–475.

Pickford, G. E., and Atz, J. W. (1957). "The Physiology of the Pituitary Gland of Fishes." Zool. Soc., New York.

Pierce, J. G., Liao, T. H., Howard, S. M., Shome, B., and Cornell, J. S. (1971). Studies on the structure of thyrotropin: its relationship to LH. *Recent Prog. Horm. Res.* **27**, 165–212.

Ponse, K. (1966). LH action on the ovary. *Eur. Rev. Endocrin. Suppl.* **2**, 13–44.

Popa, G. T., and Fielding, U. (1930). A portal circulation from the pituitary to the hypothalamic region. *J. Anat.* **65**, 88–91.

Popa, G. T., and Fielding, U. (1933). Hypophysio-portal vessels and their colloid accompaniment. *J. Anat.* **67**, 227–232.

Presl, J., Röhling, S., Horský, J., and Herzmann, J. (1970). Changes in uptake of ³H-estradiol by the female rat brain and pituitary from birth to sexual maturity. *Endocrinology* **86**, 899–902.

Price, D., and Ortiz, E. (1944). The relation of age to reactivity in the reproductive system in the rat. *Endocrinology* **34**, 215–239.

Purves, H. D. (1961). Morphology of the hypophysis related to its function. *In* "Sex and Internal Secretions" (W. C. Young, ed.), 3rd ed., Vol. 1, pp. 161–239. Williams & Wilkins, Baltimore, Maryland.

Quijada, M., Krulich, L., Fawcett, C. P., Sundberg, K., and McCann, S. M. (1971). Localization of TSH-releasing factor (TRF), LH-RF and FSH-RF in rat hypothalamus. *Fed. Proc. Fed. Am. Soc. Exp. Biol.* **30**, 197.

Rajaniemi, H., and Vanha-Perttula, T. (1971). LH-receptor in the luteal cell membrane. *Acta Endocrinol. (Copenhagen)* **67**, Suppl. **155**, 53.

Rajaniemi, H., Tuohimaa, P., and Niemi, M. (1970). Enzymic radioiodination of pituitary gonadotrophins for radioautography. *Histochemie* **23**, 342–348.

Ralph, C. L. (1959). Some effects of hypothalamic lesions on the gonadotrophin release in the hen. *Anat. Rec.* **134**, 411–431.

Ralph, C. L., and Fraps, R. M. (1959a). Long term effects of diencephalic lesions on the ovary of the hen. *Am. J. Physiol.* **197**, 1279–1283.

Ralph, C. L., and Fraps, R. M. (1959b). Effect of hypothalamic lesions on progesterone-induced ovulation in the hen. *Endocrinology* **65**, 819–824.

Ralph, C. L., and Fraps, R. M. (1960). Induction of ovulation in the hen by injection of progesterone in the brain. *Endocrinology* **66**, 269–272.

Ramaswami, L. S. (1962). Endocrinology of reproduction in fish and frog. *Gen. Comp. Endocrinol., Suppl.* **1**, 286–299.

Ramaswami, L. S., and Lakshman, A. B. (1960). Action of enzyme digested pituitary glands of the skipper frog on the ovulation of the same. *Acta Endocrinol. (Copenhagen)* **33**, 255–260.

Ramirez, V. D., and Sawyer, C. H. (1965). Fluctuations in hypothalamic LH-RF (luteinizing hormone-releasing factor) during the rat estrous cycle. *Endocrinology* **76**, 282–289.

Ramirez, V. D., and Sawyer, C. H. (1966). Changes in hypothalamic luteinizing hormone releasing factor (LHRF) in the female rat during puberty. *Endocrinology* **78**, 958–964.

Rastogi, R. K., and Chieffi, G. (1970a). A cytological study of the *pars distalis* of the pituitary gland of normal, gonadectomized, and gonadectomized steroid hormone treated green frog, *Rana esculenta* L. *Gen. Comp. Endocrinol.* **15**, 247–263.

Rastogi, R. K., and Chieffi, G. (1970b). Cytological changes in the pars distalis of pituitary of the green frog, *Rana esculenta* L. during the reproductive cycle. *Z. Zellforsch. Mikrosk. Anat.* **111**, 505–518.

Rayford, P. L., Brinkley, H. J., and Young, E. P. (1971). Radioimmunoassay determination of LH concentration in the serum of female pigs. *Endocrinology* **88**, 707–713.

Reeves, J. J., Arimura, A., and Schally, A. V. (1970). Serum levels of prolactin and luteinizing hormone (LH) in the ewe at various stages of the estrous cycle. *Proc. Soc. Exp. Biol. Med.* **134**, 938–942.

Reeves, J. J., Arimura, A., and Schally, A. V. (1971). Pituitary responsiveness to purified luteinizing hormone-releasing hormone (LHRH) at various stages of the estrous cycle in sheep. *J. Anim. Sci.* **32**, 123–126.

Reichert, L. E., Jr. (1966). Measurement of luteinizing hormone by the hyperemia and ovarian ascorbic acid depletion assays. *Endocrinology* **78**, 815–818.

Reichert, L. E., Jr. (1967). Further studies on species differences in follicle stimulating hormone as revealed by the slope of the Steelman-Pohley assay. *Endocrinology* **80**, 1180–1181.

Reichert, L. E., Jr., and Bhalla, V. K. (1974). Development of a radioligand tissue receptor assay for human FSH. *Endocrinology* **94**, 483–491.

Reichert, L. E., Jr., and Ward, D. N. (1974). On the isolation and characterization of the alpha and beta subunits of human pituitary follicle stimulating hormone. *Endocrinology* **94**, 655–667.

Reichert, L. E., Jr., Rasco, M. A., Ward, D. N., Niswender, G. D., and Midgley, A. R., Jr. (1969). Isolation and properties of subunits of bovine pituitary LH. *J. Biol. Chem.* **244**, 5110–5117.

Reichert, L. E., Jr., Midgley, A. R., Jr., Niswender, G. D., Jr., and Ward, D. N.

(1970). Formation of a hybrid molecule from subunits of human and bovine LH. *Endocrinology* **87**, 534–541.

Reichert, L. E., Jr., Leidenberger, F., and Trowbridge, C. G. (1973). Studies on LH and subunits: development and application of a radioligand-receptor assay and properties of the hormone receptor interaction. *Recent Prog. Horm. Res.* **29**, 497–532.

Réthelyi, M., and Halász, B. (1970). Origin of the nerve endings in the surface zone of the median eminence of the rat hypothalamus. *Exp. Brain Res.* **11**, 145–158.

Richards, J. S. (1975). Estradiol receptor content in rat granulosa cells during follicular development, modification by estradiol and gonadotropins. *Endocrinology* **97**, 1174–1184.

Riddle, O. (1963). Prolactin or progesterone as key to parental behaviour—a review. *Anim. Behav.* **11**, 419–432.

Riddle, O., and Lahr, E. L. (1944). On broodiness of ring doves following implants of certain steroid hormones. *Endocrinology* **35**, 255–260.

Riddle, O., Bates, R. W., and Lahr, E. L. (1935). Prolactin induces broodiness in fowl. *Am. J. Physiol.* **111**, 352–360.

Rifkind, A. B., Kulin, H. E., Rayford, P. L., Cargille, C. M., and Ross, G. T. (1970). 24-hour urinary luteinizing hormone (LH) and follicle stimulating hormone (FSH) excretion in normal children. *J. Clin. Endocrinol. Metab.* **31**, 517–525.

Riley, G. M., and Fraps, R. M. (1942). Relationships of gonad-stimulating activity of female domestic fowl pituitaries to reproductive conditions. *Endocrinology* **30**, 537–541.

Robertson, H. A. (1969). Endogenous control of estrus and ovulation in sheep, cattle and swine. *Vitam. Horm. (N.Y.)* **27**, 91–130.

Robertson, H. A., and Hutchinson, J. S. M. (1962). The levels of FSH and LH in the pituitary of the ewe in relation to follicle growth and ovulation. *J. Endocrinol.* **24**, 143–151.

Robertson, J. E., and Baker, R. D. (1969). Role of female sex steroids as possible regulators of oocyte maturation. *2nd Annu. Meet., Soc. Study Reprod.* p. 29.

Robertson, O. H., and Wexler, B. C. (1962a). Histological changes in the pituitary of the rainbow trout (*Salmo gairdnerii*) accompanying sexual maturation and spawning. *J. Morphol.* **110**, 157–170.

Robertson, O. H., and Wexler, B. C. (1962b). Histological changes in the pituitary gland of the Pacific salmon (*Genus Oncorhynchus*) accompanying sexual maturation and spawning. *J. Morphol.* **110**, 171–186.

Roche, J. F., Foster, D. L., Karsch, F. J., Cook, B., and Dziuk, P. J. (1970). Levels of luteinizing hormone in sera and pituitaries of ewes during the estrous cycle and anestrus. *Endocrinology* **86**, 568–572.

Rondell, P. (1970). Follicular processes in ovulation. *Fed. Proc. Fed. Am. Soc. Exp. Biol.* **29**, 1875–1879.

Rondell, P. (1974). Role of steroid synthesis in the process of ovulation. *Biol. Reprod.* **10**, 199–215.

Rosemberg, E. (1967). Discussion on 'Mechanism of action of gonadotrophins.' *In* "Recent Research on Gonadotrophic Hormones" (E. T. Bell and J. A. Loraine, eds.), pp. 209–210. Livingstone, Edinburgh.

Rosemberg, E., and Joshi, S. R. (1968). Effect of human urinary follicle-stimulating hormone and luteinizing hormone on uterine growth in intact immature mice.

In "Gonadotropins 1968" (E. Rosemberg, ed.), pp. 91–102. Geron-X Inc., Los Altos, California.

Rosemberg, E., Solod, E. A., and Albert, A. (1964). Luteinizing hormone activity of human pituitary gonadotropin as determined by the ventral prostate weight and the ovarian ascorbic acid depletion methods of assay. *J. Clin. Endocrinol. Metab.* **24**, 714–728.

Rosemberg, E., Joshi, S. R., and Nwe, T. T. (1968). Recovery of exogenously administered gonadotropins. *In* "Gonadotropins 1968" (E. Rosemberg, ed.), pp. 139–145. Geron-X, Inc., Los Altos, California.

Ross, G. T. (1975). The nature of gonadotrophins. *J. Endocrinol.* **66**, 6P–7P.

Ross, G. T., Cargille, C. M., Lipsett, M. B., Rayford, P. L., Marshall, J. R., Strott, C. A., and Rodbard, D. (1970). Pituitary and gonadal hormones in women during spontaneous and induced ovulatory cycles. *Recent Prog. Horm. Res.* **26**, 1–48.

Rothchild, I. (1966). The nature of the luteotrophic process. *J. Reprod. Fertil., Suppl.* **1**, 49–62.

Rowlands, I. W., and Parkes, A. S. (1966). Hypophysectomy and the gonadotrophins. *In* "Marshall's Physiology of Reproduction" (A. S. Parkes, ed.), 3rd ed., Vol. 3, pp. 26–146. Longmans Green, London and New York.

Rowson, L. E. A. (1970). The evidence for luteolysin. *Br. Med. Bull.* **26**, 14–16.

Roy, B. B. (1964). Production of corticosteroids *in vitro* in some Indian fishes with experimental, histological and biochemical studies of adrenal cortex together with general observations on gonads after hypophysectomy in *O. punctatus*. *Calcutta Med. J.* **61**, 223–244.

Rugh, R. (1935). Ovulation in the frog. I. Pituitary relations in induced ovulation. *J. Exp. Zool.* **71**, 149–162.

Ryle, M. (1969). The duration of an FSH effect *in vitro*. *J. Reprod. Fertil.* **19**, 349–351.

Ryle, M. (1971). The time factor in responses to pituitary gonadotrophins by mouse ovaries *in vitro*. *J. Reprod. Fertil.* **25**, 61–74.

Saeki, Y., Himeno, K., Tanabe, Y., and Katsuragi, T. (1956). Comparative gonadotrophic potency of anterior pituitaries from cocks, laying hens and non-laying hens in molt. *Endocrinol. Jpn.* **3**, 87–91.

Sairam, M. R., and Papkoff, H. (1974). Chemistry of pituitary gonadotrophins. *In* "Handbook of Physiology" (Am. Physiol. Soc., J. Field, ed.), Sect. 7, Vol. IV, pp. 111–131. Williams & Wilkins, Baltimore, Maryland.

Sar, M., and Meites, J. (1967). Changes in pituitary prolactin release and hypothalamic PIF content during the estrous cycle of rats. *Proc. Soc. Exp. Biol. Med.* **125**, 1018–1021.

Sasamoto, S., and Kennan, A. L. (1971). Effect of anti-PMS serum (APS) on follicular ovulability in hypophysectomized immature rats pretreated with PMS. *Program, 53rd Meet. U.S. Endocrine Soc.* p. 119.

Savard, K., Marsh, J. M., and Rice, B. F. (1965). Gonadotropins and ovarian steroidogenesis. *Recent Prog. Horm. Res.* **21**, 285–356.

Sawyer, C. H. (1970). Electrophysiological correlates of release of pituitary ovulating hormones. *Fed. Proc. Fed. Am. Soc. Exp. Biol.* **29**, 1895–1899.

Sawyer, W. H. (1975). Some recent developments in brain-pituitary-ovarian physiology. *Neuroendocrinology* **17**, 97–124.

Sawyer, C. H., and Kawakami, M. (1961). Interactions between the central nervous system and hormones influencing ovulation. *In* "Control of Ovulation" (C. A. Villee, ed.), pp. 79–97. Pergamon, Oxford.

Sawyer, W. H., and Pickford, G. E. (1963). Neurohypophyseal principles of *Fundulus heteroclitus:* characteristics and seasonal changes. *Gen. Comp. Endocrinol.* **3,** 439–445.

Saxena, B. B., Leyendecker, G., Chen, W., Gandy, H. M., and Peterson, R. E. (1969). Radioimmunoassay of follicle-stimulating (FSH) and luteinizing (LH) hormones by chromatoelectrophoresis. *Acta Endocrinol. (Copenhagen), Suppl.* **142,** 185–203.

Scaramuzzi, R. J., Caldwell, B. V., and Moor, R. M. (1970). Radioimmunoassay of LH and estrogen during the estrous cycle of the ewe. *Biol. Reprod.* **3,** 110–119.

Schally, A. V., and Kastin, A. J. (1970). The role of sex steroids, hypothalamic LH-releasing hormone and FSH-releasing hormone in the regulation of gonadotropin secretion from the anterior pituitary gland. *Adv. Steroid Biochem. Pharmacol.* **2,** 41–69.

Schally, A. V., Arimura, A., Bowers, C. Y., Kastin, A. J., Sawano, S., and Redding, T. W. (1968). Hypothalamic neurohormones regulating anterior pituitary function. *Recent Prog. Horm. Res.* **24,** 497–581.

Schally, A. V., Arimura, A., Bowers, C. Y., Wakabayashi, I., Kastin, A. J., Redding, T. W., Mittler, J. C., Nair, R. M. G., Pizzolato, P., and Segal, A. J. (1970a). Purification of hypothalamic releasing hormones of human origin. *J. Clin. Endocrinol. Metab.* **31,** 291–300.

Schally, A. V., Arimura, A., Kastin, A. J., Reeves, J. J., Bowers, C. Y., Baba, Y., and White, W. F. (1970b). Hypothalamic LH-releasing hormone: chemistry, physiology and effect in humans. *In* "Mammalian Reproduction" (H. Gibian and E. J. Plotz, eds.), pp. 45–83. Springer-Verlag, Berlin and New York.

Schally, A. V., Arimura, A., Baba, Y., Nair, R. M. G., Matsuo, H., Redding, T. W., and Debeljuk, L. (1971a). Isolation and properties of the FSH and LH-releasing hormone. *Biochem. Biophys. Res. Commun.* **43,** 393–399.

Schally, A. V., Arimura, A., Kastin, A. J., Matsuo, H., Baba, Y., Redding, T. W., Nair, R. M. G., Debeljuk, L., and White, W. F. (1971b). Gonadotropin-releasing hormone: one polypeptide regulates secretion of luteinizing and follicle-stimulating hormones. *Science* **173,** 1036–1038.

Schally, A. V., Arimura, A., Redding, T. W., Debeljuk, L., Carter, W., Duport, A., and Vilchez-Martinez, J. A. (1976). Re-examination of porcine and bovine hypothalamic fractions for additional luteinizing hormone and follicle-stimulating hormone releasing activities. *Endocrinology* **98,** 380–391.

Schomberg, D. W. (1969). The concept of a uterine luteolytic hormone. *In* "The Gonads" (K. W. McKerns, ed.), pp. 383–414. Appleton, New York.

Schuetz, A. W. (1967). Action of hormones on germinal vesicle breakdown in frog (*Rana pipiens*) oocytes. *J. Exp. Zool.* **166,** 347–354.

Schuetz, A. W. (1972). Induction of structural alterations in the preovulatory amphibian ovarian follicle by hormones. *Biol. Reprod.* **6,** 67–77.

Schuetz, A. W. (1974). Role of hormones in oocyte maturation. *Biol. Reprod.* **10,** 150–178.

Schwartz, N. B. (1969). Model for the regulation of ovulation in the rat. *Recent Prog. Horm. Res.* **25,** 1–43.

Schwartz, N. B. (1974). Role of FSH and LH and their antibodies on follicle growth and on ovulation. *Biol. Reprod.* **10,** 236–272.

Schwartz, N. B., Krone, K., and Ely, C. A. (1971). Antiserum to sheep FSH on rat estrous cycle. *Program, 53rd Meet. U.S. Endocrine Soc.* p. 79.

Scruggs, W. M. (1951). The epithelial components and their seasonal changes in the pituitary gland of the carp (*Cyprinus carpio* L.) and goldfish (*Carassius auratus* L). *J. Morphol.* **88,** 441–469.

Segaloff, A. (1966). The physiology of luteinizing hormone. *In* "The Pituitary Gland" (G. W. Harris and B. T. Donovan, eds.), Vol. 1, pp. 518–526. Univ. of California Press, Berkeley.

Sétáló, G., Vigh, S., Schally, A. V., Arimura, A., and Flerkó, B. (1975). LH-RH-containing neural elements in the rat hypothalamus. *Endocrinology* **96,** 135–142.

Sharp, P. J. (1972). Tanycyte and vascular patterns in the basal hypothalamus of Coturnix quail with reference to their possible neuroendocrine significance. *Z. Zellforsch. Mikrosk. Anat.* **127,** 552–569.

Sharp, P. J., and Follett, B. K. (1970). The adrenergic supply within the avian hypothalamus. *In* "Aspects of Neuroendocrinology" (W. Bargmann and B. Scharrer, eds.), pp. 95–103. Springer-Verlag, Berlin and New York.

Shirley, H. V., and Nalbandov, A. V. (1956). Effects of transecting hypophyseal stalks in laying hens. *Endocrinology* **58,** 694–700.

Short, R. V. (1967). Reproduction. *Annu. Rev. Physiol.* **29,** 373–400.

Shute, C. C. D. (1970). Distribution of cholinesterase and cholinergic pathways. *In* "The Hypothalamus" (L. Martini, M. Motta, and F. Fraschini, eds.), pp. 167–179. Academic Press, New York.

Simpson, M. E., and van Wagenen, G. (1958). Experimental induction of ovulation in the macaque monkey. *Fertil. Steril.* **9,** 386–399.

Simpson, M. E., Li, C. H., and Evans, H. M. (1942). Biological properties of pituitary interstitial cell-stimulating hormone (ICSH). *Endocrinology* **30,** 969–976.

Simpson, M. E., van Wagenen, G., and Carter, F. (1956). Hormone content of anterior pituitary of monkey (*Macaca mulatta*) with special reference to gonadotrophins. *Proc. Soc. Exp. Biol. Med.* **91,** 6–11.

Singh, T. P. (1970). Seasonal variations in the cyanophils and the gonadotropic potency of the pituitary in relation to gonadal activity in the catfish *Mystus vittatus. Endokrinologie* **56,** 292–303.

Sinha, Y. N., and Tucker, H. A. (1969). Mammary development and pituitary prolactin level of heifers from birth through puberty and during the estrous cycle. *J. Dairy Sci.* **52,** 507–512.

Smith, C. L. (1955). Reproduction in female amphibia. *Mem. Soc. Endocrinol.* **4,** 39–56.

Smith, P. E. (1916). Experimental ablation of the hypophysis in the frog embryo. *Science* **44,** 280–282.

Smith, P. E. (1926a). Ablation and transplantation of the hypophysis in the rat. *Anat. Rec.* **32,** 221.

Smith, P. E. (1926b). Hastening development of female genital system by daily homoplastic pituitary transplants. *Proc. Soc. Exp. Biol. Med.* **24,** 131–132.

Snook, R. B., Saatman, R. R., and Hansel, W. (1971). Serum progesterone and luteinizing hormone levels during the bovine estrous cycle. *Endocrinology* **88,** 678–686.

Sonenberg, M., Money, W. L., Keston, A. S., Fitzgerald, P. J., and Godwin, J. T.

(1951). Localization of radioactivity after administration of labeled prolactin preparations to the female rat. *Endocrinology* **49**, 709–719.

Spies, H. G., and Niswender, G. D. (1971). Levels of prolactin, LH and FSH in the serum of intact and pelvic-neurectomized rats. *Endocrinology* **88**, 937–943.

Stahl, A., Seite, R., and Leray, C. (1960). Cytologie adénohypophysaire en fonction du cycle sexuel chez le poissons. L'hypophyse du Mugilides. *C. R. Soc. Biol.* **154**, 1455–1458.

Steelman, S. L., and Pohley, F. M. (1953). Assay of the follicle stimulating hormones based on the augmentation with human chorionic gonadotrophin. *Endocrinology* **53**, 604–616.

St. Girons, H. (1967). Morphologie comparée de l'hypophyse chez les squamata. *Ann. Sci. Nat., Zool. Biol. Anim.* [12] **9**, 229–308.

Stockell Hartree, A., and Cunningham, F. J. (1969). Purification of chicken pituitary follicle-stimulating hormone and luteinizing hormone. *J. Endocrinol.* **43**, 609–616.

Subtelny, S., Smith, L. D., and Ecker, R. E. (1968). Maturation of ovarian frog eggs without ovulation. *J. Exp. Zool.* **168**, 39–48.

Sundararaj, B. I., Anand Kumar, T. C., and Donaldson, E. M. (1972). Effects of partially purified salmon pituitary gonadotropin on ovarian maintenance, ovulation and vitellogenesis in the hypophysectomized catfish, *Heteropneustes fossilis* (Bloch). *Gen. Comp. Endocrinol.* **18**, 102–114.

Swanson, L. V., and Hafs, H. D. (1970). Luteinizing hormone and prolactin in blood serum throughout estrus in heifers. *J. Dairy Sci.* **53**, 652–653.

Swanson, L. V., Hafs, H. D., and Morrow, D. A. (1970). Blood prolactin and LH during puberty in heifers. *J. Anim. Sci.* **31**, 232.

Swerdloff, R. S., and Jacobs, H. S. (1971). Positive and negative feedback control of gonadotrophin secretion during puberty in female rats. *Program, 53rd Meet. U.S. Endocrine Soc.* p. 77.

Swift, D. R., and Pickford, G. E. (1965). Seasonal variations in the hormone content of the pituitary gland of the perch. *Perca fluviatilis* L. *Gen. Comp. Endocrinol.* **5**, 354–365.

Szego, C. M., and Gitin, E. S. (1964). Ovarian histamine depletion during acute hyperaemic response to luteinizing hormone. *Nature (London)* **201**, 682–684.

Szentágothai, J., Flerkó, B., Mess, B., and Halász, B., (1968). "Hypothalamic Control of the Anterior Pituitary," 3rd ed. Akadémiai Kiadó, Budapest.

Szontágh, F., Uhlarik, S., and Jakobovits, A. (1963). A nöstény patkány adenohypophysisének ICSH tartalma serumgonadotropin PMS kezelés után. *Kiserl. Orvostud.* **15**, 526–529.

Taber, E. (1948). The relation between ovarian growth and sexual characters in brown Leghorn chicks treated with gonadotrophins. *J. Exp. Zool.* **107**, 65–108.

Taber, E., Clayton, M., Knight, J., Gambrell, D., Flowers, J., and Ayres, C. (1958). Ovarian stimulation in the immature fowl by desiccated avian pituitaries, *Endocrinology* **62**, 84–89.

Talwalker, P. K., Ratner, A., and Meites, J. (1963). *In vitro* inhibition of pituitary prolactin synthesis and release by hypothalamic extract. *Am. J. Physiol.* **205**, 213–218.

Tanaka, K., and Yoshioka, S. (1967). Luteinizing hormone activity of the hen's pituitary during the egg-laying cycle. *Gen. comp. Endocrinol.* **9**, 374–379.

Tanaka, K., Kamiyoshi, M., and Tagami, M. (1969). *In vitro* demonstration of luteinizing hormone releasing activity in the hypothalamus of the hen. *Poult. Sci.* **48**, 1985-1987.

Tejasen, T., and Everett, J. W. (1967). Surgical analysis of the preopticotuberal pathway controlling ovulatory release of gonadotropins in the rat. *Endocrinology* **81**, 1387-1396.

Te Winkel, L. E. (1969). Specialized basophilic cells in the ventral lobe of the pituitary of the smooth dogfish, *Mustelus canis. J. Morphol.* **127**, 439-452.

Thibault, C., and Mauléon, P. (1964). Quelques problèmes neuroendocriniens intéressants la reproduction des mammifères domestiques. *Proc. Int. Congr. Anim. Reprod. & Artif. Insem., 5th, 1963,* Vol. 7, pp. 427-470.

Thomas, K., Walckiers, R., and Ferin, J. (1970). Biphasic pattern of luteinizing hormone midcycle discharge. *J. Clin. Endocrinol. Metab.* **30**, 269-272.

Thompson, K. W. (1941). Antihormones. *Physiol. Rev.* **21**, 588-631.

Thomson, A. P. D., and Zuckerman, S. (1953). Functional relations of the adenohypophysis and hypothalamus. *Nature (London)* **171**, 970.

Thomson, A. P. D., and Zuckerman, S. (1954). The effect of pituitary-stalk section on light-induced oestrus in ferrets. *Proc. R. Soc. London, Ser. B*, **142**, 437-451.

Thorner, M. O., Besser, G. M., Jones, A. E., Dacie, J., and Jones, A. E. (1975). Bromocriptine treatment of female infertility: report of 13 pregnancies. *Br. Med. J.* **4**, 694-697.

Thornton, V. F., and Evennett, P. J. (1969). Endocrine control of oocyte maturation and oviducal jelly release in the toad *Bufo bufo* (L). *Gen. Comp. Endocrinol.* **13**, 268-274.

Thornton, V. F., and Geschwind, I. I. (1974). Hypothalamic control of gonadotrophin release in Amphibia: evidence from studies of gonadotrophin release *in vitro* and *in vivo. Gen. Comp. Endocrinol.* **23**, 294-301.

Tindal, J. S., Knaggs, G. S., and Turvey, A. (1967). Central nervous control of prolactin secretion in the rabbit: effect of local oestrogen implants in the amygdaloid complex. *J. Endocrinol.* **37**, 279-287.

Tixier-Vidal, A., and Follett, B. K. (1973). The adenohypophysis. *In* "Avian Physiology" (D. S. Farner and J. R. King, eds.), Vol. 3, pp. 110-182. Academic Press, New York.

van Dongen, W. J., Barker Jørgensen, C., Larsen, L. O., Rosenkilde, P., Lofts, B., and van Oordt, P. G. W. J. (1966). Function and cytology of the normal and autotransplanted pars distalis of the hypophysis in the toad *Bufo bufo* (L). *Gen. Comp. Endocr.* **16**, 491-518.

van Oordt, P. G. W. J. (1968). The analysis and identification of the hormone-producing cells of the adenohypophysis. *In* "Perspectives in Endocrinology" (E. J. W. Barrington and C. Barker Jørgensen, eds.), pp. 405-468. Academic Press, New York.

van Rees, G. P., van Dieten, J. A. M. J., Bijleveld, E., and Muller, E. R. A. (1968). Induction of ovulation during pseudopregnancy in the rat. *Neuroendocrinology* **3**, 220-228.

van Tienhoven, A., and Schally, A. V. (1972). Mammalian luteinizing hormone-releasing hormone induces ovulation in the domestic fowl. *Gen. Comp. Endocrinol.* **19**, 594-595.

van Tienhoven, A., Nalbandov, A. V., and Norton, H. W. (1954). Effect of dibenamine on progesterone-induced and "spontaneous" ovulation in the hen. *Endocrinology* **54**, 605-611.

van Wagenen, G., and Simpson, M. E. (1965). "Embryology of the Ovary and Testis—*Homo sapiens* and *Macaca mulatta*." Yale Univ. Press, New Haven, Connecticut.

Vaugien, L. (1957). La réaction ovarïenne provoquée chez le serin des jardins par la gonadotrophine équine est liée à l'état initial de la gonade. *C. R. Acad. Sci. (Paris), Series D* **245**, 1268–1271.

Velasco, M. E., and Taleisnik, S. (1969a). Release of gonadotropins induced by amygdaloid stimulation in the rat. *Endocrinology* **84**, 132–139.

Velasco, M. E., and Taleisnik, S. (1969b). Effect of hippocampal stimulation on the release of gonadotropin. *Endocrinology* **85**, 1154–1159.

Vellano, C., Bona, A., Mazzi, V., and Collucci, D. (1974). The effect of synthetic luteinizing hormone releasing hormone on ovulation in the crested newt. *Gen. Comp. Endocrinol.* **24**, 338–340.

Vijayakumar, S., Jørgenson, C. B., and Kjaer, K. (1971). Regulation of ovarian cycle in the toad *Bufo bufo bufo* (L.): effects of auto grafting pars distalis of the hypophysis, of extirpating gonadotropic hypothalamic region and of partial ovariectomy. *Gen. Comp. Endocrinol.* **17**, 432–443.

Vincent, D. S., and Anand Kumar, T. C. (1969). Electron microscopic studies on the pars intermedia of the ferret. *Z. Zellforsch. Mikrosk. Anat.* **99**, 185–197.

Vivien, J. H. (1939). Rôle de l'hypophyse dans le déterminisme du cycle genitale femelle d'un téléostéen, *Gobius paganellus* L. *C. R. Acad. Sci. (Paris), Ser. D* **208**, 948–949.

Vivien, J. H. (1941). Contribution à l'étude de la physiologie hypophysaire dans ses relations avec l'appareil génetal la thyroide et les corps suprarénaux chez les poissons sélaciens et téléostéens. *Bull. Biol. Fr. Belg.* **75**, 257–309.

Voogt, J. L., and Meites, J. (1971). Effects of an implant of prolactin in median eminence of pseudopregnant rats on serum and pituitary LH, FSH and prolactin. *Endocrinology* **88**, 286–292.

Voogt, J. L., Chen, C. L., and Meites, J. (1970). Serum and pituitary prolactin levels before, during and after puberty in female rats. *Am. J. Physiol.* **218**, 396–399.

Watanabe, S., and McCann, S. M. (1968). Localization of FSH releasing factor in the hypothalamus and neurohypophysis as determined by an *in vitro* assay. *Endocrinology* **82**, 664–673.

Watanabe, S., and McCann, S. M. (1969). Alterations in pituitary follicle stimulating hormone (FSH) and hypothalamic FSH releasing factor (FSH-RF) during puberty. *Proc. Soc. Exp. Biol. Med.* **132**, 195–201.

Weiner, R. I., Pattou, E., Kerdelhué, B., and Kordon, C. (1975). Differential effects of hypothalamic deafferentation upon luteinizing hormone releasing hormone in the median eminence and organum vasculosum of the lamina terminalis. *Endocrinology* **97**, 1597–1600.

Weisz, J., and Ferin, M. (1970). Pituitary gonadotrophins and circulating LH in immature rats. A comparison between normal females and males and females treated with testosterone in neonatal life. *In* "Gonadotrophins and Ovarian Development" (W. R. Butt, A. C. Crooke, and M. Ryle, eds.), pp. 339–350. Livingstone, Edinburgh.

Weitzman, E. D. (1974). Temporal organization of neuro-endocrine function in relation to the sleep-waking cycle in man. *In* "Recent Studies of Hypothalamic Function" (K. Lederis and K. E. Cooper, eds.), pp. 26–38. Karger, Basel.

White, W. F. (1970). On the identity of the LH- and FSH-releasing hormones. *In* "Mammalian Reproduction" (H. Gibian and E. J. Plotz, eds.), pp. 84–87. Springer-Verlag, Berlin and New York.

Wilhelmi, A. E., Pickford, G. E., and Sawyer, W. H. (1955). Initiation of the spawning reflex response in *Fundulus* by the administration of fish and mammalian neurohypophysial preparations and synthetic oxytocin. *Endocrinology* **57**, 243–252.

Williams, P. C. (1944). Ovarian stimulation by oestrogens: effects in immature hypophysectomized rats. *Proc. R. Soc. London, Ser. B*, **132**, 189–199.

Wilson, S. C., and Sharp, P. J. (1975). Effects of progesterone and synthetic luteinizing hormone releasing hormone on the release of luteinizing hormone during sexual maturation in the hen (*Gallus domesticus*). *J. Endocrinol.* **67**, 456–469.

Witschi, E. (1955). Vertebrate gonadotrophins. *Mem. Soc. Endocrinol.* **4**, 149–165.

Witschi, E., and Riley, G. M. (1940). Quantitative studies on the hormones of human pituitaries. *Endocrinology* **26**, 565–576.

Woods, M. C., and Simpson, M. E. (1961). Characterization of the anterior pituitary factor which antagonizes gonadotrophins. *Endocrinology* **68**, 647–661.

Woronzowa, M. A., and Blacher, L. J. (1930). Die Hypophyse und die Geschlecttsdrüsen der Amphibien. I. Der Einfluss der Hypophysenextirpation auf die Geschlechtsdrüse bei Urodelen. *Arch. Entwicklungs. Mech. Org.* **121**, 327–344.

Wright, P. A. (1961). Induction of ovulation *in vitro* in *Rana pipiens* with steroids. *Gen. Comp. Endocrinol.* **1**, 20–23.

Wurtman, R. J. (1964). An effect of LH on the fractional perfusion of the rat ovary. *Endocrinology* **75**, 927–933.

Yamamoto, K., and Yamazaki, F. (1967). Hormonal control of ovulation and spermiation in goldfish. *Gunma Symp. Endocrinol.* **4**, 131–145.

Yamazaki, F. (1965). Endocrinological studies on the reproduction of the female goldfish, *Carassius auratus* L with special reference to the function of the pituitary gland. *Mem. Fac. Fish., Hokkaido Univ.* **13**, 1–64.

Yamazaki, F., and Donaldson, E. M. (1968). The effects of partially purified salmon pituitary gonadotropin on spermatogenesis, vitellogenesis and ovulation in hypophysectomized goldfish (*Carassius auratus*). *Gen. Comp. Endocrinol.* **11**, 292–299.

Yang, K. P., Samaan, N. A., and Ward, D. N. (1976). Characterization of an inhibitor for luteinizing hormone receptor site binding. *Endocrinology* **98**, 233–241.

Yen, S. S. C., Llerena, O., Little, B., and Pearson, O. H. (1968). Disappearance rates of endogenous luteinizing hormone and chorionic gonadotropin in man. *J. Clin. Endocrinol. Metab.* **28**, 1763–1767.

Yen, S. S. C., Lasley, B. L., Wang, C. F., Leblanc, H., and Siler, T. M. (1975). The operating characteristics of the hypothalamic-pituitary system during the menstrual cycle and observations of biological action of somatostatin. *Recent Prog. Horm. Res.* **31**, 321–398.

Ying, S.-Y., and Greep, R. O. (1971). Effect of age of rat and dose of a single injection of estradiol benzoate (EB) on ovulation and the facilitiation of ovulation by progesterone (P). *Program, 53rd Meet. U.S. Endocrine Soc.* p. 120.

Young, W. C. (1961). The mammalian ovary. *In* "Sex and Internal Secretions" (W. C. Young, ed.), 3rd ed., Vol. 1, pp. 449–496. Williams & Wilkins, Baltimore, Maryland.

Zachariae, F. (1958). Studies on the mechanism of ovulation. Permeability of the blood-liquor barrier. *Acta Endocrinol.* (*Copenhagen*) **27**, 339–342.

Zanartu, J., Dabancens, A., Kastin, J. A., and Schally, A. V. (1974). Effect of synthetic hypothalamic gonadotrophin-releasing hormone (FSH/LH-RH) in anovulatory sterility. *Fertil. Steril.* **25**, 160–169.

Zarrow, M. X., and Bastian, J. W. (1953). Blockade of ovulation in the hen with adrenolytic and parasympatholytic drugs. *Proc. Soc. Exp. Biol. Med.* **84**, 457–459.

Zarrow, M. X., and Dinius, J. (1971). Regulation of pituitary ovulating hormone concentration in the immature rat treated with pregnant mare serum. *J. Endocrinol.* **49**, 387–392.

Zarrow, M. X., and Quinn, D. L. (1963). Superovulation in the immature rat following treatment with PMS alone and inhibition of PMS induced ovulation. *J. Endocrinol.* **26**, 181–188.

Zarrow, M. X., and Wilson, E. D. (1961). The influence of age on superovulation in the immature rat and mouse. *Endocrinology* **69**, 851–855.

Zeleznik, A. J., Midgley, A. R., Jr., and Reichert, L. E. Jr., (1974). Granulosa cell maturation in the rat: increased binding of human chorionic gonadotropin following treatment with follicle stimulating hormone *in vivo. Endocrinology* **96**, 818–825.

Zondek, B., and Aschheim, S. (1927). Das Hormon des Hypophysenvorderlappens. 1. Testobjekt zum Nachweis des Hormons. *Klin. Wochenschr.* **6**, 248–252.

Zuber-Vogeli, M. (1968). Les variations cytologiques de l'hypophyse distale des femelles de *Nectophrynoides occidentalis. Gen. Comp. Endocrinol.* **11**, 495–514.

Zuckerman, S. (1954). The secretions of the brain: relations of hypothalamus to pituitary gland. *Lancet* **1**, 789–795.

Zuckerman, S. (1955). The possible functional significance of the pituitary portal vessels. *Ciba Found. Colloq. Endocrinol.* [*Proc.*] **8**, 551–586.

Zuckerman, S. (1970). The new secretions of the brain. *In* "Beyond the Ivory Tower: The Frontiers of Public and Private Science" pp. 46–60. Weidenfeld & Nicolson, London.

5

Synthesis and Secretion of Steroid Hormones by the Ovary in Vivo

D. T. Baird

I. INTRODUCTION

The study of the factors determining the synthesis and secretion of steroids by the ovary *in vivo* is complicated by the fact that its different cellular elements are continually changing so that numerous serial measurements are required throughout at least one complete ovarian cycle. Within any one ovary the different structures, follicle, corpus luteum, and interstitial or stromal cells, secrete their characteristic steroid hormones.

305

Production of steroids by the individual cell types in the ovary can be studied only indirectly *in vivo,* e.g., by comparing the steroid secretion from an ovary containing a corpus luteum with the secretion from the contralateral ovary which has no corpus luteum, or by correlating levels of circulating hormones with development of structures within the ovary. *In vitro* techniques, such as incubation (Ryan and Smith, 1965) or tissue culture (Channing, 1969), are necessary for the study of biosynthesis by isolated cell types.

Early endocrinologists investigated the secretion of ovarian steroids indirectly by observing the effects of castration and replacement therapy on various biological signs, such as vaginal cornification, estrus, and the maintenance of pregnancy. The extraction and chemical identification of estrone (Doisy *et al.,* 1929; Butenandt, 1929) was the first step leading to specific chemical methods, the results of which confirmed those of earlier bioassays.

The pattern of steroids synthesized and secreted by the ovary is controlled by gonadotropins released from the anterior pituitary. The way in which ovarian steroids influence the release of FSH, LH, and prolactin by their feedback at the hypothalamus and pituitary is discussed in Volumes III, Chapters 2 and 4. In some mammals the uterus plays a major role in controlling ovarian function. In the ewe, for example, prostaglandin $F_{2\alpha}$ of uterine origin reaches the ovary by some form of local diffusion involving the utero-ovarian vein and ovarian artery, and is responsible for terminating the life span of the corpus luteum of the cycle (McCracken *et al.,* 1972; Pharriss *et al.,* 1972).

Some actions of gonadotropins on the ovarian cells are better defined than others; e.g., FSH stimulates the growth and development of Graafian follicles by influencing the rate of oxygen uptake and metabolism (Ahrén *et al.,* 1969), and LH stimulates steroid synthesis and secretion by all cell types in the ovary, probably by increasing the conversion of cholesterol to pregnenolone (Savard *et al.,* 1965). Under the influence of FSH there is an increase in the number of the specific receptor sites for LH on the surface of the granulosa cell within the Graafian follicle (Channing and Kammerman, 1973). Although there are superficial similarities between the action of ACTH on the adrenal and that of LH on the ovary, the physiological function and control of the two glands is very different. It would not be advantageous to the animal if the secretion of progesterone by the corpus luteum was as sensitive to changes in concentration of LH as is the secretion of cortisol to minor alterations in ACTH. The secretion of progesterone by the corpus luteum *in vivo* is, in fact, altered very little, if at all, by marked increases in the concentration of LH above a certain minimal level (Short *et al.,* 1963; Collett *et al.,* 1973; Land *et al.,* 1974).

It must be assumed that the only way gonadotropins can reach the ovary is via the arterial blood. The ovarian blood flow per unit weight is high relative to other organs (Goding *et al.,* 1971; Reynolds, 1973). In the rabbit, the ovarian vessels appear to be without autoregulatory control and hence flow is almost totally related to perfusion pressure (Janson and Albrecht, 1975). LH stimulates a marked increase in ovarian blood flow, a property which it shares with many other tropic hormones (Wurtman, 1964). Very little is known about the factors controlling blood flow to the ovary, although the growth of follicles and the formation of a corpus luteum is dependent on the development of an adequate blood supply (see Reynolds, 1973, for review).

Although there is much information about the biochemistry of synthesis of steroids, very little is known about the mechanism of secretion by ovarian cells *in vivo* (Eik-Nes and Hall, 1965). The enzymes responsible for the synthesis of steroids are situated in the cytoplasm and the mitochondria with the intermediates being transported back and forth for specific enzymatic steps (see Volume 3, Chapter 6). The turnover of synthesized steroids in the corpus luteum, at least, appears to be very rapid (Short, 1964). In contrast to the secretion of protein hormones, there is very little steroid hormone stored in the ovary and any change in secretion usually represents increased synthesis (Aakvaag and Eik-Nes, 1969; Rado *et al.,* 1970). Preliminary reports suggest that progesterone may be stored and secreted in the lutein cell in the form of granules (Gemmell *et al.,* 1974).

Since the first edition of this book, much quantitative information about the secretion of steroids from the ovary has become available. This has been due to: (1) the development of microanalytical techniques sensitive enough to measure the concentration of steroids present in urine, blood, and lymph (see Gray and Bacharach, 1967; Diczfalusy, 1970); (2) the development of experimental models for perfusion of the ovary either *in situ* (Aakvaag and Eik-Nes, 1969) or autotransplanted to a more accessible site (Goding *et al.,* 1967); and (3) the application of techniques of isotope dilution to the study of secretion, production, and metabolism of ovarian steroid hormones, made possible by the availability of radioactive isotopes of steroids of high specific activity.

It has been demonstrated that the ovary secretes a variety of steroids in addition to the major sex steroids, estradiol-17β and progesterone, although the precise biological significance of many of these other substances is not yet understood. This chapter will be confined, where possible, to studies in which a quantitative assessment of the secretion of ovarian hormones has been made by using specific chemical methods.

The secretion of a hormone from an endocrine gland is demonstrated by a difference in concentration between the afferent arterial and the efferent venous blood, and the sensitivity of the techniques used limits the determination of secretory activity. Microanalytical techniques, particularly those of radioimmunoassay and protein binding, have permitted analysis of steroids in small volumes of peripheral blood and sequential measurements in the same animal have become possible.

II. METHODS OF STUDYING SECRETION OF STEROIDS *IN VIVO*

A. Direct Approach

The secretion rate of a hormone from an endocrine gland can be defined as the net release of hormone per unit time. It is usually calculated from the product of the blood flow through the gland (O.B.F.) and the difference in concentration of hormone in afferent arterial blood (i^A) and efferent venous blood (i^V):

$$S = \text{O.B.F.} \times (i^V - i^A) \tag{1}$$

Hormones from the ovary are secreted not only into the venous effluent blood but also into follicular fluid (Short, 1964) and the ovarian lymphatics (Lindner *et al.*, 1964). Drainage by the lymphatics may be considerable, but these other routes of secretion are usually ignored and Eq. (1) is used as a basis for the calculation of secretion rates of ovarian hormones. With the analytical techniques now available, it is possible to measure i^A and i^V, although ovarian venous blood can usually only be collected at operation when the animal may be stressed by anesthesia and surgical manipulation. Furthermore, in many animals, for example, women and pigs, venous drainage from the ovary is not by a single vein but by a plexus of interconnecting veins making representative sampling extremely difficult (Reynolds, 1973). Therefore, although studies involving collection of venous effluent blood from the ovary *in situ* have been used extensively to obtain qualitative information about the secretion of ovarian steroids, quantitative measurements are much more difficult (Baird, 1970).

Because of the difficulty of access to the ovary, attempts have been made to perfuse the ovary *in vivo* and *in vitro*. Although the latter techniques are not strictly within the scope of this chapter, there are gradations from the *in vivo* situation to a complete *in vitro* perfusion (Ahrén *et al.*, 1971). Perfusion systems *in vivo* have the advantage of being more "physiological" than equivalent preparations *in vitro,* but the disadvantage of having more uncontrolled variables. Ahrén *et al.* (1971) classified *in vivo* perfusions into Type I: those involving cannulation of the ovarian vein(s) and repeated or continuous collection of ovarian venous blood; Type II: experiments involving collection of ovarian venous blood and cannulation of the ovarian artery with the circulation intact; Type III: experiments in which the ovarian artery has been ligated and the ovary is perfused exclusively by a perfusion pump. Type III experiments exclude blood-borne endogenous hormones but the ovary is still exposed to nervous factors and probably also to hormones reaching it by lymphatics and small vascular anastomoses. For technical reasons, *in vivo* perfusions of Type I have been the most widely used. Type II perfusions have been carried out in sheep by cannulating the tubal branch of the ovarian artery and by retrograde perfusion with gonadotropins (Domanski *et al.,* 1967). This technique has also been used in women (Douss and Desphpande, 1968; Oertel *et al.,* 1968) and dogs (De Paoli and Eik-Nes, 1963), in which labeled precursors have been injected into the ovarian artery *in situ.* Type III perfusions have been extensively used in the rabbit by Hilliard *et al.* (1969); the ovary *in situ* is perfused with donor blood through an infusion pump and a cannula in the ovarian artery. A semi *in vivo*:semi *in vitro* system in the dog involves removal of the ovary and then perfusion via a cannula in the ovarian artery with the dog's own blood from the femoral artery (Engels *et al.,* 1968).

All these *in vivo* perfusion systems have the disadvantage that the experiments must be performed on anesthetized animals. It is possible that the anesthetic agents and/or the stress of surgery may disturb the function of the ovary. Attempts have been made, with varied success, to measure secretion rates directly in the conscious animal. Lindner *et al.* (1964) cannulated the ovarian vein and major lymphatic vessels in the ewe, and were able to sample ovarian venous blood in a small number of conscious ewes for several days after surgery by exteriorizing the cannulae through the flank. This technique has not been widely applied, however, because clotting occurred in the venous cannula. The cannulation of the ovarian lymph vessels has been more successful (Lindner *et al.,* 1964). In seven conscious ewes, ovarian lymph was collected over periods varying from 3 to 8 days, the rates of flow (1.5–8.6 ml/hour/ovary) being higher per unit weight than from any other organ studied. A similar technique of chronic catheterization of the utero-ovarian vein has been used to study the secretion of ovarian steroids and uterine prostaglandins in the sheep (Thorburn *et al.,* 1972, 1973), cow (Nancarrow *et al.,* 1973), and pig (Gleeson and Thorburn, 1973). These preparations usually last only 7–14 days when infection and/or clotting of the veins makes further collections impossible.

Autotransplantation of organs to a superficial site in the body in order to gain access to their blood supply was first described by Lockett *et al.* (1942) for the dog kidney. The mammary gland of the goat (Linzell, 1957) and the thyroid gland of the dog (Falconer, 1963) have been transplanted to increase the accessibility of their blood supply. A refinement of this technique was the autotransplantation of the adrenal gland of the sheep, together with its blood supply, to a skin loop in the neck (McDonald *et al.*, 1958; Goding *et al.*, 1971). A similar transplant of the ovary in sheep (Goding *et al.*, 1967) has been widely used (see Baird *et al.*, 1973a, for review). The left ovary, with its vascular pedicle, is transplanted to a skin loop in the neck which contains the carotid artery and jugular vein (McCracken and Baird, 1969; McCracken *et al.*, 1971) (Fig. 1A). The ovarian circulation is maintained by

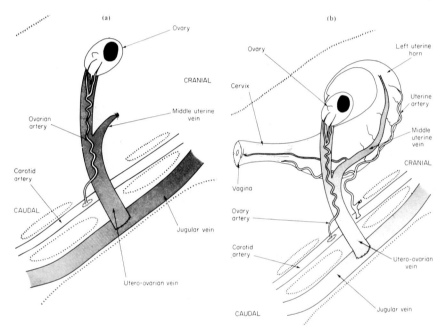

Fig. 1. Diagram to show the technique of autotransplantation of (A) the ovary and (B) the ovary and uterus of the sheep.

(A) The left ovary with the ovarian vessels attached has been relocated in a skin pouch in the neck. The ovarian artery has been anastomozed to the carotid artery and the utero-ovarian vein to the jugular vein. When the jugular vein is occluded cranial and caudal to the site of anastomosis, the total ovarian venous drainage enters the jugular vein between the occlusion points and it can there be collected through a catheter. When the carotid artery is occluded cranial to the origin of the ovarian artery, the carotid artery supplies the ovary exclusively. In this way, substances can be perfused through the ovary via a cannula in the carotid artery.

(B) The left ovary, Fallopian tube, left uterine horn, body, cervix, and vagina have been relocated in a skin pouch in the neck. The right ovary and uterine horn have been excised and discarded. The arterial supply is maintained by anastomosis of the ovarian and uterine arteries to the carotid artery. The utero-ovarian vein, draining uterus and ovary, is anastomosed to the jugular vein. The vagina is anastomosed to the skin of the neck to form a short vaginal pocket for the cervix. Utero-ovarian venous samples can be collected via a catheter in the jugular vein as described for (A). These ewes continue to show cyclic ovarian activity due to preservation of the contiguity of the uterus and ovary.

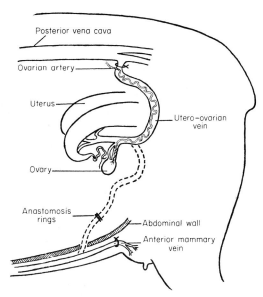

Fig. 2.   Diagram of the anastomosis technique of Thorburn and Mattner (1971) of the utero-ovarian vein to the anterior mammary vein in the sheep.

anastomosis of the ovarian artery and utero-ovarian vein to the carotid artery and jugular vein, respectively. The right ovary is removed so that all the ovarian secretions enter the jugular vein in the skin loop. Blood can be withdrawn repeatedly from the ovarian artery and vein in the conscious unstressed animal by percutaneous cannulation of the carotid artery and jugular vein. Transplantation of the ovary alone results in prolongation of the life span of the corpus luteum (Baird *et al.*, 1968a), making this preparation unsuitable for the study of cyclic ovarian function. However, the ovary functions normally if the ovary and the uterus are transplanted to the neck (Harrison *et al.*, 1968; Baird *et al.*, 1976a) (Fig. 1B). In this preparation, the uterine and ovarian arteries are anastomosed to the carotid artery and the combined venous drainage of uterus and ovary is maintained by the utero-ovarian vein. Normal cyclic function of the ovary then occurs and the secretion of steroids from the ovary throughout the estrous cycle is measured (McCracken *et al.*, 1969; Baird *et al.*, 1976b).

Thorburn and Mattner (1969) anastomosed the utero-ovarian vein to the mammary vein in the ewe (Fig. 2), and samples of utero-ovarian blood were collected throughout the estrous cycle (Thorburn and Mattner, 1971). This preparation has the advantage that the ovary and uterus remain *in situ,* but it is not clear how retrograde drainage of ovarian venous blood is prevented. Measurements of ovarian blood flow may therefore be unreliable.

Jacobs (1969) has described catheterization of the adrenal and ovarian veins in conscious women. The catheterizations are performed percutaneously via the femoral vein using a Müller-guided catheter system under fluoroscopy with an image intensifier. The catheters are wedged into each adrenal and ovarian vein in turn; the entire procedure takes about 60 minutes. The source of excess androgen in hirsute women has been investigated by this method (Kirschner and Jacobs, 1971). A similar technique has been described for adrenal vein catheterization (Kahn, 1967).

The collection of follicular fluid from unanesthetized mares has been described by Short (1961), who punctured the follicular cavity through the wall of the rectum. Although this enables the concentration of steroids in follicular fluid to be measured, there appears to be very poor correlation between follicular fluid content and the secretion of steroids into ovarian venous blood (Aakvaag and Fylling, 1968).

B. Indirect Approach

It is possible to study the secretion of ovarian hormones indirectly by measuring the concentration of the secreted steroid or its metabolite in peripheral blood and/or in urine. For example, by correlating the excretion of estriol conjugates in the urine with the stage of the menstrual cycle, it was suggested that both the developing follicle and the corpus luteum secrete an estrogen which is metabolized to estriol (Brown, 1955). However, estrogen is also secreted by the adrenal glands (Baird et al., 1969a) and it is possible that the excreted estriol is not derived exclusively from the ovary. The type of estrogen secreted cannot be identified because estrone, estradiol-17β, and estriol are all metabolized in the liver and kidneys and excreted as estriol conjugates (Brown and Matthew, 1962). The concentrations of estrone and estradiol-17β in peripheral venous blood are a more direct measure of the secreted hormones (Baird, 1968), although different rates of metabolism or clearance may distort the actual patterns of secretion (Baird, 1971). Some steroids circulating in peripheral blood (e.g., estrone) are produced by conversion in peripheral tissues of other secreted steroids (precursors) in addition to being secreted from more than one endocrine gland (Baird et al., 1969b).

In spite of all these potential sources of error, much information about the pattern of hormones secreted by the ovary has been obtained by these indirect methods because the ovary is the major source of estradiol-17β and progesterone in most species throughout reproductive life (see Section IV).

C. Isotope-Dilution Techniques

The concentration of a steroid in peripheral blood or the amount excreted in the urine is a function not only of its rate of secretion but also its total production from all sources in the body and its rate of metabolism. Differences in the concentration of a steroid hormone in peripheral blood, or its metabolite in urine, may represent differences in metabolism rather than differences in rates of secretion. Isotope-dilution techniques have been applied to problems of the metabolism of steroids (Vande Wiele et al., 1963; Tait, 1963), and complete accounts of the application of "steroid dynamics" to ovarian steroids may be found in several comprehensive reviews (Vande Wiele et al., 1963; Lipsett et al., 1966; Baird et al., 1969b). A tracer amount of the steroid, labeled with a radioactive isotope, is injected into the same compartment (usually blood) as the endogenous hormone, and its fate in blood and/or urine is followed by measuring the dilution of the injected tracer with endogenous hormone. In the "urinary" method, urine is collected after injection of the tracer until no further radioactive substance is excreted (usually about 4 days). The production of hormone can then be calculated from the cumulative specific activity of the metabolite of the hormone in urine.

If the production rate of hormone measured in urine is P_u and the total amount of radioactive material injected is R, the specific activity of hormone excreted in time $t = \mathrm{dpm}/\mu\mathrm{g} \times t$ and $P_u = R/(\mathrm{dpm}/\mu\mathrm{g} \times t)$.

The production rate measured in urine will represent secretion only if the hormone (e.g., cortisol) arises solely from glandular secretion and the urinary metabolite chosen for the measurement of specific activity is unique for that hormone (e.g., tetrahydrocortisol). If there is significant extraglandular production, e.g., testosterone in the human female, the total urinary production rate will represent the sum of glandular secretion and the total irreversible production of the steroid from precursors in organs such as the liver outside the endocrine glands. When there is significant production of this steroid in a compartment which is not in equilibrium with blood, the production rate as measured in urine may give a misleading impression of the amount of steroid which is biologically active (Korenman and Lipsett, 1965). Diurnal variations in excretion are overcome by collecting urine for several days, but the method is strictly applicable only when there is no significant alteration in secretion rate or metabolism during collection. The method is therefore of limited use in study of the ovarian secretions which change from day to day. The metabolism of the steroid (e.g., cortisol) may also vary with the time of injection of the tracer, and the calculation of a true secretion rate is then impossible (Kelly, 1970; Fukushima et al., 1970). If this were found to apply to other steroids, it would cast doubt on the urinary method of determining secretion rates.

The so called "blood" method involves the determination of the concentration of tracer (dpm/liter plasma), or the steroid metabolite, in samples of blood withdrawn at various times after the single injection of the radioactive steroid. The plot of concentration against time can usually be expressed as one or more exponential curves which can be analyzed either graphically by log conversion or by computer analysis. The rate of disappearance of steroid can be described by integrating the area under the disappearance curve. The metabolic clearance rate (MCR), introduced by Tait (1963), is defined as the volume of blood completely and irreversibly cleared of steroid per unit time. Thus, for a given steroid concentration (i), in steady state conditions, the total amount of steroid removed from the circulation will be the product of the MCR and (i). If the plasma concentration is constant, the amount of steroid removed from the circulation must be balanced by the amount entering, i.e., the production rate (P_B), and P_B = $MCR \times i$, or $MCR = P_B/i$. The MCR can be determined experimentally by injection of a radioactively labeled isotope (Tait and Burstein, 1964):

$$MCR = R/x_1 \times dt$$

where R = the amount of radioactivity injected, and x_1 = the concentration of labeled steroid.

One difficulty with the single injection technique is that assumptions must be made about the number of exponentials required to describe the curve. If the labeled steroid is infused at a constant rate until equilibrium is established, then integration is automatic (Tait, 1963), and $MCR = R/x_1$, where R = the rate of infusion, and x_1 = the concentration of radioactive steroid at the steady state. The constant infusion technique has been applied successfully to the study of ovarian secretions (Baird *et al.*, 1969b). However, the values only apply to the specific time when the infusion is performed, and can be extrapolated to 24 hours only if there is no diurnal change in secretion rate.

In women, about 60% of the testosterone in the blood is derived from the extraglandular conversion of androstenedione, which is secreted by the ovaries and adrenal. The secretion rate of a steroid produced directly and by extraglandular conversion is determined by use of a "fractional" conversion factor (ρ). This factor can be obtained from the ratio of product to precursor at the steady state after constant infusion of precursor. Then the sum of the contributions from all possible precursors is subtracted from the total production rate in the blood to give the steroid secretion rate (Baird *et al.*, 1969b):

$$S^{\text{PRO}} = [P_B{}^{\text{PRO}} - P_B{}^{\text{PRE}} \times (\rho)_{\text{BB}}{}^{\text{PRE-PRO}}]/[1 - (\rho)_{\text{BB}}{}^{\text{PRE-PRO}} \times (\rho)_{\text{BB}}{}^{\text{PRO-PRE}}]$$

where $1 - (\rho)_{\text{BB}}{}^{\text{PRE-PRO}} \times (\rho)_{\text{BB}}{}^{\text{PRO-PRE}}$ is a factor used to correct for back conversion of product to precursor. In practice, this factor is negligible for all steroids which have been studied and the secretion rate can be derived from the numerator:

$$S^{\text{PRO}} = P_B{}^{\text{PRO}} - P_B{}^{\text{PRE}} \times (\rho)_{\text{BB}}{}^{\text{PRE-PRO}}$$

It is clear that the rate of secretion from the endocrine gland(s) can be estimated by this method only if contributions from all possible precursors have been determined.

As a check for precursor conversions and to determine the endocrine gland responsible for the secretion, indirect measurements should be confirmed by direct sampling (Eik-Nes and Hall, 1965). This has been done for cortisol secretion in the sheep by comparison of the blood production rate, derived from isotope-dilution studies, with the secretion rate measured simultaneously from the autotransplanted adrenal gland (Paterson and Harrison, 1967).

It is difficult to determine the blood flow from endocrine glands *in situ* (see Section II,A), but a reasonable estimate of glandular secretion can be obtained from a combination of isotope-dilution techniques and direct sampling (Baird, 1970). For a steroid such as cortisol, which enters the blood solely as a secretion from the adrenal (Luetscher and Cheville, 1968), the blood production rate (P_B^1) as measured by isotope-dilution techniques will equal the secretion rate (S_1). Hence

$$S_1 = P_B^1 = MCR_1 \times i_1^A$$

where MCR_1 = metabolic clearance i_1^A = the concentration of such a steroid in arterial blood. By definition, $S_1 = $ O.B.F. $\times (i_1^V - i_1^A)$ (see Section II,A). Hence

$$S_1 = \text{O.B.F.} \times (i_1^V - i_1^A) = MCR_1 \times i_1^A$$

or

$$\text{O.B.F.} = (MCR_1 \times i_1^A)/(i_1^V - i_1^A) = MCR_1/(r_1 - 1)$$

when $r_1 = i_1^V/i_1^A$. The secretion rate (S_2) of any other steroid from that endocrine gland can be calculated if its concentrations in venous effluent (i_2^V) and arterial blood (i_2^A) are measured (Baird, 1970):

$$S_2 = \text{O.B.F.} \times (i_2^V - i_2^A) = [MCR_1 \times (i_2^V - i_2^A)]/(r_1 - 1)$$

III. METHODS FOR STUDYING THE BIOSYNTHESIS OF STEROIDS *IN VIVO*

The biosynthesis of steroid hormones by the ovary has been studied mainly by incubation of slices or minces of ovarian tissue. The results of investigations on the whole ovary *in vivo* are largely confirmatory, in spite of the potential sources of error inherent in the *in vivo* and *in vitro* techniques.

The most widely used technique has been that of ovarian perfusion. Radioactive precursors are added to the afferent arterial blood in the ovarian artery and the conversion products in the venous effluent blood or in the ovary at the end of the experiment are measured. The perfusion of whole organs via the arterial blood supply is advantageous because the cells are supplied with the optimal concentrations of oxygen and nutrients that are necessary for normal enzymatic activity (Goding *et al.,* 1971). However, steroids are probably synthesized in the ovary from precursors made by other enzyme systems within the cell, and not from blood-borne substances. Thus the conversion, or lack of conversion, of a precursor by the perfused ovary may measure the ability of the precursor to leave the blood and penetrate the cell wall, as well as the activity of biosynthetic enzymes (Eik-Nes and Hall, 1965). Even when the ovary *in vivo* may use a plasma-borne precursor such as cholesterol, it is difficult to ensure that the labeled precursor is fully equilibrated with the endogenous substance in plasma (Bolte *et al.,* 1974). These reservations are applicable also to minces and slices of tissue *in vitro,* and to preparations of subcellular organelles (Matsumoto and Samuels, 1969).

It has not yet been demonstrated that acetate can be converted to steroid hormones by the ovary *in vivo,* although such synthesis has been found in the bovine ovary perfused *in vitro* (Bartosik and Romanoff, 1969). When [^{3}H]cholesterol was perfused via the ovarian artery through the ovary of women, it was converted to progesterone and pregnenolone. Progesterone was formed only after injection of LH (Dell'Acqua *et al.,* 1971). The biosynthesis of steroids by the canine ovary has been reported by Eik-Nes and his colleagues. The dogs were pretreated for 10 days with HCG or FSH, and the dogs' own blood was used to perfuse the ovary *in vitro.* Pregnenolone, dehydroepiandrosterone, dehydroepiandrosterone sulfate, and 17β-hydroxyprogesterone were converted to androstenedione, testosterone, estrone, and estradiol-17β (Aakvaag and Eik-Nes, 1969).

The biosynthesis of estrogens *in vivo* by the autotransplanted ovary of the ewe has been reported (Rado *et al.,* 1970). Follicular development was induced by pretreating the sheep with 750 IU PMSG, 3 days before the experiment. Both androstenedione and testosterone were converted to estrone and estradiol-17β, although the conversion was ten times greater from androstenedione. No estriol or estradiol-17α was formed by the ovary from these precursors.

YoungLai and Short (1970) injected labeled precursors into the follicular cavity of a mare at estrus, and the conversion products were isolated from ovarian venous blood and follicular fluid. It was concluded that precursors in the follicular fluid contribute very little to the biosynthesis of those steroids that are secreted into the ovarian venous blood.

In summary, the few studies of steroid biosynthesis by the ovary *in vivo* have confirmed the results of *in vitro* experiments. However, there is a need for additional work, especially on ovaries which are not stimulated by tropic hormones.

IV. SECRETION OF OVARIAN STEROIDS IN DIFFERENT MAMMALS

There are very great differences between species in their ovarian function and each species will therefore be considered separately. Evidence of secretion by direct sampling of ovarian venous blood is quoted when possible. Other indirect evidence, such as excretion of a steroid or its metabolite in urine, its concentration in peripheral blood, ovarian tissue, or follicular fluid, and isotope dilution will also be discussed.

A. Man

1. During Reproductive Life

In 1958, Zander isolated progesterone, 20α-dihydroprogesterone, and 17α-hydroxyprogesterone from preovulatory follicles and corpora lutea of the human ovary. The following year, estrone and estradiol-17β were extracted and identified from human ovaries (Zander *et al.*, 1959). Since then, at least seven steroids have been found in higher concentrations in ovarian than in peripheral venous blood: progesterone and 17α-hydroxyprogesterone (Mikhail, 1967), dehydroepiandrosterone (Rivarola *et al.*, 1967), androstenedione and testosterone (Horton *et al.*, 1966), and estrone and estradiol-17β (Varangot and Cédard, 1959; Schild, 1966). Other steroids, such as pregnenolone (Lloyd *et al.*, 1971), 20α-dihydroprogesterone (Mikhail, 1967), and dehydroepiandrosterone sulfate (Rivarola *et al.*, 1967; Aakvaag and Fylling, 1968), have been measured in ovarian venous plasma, but their secretion by the ovary has not been established by the demonstration of a higher concentration in ovarian venous than in peripheral arterial or venous blood in normal subjects.

a. Estrogens. Estrone and estradiol-17β have both been identified in ovarian venous blood, the latter being present in higher concentrations (Table I). Varangot and Cédard (1959) and Schild (1966) measured total

TABLE I

Concentration (ng/ml) of Estrone and Estradiol-17β in Human Ovarian Venous Plasma[a]

Active ovary			Inactive ovary			
Estrone	Estradiol	N	Estrone	Estradiol	N	Reference
17 ± 3	20 ± 13	?				Varangot and Cedard (1959)
10.1	15.9	2	1.9	1.2	2	Schild (1966)
<2.0	11–28	3				Mahesh (1965)
0.65–1.72	4.03–17.6	4	0.71–0.80	1.93–3.59	2	Mikhail (1967)
1.22 ± 0.33	11.4 ± 3.3	13	0.24 ± 0.05	0.93 ± 0.31	7	Lloyd *et al.* (1971)
Follicular phase						
1.93 ± 0.73	14.14 ± 3.25	11	0.44 ± 0.18	0.65 ± 0.14	7 ⎫	
Early luteal phase						
0.55 ± 0.13	4.49 ± 0.86	8	0.21 ± 0.08	0.93 ± 0.38	4 ⎬ Baird and Fraser (1975)	
Midluteal phase						
1.01 ± 0.29	8.12 ± 2.16	10	0.15 ± 0.04	0.50 ± 0.10	7 ⎭	

[a] Values are means ± S.E.M. The "active" ovary contains either the preovulatory follicle or the corpus luteum and secretes twenty times more estradiol-17β than the contralateral ("inactive") ovary.

hydrolyzable estrogens and found more estrone than did those workers who measured unconjugated estrogens (Mahesh, 1965; Mikhail, 1967; Lloyd *et al.*, 1971; Baird and Fraser, 1975), suggesting that estrone may be secreted by the ovary as a conjugate such as estrone sulfate. Although estriol has been isolated from the human corpus luteum (Wotiz *et al.*, 1956) and isotope dilution studies suggest that it could be secreted during the luteal phase of the cycle (Barlow and Logan, 1966), its secretion by the ovary has never been established (Schild, 1966).

Estradiol-17β is metabolized in the liver and elsewhere to estrone and estriol, which are excreted in the urine as water-soluble conjugates of glucuronic and sulfuric acid. The urinary excretion of the three classical estrogens shows a biphasic pattern (Brown, 1955; Brown and Matthew, 1962). These two peaks at midcycle and during the luteal phase reflect the secretion of estradiol by the preovulatory follicle and corpus luteum, respectively. The concentration of estrogens in peripheral plasma has a similar biphasic pattern throughout the menstrual cycle (Baird 1968; Baird and Guevara, 1969; Korenman *et al.*, 1969; Corker *et al.*, 1969), but the ratio of the concentrations of estradiol and estrone changes significantly from 0.67 ± 0.13 during menstruation to a maximum of 1.90 ± 0.12 at the preovulatory peak suggesting that, as the Graafian follicle develops, it secretes increasing amounts of estradiol.

Several groups of workers have measured the production and secretion rates of estrogens with urinary isotope dilution techniques (Gurpide *et al.*,

1962; Fishman *et al.,* 1962; Morse *et al.,* 1963; Barlow and Logan, 1966; Eren *et al.,* 1967; Kirschner and Taylor, 1972). The results agree with the data on the recovery of injected estrogen (Brown, 1957) and with direct measurements in ovarian vein blood (Mikhail, 1970; Lloyd *et al.,* 1971), and show that the secretion of estradiol is much greater than that of estrone. The blood production rate of estradiol increases from 36 μg/24 hours during menstruation to <400 μg/24 hours just before ovulation (Fig. 3a) (Baird *et al.,* 1969b; Baird, 1973a; Baird and Fraser, 1974). The smaller peak of peripheral estrogen concentration during the luteal phase corresponds with the period of maximal activity of the corpus luteum.

The concentration of estradiol in plasma draining the ovary containing the preovulatory follicle or the corpus luteum ("active ovary") is almost 25 times greater than that draining the contralateral ("inactive") ovary (Baird, 1973a). Since the secretion of estradiol by the adrenal glands is negligible (Baird *et al.,* 1969a), and the extraglandular production of estradiol from estrone and testosterone in menstruating women is small (Baird *et al.,* 1969b), in the absence of bilateral ovulation, the "active ovary" contributes approximately 95% of the total blood production of estradiol (Baird, 1971). Thus, in reproductively active women, estradiol is secreted almost exclusively from the preovulatory follicle or the corpus luteum, and the blood production rate equals the secretion rate.

Most of the circulating estrone is not directly secreted (Fig. 3b). The combined secretion from the ovaries (Mikhail, 1970; Baird, 1970) and the adrenals (Baird *et al.,* 1969a) contributes less than 50% of the production rate in blood. Estrone is cleared more rapidly from the blood than is estradiol (Longcope *et al.,* 1968; Hembree *et al.,* 1969) and the lower estradiol:estrone ratio in peripheral than in ovarian venous plasma reflects a significant extraglandular production of estrone (Baird and Guevara, 1969). The amount of estrone produced by peripheral conversion of secreted estradiol varies with the stage of the cycle and represents 20–35% of the blood production rate (Baird *et al.,* 1969b). When the ovarian secretion of estradiol is low, as during the first few days after ovulation and in postmenopausal and castrated women, most of the circulating estrogen may arise from androstenedione (Baird *et al.,* 1968b). The fractional conversion from androstenedione to estrone is low (0.007: Longcope *et al.,* 1969), but the blood production rate of androstenedione (almost 3 mg/day) is so much higher than that of estrone (about 80 to 300 μg/day) that an appreciable proportion of circulating estrone is derived from this steroid (MacDonald *et al.,* 1967).

Estrogens are also present in high concentrations in follicular fluid (Smith, 1960; Smith and Ryan, 1962; Giorgi, 1965; Baird and Fraser, 1975), and the concentration of estradiol is high in ovarian venous plasma draining

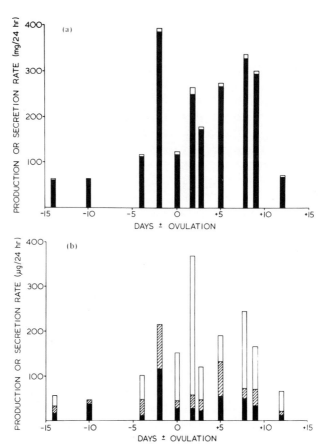

Fig. 3. The dynamics of (a) estradiol and (b) estrone production throughout the menstrual cycle in eleven women. The blood production rates and ovarian secretion rates were measured by a constant infusion technique combined with direct sampling from the ovarian veins. The subjects are grouped around the estimated day of ovulation (day 0). The blood production rate for each subject is indicated by the total height of each column and the individual components by the corresponding codes. (a) Ovarian secretion of estradiol (solid column) and estradiol produced by conversion from estrone (open column). (b) Ovarian secretion of estrone (solid column), estrone produced by conversion from estradiol (hatched column), and estrone derived from other sources such as conversion from androstenedione and adrenal secretion (open columns). (Reproduced by permission from Baird and Fraser, 1974.)

an ovary containing a follicle in which the concentration of estradiol exceeds 400 ng/ml (Baird and Fraser, 1975). The concentration of estradiol is higher than that of estrone throughout the ovarian cycle (Fig. 4), and the estradiol:estrone ratio of about 20 is greater in preovulatory follicles than in smaller follicles in the contralateral ovary or during the luteal phase

(Edwards *et al.*, 1972; Sanyal *et al.*, 1974; Baird and Fraser, 1975). The high ratio of estradiol to estrone may be due to the preferential binding of estradiol by a protein of high affinity which is present in follicular fluid (Takikawa, 1966; Giorgi, 1967). The physiological importance of estrogen in follicular fluid is unknown, although it may influence development of the oocyte and/or granulosa cells within the follicular cavity or the transport of the former through the Fallopian tube after ovulation.

b. Progestagens. Progesterone is present in high concentrations in ovarian venous blood from the ovary containing the corpus luteum (Table II) (Mikhail, 1970; Lloyd *et al.*, 1971). The tenfold difference between the production rate of progesterone during the luteal and follicular phases of the cycle is caused entirely by secretion from the corpus luteum (Little and Billiar, 1969; Lin *et al.*, 1972) (Fig. 5). At midcycle, the mature follicle starts secreting progesterone as the granulosa cells start to luteinize before ovulation (Hertig, 1967; Delforge *et al.*, 1972). The concentrations of progesterone in ovarian venous plasma (Mikhail, 1970; de Jong *et al.*, 1974), peripheral plasma (Johansson and Wide, 1969; Yussman and Taymor, 1970), and in follicular fluid (Sanyal *et al.*, 1974; McNatty *et al.*, 1975) rise sharply at the same time as the preovulatory surge of LH.

During the luteal phase, over 90% of the production rate of progesterone (25 mg) is derived from direct secretion by the corpus luteum. In this phase

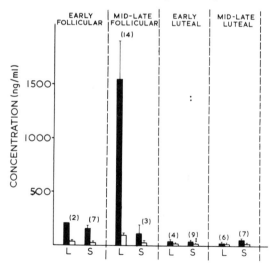

Fig. 4.   The concentrations of estradiol (solid columns) and estrone (open columns) in follicular fluid collected from the ovaries *in situ* of women at different stages of the menstrual cycle. The height of each bar represents the mean ±S.E. of the number of observations indicated within parentheses. L, follicles with diameter >1 cm; S, follicles with diameter <1 cm. (Reproduced by permission from Baird and Fraser, 1975.)

of the cycle, therefore, the production rate of progesterone, based on the cumulative specific activity of pregnanediol, is very similar to the secretion rate from the ovary (Tagatz and Gurpide, 1973). In the follicular phase, such measurements overestimate the secretion of progesterone because of the significant extraglandular production of progesterone and the production of pregnanediol in the liver and elsewhere from pregnenolone and pregnenolone sulfate (Arcos *et al.*, 1964).

It is now clear that the human ovary secretes 17α-hydroxyprogesterone in addition to progesterone (Table II) (Mikhail *et al.*, 1963). The concentration of this hormone in ovarian (Mikhail, 1970; Lloyd *et al.*, 1971) and in peripheral venous plasma (Strott *et al.*, 1969; Stewart-Bentley and Horton, 1971) is highest in the immediate preovulatory period and during the luteal phase of the cycle, indicating that the follicle and the corpus luteum are important sources (Fig. 6). In the follicular phase of the cycle, much of the circulating 17α-hydroxyprogesterone is formed by the extraglandular conversion of 17α-hydroxypregnenolone that is secreted by the adrenal. In the luteal phase, the ovarian secretion comprises almost all the production rate in blood (4 mg) (Strott *et al.*, 1970). It has been reported that the preovulatory peak of 17α-hydroxyprogesterone in peripheral plasma occurs at the same time as the peak of estradiol (Ross *et al.*, 1970), but simultaneous measurements of both hormones in the same plasma samples suggest that

TABLE II

Concentration (ng/ml) of Progesterone and 17α-Hydroxyprogesterone in Human Ovarian Venous Plasma[a]

	Luteal phase			
Follicular phase	Ovary + corpus luteum	Inactive ovary	N	Reference
Progesterone				
3.9–15.5	604 ± 167	26	11	Mikhail *et al.* (1963); Mikhail (1967)
ND[b]–81				Aakvaag and Fylling (1968)
1.7–15.0	549 ± 220	37	8	Lloyd *et al.* (1971)
17α-Hydroxyprogesterone				
20.0–41.1	70.9 ± 17.6	—	8	Mikhail (1967)
10.0–18.3	222 ± 94	<10.0–49.0		Lloyd *et al.* (1971)

[a] Values are means ± S.E.M. The concentration of progesterone in blood draining the ovary containing the corpus luteum is about twenty times greater than that in plasma draining the contralateral ("inactive") ovary.

[b] ND, not detectable.

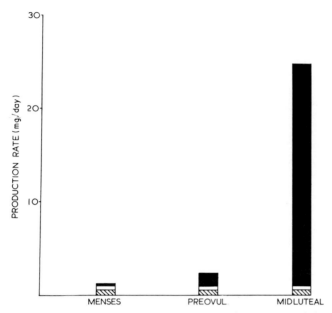

Fig. 5. The dynamics of progesterone production at three stages of the menstrual cycle. The plasma concentrations reported by Johansson (1969) have been used in calculating the blood production rate; the MCR and transfer constants are from Lin *et al.* (1972). The height of each bar represents the total blood production rate. The contributions from other sources are: ovarian secretion (solid column), adrenal secretion (open column), and extraglandular conversion of pregnenolone (hatched column). Note the increased secretion of progesterone from the preovulatory follicle. (Reproduced by permission from Baird, 1974.)

the maximum concentration of 17α-hydroxyprogesterone occurs 12–24 hours after the peak of estradiol (Abraham *et al.*, 1972; Thorneycroft *et al.*, 1972, 1974). Pregnanetriol is a metabolite of 17α-hydroxyprogesterone and its excretion pattern in the urine is biphasic like that of estrogen (Fotherby, 1962).

There is presumptive evidence that the ovary secretes pregnenolone, 20α-dihydroprogesterone and, possibly, 20β-dihydroprogesterone. 20α-Dihydroprogesterone has been isolated and identified from human ovaries (Zander, 1958) and the concentrations found in ovarian venous plasma during the luteal phase, up to 66 ng/ml (Mikhail, 1967; Lloyd *et al.*, 1971), far exceed those (2.5 ng/ml) reported in peripheral venous blood at this stage of the cycle (van der Molen, 1969; Wu *et al.*, 1974). Proof of the ovarian secretion of pregnenolone is also lacking, although this hormone has reached concentrations as high as 77 ng/ml in ovarian venous plasma (Lloyd *et al.*, 1971) and the concentration of pregnenolone in peripheral plasma of

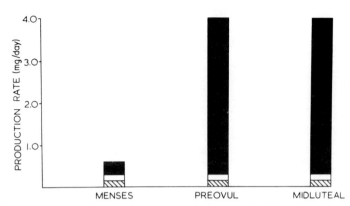

Fig. 6. The dynamics of 17α-hydroxyprogesterone production at three stages of the menstrual cycle. The plasma concentrations reported by Strott *et al.* (1969) have been used in calculating the blood production rate. The MCR and transfer constants are those of Strott *et al.* (1970). The height of each column represents the blood production rate of 17α-hydroxyprogesterone. The contributions from other sources are: ovarian secretion (solid column), adrenal secretion (open column), and extraglandular conversion of 17α-hydroxypregnenolone (hatched column).

reproductively active women is higher than in postmenopausal women (McKenna and Brown, 1974). The concentration is lowered by suppression with dexamethasone or estrogen, suggesting that it arises from the ovary and adrenal. There is some doubt as to whether the concentration changes throughout the menstrual cycle (McKenna and Brown, 1974; Bermudez *et al.*, 1970; Abraham *et al.*, 1973; Di Pietro *et al.*, 1972). Although 20β-dihydroprogesterone has been isolated from human ovaries (Zander *et al.*, 1959), it has not been detected in ovarian venous blood at any stage of the cycle (Mikhail, 1970), and its ovarian origin in women remains in doubt.

c. Androgens. Although testosterone (Table III) and dehydroepiandrosterone are present in higher concentrations in ovarian than in peripheral venous blood, androstenedione is the main androgen secreted by the ovary (Mikhail, 1970). The ratio of testosterone and dehydroepiandrosterone in ovarian and peripheral blood is only about 2:1 while that of androstenedione is at least 20:1. An ovarian origin of androstenedione is suggested by its low concentration in peripheral plasma in patients with ovarian dysgenesis (0.47 ng/ml), castrated women (0.88 ng/ml), and prepubertal females (0.6 ng/ml) (Gandy and Peterson, 1968).

Several authors have found that the concentration of dehydroepiandrosterone is higher in ovarian than in peripheral venous plasma. However, the calculated ovarian secretion (0.3–0.6 mg/day) is insignificant compared with the adrenal secretion of 6 to 8 mg/day (Abraham and Chakmakjian, 1973). Although dehydroepiandrosterone sulfate is present in ovarian venous blood, its concentration is not significantly different from that in peripheral blood, and it is probably not secreted by the normal ovary (Rivarola *et al.*, 1967; Gandy and Peterson, 1968; Kalliala *et al.*, 1970; de Jong *et al.*, 1974).

The concentration of androstenedione in adrenal (Wieland *et al.*, 1965) and ovarian venous plasma (Mikhail, 1970) is relatively high. About 15% of the blood production rate of androstenedione arises from peripheral conversion of dehydroepiandrosterone (Horton and Tait, 1967), with the remainder being secreted directly from the adrenal and ovary (Horton and Tait, 1966) (Fig. 7). The adrenal secretion of androstenedione shows a diurnal rhythm comparable to that of cortisol (Tunbridge *et al.*, 1973; Lobotsky and Lloyd, 1973), but remains relatively constant throughout the menstrual cycle at about 1.2 mg/day (Baird, 1974). The rise in the concentration of androstenedione in peripheral plasma at midcycle and during the luteal phase of the cycle (Judd and Yen, 1973; Dupon *et al.*, 1973; Baird *et al.*, 1973b) is due to increased ovarian secretion (Abraham *et al.*, 1969). The concentration in plasma draining the ovary containing the

TABLE III

Concentration (ng/ml) of Testosterone, Dehydroepiandrosterone, and Androstenedione in Human Ovarian Venous Plasma

Mean ± SE	Range	N	Reference
Testosterone			
0.74 ± 0.06	0.54–0.95	5	Horton *et al.* (1966)
2.09 ± 0.14	1.59–2.42	5	Mikhail (1967)
1.30 ± 0.25	0.52–2.05	5	Rivarola *et al.* (1967)
3.3 ± 0.9	1.2–10.4	9	Gandy and Peterson (1968)
5.14 ± 1.5 (A)[a]	0.83–20.4	12	Lloyd *et al.* (1971)
2.36 ± 0.6 (I)[a]		7	
Dehydroepiandrosterone			
18.7 ± 3.1	7.2–25.1	5	Rivarola *et al.* (1967)
11.5 ± 0.9	7.2–17.4	11	Gandy and Peterson (1968)
Androstenedione			
48.8 ± 10.5	15–98	10	Mikhail (1967)
8.22	4.5–12.0	5	Horton *et al.* (1966)
16.3 ± 5.6	4.7–35.6	5	Rivarola *et al.* (1968)
20.4 ± 5.1	5.0–59.7	11	Gandy and Peterson (1968)
79.5 ± 21.2	15–307	17	Lloyd *et al.* (1971)
33.2 ± 9.0 (A)	3.4–65.0	7	Baird *et al.* (1974)
10.1 ± 1.6 (I)		13	

[a] A, the active ovary containing the corpus luteum; I, the contralateral, "inactive" ovary.

preovulatory follicle or corpus luteum is significantly higher than in that draining the contralateral ovary, indicating that follicles and corpora lutea, as well as stromal cells, secrete androstenedione (Baird *et al.*, 1974). In the early follicular phase of the cycle, ovarian secretion of androstenedione (0.9 mg/day) represents 30% of the total blood production rate (2.8 mg/day), and, at midcycle, 3.1 mg (60%) of the production rate is ovarian in origin (Baird, 1974; Abraham, 1974).

Testosterone in women also arises from several sources (Fig. 8). Testosterone is present in adrenal (Wieland *et al.*, 1965; Baird *et al.*, 1969a) and ovarian venous plasma (Table III), but in contrast to androstenedione the combined glandular secretion amounts to less than half the total production rate (Fig. 8) (Horton and Tait, 1966; Tait and Horton, 1966; Baird *et al.*, 1969b). The extraglandular conversion of dehydroepiandrosterone and androstenedione contributes significantly to the total production rate at all stages of the cycle, and the increased production at midcycle (Judd and Yen, 1973; Abraham, 1974), the physiological significance of which is unknown, is partly related to the increased amount of androstenedione

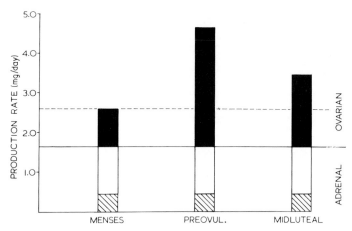

Fig. 7. The dynamics of production and secretion of androstenedione throughout the menstrual cycle. The plasma concentrations reported by Baird *et al.* (1974) have been used in calculating the blood production rate. The MCR and transfer constants are those quoted by Baird *et al.* (1968b). At each stage of the cycle the blood production rate is indicated by the total height of the column. Contributions from other sources are: ovarian secretion (solid column), adrenal secretion (open column), and extraglandular conversion of dehydroepiandrosterone (hatched column). The total amount arising from adrenal sources (below the horizontal continuous line) remains constant throughout the cycle. The ovarian secretion from the preovulatory Graafian follicle and the corpus luteum is indicated above the dotted horizontal line. (Reproduced by permission from Baird, 1974.)

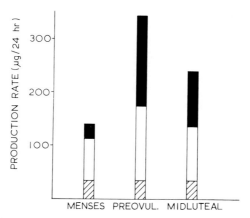

Fig. 8. The dynamics of production and secretion of testosterone throughout the menstrual cycle. The plasma concentrations reported by Abraham (1974) have been used in calculating the blood production rate. The MCR and transfer constants are those quoted by Baird *et al.* (1968b). At each stage of the cycle the blood production rate is indicated by the total height of the column. Contributions from other sources are: secretion by ovary and adrenal (solid column), extraglandular conversion of androstenedione (open column), and extraglandular conversion of dehydroepiandrosterone (hatched column). Note that the glandular secretion of testosterone accounts for less than one-half the blood production rate at all stages of the cycle.

secreted from the ovary (Fig. 8). In rhesus monkeys, androgens appear to be important for female libido (Everitt and Herbert, 1971), and the relatively high concentrations of androgens throughout the cycle may help in maintaining sexual activity in female primates (Herbert, 1972; see Volume II, Chapter 10).

The physiological significance of the production of the biologically potent androgen, testosterone, from the weak androgen, androstenedione, is discussed in detail by Baird *et al.* (1968b, 1969b). Androstenedione has been described as a prehormone, i.e., a hormone with little intrinsic biological activity but convertible in peripheral tissues to a more potent compound. It provides one means of achieving high concentrations of a hormone in the target organ without concentrations in peripheral blood being high enough to cause undesirable side effects. Peripheral conversion of androstenedione is responsible for over 50% of the blood testosterone, dihydrotestosterone (Ito and Horton, 1970), and almost all blood estrogens in postmenopausal women (see below). The sites of extraglandular conversion of androstenedione are unknown, although the hypothalamus can convert androstenedione to estrone locally (Flores *et al.*, 1973), and it may have a role in the feedback control of gonadotropins (Martenz *et al.*, 1975).

2. After the Menopause

Estrogens are still detectable after the menopause in blood (Baird and Guevara, 1969; Korenman *et al.*, 1969; Tulchinsky and Korenman, 1970; Longcope, 1971) and in urine (Brown and Matthew, 1962). It is unlikely, however, that the ovaries are normally the source of this estrogen because the concentration of estrogens in ovarian venous plasma is negligible (Mikhail, 1970) and the production rates of estradiol (Barlow *et al.*, 1969) and estrone (MacDonald *et al.*, 1967) are not significantly altered by removal of postmenopausal ovaries. After the menopause, the proportions of the circulating estrogens change; the concentration of estrone is reduced less than that of estradiol (Baird and Guevara, 1969; Nagai and Longcope, 1971). The ratio of estradiol:estrone in the peripheral plasma of postmenopausal women (0.27 ± 0.07 S.E.) is similar to that in castrated women (0.42 ± 0.07) and in men (0.35 ± 0.04) in whom most of the estradiol is formed by extraglandular conversion of steroid precursors (Baird *et al.*, 1969b). The entire production of estradiol in postmenopausal women can be accounted for by extraglandular conversion of circulating estrone (MacDonald *et al.*, 1971).

The production rate of estrone, as measured by isotope-dilution techniques in blood (Longcope, 1971) or in urine (MacDonald *et al.*, 1967), is about 40 μg/day (Fig. 9). In reproductively active women, estrone is directly secreted by the ovaries (Lloyd *et al.*, 1971) and adrenals (Baird *et al.*, 1969a), as well as being produced by the peripheral conversion of secreted estradiol and androstenedione (Baird *et al.*, 1969b). In postmenopausal women, however, small amounts of estrone are secreted directly from the adrenal glands (Baird *et al.*, 1969a) and estrone is formed almost entirely by peripheral conversion of androgen precursors. Androstenedione is a precursor for estrone and is almost certainly of adrenal origin because the production of estrone is reduced after adrenalectomy, or administration of dexamethasone, but not after castration (MacDonald *et al.*, 1967). It is unlikely, therefore, that the small quantities of androstenedione secreted into the ovarian vein after the menopause (Mikhail, 1970) contribute significantly to the total blood production of androstenedione, which falls to 50% of the level found in menstruating women (MacDonald *et al.*, 1967; Abraham *et al.*, 1969).

The site of conversion of androstenedione to estrone is extraglandular because the efficiency of conversion (about 1.2%) is similar in menstruating, postmenopausal, oophorectomized, or adrenalectomized subjects (MacDonald *et al.*, 1967). The rate of conversion varies considerably in postmenopausal subjects (from less than 1–20%), and is probably related to the degree of obesity as well as increasing age (MacDonald *et al.*, 1971;

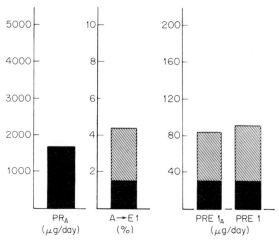

Fig. 9.   Estrone production in eight women studied more than 10 years after the menopause, all but one of whom had symptoms of excess estrogen production (mean age = 66.8 years; mean weight = 204.8 lb). The solid portion of each column represents the normal value and the total height of each column represents the mean for the group. Although the total production rate of androstenedione/day (PR_A) is similar in both groups, the proportion of androstenedione converted to estrone (A $\rightarrow$ E_1) is more than double in the group with symptoms of excessive estrogen production. Thus the amount of estrone produced from androstenedione (PRE 1_A) is greatly increased in the women with symptoms. Production of estrone from androstenedione (PRE 1_A) accounts for virtually all the total production of estrone (PRE 1). (Reproduced by permission from Siiteri and MacDonald, 1973.)

Hemsell *et al.*, 1974). Although fat tissue from postmenopausal women will convert androstenedione to estrone when incubated *in vitro* (Schindler *et al.*, 1972) the site of conversion *in vivo* is unknown. The amount of estrone produced depends on the rate of secretion of androstenedione from the adrenal, the efficiency of peripheral conversion, and the release of estrone into the circulation from the peripheral sites. These factors have been related to the degree of stimulation of the endometrium, and it has been calculated that postmenopausal bleeding usually occurs when the amount of estrone entering the blood exceeds 40 μg/day (MacDonald *et al.*, 1969; Siiteri and MacDonald, 1973).

B. Monkey

Because of the relatively small ovarian vessels and insensitivity of the analytical methods, early attempts to demonstrate the secretion of steroids by the ovary of the monkey were restricted to animals which had been treated with gonadotropins. Progesterone, 20α-dihydroprogesterone, and 17α-hydroxyprogesterone were identified in the ovarian vein of rhesus monkeys (*Macaca mulatta*) which had been injected for several days with FSH and HCG (Hayward *et al.*, 1963). A similar study in the squirrel monkey (*Saimiri sciureus*) demonstrated the ovarian secretion of pregnenolone as well as progesterone (Fajer and Bechini, 1971). It is difficult to assess how these findings in stimulated ovaries relate to the physiological situation.

Indirect evidence of the ovarian secretion of progesterone in the rhesus monkey is available from the pattern of concentration in peripheral plasma throughout the menstrual cycle (Johansson *et al.*, 1968; Munroe *et al.*, 1970; Hopper and Tullner, 1970). The marked rise in the second half of the cycle is very similar to that found in women and strongly suggests that the corpus luteum secretes progesterone. The ovary in this species may also secrete progesterone in the follicular phase, the concentration in ovarian venous plasma draining the preovulatory follicle rising to 18.6 ng/ml (Koering *et al.*, 1968). After ovulation the concentration rises further to over 100 ng/ml in plasma draining the ovary containing the corpus luteum (Weiss *et al.*, 1973).

There is little doubt that estradiol-17β is the main estrogen secreted by the Graafian follicle in the rhesus monkey. The concentrations of estrone and estradiol-17β are higher in venous plasma (Weiss *et al.*, 1973; Bosu and Johansson, 1974). The concentration of estradiol in peripheral plasma exceeds that of estrone in the follicular phase of the cycle (Bosu *et al.*, 1973) and reaches a peak at midcycle (Hotchkiss *et al.*, 1971; Weick *et al.*, 1973).

It is still not clear whether the corpus luteum of the rhesus monkey secretes significant amounts of estrogen. The concentration in peripheral plasma of estradiol or estrone does not rise during the luteal phase as does that of progesterone and 17α-hydroxyprogesterone (Knobil, 1973; Bosu *et al.*, 1973). After enucleation of the corpus luteum or in the few days before menstruation, the concentration of estrogens (in marked contrast to the concentration of progesterone) may rise (Bosu *et al.*, 1973). Although estradiol is apparently present in relatively high concentrations in luteal tissue (Karsch *et al.*, 1973; Knobil, 1974), the concentration in blood draining the contralateral ovary is similar to that in blood draining the ovary containing the corpus luteum (Weiss *et al.*, 1973; Bosu and Johansson, 1974). These findings indicate that it is unlikely that the corpus luteum in the rhesus monkey is a major source of the estrogen secreted by the ovary, but the corpus luteum may synthesize estrogen and retain it locally within the gland.

In contrast to the rhesus monkey and the baboon (Stevens *et al.*, 1970), the chimpanzee (Graham *et al.*, 1972) and the orangutan (Collins *et al.*,

1973) have a biphasic pattern of estrogen excretion throughout the menstrual cycle, similar to that found in women, indicating that in these primates the corpus luteum secretes estrogens as well as progesterone.

Although the rhesus monkey excretes all three "classical" estrogens in the urine during the menstrual cycle, the quantity of estrone is much greater than that of estradiol-17β or estriol (Hopper and Tullner, 1970). The production rate of estradiol-17β in nonpregnant rhesus monkeys at different stages of the menstrual cycle has been measured by isotope-dilution techniques (Laumas, 1969): the secretion rate of estrone and estradiol-17β on day 8 of the menstrual cycle was 21 and 124 μg/24 hours, respectively.

C. Sheep

The ovarian cycle in the ewe lasts 16 or 17 days with ovulation occurring toward the end of estrus (= day 0) (Cupps et al., 1969). During the luteal phase, the corpus luteum secretes progesterone almost exclusively (Edgar and Ronaldson, 1958; Stormshak et al., 1963; Short et al., 1963), although smaller quantities of 20α-dihydroprogesterone and 17α-hydroxyprogesterone are also produced (Lindner et al., 1964; Baird et al., 1968a). The main estrogen secreted by the follicle is estradiol-17β, but some estrone is found (Short et al., 1963; Lindner et al., 1964; Baird et al., 1968a; Moore et al., 1969; Smith and Robinson, 1970; Scaramuzzi et al., 1970). The estrogens secreted throughout the luteal phase of the cycle (Baird et al., 1968a; Scaramuzzi et al., 1970) probably arise from developing Graafian follicles rather than the corpus luteum (Holst et al., 1972; Baird et al., 1973a). There are three peaks of secretion of estradiol-17β into the uteroovarian vein of ewes in which this vein is anastomosed with the mammary vein (Cox et al., 1971a,b). As well as the preovulatory peak, there is a similar peak on day 2 or 3, and smaller peaks which occur at random in the luteal phase (McCracken et al., 1969; Baird and Scaramuzzi, 1975; Baird et al., 1976b) (Fig. 10).

Androstenedione and testosterone are secreted by the sheep ovary in situ (Baird et al., 1973a) and when autotransplanted to the neck (Baird et al., 1968a). In both situations, androstenedione is secreted in significant amounts, indicating that the interstitial cells as well as the follicles are its source, as in other species (Savard et al., 1965). Androgens are probably also secreted by the follicle because the concentrations of androstenedione and testosterone increase in parallel with that of estradiol-17β after induction of follicular development with PMSG (Baird et al., 1968a, 1976b).

The concentration of estrogens in peripheral plasma of nonpregnant ewes is very low and is not detectable by existing methods (Cox et al., 1971a).

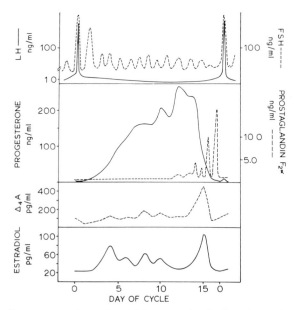

Fig. 10. The concentrations of steroids, prostaglandin F, and gonadotropins in utero-ovarian venous plasma throughout the estrous cycle of the ewe. The concentrations of estradiol, androstenedione, progesterone, and prostaglandin $F_{2\alpha}$ were measured in serial samples of utero-ovarian venous plasma collected from three ewes with cervical-utero-ovarian transplants. The values are the mean of eight cycles adjusted for differences in cycle length by dating back 5 days from the onset of estrus (day 0). The concentrations of LH and FSH, expressed in nanograms of LH-M1 (potency × 1.5 NIH-LH-S1) and NIH-FSH-S6/ml are redrawn from the data of Land *et al.* (1973) and Salamonsen *et al.* (1973), respectively. The magnitude of the proestrous rise of estradiol (and androstenedione) is obscured by the twofold increase in uterine blood flow at this time. (Reproduced by permission from Baird and Scaramuzzi, 1975).

From a knowledge of the metabolic clearance rate of estrogens (Challis *et al.*, 1973), it can be calculated that the concentration is unlikely to exceed 5 pg/ml plasma (Baird *et al.*, 1973a) even during the maximum preovulatory secretion of about 12 μg/day. The values of 9500 pg/ml (Norman *et al.*, 1968) and 50–140 pg/ml (Obst *et al.*, 1971) are probably overestimates due to relative nonspecificity of the methods used. Terqui *et al.* (1973), using an extremely sensitive radioimmunoassay, have reported a peak concentration in peripheral plasma of between 30 and 50 pg/ml 3 days before the onset of estrus.

The biosynthesis of estrogens has been studied in the sheep ovary auto-transplanted to the neck; PMSG was used to induce follicular development (Rado *et al.*, 1970). It was concluded that androstenedione was the major precursor for the synthesis of estradiol *in vivo*.

D. Cow

In the cow, the main steroid secreted into ovarian venous blood by the corpus luteum is progesterone (Edgar and Ronaldson, 1958; Gomes *et al.,* 1963). However, Gomes *et al.* (1963) concluded that the peripheral plasma progesterone levels bear no consistent relationship to the stages of the estrous cycle. This conclusion probably reflects the insensitivity of the method used because numerous subsequent studies have shown an excellent correlation between the function of the corpus luteum and concentration of progesterone in peripheral plasma (Pope *et al.,* 1969; Stabenfeldt *et al.,* 1969; Shemesh *et al.,* 1971). Dobrowolski *et al.* (1968) studied the concentration of progesterone in ovarian venous blood throughout the estrous cycle by inserting a silicone catheter into the ovarian vein via the uterine vein. The catheter was kept patent for 4 to 9 days by flushing with a solution of heparin every 2 hours, and samples were withdrawn every 6 hours for serial measurements of hormone concentration in the same animal. The concentration of progesterone in ovarian venous plasma was lowest on day 1 (56 ng/ml) and rose to a maximum about day 8 to 15 (1250 to 1800 ng/ml).

Estrone, estradiol-17α, and estradiol-17β are excreted in the urine, and it is likely that the latter is the hormone secreted by the ovary (Mellin *et al.,* 1966). Estradiol-17β has been identified in the follicular fluid of cows (Velle, 1959), but its secretion into ovarian venous blood has not been confirmed. The concentrations of estrone, estradiol-17β, and estradiol-17α in peripheral plasma rise 3 days before the onset of estrus to reach a peak coincidentally with the maximum development of the preovulatory follicle (Shemesh *et al.,* 1972; Hansel and Echternkamp, 1972; Glencross *et al.,* 1973; Dobson and Dean, 1974). The hormonal events occurring around the time of estrus have been elucidated by results of studies in which samples of utero-ovarian (Nancarrow *et al.,* 1973) and peripheral venous plasma (Lemon *et al.,* 1975) were collected every 2–3 hours. As the secretion of progesterone declines 3–4 days before the onset of estrus, the secretion of estradiol-17β rises to reach its peak at about the onset of estrus. The temporal relationships of these events make it virtually certain that the follicle secretes estradiol-17β in the cow.

E. Horse

The mare has an ovarian cycle lasting 22 days with a 6-day follicular or estrous phase. Follicular development continues throughout the luteal phase although estrus is not displayed and ovulation fails to occur (Aitken, 1927).

The mature follicle secretes large quantities of estradiol-17β and androstenedione into the ovarian vein at estrus (Short, 1964), and this is reflected by an increase in the concentration of estradiol in peripheral plasma at this time (Pattison *et al.*, 1974). In the preovulatory follicle the cytoplasm of the thecal cells is foamy and very well supplied with blood capillaries, and these cells are probably the major source of steroids from the follicle. Aspiration of follicular fluid does not affect the secretion of steroids from the follicle into the ovarian vein for at least a short period (Short, 1964). It must be assumed, therefore, that the granulosa cells within the follicle or steroids in follicular fluid play little part in the synthesis of steroids secreted into the ovarian vein.

The pattern of steroids present in follicular fluid differs from that secreted into the ovarian vein (Short, 1960a; Knudsen and Velle, 1961; YoungLai, 1971). The concentration of estradiol is extremely high (about 1500 ng/ml), but smaller amounts of estrone, androstenedione, testosterone, progesterone, 20α-dihydroprogesterone, 17α-hydroxyprogesterone, epitestosterone, and 19-norandrostenedione are also present (Short, 1964). 19-Norandrostenedione is an important intermediate in the biosynthesis of estrogens and the follicular fluid of mares at estrus is the only site in which it has been demonstrated *in vivo* (Short, 1961; YoungLai, 1971). It was suggested on the basis of the different patterns of steroids in ovarian venous plasma and follicular fluid that the synthesis of steroids in ovarian venous plasma and follicular fluid involves a "two-cell" process (Short, 1962, 1964). The granulosa cells *in vivo* have a weak 17α-hydroxylase system and little or no 17-desmolase, and appear to play little part in the synthesis of steroids present in follicular fluid which originate in the theca cell layer and are transported into the follicular cavity (YoungLai and Short, 1970).

After ovulation the luteinized granulosa cells become vascularized and secrete large quantities of progesterone. This is reflected by the marked rise in concentration of this steroid in luteal tissue (Short, 1962), ovarian venous plasma (Short, 1964), and peripheral plasma (Short, 1959; Smith *et al.*, 1970). The corpus luteum also secretes 20α-dihydroprogesterone, 17α-hydroxyprogesterone, and pregnenolone in smaller amounts (Short, 1962; YoungLai, 1971). Traces of estradiol-17β were also found in luteal tissue but rigorous chemical identification was not possible (YoungLai, 1971).

F. Pig

Progesterone levels have been measured in the ovarian venous effluent of the sow (Gomes *et al.*, 1965; Masuda *et al.*, 1967; Brinkley and Young, 1968). Masuda *et al.* (1967) collected blood by catheterization of the ovarian vein and found that the concentration of progesterone in ovarian venous plasma correlated well with the weight of all luteal tissue throughout the estrous cycle. Brinkley and Young (1968) ligated all the vessels which anastomosed with the ovarian vein. Between days 7 and 13 after ovulation, the secretion rate of progesterone rose from 7.2 μg/minute/ovary to 13.4 μg/minute/ovary, and, correlated with the weight of the corpus luteum, so that the secretion rate remained fairly constant at 2.74 to 3.94 μg/minute/gm luteal tissue. The secretion of progesterone declines sharply in the few days before the onset of estrus and is accompanied by marked release of prostaglandin F from the uterus (Gleeson and Thorburn, 1973).

Estradiol is the major estrogen present in the pig ovary (Westerfeld *et al.*, 1938). Estrogen is secreted throughout the luteal phase, but a marked rise occurs as the corpus luteum regresses at the end of the cycle. The urinary excretion of estrogens (Liptrap and Raeside, 1966) and the concentration of estradiol in plasma (Henricks *et al.*, 1971) reach a maximum 24-48 hours before the onset of estrus, coincidental with the maximum development of the preovulatory follicles.

G. Rabbit

Progesterone has been identified as a secretory product of the rabbit ovary before ovulation and during pseudopregnancy (Hilliard *et al.*, 1963) and pregnancy (Mikhail *et al.*, 1961). The corpus luteum is the main source of progesterone (Hilliard *et al.*, 1973), but this hormone and 20α-dihydroprogesterone are produced in addition by the interstitial cells between copulation and ovulation when the secretion rate may be over 400 μg/gm ovary/hour (Hilliard *et al.*, 1967). The synthesis and secretion of 20α-dihydroprogesterone (and progesterone) is further stimulated by the preovulatory LH surge which occurs in response to coitus in the rabbit and is thought to prolong the release of LH from the pituitary (Hilliard *et al.*, 1969).

On the basis of their relative biological potencies, it has been postulated that estradiol-17β rather than estradiol-17α was the main estrogen secreted by the rabbit ovary (Saldarini *et al.*, 1970). Using a specific and extremely sensitive competitive protein-binding method (Korenman *et al.*, 1969) it has been confirmed that, as in other species, the rabbit ovary secretes estradiol-17β (Eaton and Hilliard, 1971; Hilliard and Eaton, 1971). The secretion of estradiol-17β rises at estrus to 17.7 $\pm$ 1.7 ng/ovary/hour. During the 4 hours after coitus, the secretion rate of estradiol-17β reaches a maximum of about 60 ng/ovary/hour and parallels the increased rate of secretion of 20α-dihydroprogesterone and testosterone. The secretion of both these steroids is almost certainly stimulated by the increased concentration of LH in serum which occurs at this time. The secretion of progesterone, 20α-dihydroprogesterone, and estradiol-17β is very low 24–72 hours after mating but increases before implantation (Hilliard and Eaton, 1971). Hilliard and Eaton (1971) demonstrated, by destroying the follicles by cautery, that estradiol-17β is secreted almost exclusively by the follicle, 20α-dihydroprogesterone is derived from the interstitial gland, and progesterone from the corpus luteum. The fact that secretion of these three steroids can be stimulated by the appropriate dose of LH (0.5–2.0 μg/kg) supports the view that LH increases steroidogenesis in all cell types in the ovary.

The rabbit ovary secretes large amounts of androgen in addition to estrogens and progesterone (Mills and Savard, 1973; Hilliard *et al.*, 1973, 1974a). The secretion of testosterone is ten times that of estradiol and the change in the pattern of secretion after coitus or administration of LH is similar for both steroids (Hilliard *et al.*, 1974a). The secretion rates of steroids from each ovary were found to be similar by Hilliard *et al.* (1974b), but Shaikh and Harper (1972) found marked differences in secretion rates between the two ovaries. The technique of cannulation used by Shaikh and Harper involved ligation of the ovarian artery and their results, therefore, must be regarded as unphysiological.

By selectively cauterizing the follicles in one ovary, Hilliard *et al.* (1974a) demonstrated that, although estradiol was secreted exclusively by the follicle, significant amounts of testosterone are secreted by the interstitial tissue as well as the follicle. Further studies of the secretion of androgen by the rabbit ovary are needed to determine its function (if any) and whether testosterone is the dominant androgen in this species (in contrast to the woman, mare, ewe, and bitch).

The mechanism of action of LH in increasing steroid synthesis *in vivo* has been investigated in the rabbit (Solod *et al.*, 1966; Armstrong *et al.* 1969); LH mobilizes cholesterol in the ovary and the store of cholesterol is very rapidly depleted following injection of LH. The endogenous pools of cholesterol within the ovary were labeled by injecting [7α-^{3}H]cholesterol intravenously 24 hours before the experiment. The secretion rate of progesterone and 20α-dihydroprogesterone, and the incorporation of radioactivity into the progestagens, was measured *in vivo* after cannulation of the ovarian vein. Intravenous injection of 100 μg LH was followed by a marked increase in the secretion rate of both hormones. The specific activity of the progestagens remained similar, indicating that the increased secretion represented increased synthesis of steroid from acetate.

H. Rat

Most of the information about the secretion of ovarian hormones in the rat has been obtained by investigation of the changes occurring in target organs such as the uterus. However, in spite of the technical difficulties presented by the size of the blood vessels, the secretion of steroids by the ovary has also been studied directly by cannulating the ovarian vein *in situ* (Eto *et al.*, 1962; Telegdy and Endroczi, 1963; Yoshinaga *et al.*, 1967). Progesterone is not secreted in significant amounts by the corpus luteum in the unmated rat (Telegdy and Endroczi, 1963), although 20α-dihydroprogesterone does appear in ovarian venous blood in increasing concentrations, especially at proestrus when progesterone secretion is at a minimum (Cortes *et al.*, 1971; Piaczek *et al.*, 1971). The corpus luteum of pregnancy or pseudopregnancy secretes large amounts of progesterone (Fajer and Barraclough, 1967) and little 20α-dihydroprogesterone. It has been suggested that reduction of progesterone to 20α-dihydroprogesterone is important in the regulation of progestational effects (Wiest and Kidwell, 1969).

Estrogen has been measured in ovarian venous blood throughout the estrous cycle and during pregnancy by bioassay (Yoshinaga *et al.*, 1969; Hori *et al.*, 1968) and by radioimmunoassay (Shaikh, 1971). The secretion of estrogen is low during diestrus but rises rapidly on the morning of proestrus to reach a maximum at about 11.00 hours. There is then a rapid fall that evening following the preovulatory surge of LH (Munroe *et al.*, 1969). The concentration of estradiol-17β has been measured in peripheral plasma by competitive protein binding (Brown-Grant *et al.*, 1970) and by radioimmunoassay (Shaikh and Abraham, 1969) and the changes are in accord with those reported for ovarian secretion of estrogen.

Various other neutral steroids, such as 5α-pregnane-3,20-dione, progesterone, 3α-hydroxy-5α-pregnan-20-one, 20α-hydroxy-5α-pregnan-3-one, 20α-hydroxy-4-ene-3-one, and 5α-pregnane-3α-20α-diol have been identified chemically in the ovarian vein of rats stimulated by administration of LH (Hirai *et al.*, 1968; Ichikawa *et al.*, 1971). It is not known if all these compounds are normally secreted by the rat ovary.

I. Hamster

The hamster has a 4-day estrous cycle, similar to that of the rat; ovulation occurs between 13.00 and 15.00 hours on the day of estrus (day 1). In spite of the small size of the vessels, the ovarian vein has been cannulated (Eto *et al.*, 1962; Shaikh, 1972; Labhsetwar *et al.*, 1972).

The secretion of estradiol rises on the afternoon of day 3 and reaches a maximum at 15.00 hours on day 4 (proestrus) (Baranczuk and Greenwald, 1973). The amount of estrone secreted is about one-half that of estradiol and remains unchanged throughout the estrous cycle. Late on day 4, and synchronously with the preovulatory surge of LH, the secretion of progesterone starts to rise and reaches a maximum value on day 1. Unless pregnancy or pseudopregnancy occurs, the corpus luteum only secretes progesterone for 1 or 2 days as in the rat (Hoffman and Fajer, 1970).

J. Dog

Progesterone has been identified in the ovarian venous blood of dogs at different stages of the estrous cycle (Romanoff *et al.,* 1962; Telegdy *et al.,* 1963). The secretion of progesterone and androstenedione increased immediately following the administration of HCG or LH (Telegdy and Huszar, 1962; Nishizawa and Eik-Nes, 1964). Eik-Nes and his group have studied the biosynthesis of ovarian steroids in dogs treated previously for several days with HCG (see Aakvaag and Eik-Nes, 1969). The results may not be comparable with events in normal animals; [^{3}H]pregnenolone was metabolized to 17α-hydroxypregnenolone, dehydroepiandrosterone, progesterone, 17α-hydroxyprogesterone, androstenedione, testosterone, and estradiol-17α (De Paoli and Eik-Nes, 1963; Aakvaag *et al.,* 1964; Aakvaag and Eik-Nes, 1965).

When [7α-^{3}H]dehydroepiandrosterone and [^{14}C]17α-hydroxyprogesterone were infused through the ovarian artery, the conversion of dehydroepiandrosterone to androstenedione and estradiol-17β was much greater than that from 17β-hydroxyprogesterone. Similar experiments with labeled androstenedione and testosterone confirm the view that the pathway for estrogen biosynthesis is by way of androstenedione (Aakvaag *et al.,* 1964), and the results agree with those obtained from incubations of ovarian tissue (Ryan and Smith, 1965).

When [^{14}C]androstenedione was perfused through the ovarian artery in a semi *in vivo* system, its conversion to [^{14}C]estradiol-17β was much greater than that to [^{14}C]estrone (Engels *et al.,* 1968). Clomiphene was found to stimulate the conversion of androstenedione to estrogens, confirming that this substance, in addition to its effect on the hypothalamic–pituitary axis, has a direct effect on the aromatizing enzyme (Hagerman *et al.,* 1966).

K. Other Species

With the technological improvements for measuring substances in small amounts of blood, it has become more common for reports to contain not only descriptions of the structural changes in the ovaries but also the concomitant hormone levels in ovarian tissues and ovarian venous and peripheral plasma. Some of these studies have been less extensive than others because of lack of suitable animals, but data for many species are accumulating—mainly on peripheral plasma levels of hormones.

The animals studied in most detail have been those in which some peculiarities of the reproductive cycle are known (see Volume II, Chapter 6). For example, roe deer (Short and Hay, 1966), badgers (Canivenc, 1966), armadillos (Talmage and Buchanan, 1954; Brinck-Johnson et al., 1967), and spotted skunks (Mead and Eik-Nes, 1969) experience a delay of implantation, but as yet there appears to be no common pattern of steroid secretion. The mink also has a delay of implantation and complex multiple mating requirements (Moller, 1973a, 1974); plasma progesterone levels have also been studied in the ferret (Heap and Hammond, 1974) and the blue fox (Møller, 1973b). Like mink, alpaca are commercially valuable animals for which reproductive parameters are needed (Fernandez-Baca et al., 1970). Of the more usual domesticated species, goats (Linzell and Heap, 1968) and horses (Short, 1959; Savard, 1961) have also been investigated. All equids seem to have unusual estrogens–equilin and equilenin (see Rowlands et al., 1975). The South American hystricomorph rodents have been studied because of their long gestation periods, but again the progesterone levels in pregnancy do not show a similar pattern in the different species (guinea pig: Heap and Deanesly, 1966, 1967; Heap et al., 1967; Illingworth et al., 1970; coypu: Rowlands and Heap, 1966; acouchi: Rowlands et al., 1970; chinchilla and cuis: Tam, 1974). Although the concentration of progesterone rises progessively throughout pregnancy in hystricomorphs, as in women, this does not represent increased production of progesterone. The dynamics of progesterone metabolism are unusual in that the *MCR* decreases dramatically due to a marked increase in the concentration of a specific binding globulin (Heap and Illingworth, 1974); the source of the globulin has not yet been established. There have been several studies on African elephants (Short and Buss, 1965; Hanks and Short, 1972; Ogle et al., 1973), but peripheral progesterone levels and the progesterone concentrations in the corpus luteum have been exceedingly low. Recent studies on the hyrax (believed to be related to elephants) show that plasma progesterone levels are very low, although luteal progesterone concentrations are comparable to those of many other mammals (R. B. Heap, S. Gombe, and J. B. Sale, personal communication). Progesterone levels are also very low at all stages of reproduction in the tammar wallaby, but are higher during pregnancy than during the estrous cycle, although the gestation period is shorter (Lemon, 1972).

V CONCLUSIONS

From the information now available concerning the secretion of ovarian steroids, certain generalizations can be made.

The Graafian follicle of every species so far studied secretes estradiol-17β as the major estrogen. The follicle appears to be the only source of this estrogen in the follicular phase of the cycle, although appreciable quantities of estrone are produced from other sources. The preovulatory rise in the secretion of estrogen is responsible for the manifestation of estrus and provokes the ovulatory discharge of LH from the anterior pituitary (Harris and Naftolin, 1970). Thus the stimulus for release of LH can only be evoked by a Graafian follicle which has reached the last stages of maturation and contains the cellular apparatus for conversion into a corpus luteum. The correct synchronization of these events is important in ensuring that coitus occurs at a time of the ovarian cycle when a mature ovum is available for fertilization.

In addition to estradiol-17β, the follicle of many species secretes androstenedione, testosterone, 17α-hydroxyprogesterone, and estrone. The possible physiological significance of the secretion of these steroids by the follicle has been discussed (Section IV). Very little is known about the functions of androgens in the female, although it has been suggested that they are concerned with sexual libido and/or synthesis of protein (Herbert, 1972).

The granulosa cells of the mature follicle undergo histological changes suggestive of early luteinization a short time before ovulation. These changes occur simultaneously with, and are almost certainly dependent on, the preovulatory discharge of LH. A preovulatory secretion of a progestin has been demonstrated in women (Johansson and Wide, 1969; Yussman and Taymor, 1970), rhesus monkey (Weick *et al.,* 1973), sheep (Wheeler *et al.,* 1975), rabbit (Hilliard *et al.,* 1969), rat (Cortes *et al.,* 1971), hamster (Baranczuk and Greenwald, 1973), and guinea pig (Feder *et al.,* 1968). In the rabbit, 20α-dihydroprogesterone apparently reinforces and prolongs the ovulatory discharge of LH (Hilliard *et al.,* 1969), and in other species preovulatory progesterone may facilitate the positive feedback effect of estradiol (Harris and Naftolin, 1970).

In all species, progesterone is the major steroid secreted by the corpus luteum (Yoshinaga, 1973). In the rat, which, like the hamster, does not develop a functional corpus luteum in the absence of mating (see Volume II, Chapter 6), the corpus luteum changes rapidly from the secretion of progesterone to 20α-dihydroprogesterone. The primate corpus luteum is unusual in its ability to synthesize and secrete androgens and estrogens as well as gestagens, and this may be because of the persistence of prominent theca cells within the corpus luteum (Corner, 1956). In other species, estradiol appears to originate from the follicles which continue to develop throughout the luteal phase. It has been suggested that the cellular compartments in the ovary responsible for the secretion of estrogen in the luteal phase have a critical influence on the length of the follicular phase (Baird *et al.*, 1975), and possibly on the mechanism of luteal regression (Baird, 1973b). Further careful studies on the secretory activity of the corpus luteum in a variety of species is required.

The function of many steroids known to be secreted by the ovary (and the adrenal) is unknown. They may represent a "leak" of key biosynthetic intermediates (Short, 1960b). In some instances, they act as prehormones and may provide a constant source of biological activity (Baird *et al.*, 1968b).

REFERENCES

Aakvaag, A., and Eik-Nes, K. B. (1965). Metabolism *in vivo* of steroids in the canine ovary. *Biochim. Biophys. Acta* **111**, 273–285.

Aakvaag, A., and Eik-Nes, K. B. (1969). Biosynthesis and secretion in vivo of ovarian steroids in the canine. *In* "The Gonads" (K. W. McKerns, ed.), pp. 93–118. Appleton, New York.

Aakvaag, A., and Fylling, P. (1968). A method for the simultaneous determination of progesterone, androstenedione, testosterone, and dehydroepiandrosterone sulphate in biological fluid. Its application in the analysis of venous plasma and cyst fluid from human ovaries *in situ. Acta Endocrinol. (Copenhagen)* **57**, 447–456.

Aakvaag, A., Hagen, A. A., and Eik-Nes, K. B. (1964). Biosynthesis *in vivo* of testosterone and $^4\Delta$ androstenedione from dehydroepiandrosterone sodium sulphate by the canine testis and ovary. *Biochim. Biophys. Acta* **86**, 622–627.

Abraham, G. E. (1974). Ovarian and adrenal contribution to peripheral androgens during the menstrual cycle. *J. Clin. Endocrinol. Metab.* **39**, 340–346.

Abraham, G. E., and Chakmakjian, Z. H. (1973). Serum steroid levels during the menstrual cycle in a bilaterally adrenalectomized woman. *J. Clin. Endocrinol. Metab.* **37**, 581–587.

Abraham, G. E., Lobotsky, J. H., and Lloyd, C. W. (1969). Metabolism of testosterone and androstenedione in normal and ovariectomized women. *J. Clin. Invest.* **48**, 696–703.

Abraham, G. E., Odell, W., Swerdloff, R., and Hopper, K. (1972). Simultaneous radioimmunoassay of plasma FSH, LH, progesterone, 17-hydroxyprogesterone, and estradiol-17β during the menstrual cycle. *J. Clin. Endocrinol. Metab.* **34**, 312–318.

Abraham, G. E., Bister, J. E., Kyle, F. W., Corrales, P. C., and Teller, R. C. (1973). Radioimmunoassay of plasma pregnenolone. *J. Clin. Endocrinol. Metab.* **37**, 40–45.

Ahrén, K., Hamberger, L., and Rubinstein, L. (1969). Acute in vivo effects of gonadotrophins on the metabolism of the rat ovary. *In* "The Gonads" (K. W. McKerns, ed.), pp. 327–350. Appleton, New York.

Ahrén, K., Janson, P. O., and Selstam, G. (1971). Perfusion of ovaries *in vitro* and *in vivo. Acta Endocrinol. (Copenhagen), Suppl.* **158**, 285–303.

Aitken, W. A. (1927). Some observations on the oestrous cycle and reproductive phenomenon of the mare. *J. Amer. Vet. Med. Assoc.* **23**, 481–491.

Arcos, M., Gurpide, E., Vande Wiele, R. L., and Lieberman, S. (1964). Precursors of urinary pregnanediol and their influence on the determination of the secretory rate of progesterone. *J. Clin. Endocrinol. Metab.* **24,** 237.

Armstrong, D. T., Jackanicz, T. M., and Keyes, P. L. (1969). Regulation of steroidogenesis in the rabbit ovary. *In* "The Gonads" (K. W. McKerns, ed.), pp. 3–21. Appleton, New York.

Baird, D. T. (1968). A double isotope derivative method for the estimation of oestrone and oestradiol-17β in peripheral human blood, ovarian and adrenal venous blood of sheep and other biological fluids using (^{35}S) pipsyl chloride. *J. Clin. Endocrinol. Metab.* **28,** 244–258.

Baird, D. T. (1970). The secretion of androgens and oestrogens from the ovary and adrenal gland. *In* "Reproductive Endocrinology" (W. J. Irvine, ed.), pp. 95–98. Livingstone, Edinburgh.

Baird, D. T. (1971). Steroids in blood reflecting ovarian function. *In* "Control of Gonadal Steroid Secretion" (D. T. Baird and J. A. Strong, eds.), pp. 175–189. Edinburgh Univ. Press, Edinburgh.

Baird, D. T. (1973a). The secretion of oestrogens from the ovary in normal and abnormal menstrual cycles. *Proc. 4th Int. Congr. Endocrinol., 1972,* pp. 851–856.

Baird, D. T. (1973b). Steroid synthesis and secretion in the compartments of the ovary. *In* "Le Corps Jaune" (R. Denamur and A. Netter, eds.), pp. 33–38. Masson, Paris.

Baird, D. T. (1974). The endocrinology of ovarian steroid secretion. *Eur. J. Obstet. Gynecol. Reprod. Biol.* **4,** 31–39.

Baird, D. T., and Fraser, I. S. (1974). Blood production and ovarian secretion rates of oestradiol-17β and oestrone in women throughout the menstrual cycle. *J. Clin. Endocrinol. Metab.* **38,** 779–787.

Baird, D. T., and Fraser, I. S. (1975). Concentration of oestrone and oestradiol-17β in follicular fluid and ovarian venous blood of women. *Clin. Endocrinol.* **4,** 259–266.

Baird, D. T., and Guevara, A. (1969). Concentration of unconjugated oestrone and oestradiol in peripheral plasma in non-pregnant women throughout the menstrual cycle, castrate and post-menopausal women and in men. *J. Clin. Endocrinol. Metab.* **29,** 149–156.

Baird, D. T., and Scaramuzzi, R. J. (1975). Prostaglandin F2α and luteal regression in the ewe: comparison with 16-aryloxy-prostaglandin (I.C.I. 80,996). *Ann. Biol. Anim. Biochim. Biophys.* **15,** 161–174.

Baird, D. T., Goding, J. R., Ichikawa, Y., and McCracken, J. A. (1968a). The secretion of steroids from the autotransplanted ovary in the ewe spontaneously and in response to systemic gonadotrophin. *J. Endocrinol.* **42,** 283–299.

Baird, D. T., Horton, R., Longcope, C., and Tait, J. F. (1968b). Steroid prehormones. *Perspect. Biol. Med.* **11,** 384–421.

Baird, D. T., Uno, A., and Melby, J. C. (1969a). Adrenal secretion of androgens and oestrogens. *J. Endocrinol.* **45,** 135–136.

Baird, D. T., Horton, R., Longcope, C., and Tait, J. F. (1969b). Steroid dynamics under steady state conditions. *Recent Progr. Horm. Res.* **25,** 611–656.

Baird, D. T., McCracken, J. A., and Goding, J. R. (1973a). Studies in steroid synthesis and secretion with the autotransplanted sheep ovary and adrenal. *In* "The Endocrinology of Pregnancy and Parturition" (C. G. Pierrepoint, ed.), pp. 5–16. Alpha Omega Alpha Publ., Cardiff, Wales.

Baird, D. T., Burger, P. E., Heavon-Jones, G. D., and Scaramuzzi, R. J. (1973b). Androstenedione: a secretory product of the Graafian follicle. *J. Endocrinol.* **59**, xxxvi (abstr.).

Baird, D. T., Burger, P. E., Heavon-Jones, G. D., and Scaramuzzi, R. J. (1974). The site of secretion of androstenedione in non-pregnant women. *J. Endocrinol.* **63**, 201–202.

Baird, D. T., Baker, T. G., McNatty, K. P., and Neal, P. (1975). Relationship between the secretion of the corpus luteum and the length of the follicular phase in the ovarian cycle. *J. Reprod. Fertil.* **45**, 611–619.

Baird, D. T., Land, R. B., Scaramuzzi, R. J. and Wheeler, A. G. (1976a). Functional assessment of the autotransplanted uterus and ovary in the ewe. *Proc. R. Soc. London, Ser B.* **192**, 463–474.

Baird, D. T., Land, R. B., Scaramuzzi, R. J., and Wheeler, A. G. (1976b). Endocrine changes associated with lateral regression in the ewe: the secretion of ovarian oestradiol, progesterone and androstenedione and uterine prostaglandin F2α throughout the oestrous cycle. *J. Endocrinol.* **69**, 275–286.

Baranczuk, R., and Greenwald, G. S. (1973). Peripheral levels of oestrogen in the cyclic hamster. *Endocrinology* **92**, 805–812.

Barlow, J. J., and Logan, C. H. (1966). Oestrogen secretion, biosynthesis and metabolism; their relationship to the menstrual cycle. *Steroids* **7**, 309–314.

Barlow, J. J., Emerson, K., and Saxena, B. N. (1969). Oestradiol production after ovariectomy for carcinoma of the breast. *N. Engl. J. Med.* **280**, 633–637.

Bartosik, D. B., and Romanoff, E. B. (1969). The luteotrophic process: effects of prolactin and LH on sterol and progesterone metabolism in bovine luteal ovaries perfused in vitro. *In* "The Gonads" (K. W. McKerns, ed.), pp. 211–244. Appleton, New York.

Bermudez, J. A., Doerr, P., and Lipsett, M. B. (1970). Measurement of pregnenolone in blood. *Steroids* **16**, 505–515.

Bolte, E., Coudert, S., and Lefebvre, Y. (1974). Steroid production from plasma cholesterol. II. In vivo conversion of plasma cholesterol to ovarian progesterone and adrenal C_{19} and C_{21} steroids in humans. *J. Clin. Endocrinol. Metab.* **38**, 394–400.

Bosu, W. T. K., and Johansson, E. D. B. (1974). Effect of HCG on plasma levels of oestrogens and progesterone during the luteal phase of the menstrual cycle in rhesus monkeys (*Macaca mulatta*). *Int. J. Fertil.* **19**, 28–32.

Bosu, W. T. K., Johansson, E. D. B., and Gemzell, C. (1973). Peripheral plasma levels of oestrone, oestradiol-17β and progesterone during ovulatory menstrual cycles in the rhesus monkey with special reference to the onset of menstruation. *Acta Endocrinol. (Copenhagen)* **74**, 732–742.

Brink-Johnson, T., Benirschke, K., and Brink-Johnsen, K. (1967). Hormonal steroids in the armadillo, *Dasypus novemcinctus*. *Acta Endocrinol. (Copenhagen)* **56**, 675–690.

Brinkley, H. J., and Young, E. P. (1968). Determination of the in vivo rate of progesterone secretion by the ovary of the pig during the luteal phase of the estrous cycle. *Endocrinology* **82**, 203–208.

Brown, J. B. (1955). A chemical method for the determination of oestriol, oestrone and oestradiol in human urine. *Biochem. J.* **60**, 185–193.

Brown, J. B. (1957). The relationship between urinary oestrogens and oestrogens produced in the body. *J. Endocrinol.* **16**, 202–212.

Brown, J. B., and Matthew, G. D. (1962). The application of urinary oestrogen measurements to problems in gynaecology. *Recent Progr. Horm. Res.* **18**, 337–373.

Brown-Grant, K., Exley, D., and Naftolin, F. (1970). Peripheral plasma oestradiol and luteinizing hormone concentrations during the oestrous cycle of the rat. *J. Endocrinol.* **48**, 295–296.

Butenandt, A. (1929). Über "Progynon" ein Krystallisieres Weibliches Sexual hormon. *Naturwissenschaften* **17**, 879.

Canivenc, R. (1966). A study of progestation in the European badger (*Meles meles* L.). *Symp. Zool. Soc. London* **15**, 15–26.

Challis, J. R. G., Harrison, F. A., and Heap, R. B. (1973). The kinetics of oestradiol-17β metabolism in the sheep. *J. Endocrinol.* **57**, 97–110.

Channing, C. P. (1969). The use of tissue culture of granulosa cells as a method of studying the mechanism of luteinization. *In* "The Gonads" (K. W. McKerns, ed.), pp. 245–276. Appleton, New York.

Channing, C. P., and Kammerman, S. (1973). Characteristics of gonadotrophin receptors of porcine granulosa cells during follicle maturation. *Endocrinology* **92**, 531–539.

Collett, R. A., Land, R. B., and Baird, D. T. (1973). The pattern of progesterone secretion by the autotransplanted ovary of the ewe in response to ovine luteinizing hormone. *J. Endocrinol.* **56**, 403–411.

Collins, D. C., Graham, C. E., and Preedy, J. R. K. (1973). Urinary levels of oestrogens, pregnanediol and androsterone during the menstrual cycle of the orangutan. *Biol. Reprod.* **9**, 107 (abstr.).

Corker, C., Naftolin, F., and Exley, D. (1969). Menstrual cycle—midcycle rise in plasma oestradiol and luteinizing hormone. *Nature* (*London*) **222**, 1063.

Corner, J. W., Jr. (1956). The histological dating of the corpus luteum of menstruation. *Amer. J. Anat.* **98**, 377–401.

Cortes, V., McCracken, J. A., Lloyd, C. W., and Weisz, J. (1971). Progestin production by the ovary of the testosterone-sterilized rat treated with an ovulatory dose of LH, and the normal, proestrous rat. *Endocrinology* **89**, 878–885.

Cox, R. I., Mattner, P. E., Shutt, D. A., and Thorburn, G. D. (1971a). Ovarian secretion of oestradiol-17β during the oestrous cycle in the ewe. *J. Reprod. Fertil.* **24**, 133–134.

Cox, R. I., Mattner, P. E., and Thorburn, G. D. (1971b). Changes in ovarian secretion of oestradiol-17β around oestrus in the sheep. *J. Endocrinol.* **49**, 345–346.

Cupps, P. T., Anderson, L. L., and Cole, H. H. (1969). The estrous cycle. *In* "Reproduction in Domestic Animals" (H. H. Cole and P. T. Cupps, eds.), pp. 217–250. Academic Press, New York.

de Jong, F. A., Baird, D. T., and van der Molen, H. J. (1974). Ovarian secretion rates of oestrogens, androgens and progesterone in normal women and in women with persistent ovarian follicles. *Acta Endocrinol.* (*Copenhagen*) **77**, 575–587.

Delforge, J. R., Thomas, K., Roux, F., Carneiro De Seiquira, J., and Ferin, J. (1972). Time relationship between granulosa cells growth and luteinization, and plasma luteinizing hormone discharge in human. 1. A morphometric analysis. *Fertil. Steril.* **23**, 1–11.

Dell'Acqua, S., Mancuso, S., Catelli, M. G., and Bompiani, A. (1971). Conversion of [³H] cholesterol to neutral steroids by human ovaries perfused *in vivo*. *Proc. 3rd Int. Congr. Horm. Steroids, 1970* Excerpta Med. Found. Int. Congr. Ser. No. 210, Abstr. 240.

De Paoli, J. C., and Eik-Nes, K. B. (1963). Metabolism in vivo of ($7\alpha^3$H) pregnenolone by the dog ovary. *Biochim. Biophys. Acta* **78**, 457–465.

Diczfalusy, E. (1970). Steroid assay by protein binding. *Acta Endocrinol. (Copenhagen), Suppl.* **147**.

Di Pietro, D. L., Brown, R. D., and Strott, C. A. (1972). A pregnenolone radioimmunoassay utilising a new fractionation technique for sheep antiserum. *J. Clin. Endocrinol. Metab.* **35**, 729–735.

Dobrowolski, W., Stupnicka, E., and Domanski, E. (1968). Progesterone levels in ovarian venous blood during the oestrous cycle of the cow. *J. Reprod. Fertil.* **15**, 409–414.

Dobson, H., and Dean, P. D. G. (1974). Radioimmunoassay of oestrone, oestradiol-17α and -17β in bovine plasma during the oestrous cycle and last stages of pregnancy. *J. Endocrinol.* **61**, 479–486.

Doisy, E. A., Veler, C. D., and Thayer, S. A. (1929). Folliculin from urine of pregnant women. *Amer. J. Physiol.* **90**, 329–330 (abstr.).

Domanski, E., Skrzeczkowski, L., Stupnicka, E., Fitko, R., and Dobrowolski, W. (1967). The effect of gonadotrophins on the secretion of progesterone and oestrogens by the sheep ovary perfused *in situ. J. Reprod. Fertil.* **14**, 365–372.

Douss, T. N., and Desphpande, N. (1968). In vivo perfusion of the human adrenal gland and ovary in patients with mammary cancer. *Brit. J. Surg.* **55**, 673–677.

Dupon, C., Hosseinian, A., and Kim, M. H. (1973). Simultaneous determination of plasma oestrogens, androgens and progesterone during the human menstrual cycle. *Steroids* **22**, 47–61.

Eaton, L. W., and Hilliard, J. (1971). Oestradiol-17β, progesterone and 20α-OH-pregn-4-ene-3-one in rabbit ovarian venous plasma. 1. Steroid secretion from paired ovaries with and without corpora lutea: effect of LH. *Endocrinology* **89**, 105–111.

Edgar, D. G., and Ronaldson, J. W. (1958). Blood levels of progesterone in the ewe. *J. Endocrinol.* **16**, 378–384.

Edwards, R. G., Steptoe, P. C., Abraham, G. E., Walters, E., Purdy, J. M., and Fotherby, K. (1972). Steroid assays and pre-ovulatory follicular development in human ovaries primed with gonadotrophin. *Lancet* **2**, 611–615.

Eik-Nes, K. B., and Hall, P. F. (1965). Secretion of steroid hormones *in vivo. Vitam. Horm. (N.Y.)* **23**, 153–202.

Engels, J. A., Friedlander, R. L., and Eik-Nes, K. B. (1968). An effect in vivo of clomiphene in the rate of conversion of androstenedione C^{14} to estrone-C^{14} and estradiol-C^{14} by the canine ovary. *Metab. Clin. Exp.* **27**, 189–198.

Eren, S., Reynolds, G. H., Turner, M. B., Schmidt, F. H., Mackay, J. H., Howard, C. M., and Preedy, J. R. K. (1967). Oestrogen metabolism in the human. III. A comparison between females, studied during the first and second halves of the menstrual cycle, and males. *J. Clin. Endocrinol. Metab.* **27**, 1451–1462.

Eto, T., Masuda, H., Suzuki, Y., and Hosi, T. (1962). Progesterone and preg-4-en-20α-ol-3-one in rat ovarian venous blood at different stages in reproductive cycle. *Jpn. J. Anim. Reprod.* **8**, 39–40.

Everitt, B. J., and Herbert, J. (1971). The effects of dexamethasone and androgens on sexual receptivity of female monkeys. *J. Endocrinol.* **51**, 575–588.

Fajer, A. B., and Barraclough, C. A. (1967). Ovarian secretion of progesterone and 20α-hydroxypregn-4-ene-3-one during pseudopregnancy and pregnancy in rats. *Endocrinology* **81**, 617–622.

Fajer, A. B., and Bechini, D. (1971). Pregnenolone and progesterone concentrations in ovarian venous blood of the artificially ovulated squirrel monkey (*Saimiri sciureus*). *J. Reprod. Fertil.* **27**, 193–200.

Falconer, I. R. (1963). The exteriorization of the thyroid gland and measurement of its function. *J. Endocrinol.* **26**, 241–247.

Feder, H. H., Resko, J. A., and Goy, R. W. (1968). Progesterone concentrations in the arterial plasma of guinea pigs during the oestrous cycle. *J. Endocrinol.* **40**, 505–513.

Fernandez-Baca, S., Hansel, W., and Novoa, C. (1970). Corpus luteum function in the alpaca. *Biol. Reprod.* **3**, 252–261.

Fishman, J., Brown, J. B., Hellman, L., Zumoff, B., and Gallagher, T. F. (1962). Estrogen metabolism in normal and pregnant women. *J. Biol. Chem.* **237**, 1489–1494.

Flores, F., Naftolin, F., Ryan, K. J., and White, R. J. (1973). Estrogen formation by the isolated perfused uterus monkey brain. *Science* **180**, 1074–1075.

Fotherby, K. (1962). The ovarian production of a pregnanetriol precursor. *J. Endocrinol.* **25**, 19–28.

Fukushima, D. K., Bradlow, L. H., Hellman, L., and Gallagher, T. F. (1970). Cortisol metabolism in the morning and evening; relation to cortisol secretion rate measurements. *Steroids* **16**, 603–610.

Gandy, H. M., and Peterson, R. E. (1968). Measurement of unconjugated testosterone and 17-keto steroids in plasma by the double isotope dilution derivative technique. *J. Clin. Endocrinol. Metab.* **28**, 949–977.

Gemmell, R. T., Stacy, B. D., and Thorburn, G. D. (1974). Ultrastructural study of secretory granules in the corpus luteum of the sheep during the estrous cycle. *Biol. Reprod.* **11**, 447–462.

Giorgi, E. P. (1965). Steroids in cyst fluid from ovaries from normally menstruating women and women with functional uterine bleeding. *J. Reprod. Fertil.* **10**, 309–319.

Giorgi, E. P. (1967). Determination of free and conjugated oestrogens in fluid from human ovaries. *J. Endocrinol.* **37**, 219.

Gleeson, A. R., and Thorburn, G. D. (1973). Plasma progesterone and prostaglandin F concentration in the cyclic sow. *J. Reprod. Fertil.* **32**, 343–344.

Glencross, R. G., Munro, I. B., Senior, B. E., and Pope, G. S. (1973). Concentration of 17β-oestradiol, oestrone and progesterone in jugular venous plasma of cows during the oestrous cycle and in early pregnancy. *Acta Endocrinol. (Copenhagen)* **73**, 374–384.

Goding, J. R., McCracken, J. A., and Baird, D. T. (1967). The study of ovarian function in the ewe by means of a vascular autotransplantation technique. *J. Endocrinol.* **39**, 37–62.

Goding, J. R., Baird, D. T., Cumming, I. A., and McCracken, J. A. (1971). Functional assessment of autotransplanted endocrine organs. *Acta Endocrinol. (Copenhagen), Suppl.* **158**, 169–191.

Gomes, W. R., Estergreen, V. C., Frost, O. L., and Erb, R. E. (1963). Progestin levels in jugular and ovarian venous blood, corpora lutea and ovaries of the non-pregnant bovine. *J. Dairy Sci.* **46**, 553–558.

Gomes, W. R., Herschler, R. C., and Erb, R. E. (1965). Progesterone levels in ovarian venous effluent of the non-pregnant sow. *J. Anim. Sci.* **24**, 722–725.

Graham, C. E., Collins, D. C., Robinson, H., and Preedy, J. R. K. (1972). Urinary levels of estrogens and pregnanediol and plasma levels of progesterone during the menstrual cycle of the chimpanzee: relationship to the sexual swelling. *Endocrinology* **91**, 13–24.

Gray, C. H., and Bacharach, A. L., eds. (1967). "Hormones in Blood," 2nd rev. ed., Vol. 2. Academic Press, New York.

Gurpide, E., Angers, M., Vande Wiele, R. L., and Lieberman, S. (1962). Determination of secretory rates of oestrogens in pregnant and non-pregnant women from specific activities of urinary metabolites. *J. Clin. Endocrinol. Metab.* **22**, 935–945.

Hagerman, D. D., Smith, O. W., and Day, C. F. (1966). Mechanism of the stimulatory effect of Clomid® on aromatization of steroids by human placenta in vitro. *Acta Endocrinol. (Copenhagen)* **51**, 591–598.

Hanks, J., and Short, R. V. (1972). The formation and function of the corpus luteum in the African elephant, *Loxodonta africana. J. Reprod. Fertil.* **29**, 79–89.

Hansel, W., and Echternkamp, S. E. (1972). Control of ovarian function in domestic animals. *Amer. Zool.* **12**, 225–243.

Harris, G. W., and Naftolin, F. (1970). The hypothalamus and control of ovulation. *Brit. Med. Bull.* **26**, 3–9.

Harrison, F. A., Heap, R. B., and Linzell, J. L. (1968). Ovarian function in the sheep after autotransplantation of the ovary and uterus to the neck. *J. Endocrinol.* **40**, xiii.

Hayward, J. N., Hilliard, J., and Sawyer, C. H. (1963). Pre-ovulatory and post-ovulatory progestins in monkey ovary and ovarian vein blood. *Proc. Soc. Exp. Biol. Med.* **113**, 256–259.

Heap, R. B., and Deanesly, R. (1966). Progesterone in systemic blood and placentae of intact and ovariectomized pregnant guinea-pigs. *J. Endocrinol.* **34**, 417–423.

Heap, R. B., and Deanesly, R. (1967). The increase in plasma progesterone levels in the pregnant guinea-pig and its possible significance. *J. Reprod. Fertil.* **14**, 339–341.

Heap, R. B., and Hammond, J., Jr. (1974). Plasma progesterone levels in pregnant and pseudopregnant ferrets. *J. Reprod. Fertil.* **39**, 149–152.

Heap, R. B., and Illingworth, D. V. (1974). The maintenance of gestation in the guinea-pig and other hystricomorph rodents: changes in the dynamics of progesterone metabolism and the occurrence of progesterone-binding globulin (PBG). *Symp. Zool. Soc. London* **34**, 385–415.

Heap, R. B., Perry, J. S., and Rowlands, I. W. (1967). Corpus luteum function in the guinea-pig; arterial and luteal progesterone levels, and the effects of hysterectomy and hypophysectomy. *J. Reprod. Fertil.* **13**, 537–553.

Hembree, W. C., Bardin, C. W., and Lipsett, M. B. (1969). A study of estrogen metabolic clearance rates and transfer factors. *J. Clin. Invest.* **48**, 1809–1819.

Hemsell, D. L., Grodin, J. M., Brenner, P. F., Siiteri, P. K., and MacDonald, P. C. (1974). Plasma precursors of estrogen. II. Correlation of the extent of conversion of plasma androstenedione to estrone with age. *J. Clin. Endocrinol. Metab.* **38**, 476–479.

Henricks, D. M., Guthrie, H. D., and Handlin, D. L. (1971). Plasma estrogen, progesterone and luteinizing hormone levels during the estrous cycle in pigs. *Biol. Reprod.* **6**, 210–218.

Herbert, J. (1972). Behavioural patterns. *In* "Reproductive Patterns" (C. R. Austin and R. V. Short, eds.), pp. 34–68. Cambridge Univ. Press, London and New York.

Hertig, A. (1967). Morphological determinants of placentation. *In* "Fetal Homeostasis" (R. M. Wynn, ed.), Vol. II, pp. 98–109. N.Y. Acad. Sci., New York.

Hilliard, J., and Eaton, L. W. (1971). Estradiol-17β, progesterone, and 20α-hydroxypregn-4-ene-3-one in rabbit ovarian venous plasma. II. From mating through implantation. *Endocrinology* **89,** 522–527.

Hilliard, J., Archibald, D., and Sawyer, C. H. (1963). Gonadotropic activation of pre-ovulatory synthesis and release of progestin in the rabbit. *Endocrinology* **72,** 59–66.

Hilliard, J., Penardi, R., and Sawyer, C. H. (1967). Functional role of 20α-hydroxypregn-4-ene-3-one in the rabbit. *Endocrinology* **80,** 901–909.

Hilliard, J., Spies, H. G., and Sawyer, C. H. (1969). Hormonal factors regulating ovarian cholesterol mobilization and progestin secretion in intact and hypophysectomized rabbits. *In* "The Gonads" (K. W. McKerns, ed.), pp. 55–85. Appleton, New York.

Hilliard, J., Scaramuzzi, R. J., Penardi, R., and Sawyer, C. H. (1973). Progesterone, estradiol and testosterone levels in ovarian venous blood of pregnant rabbits. *Endocrinology* **93,** 1235–1238.

Hilliard, J., Scaramuzzi, R. J., Pang, C.-N., Penardi, R., and Sawyer, C. H. (1974a). Testosterone secretion by the rabbit ovary *in vivo*. *Endocrinology* **94,** 267–271.

Hilliard, J., Pang, C.-N., Scaramuzzi, R. J., Penardi, R., and Sawyer, C. H. (1974b). Secretion rates of estradiol, testosterone and progesterone from right and left rabbit ovaries cannulated concurrently or successively. *Biol. Reprod.* **10,** 364–369.

Hirai, M., Morita, Y., and Nakao, T. (1968). Isolation of 4-pregnen-20α-ol-3-one from rats' ovarian venous blood. *Perspect. Biol. Med.* **11,** 427–440.

Hoffman, D. C., and Fajer, A. B. (1970). Progesterone concentration in the ovarian venous blood of the hamster during the estrous cycle and pregnancy. *Fed. Proc. Fed. Amer. Soc. Exp. Biol.* **9,** 250.

Holst, P. J., Braden, A. W. H., and Mattner, P. E. (1972). Oestradiol-17β secretion from the ewe ovary and related ovarian morphology on Days 2 and 3 of the cycle. *J. Reprod. Fertil.* **28,** 136.

Hopper, B., and Tullner, W. W. (1970). Urinary estrone and plasma progesterone levels during the menstrual cycle of the Rhesus monkey. *Endocrinology* **86,** 1225–1230.

Hori, T., Ide, M., and Miyaket, T. (1968). Ovarian estrogen secretion during the estrous cycle and under the influence of exogenous gonadotrophins in rats. *Endocrinol. Jpn.* **15,** 215–222.

Horton, R., and Tait, J. F. (1966). Androstenedione production and inter-conversion rates measured in peripheral blood and studies on the possible site of its conversion to testosterone. *J. Clin. Invest.* **45,** 301–313.

Horton, R., and Tait, J. F. (1967). In vivo conversion of dehydroisoandrosterone to plasma androstenedione and testosterone in man. *J. Clin. Endocrinol. Metab.* **27,** 79–88.

Horton, R., Romanoff, E., and Walker, J. (1966). Androstenedione and testosterone in ovarian venous and peripheral plasma during ovariectomy for breast cancer. *J. Clin. Endocrinol. Metab.* **26,** 1267–1269.

Hotchkiss, J., Atkinson, L. E., and Knobil, E. (1971). Time course of serum estrogen and luteinizing hormone (LH) concentrations during the menstrual cycle of the Rhesus monkey. *Endocrinology* **89,** 177–183.

Ichikawa, S., Morioka, H., and Sawada, T. (1971). Identification of the neutral steroids in the ovarian venous plasma of LH-stimulated rats. *Endocrinology* **88,** 372–383.

Illingworth, D. V., Heap, R. B., and Perry, J. S. (1970). Changes in the metabolic clearance rate of progesterone in the guinea-pig. *J. Endocrinol.* **48,** 409–417.

Ito, T., and Horton, R. (1970). Dihydrotestosterone in human peripheral plasma. *J. Clin. Endocrinol. Metab.* **31,** 362–368.

Jacobs, J. B. (1969). Selective gonadal venography. *Radiology* **92,** 885–888.

Janson, P. O., and Albrecht, I. (1975). Methodological aspects of blood flow measurements in ovaries containing corpora lutea. *J. Appl. Physiol.* **38,** 288–293.

Johansson, E. D. B. (1969). Progesterone levels in peripheral plasma during the luteal phase of the normal human menstrual cycle measured by a competitive protein binding technique. *Acta Endocrinol. (Copenhagen)* **61,** 592–606.

Johansson, E. D. B., and Wide, L. (1969). Periovulatory levels of plasma progesterone and luteinizing hormone in women. *Acta Endocrinol. (Copenhagen)* **62,** 82–88.

Johansson, E. D. B., Neill, J. D., and Knobil, E. (1968). Periovulatory progesterone concentration in the peripheral plasma of the Rhesus monkey with a methodologic note on the detection of ovulation. *Endocrinology* **82,** 143–148.

Judd, H. L., and Yen, S. S. C. (1973). Serum androstenedione and testosterone levels during the menstrual cycle. *J. Clin. Endocrinol. Metab.* **36,** 475–481.

Kahn, P. C. (1967). Radiological identification of functioning ovarian tumors. *Clin. Radiol. North Am.* **5,** 221–230.

Kalliala, K., Laatikainen, T., Luukkainen, T., and Vihko, R. (1970). Neutral steroid sulphates in human ovarian blood. *J. Clin. Endocrinol. Metab.* **30,** 533–535.

Karsch, F. J., Krey, L. C., Weick, R. F., Dierschke, D. J., and Knobil, E. (1973). Functional luteolysis in the Rhesus monkey: the role of estrogen. *Endocrinology* **92,** 1148–1152.

Kelly, W. G. (1970). Questions concerning the validity of one of the assumptions underlying the determination of the secretory rate of cortisol. *Steroids* **16,** 579–602.

Kirschner, M. A., and Jacobs, J. R. (1971). Combined ovarian and adrenal vein catheterization to determine the site(s) of androgen over-production in hirsute women. *J. Clin. Endocrinol. Metab.* **33,** 199–209.

Kirschner, M. A., and Taylor, J. P. (1972). Urinary estrogen production rates in normal and endocrine-ablated subjects. *J. Clin. Endocrinol. Metab.* **35,** 513–521.

Knobil, E. (1973). On the regulation of the primate corpus luteum. *Biol. Reprod.* **8,** 246–258.

Knobil, E. (1974). On the control of gonadotrophin secretion in the Rhesus monkey. *Recent Progr. Horm. Res.* **30,** 1–36.

Knudsen, O., and Velle, W. (1961). Ovarian oestrogen levels in the non-pregnant mare: relationship to histological appearance of the uterus and clinical status. *J. Reprod. Fertil.* **2,** 130–137.

Koering, M., Resko, J., Phoenix, C. H., and Goy, R. W. (1968). Ovarian morphology and progesterone levels throughout the menstrual cycle in *Macaca mulatta. Anat. Rec.* **160,** 378 (abstr.).

Korenman, S. G., and Lipsett, M. B. (1965). Direct peripheral conversion of dehydroepiandrosterone to testosterone glucuronide. *Steroids* **5,** 509–517.

Korenman, S. G., Perrin, L. E., and McCallam, T. P. (1969). A radioligand binding assay system for estradiol measurement in human plasma. *J. Clin. Endocrinol. Metab.* **29**, 879–883.

Labhsetwar, A. R., Joshi, H. S., and Watson, D. (1972). Temporal relationship between estradiol, estrone and progesterone secretion in the ovarian venous blood and LH in the peripheral plasma of cyclic hamsters. *Biol. Reprod.* **8**, 321–326.

Land, R. B., Pelletier, J., Thimonier, J., and Mauléon, P. (1973). A quantitative study of genetic differences in the incidence of oestrus, ovulation and plasma luteinizing hormone concentration in the sheep. *J. Endocrinol.* **58**, 305–317.

Land, R. B., Collett, R. A., and Baird, D. T. (1974). Variation in the sensitivity of the autotransplanted ovary of the ewe to ovine luteinizing hormone. *J. Endocrinol.* **62**, 165–166.

Laumas, K. R. (1969). Estrogen production and metabolism in *Macaca mulatta. Gen. Comp. Endocrinol., Suppl.* **2**, 141–146.

Lemon, M. (1972). Peripheral plasma progesterone during pregnancy and the oestrous cycle in the tammar wallaby, *Macropus eugenii. J. Endocrinol.* **55**, 63–71.

Lemon, M., Pelletier, J., Saumande, J., and Signoret, J. P. (1975). Peripheral plasma concentrations of progesterone, oestradiol-17β and luteinizing hormone around oestrus in the cow. *J. Reprod. Fertil.* **42**, 137–140.

Lin, T. J., Billiar, R. B., and Little, B. (1972). Metabolic clearance rate of progesterone in the menstrual cycle. *J. Clin. Endocrinol. Metab.* **35**, 879–886.

Lindner, H. R., Sass, M. B., and Morris, B. (1964). Steroids in the ovarian lymph and blood of conscious ewes. *J. Endocrinol.* **30**, 361–376.

Linzell, J. L. (1957). Measurement of udder blood flow in the conscious goat. *J. Physiol. (London)* **137**, 75–76.

Linzell, J. L., and Heap, R. B. (1968). A comparison of progesterone metabolism in the pregnant sheep and goat: sources of production and an estimation of uptake by some target organs. *J. Endocrinol.* **41**, 433–438.

Lipsett, M. B., Wilson, H., Kirschner, M. A., Korenman, S. G., Fishman, L. W., Sarfaty, G. A., and Bardin, C. W. (1966). Studies on Leydig cell physiology and pathology: secretion and metabolism of testosterone. *Recent Progr. Horm. Res.* **22**, 245–271.

Liptrap, R. M., and Raeside, J. I. (1966). Luteinizing hormone activity in blood and urinary oestrogen excretion by the sow at oestrus and ovulation. *J. Reprod. Fertil.* **11**, 439–446.

Little, B., and Billiar, R. B. (1969). Progesterone production. *In* "Progress in Endocrinology" (C. Gual, ed.), pp. 871–879. Excerpta Medica Found., Amsterdam.

Lloyd, C. W., Lobotsky, J., Baird, D. T., McCracken, J. A., Weisz, J., Pupkin, M., Zanartu, J., and Puga, J. (1971). Concentration of unconjugated estrogens, androgens and gestogens in ovarian and peripheral venous plasma of women: the normal menstrual cycle. *J. Clin. Endocrinol. Metab.* **32**, 155–166.

Lobotsky, J., and Lloyd, C. W. (1973). Variations in the concentration of androstenedione in peripheral plasma of women during the day and during the menstrual cycle. *Steroids* **22**, 133–138.

Lockett, M. F., O'Connor, W. J., and Verney, E. B. (1942). Renal artery loop in the dog. *Q. J. Exp. Physiol. Cogn. Med. Sci.* **31**, 333–336.

Longcope, C. (1971). Metabolic clearance and blood production rates of estrogens in post-menopausal women. *Amer. J. Obstet. Gynecol.* **111**, 779–780.

Longcope, C., Layne, D. S., and Tait, J. F. (1968). Metabolic clearance rates and interconversion of estrone and 17β-estradiol in normal males and females. *J. Clin. Invest.* **47**, 93–106.

Longcope, C., Kato, T., and Horton, R. J. (1969). Conversion of blood androgens to estrogens in normal adult men and women. *J. Clin. Invest.* **48**, 2191–2201.

Luetscher, J. A., and Cheville, R. A. (1968). Measurement of secretion rates of adrenal cortical hormones. *Clin. Endocrinol.* **2**, 444–455.

McCracken, J. A., and Baird, D. T. (1969). The study of ovarian function by means of transplantation of the ovary in the ewe. *In* "The Gonads" (K. W. McKerns, ed.), pp. 175–209. Appleton, New York.

McCracken, J. A., Glew, M. E., and Levy, L. K. (1969). Regulation of corpus luteum function by gonadotrophins and related compounds. *Adv. Biosci.* **4**, 377–393.

McCracken, J. A., Baird, D. T., and Goding, J. R. (1971). Factors affecting the secretion of steroids from the transplanted ovary in the sheep. *Recent Progr. Horm. Res.* **27**, 537–582.

McCracken, J. A., Carlson, J. C., Glew, M. E., Goding, J. R., Baird, D. T., Green, K., and Samuelsson, B. (1972). Prostaglandin F2α identified as a luteolytic hormone in the sheep. *Nature (London) New Biol.* **238**, 129–134.

McDonald, I. R., Goding, J. R., and Wright, R. D. (1958). Transplantation of the adrenal gland of the sheep to provide access to its blood supply. *Aust. J. Exp. Biol. Med. Sci.* **36**, 83–96.

MacDonald, P. C., Rombaut, R. P., and Siiteri, P. K. (1967). Plasma precursors of estrogen. 1. Extent of conversion of plasma Δ⁴-androstenedione to estrone in normal males and non-pregnant normal, castrate and adrenalectomized females. *J. Clin. Endocrinol. Metab.* **27**, 1103–1111.

MacDonald, P. C., Grodin, J. M., and Siiteri, P. K. (1969). The utilization of plasma androstenedione for estrone production in women. *In* "Progress in Endocrinology" (C. Gual, ed.), pp. 770–776. Excerpta Med. Found., Amsterdam.

MacDonald, P. C., Grodin, J. M., and Siiteri, P. K. (1971). Dynamics of androgen and oestrogen secretion. *In* "Control of Gonadal Steroid Secretion" (D. T. Baird and J. A. Strong, eds.), pp. 157–174. Edinburgh Univ. Press, Edinburgh.

McKenna, T. J., and Brown, R. D. (1974). Pregnenolone in man: plasma levels in states of normal and abnormal steroid oogenesis. *J. Clin. Endocrinol. Metab.* **38**, 480–485.

McNatty, K. P., Hunter, W. M., McNeilly, A. S., and Sawers, R. S. (1975). Changes in the concentrations of pituitary and steroid hormones in the follicular fluid of human Graafian follicles throughout the menstrual cycle. *J. Endocrinol.* **64**, 555–571.

Mahesh, V. B. (1965). Steroid secretions of polycystic ovaries. *Proc. 2nd Int. Congr. Endocrinol., 1964* Excerpta Med. Found. Int. Congr. Ser. No. 83, p. 945.

Martenz, N. D., Scaramuzzi, R. J., van Look, P., and Baird, D. T. (1975). A physiological role for ovarian androstenedione. *J. Endocrinol.* **69**, 227–237.

Masuda, H., Anderson, L. L., Henricks, D. M., and Melampy, R. M. (1967). Progesterone in ovarian venous plasma and corpora lutea of the pig. *Endocrinology* **80**, 240–246.

Matsumoto, K., and Samuels, L. T. (1969). Influence of steroid distribution between microsomes and soluble fraction on steroid metabolism by microsomal enzymes. *Endocrinology* **85**, 402–409.

Mead, R. A., and Eik-Nes, K. (1969). Seasonal variations in plasma levels of progesterone in western forms of the spotted skunk. *J. Reprod. Fertil., Suppl.* **6**, 397–403.

Mellin, T. N., Erb, R. E., and Estergreen, V. L. (1966). Quantitative estimation and identification of estrogens in bovine urine. *J. Dairy Sci.* **48**, 895–902.

Mikhail, G. (1967). Sex steroids in blood. *Clin. Obstet. Gynecol.* **10**, 29–39.

Mikhail, G. (1970). Hormone secretion by human ovaries. *Gynecol. Invest.* **1**, 5–20.

Mikhail, G., Noall, M. W., and Allen, W. M. (1961). Progesterone levels in the rabbit ovarian vein blood throughout pregnancy. *Endocrinology* **69**, 504–509.

Mikhail, G., Zander, J. V., and Allen, W. M. (1963). Steroids in human ovarian vein blood. *J. Clin. Endocrinol. Metab.* **23**, 1267–1270.

Mills, T. M., and Savard, K. (1973). Steroidogenesis in ovarian follicles isolated from rabbits before and after mating. *Endocrinology* **92**, 788–791.

Møller, O. M. (1973a). The progesterone concentrations in the peripheral plasma of the mink (*Mustela vison*) during pregnancy. *J. Endocrinol.* **56**, 121–132.

Møller, O. M. (1973b). Progesterone concentrations in the peripheral plasma of the blue fox (*Alopex lagopus*) during pregnancy and the oestrous cycle. *J. Endocrinol.* **59**, 429–438.

Møller, O. M. (1974). Plasma progesterone before and after ovariectomy in unmated and pregnant mink, *Mustela vison. J. Reprod. Fertil.* **37**, 367–372.

Moore, N. W., Barrett, S., Brown, J. B., Schindler, I., Smith, M. A., and Smyth, B. (1969). Oestrogen and progesterone content of ovarian vein blood of the ewe during the oestrous cycle. *J. Endocrinol.* **44**, 55–62.

Morse, W. J., Clark, A. F., and McLeod, S. C. (1963). Size and production rate of the oestradiol and oestrone miscible body pool of a healthy woman. *J. Endocrinol.* **26**, 25–30.

Munroe, S. E., Rebar, R. W., Gay, V. L., and Midgley, A. R. (1969). Radioimmunoassay determination of luteinising hormone during the estrous cycle of the rat. *Endocrinology* **85**, 720–724.

Munroe, S. E., Atkinson, L. E., and Knobil, E. (1970). Patterns of circulating luteinizing hormone and their relationship to plasma progesterone levels during the menstrual cycle of the Rhesus monkey. *Endocrinology* **87**, 453–455.

Nagai, N., and Longcope, C. (1971). Estradiol-17β and estrone: studies on their binding to rabbit uterine cytosol and their concentration in plasma. *Steroids* **17**, 631–646.

Nancarrow, C. D., Buckmaster, J. M., Chamley, W., Cox, R. I., Cumming, I. A., Cummins, L., Drinan, J. P., Findlay, J. K., Goding, J. R., Restall, B. J., Schneider, W., and Thorburn, G. D. (1973). Hormonal changes around oestrus in the cow. *J. Reprod. Fertil.* **32**, 320–321.

Nishizawa, E. E., and Eik-Nes, K. B. (1964). On the secretion of progesterone and $^4\Delta$ androstene 3,17-dione by the canine ovary in animals stimulated with human chorionic gonadotrophins. *Biochim. Biophys. Acta* **86**, 610–621.

Norman, R. L., Eleftheriou, B. E., Spies, H. G., and Hope, P. (1968). Free plasma oestrogens in the ewe during oestrous cycle. *Steroids* **11**, 667–671.

Obst, J. M., Seamark, R. F., and Brown, J. M. (1971). Application of a competitive protein-binding assay for oestrogens to the study of ovarian function in sheep. *J. Reprod. Fertil.* **24**, 140 (Abstr.).

Oertel, G. W., Treiber, L., Wenzel, D., Knapstein, P., Wendlberger, F., and Menzel, P. (1968). In vivo perfusion of the human ovary with 4-^{14}C-dehydroepiandrosterone and 7α-^{3}H-dehydroepiandrosterone 35S-sulfate. Biosynthesis of steroid hormones in human gonads 5. *Experientia* **24**, 607–609.

Ogle, T. F., Braach, N. H., and Buss, I. O. (1973). Fine structure and progesterone concentration in the corpus luteum of the African elephant. *Anat. Rec.* **175**, 707–724.

Paterson, J. Y. F., and Harrison, F. A. (1967). The specific activity of plasma cortisol during continuous infusion of [1,2-^{3}H$_2$]cortisol, and its relation to the rate of cortisol secretion. *J. Endocrinol.* **37**, 269–277.

Pattison, M. L., Chen, C. L., Kelley, S. T., and Brandt, G. W. (1974). Luteinizing hormone and estradiol in peripheral blood of mares during estrous cycle. *Biol. Reprod.* **11**, 245–250.

Pharriss, B. B., Tillson, S. A., and Erickson, R. R. (1972). PGs in luteal function. *Recent Progr. Horm. Res.* **28**, 51–89.

Piaczek, B. E., Schneide, T. C., and Gay, V. L. (1971). Sequential study of LH and progestin secretion on the afternoon of pro-estrus in the rat. *Endocrinology* **89**, 39–45.

Pope, G. S., Gupta, S. K., and Munro, I. B. (1969). Progesterone levels in the systemic plasma of pregnant, cycling and ovariectomized ewes. *J. Reprod. Fertil.* **20**, 369–381.

Rado, A., McCracken, J. A., and Baird, D. T. (1970). The formation of oestrogens by the autotransplanted ovary of the ewe perfused *in vivo* with C_{19} steroids. *Acta Endocrinol. (Copenhagen)* **65**, 244–260.

Reynolds, S. R. M. (1973). Blood and lymph vascular systems of the ovary. *In* "Handbook of Physiology" (Am. Physiol. Soc., J. Field, ed.), Sect. 7, Vol. II, Part 1, pp. 261–316. Williams & Wilkins, Baltimore, Maryland.

Rivarola, M. A., Saez, J. M., Jones, H. W., Jones. G. S., and Migeon, C. J. (1967). The secretion of androgens by the normal, polycystic and neoplastic ovaries. *Johns Hopkins Med. J.* **121**, 82–90.

Romanoff, E. B., Deshpande, N., and Pincus, G. (1962). Rate of progesterone secretion in the dog. *Endocrinology* **70**, 532–539.

Ross, G. T., Cargille, C. M., Lipsett, M. B., Rayford, P. L., Marshall, J. R., Strott, C. A., and Rodbard, D. (1970). Pituitary and gonadal hormones in women during spontaneous and induced ovulatory cycles. *Recent Progr. Horm. Res.* **26**, 1–62.

Rowlands, I. W., and Heap, R. B. (1966). Histological observations on the ovary and progesterone levels in the coypu, *Myocastor coypus. Symp. Zool. Soc. London* **15**, 335–352.

Rowlands, I. W., Tam, W. H., and Kleiman, D. G. (1970). Histological and biochemical studies on the ovary and progesterone levels in the systemic blood of the green acouchi (*Myoprocta pratti*). *J. Reprod. Fertil.* **22**, 533–545.

Rowlands, I. W., Allen, W. R., and Rossdale, P. P. (eds.) (1975). Equine reproduction. *J. Reprod. Fertil., Suppl.* **23**.

Ryan, K. J., and Smith, O. W. (1965). Biogenesis of steroid hormones in the human ovary. *Recent Progr. Horm. Res.* **21**, 367–403.

Salamonsen, L. A., Jonas, H. A., Burger, H. G., Buckmaster, J. M., Charnley, W. A., Cumming, I. A., Findlay, J. K., and Goding, J. R. (1973). A heterologous radioimmunoassay for follicle-stimulating hormone: application to measurement of FSH in the ovine estrous cycle and in several other species including man. *Endocrinology* **93**, 610–618.

Saldarini, R. J., Hilliard, J., Abraham, G. E., and Sawyer, C. H. (1970). Relative potencies of 17α- and 17β-estradiol in the rabbit. *Biol. Reprod.* **3**, 105–109.

Sanyal, M. K., Berger, M. J., Thompson, I. E., Taymor, M. L., and Horne, H. W. (1974). Development of Graafian follicles in adult human ovary. 1. Correlation of estrogen and progesterone concentration in antral fluid with growth of follicles. *J. Clin. Endocrinol. Metab.* **38**, 828–835.

Savard, K. (1961). The estrogens of the pregnant mare. *Endocrinology* **68**, 411–416.

Savard, K., Marsh, J. M., and Rice, B. F. (1965). Gonadotrophins and ovarian steroidogenesis. *Recent Progr. Horm. Res.* **21**, 285–365.

Scaramuzzi, R. J., Caldwell, B. V., and Moor, R. M. (1970). Radioimmunoassay of LH and estrogen during the estrous cycle of the ewe. *Biol. Reprod.* **3**, 110–119.

Schild, W. (1966). Untersuchungen uber die Kozentrationen von Oestron, Ostradiol, und Ostriol in Ovarialveneblut. *Geburtshilfe Frauenheilkol.* **26**, 607–610.

Schindler, A. E., Ebert, A., and Friedrich, E. (1972). Conversion of androstenedione to estrone by human fat tissue. *J. Clin. Endocrinol. Metab.* **35**, 627–630.

Shaikh, A. A. (1971). Estrone and estradiol measurements during the rat estrous cycle and pregnancy. *Anat. Rec.* **169**, 424–425 (abstr.).

Shaikh, A. A. (1972). Estrone, estradiol, progesterone, and 17α-hydroxyprogesterone in the ovarian venous plasma during the estrous cycle of the hamster. *Endocrinology* **91**, 1136–1140.

Shaikh, A. A., and Abraham, G. E. (1969). Measurements of oestrogen surge during pseudopregnancy in rats by radioimmunoassay. *Biol. Reprod.* **1**, 378–380.

Shaikh, A. A., and Harper, M. J. J. (1972). Ovarian steroid secretion in estrous, mated and HCG treated rabbits, determined by concurrent cannulation of both ovarian veins. *Biol. Reprod.* **7**, 387–397.

Shemesh, M., Lindner, H. R., and Ayalon, N. (1971). Competitive protein binding assay of progesterone in bovine jugular venous plasma during the oestrous cycle. *J. Reprod. Fertil.* **26**, 167–174.

Shemesh, M., Ayalon, N., and Lindner, H. R. (1972). Oestradiol levels in the peripheral blood of cows during the oestrous cycle. *J. Endocrinol.* **55**, 72–76.

Short, R. V. (1959). Progesterone in blood. IV. Progesterone in the blood of mares. *J. Endocrinol.* **19**, 207–210.

Short, R. V. (1960a). Steroids present in the follicular fluid of the mare. *J. Endocrinol.* **20**, 147–156.

Short, R. V. (1960b). The secretion of sex steroids by the adrenal glands. *Biochem. Soc. Symp.* **18**, 59–84.

Short, R. V. (1961). Steroid concentrations in the follicular fluid of mares at various stages of the reproductive cycle. *J. Endocrinol.* **22**, 153–163.

Short, R. V. (1962). Steroids in the follicular fluid and the corpus luteum of the mare. A two-cell type theory of ovarian steroid synthesis. *J. Endocrinol.* **24**, 59–63.

Short, R. V. (1964). Ovarian steroid synthesis and secretion *in vivo. Recent Progr. Horm. Res.* **20**, 303–332.

Short, R. V., and Buss, I. O. (1965). Biochemical and histological observations on the corpora lutea of the African elephant, *Loxodonta africana. J. Reprod. Fertil.* **9**, 61–67.

Short, R. V., and Hay, M. F. (1966). Delayed implantation in the roe deer, *Capreolus capreolus. Symp. Zool. Soc. London* **15**, 173–194.

Short, R. V., MacDonald, M. F., and Rowson, L. E. A. (1963). Steroids in the ovarian venous blood of ewes before and after gonadotrophic stimulation. *J. Endocrinol.* **26**, 155–169.

Siiteri, P. K., and MacDonald, P. C. (1973). Role of extraglandular estrogen in human endocrinology. *In* "Handbook of Physiology" Sect. 7, (Am. Physiol. Soc., J. Field, ed.), Vol. II, pp.615–629. Williams & Wilkins, Baltimore, Maryland.

Smith, I. D., Bassett, J. M., and Williams, T. (1970). Progesterone concentrations in the peripheral plasma of the mare during the oestrous cycle. *J. Endocrinol.* **47**, 523.

Smith, J. F., and Robinson, T. J. (1970). The effect of exogenous progestin on the levels of free oestrogen in the ovarian vein plasma of the ewe. *J. Endocrinol.* **48**, 485–496.

Smith, O. W. (1960). Estrogens in the ovarian fluids of normally menstruating women. *Endocrinology* **67**, 698–707.

Smith, O. W., and Ryan, K. J. (1962). Estrogen in the human ovary. *Amer. J. Obstet. Gynecol.* **84**, 141–153.

Solod, E. A., Armstrong, D. T., and Greep, R. O. (1966). Action of luteinizing hormone on conversion of cholesterol stores to steroids secreted in vivo and synthesised in vitro by the pseudopregnant rabbit ovary. *Steroids* **7**, 607–620.

Stabenfeldt, G. H., Ewing, I. L., and McDonald, L. E. (1969). Peripheral plasma progesterone levels during the bovine oestrous cycle. *J. Reprod. Fertil.* **19**, 433–442.

Stevens, V. C., Sparks, S. J., and Powell, J. E. (1970). Levels of estrogens, progestogens and luteinizing hormone during the menstrual cycle of the baboon. *Endocrinology* **87**, 658–666.

Stewart-Bentley, M., and Horton, R. (1971). 17α-Hydroxyprogesterone in human plasma. *J. Clin. Endocrinol. Metab.* **33**, 542–544.

Stormshak, F., Inskeep, E. K., Lynn, L. E., Pope, A. L., and Casida, L. E. (1963). Progesterone levels in corpora lutea and ovarian effluent blood of the ewe. *J. Anim. Sci.* **22**, 1021–1026.

Strott, C. A., Yoshimi, Y., Ross, G. T., and Lipsett, M. B. (1969). Ovarian physiology: relationship between plasma LH and steroidogenesis by the follicle and corpus luteum: effect of HCG. *J. Clin. Endocrinol. Metab.* **29**, 1157–1167.

Strott, C. A., Bermudez, J. A., and Lipsett, M. B. (1970). Blood levels and production rate of 17-hydroxypregnenolone in man. *J. Clin. Invest.* **49**, 1999–2007.

Tagatz, G. E., and Gurpide, E., (1973). Hormone secretion by the human ovary. In "Handbook of Physiology" (Am. Physiol. Soc., J. Field, ed.), Sect. 7, Vol. II, Part 1, pp, 603–613. Williams & Wilkins, Baltimore, Maryland.

Tait, J. F. (1963). Review—the use of isotopic steroids for the measurement of production rates *in vivo. J. Clin. Endocrinol. Metab.* **23**, 1285–1297.

Tait, J. F., and Bernstein, S. (1964). *In vivo* studies of steroid dynamics in man. *In* "The Hormones" (G. Pincus, K. V. Thimann, and E. B. Astwood, eds.), Vol. 5, pp. 441–557. Academic Press, New York.

Tait, J. F., and Horton, R. (1966). The in vivo estimation of blood production and interconversion rates of androstenedione and testosterone and the calculation of their secretion rates. *In* "Steroid Dynamics" (G. Pincus, T. Nakao, and J. F. Tait, eds.), pp. 393–424. Academic Press, New York.

Takikawa, H. (1966). Binding of steroids to follicular fluid proteins. *In* "Steroid Dynamics" (G. Pincus, T. Nakao, and J. F. Tait, eds.), pp. 217–236. Academic Press, New York.

Talmage, R. V., and Buchanan, G. D. (1954). The armadillo (*Dasypus novemcinctus*). A review of its natural history, ecology, anatomy and reproductive physiology. *Rice Inst. Pam.* **41**, 1 135.

Tam, W. H. (1974). The synthesis of progesterone in some hystricomorph rodents. *Symp. Zool. Soc. London* **34**, 363 384.

Telegdy, G., and Endroczi, E. (1963). The ovarian secretion of progesterone and 20α-hydroxpregn-4-ene-3-one in rats during the oestrous cycle. *Steroids* **2**, 119 124.

Telegdy, G., and Huszar, L. (1962). The effect of FSH and HCG on the dog's ovarian progesterone and androstenedione secretion in vivo. *Acta Physiol. Acad. Sci. Hung.* **21**, 339 345.

Telegdy, G., Endroczi, E., and Lissak, K. (1963). Ovarian progesterone secretion during the oestrous cycle, pregnancy and lactation in dogs. *Acta Endocrinol. (Copenhagen)* **44**, 461 466.

Terqui, M., Dray, F., and Cotta, J. (1973). Variations de la concentration de l'oestradiol dans le sang périphérique de la Brebis au cours du cycle oestrial. *C. R. Acad. Sci., Ser. D* **277**, 1795 1798.

Thorburn, G. D., and Mattner, P. E. (1969). A method for collecting utero-ovarian venous blood in conscious ewes for assessment of ovarian and uterine steroid production. *Aust. J. Exp. Biol. Med. Sci.* **47**, p. 32.

Thorburn, G. D., and Mattner, P. E. (1971). Anastomosis of the utero-ovarian and anterios mammary veins for collection of utero-ovarian venous blood: progesterone secretion rates in cyclic ewes. *J. Endocrinol.* **50**, 307 320.

Thorburn, G. D., Cox, R. I., Currie, W. B., Restall, B. J., and Schneider, W. (1972). Prostaglandin F concentration in the utero-ovarian venous plasma of the ewe during the oestrous cycle. *J. Endocrinol.* **53**, 325 326.

Thorburn, G. D., Cox, R. I., Currie, W. B., Restall, B. J., and Schneider, W. (1973). Prostaglandin F and progesterone concentrations in the utero-ovarian venous plasma of the ewe during the oestrous cycle and early pregnancy. *J. Reprod. Fertil., Suppl.* **18**, 151 158.

Thorneycroft, I. H., Mishell, D. R., Jr., Stone, S. C., Khama, K. M., and Nakamura, R. M. (1972). The relationship of serum 17-hydroxyprogesterone and estradiol-17β levels during the human menstrual cycle. *Am. J. Obstet. Gynecol.* **111**, 947 951.

Thorneycroft, I. H., Sribyatta, B., Tom, W. K., Nakamura, R. M., and Mishell, D. R., Jr. (1974). Measurement of serum LH, FSH, progesterone, 17-hydroxyprogesterone, and estradiol-17β levels at 4 hourly intervals during the periovulatory phase of the menstrual cycle. *J. Clin. Endocrinol. Metab.* **39**, 754 758.

Tulchinsky, D., and Korenman, S. G. (1970). A radio-ligand assay for plasma estrone: normal values and variations during the menstrual cycle. *J. Clin. Endocrinol. Metab.* **31**, 76 80.

Tunbridge, D., Rippon, A. E., and James, V. H. T. (1973). Circadian relationships of plasma androgen and cortisol levels. *J. Endocrinol.* **59**, xxi.

van der Molen, H. J. (1969). Patterns of gonadal steroids in the normal human female. *In* "Progress in Endocrinology" (C. Gual, ed.), pp. 894 901. Excerpta Med. Found., Amsterdam.

Vande Wiele, R. L., MacDonald, P. C., Gurpide, E., and Lieberman, S. (1963). Studies on the secretion and interconversion of the androgens. *Recent Progr. Horm. Res.* **19**, 275 305.

Varangot, J., and Cédard, L. (1959). Dosage comparatif des oestrogenes dans le sang veineux peripherique et le sang de la veine ovarienne. *C. R. Soc. Biol.* **153,** 1701–1707.

Velle, W. (1959). "Oestradiol-17β identified in follicular fluid of cows." Thesis, Norges Veterinaerhogskole, Oslo.

Weick, R. F., Dierschke, D. J., Karsch, F. J., Butler, W. R., Hotchkiss, J., and Knobil, E. (1973). Periovulatory time courses of circulating gonadotrophic and ovarian hormones in the rhesus monkey. *Endocrinology* **93,** 1140–1147.

Weiss, G., Dierschke, D. J., Karsch, F. J., Hotchkiss, J., Butler, W. R., and Knobil, E. (1973). The influence of lactation on luteal function in the rhesus monkey. *Endocrinology* **93,** 954–959.

Westerfeld, W. W., Thayer, S. A., MacCorquadale, D. W., and Doisy, E. A. (1938). The ketonic estrogens of sow ovaries. *J. Biol. Chem.* **136,** 181–193.

Wheeler, A. G., Baird, D. T., Land, R. B., and Scaramuzzi, R. J. (1975). Increased secretion of progesterone from the ovary of the ewe during the preovulatory period. *J. Reprod. Fertil.* **45,** 519–522.

Wieland, R. B., de Courcy, C., Levy, R. P., Zala, A. P., and Hirshman, H. (1965). $C_{19}O_2$ steroids and some of their precursors in blood from normal human adrenals. *J. Clin. Invest.* **44,** 159–168.

Wiest, W. G., and Kidwell, W. R. (1969). The regulation of progesterone secretion by ovarian dehydrogenase. *In* "The Gonads" (K. W. McKerns, ed.), pp. 295–320. Appleton, New York.

Wotiz, H. H., Davis, J. W., Lemon, H. M., and Gut, M. (1956). Studies in steroid metabolism. V. Conversion of testosterone-4-C^{14} to oestrogens by human ovarian tissue. *J. Biol. Chem.* **222,** 487–495.

Wu, C.-H., Pazak, L., Flickinger, G., and Mikhail, G. (1974). Plasma 20α-hydroxypregn-4-ene-3-one in the normal menstrual cycle. *J. Clin. Endocrinol. Metab.* **39,** 536–539.

Wurtman, R. J. (1964). An effect of LH on the fractional perfusion of the rat ovary. *Endocrinology* **75,** 927–933.

Yoshinaga, K. (1973). Gonadotrophin induced hormone secretion and structural changes in the ovary during the non-pregnant reproductive cycle. *In* "Handbook of Physiology" (Am. Physiol. Soc., J. Field, ed.), Sect. 7, Vol. II, Part 1, pp. 363–388. Williams & Wilkins, Baltimore, Maryland.

Yoshinaga, K., Grieves, S. A., and Short, R. V. (1967). Steroidogenic effects of luteinizing hormone and prolactin on the rat ovary *in vivo*. *J. Endocrinol.* **38,** 423–430.

Yoshinaga, K., Hawkins, R. A., and Stocker, J. F. (1969). Estrogen secretion by the rat ovary *in vivo* during the estrous cycle and pregnancy. *Endocrinology* **85,** 103–112.

YoungLai, E. V. (1971). Steroid content of equine ovary during the reproductive cycle. *J. Endocrinol.* **50,** 589–597.

YoungLai, E. V., and Short, R. V. (1970). Pathways of steroid biosynthesis in the intact Graafian follicle of mares in oestrus. *J. Endocrinol.* **47,** 321–331.

Yussman, M. A., and Taymor, M. L. (1970). Serum levels of follicle stimulating hormone and luteinizing hormone and of plasma progesterone related to ovulation by corpus luteum biopsy. *J. Clin. Endocrinol. Metab.* **30,** 396–399.

Zander, J. V. (1958). Steroids in human ovary. *J. Biol. Chem.* **232,** 117–122.

Zander, J. V., Brendle, E., von Munstermann, A. M., Diczfalusy, E., Martinsen, B., and Tillinger, K. G. (1959). Identification and estimation of oestradiol-17β and oestrone in human ovaries. *Acta Obstet. Gynecol. Scand.* **38,** 724–736.

6

Steroidogenesis in Vitro

Jennifer H. Dorrington

I. INTRODUCTION

The development of the follicle and the formation and maintenance of the corpus luteum provide fascinating examples of differentiating systems which require the interaction of polypeptide and steroid hormones. The study of the mechanisms by which gonadotropins and steroid hormones regulate ovarian function has expanded to an extent which necessitates the selection of particular areas for review. In this chapter, I have accumulated information on four main topics: (1) the changing populations of receptors for gonadotropins and steroid hormones in the follicle and the corpus luteum; (2) the steroidogenic capacities of the isolated tissue components of the

359

ovary *in vitro*; (3) the cellular and biochemical sites of action of gonadotropins; and (4) the mechanisms of action of gonadotropins.

II. THE FOLLICLE

A. Development

Shortly after birth, the female gonad contains a finite number of oocytes which are arrested in the dictyate stage of meiosis and are surrounded by a single layer of granulosa cells. The trigger which initiates the growth of a small follicle is unknown, but in the mouse a number of follicles begin to differentiate every day. The majority of follicles are destined to undergo atresia, and only a small proportion complete the final stages of differentiation and ovulate (Pedersen, 1970). It has been proposed that estrogens and other substances (perhaps androgens) produced in response to gonadotropins may act antagonistically to determine the ultimate fate of the follicle; either the acquisition of increasing numbers of FSH, estrogen, and LH receptors followed by ovulation or the loss of receptors and follicular atresia (Harman *et al.*, 1975; Louvet *et al.*, 1975a,b; Richards and Midgley, 1976).

Both FSH and estrogens are involved in the early stages of development of the follicle. There is evidence that granulosa cells from small follicles contain FSH binding sites, and treatment of neonatal rats with FSH will stimulate follicular growth (Eshkol and Lunenfeld, 1972). Morphological observations (Ryle, 1969) and studies on the uptake of [^{3}H]thymidine (Ryle, 1971) indicated that FSH caused a progressive increase in the number of granulosa cells in cultured intact ovaries from 15-day-old mice. Estrogen has also been reported to cause increased incorporation of thymidine into granulosa cell DNA followed by granulosa cell proliferation in hypophysectomized rats (Goldenberg *et al.*, 1972b), thereby increasing the number of cells which contained FSH receptors. The number of FSH receptors/cell was influenced by FSH but not by estrogen (Richards and Midgley, 1976). It is possible, therefore, that FSH activates the aromatization enzyme system in granulosa cells and, in the presence of testosterone from differentiated thecal cells, estrogen is synthesized which stimulates granulosa cell proliferation. This would result in a cascade of events culminating in several layers of granulosa cells containing increased numbers of FSH receptors. Although FSH stimulated follicular growth, it was less effective than estrogen as a granulosa-cell mitogen but did cause antrum formation after pretreatment with estrogen (Goldenberg *et al.*, 1972b). Thus, exposure to estrogen and FSH during the early follicular phase promotes the mitotic activity of the granulosa cells, antrum formation, and consequently, a marked increase in ovarian weight.

The next stage of follicular maturation involves the appearance of LH receptors. Channing and Kammerman (1973) demonstrated that granulosa cells from large porcine follicles had a greater capacity to bind HCG than granulosa cells from smaller follicles. Granulosa cells isolated from intact and hypophysectomized rats treated with FSH bound more HCG than those isolated from saline-treated animals (Zeleznik *et al.*, 1974). FSH, therefore, appears to induce or activate LH receptors on granulosa cells and influences 3β-hydroxysteroid dehydrogenase in preparation for subsequent ovulation and steroidogenesis (Goldenberg *et al.*, 1972b; Zeleznik *et al.*, 1974; Channing and Kammerman, 1973).

Ovulation and luteinization mark the final stages of differentiation of the follicle. Exposure of the preovulatory follicle to LH (or FSH) either *in vivo* or *in vitro* triggers a series of morphological and biochemical changes. An early event, namely, elevated 3′,5′-AMP levels within the follicle, is followed by increased secretion of estrogen, androgen, and progesterone (Lindner *et al.*, 1974). The high levels of estrogen and androgen are short-lived, whereas elevated progesterone synthesis persists. Prostaglandin levels increase approximately 4 hours after exposure to LH and may play a role in follicular rupture (Armstrong and Grinwich, 1972; Tsafriri *et al.*, 1972; Lindner *et al.*, 1974). Gonadotropins cause the breakdown of the germinal vesicle in the oocyte and the resumption of meiosis. Shortly before ovulation in most mammalian species there is the completion of the first reduction division and the formation of the first polar body. Completion of the second meiotic division occurs only when a spermatozoon penetrates the oocyte. Compounds which elevate 3′,5′-AMP levels in the follicle (LH, FSH, prostaglandin E, and enterotoxin) will terminate meiotic arrest and these observations suggest that 3′,5′-AMP may play a key role in mediating this event (Lindner *et al.*, 1974). The effect of LH on the resumption of meiosis does not appear to involve steroids or prostaglandins since the process is induced in the presence of inhibitors of steroid or prostaglandin synthesis (Lipner and Greep, 1971; Lindner *et al.*, 1974). The effects of inhibitors of protein synthesis have suggested the involvement of specific protein(s) in the LH-induced meiotic process (Tsafriri *et al.*, 1973).

B. Steroidogenesis

The follicle synthesizes three classes of steroids—progestational hormones, androgens, and estrogens—the relative proportions depending upon the stage of development of the follicle. Even though the synthesis of steroids is a well-recognized function of the follicle, the roles played by theca interna and granulosa cells remain controversial.

Valuable information concerning follicular steroidogenesis has been obtained by analyzing the steroid content of ovarian venous blood during estrus, and of follicular fluid at various stages of the reproductive cycle (Short, 1964). The immature rat ovary, composed mainly of small follicles, has also been studied extensively *in vitro* (Armstrong, 1968; Villee *et al.*, 1969; Ahrén *et al.*, 1969). Although these preparations are free from corpora lutea, the interstitial tissue may contribute its characteristic pattern of steroid secretion (Savard *et al.*, 1965) and thereby complicate the interpretation. To fully assess steroidogenesis by the follicle alone, two approaches have been used *in vitro*—studies on the whole follicle, and those on isolated thecal and granulosa cells.

Studies on the isolated human follicle were first reported in 1961 (Ryan and Smith, 1961a–d; Smith and Ryan, 1961). In a series of *in vitro* experiments on follicular cyst linings from a normally menstruating woman, treated with FSH before surgery, the complete spectrum of steroids synthesized by the follicle was evaluated. The major radioactive steroids formed by the minced follicle walls incubated with [1-^{14}C]acetate were estradiol and estrone (0.03% yield). Progesterone, 17α-hydroxyprogesterone, androstenedione, pregnenolone, 17-hydroxypregnenolone, and dehydroepiandrosterone were also identified (Ryan and Smith, 1961d). The incorporation of label into the latter two steroids suggested the involvement of an alternate pathway for the synthesis of estrogens, this pathway bypassing progesterone. Additional evidence for the existence of such a pathway was obtained from studies of polycystic ovaries (Ryan and Smith, 1965; Mahesh and Greenblatt, 1964). Estrogens can be synthesized by follicular cyst linings *in vitro* from a number of intermediates, the yield depending inversely upon the number of steps involved in the conversion, i.e., 15.3% yield of estrogen from [4-^{14}C]androstenedione (Smith and Ryan, 1961) and 5.6% from [4-^{14}C]progesterone (Ryan and Smith, 1961c).

Follicles have been successfully isolated by microdissection from rabbit (Mills *et al.*, 1971; YoungLai, 1972, 1975) and rat ovaries (Stoklosowa and Nalbandov, 1972; Lindner *et al.*, 1974). Intact follicles from estrous rabbit ovaries incorporated [^{14}C]acetate into estrone and estradiol-17β (Mills *et al.*, 1971) but not into estradiol-17α. In this system, LH (0.0025–2.5 μg/ml) caused a 5- to 13-fold increase in estrogen production from labeled acetate. The stimulatory effect produced by FSH preparations was related to the degree of LH contamination. Preparations of FSH in which the level of LH contamination was reduced by treatment with either 6 M urea or anti-LH serum were ineffective in stimulating estrogen production (Mills *et al.*, 1971). Rabbit corpora lutea and interstitial tissue did not synthesize estrogen *in vitro* (Telegdy and Savard, 1966), demonstrating that the follicle is responsible for ovarian estrogen production *in vitro*. This is consistent with conclusions drawn from *in vivo* studies of estrogen synthesis by the rabbit (Keyes and Nalbandov, 1967; Hilliard *et al.,* 1974).

In addition to producing estrogens, the rabbit ovarian follicle also produces testosterone *in vitro* and the synthesis of this steroid is likewise regulated by LH (Mills and Savard, 1973; YoungLai, 1975). The increased synthesis *in vitro* of testosterone, estrogens, and progestational steroids which occurred shortly after stimulation was followed by a decline to very low levels by the time of ovulation, 10–12 hours after stimulation. The same changes have been demonstrated in the levels of steroids in the peripheral blood of rabbits during the preovulatory period (Eaton and Hilliard, 1971; Hilliard *et al.,* 1974). These similarities emphasize the value of the isolated follicle *in vitro* as a model system in which to study the changing patterns of steroid synthesis during the preovulatory phase.

Graafian follicles isolated from the rat on the morning of proestrus, i.e., before the LH surge, and maintained in culture, released considerably more estradiol than androstenedione and progesterone into the culture medium. LH increased the rate of accumulation of all three steroids during the initial 4–6 hours of incubation, after which time estradiol and androstenedione synthesis ceased. Progesterone synthesis continued and became the major product after incubation for 6 hours (Lindner *et al.,* 1974). An abrupt decline in the secretion of estradiol-17β in the ovarian venous blood of rats is observed late in proestrus. Rapid stimulation of testosterone synthesis after LH administration to immature rats is also followed by an abrupt decline. It is possible, as suggested by Lindner *et al.* (1974), that LH induces the synthesis of an inhibitor of 17α-hydroxylase or 17–20 lyase. This may be an indirect effect, however, mediated by androgen or estrogen. Additional support for this action of LH on progesterone metabolism comes from the studies of Armstrong (1974). Prepubertal rats injected with PMSG synthesized androsterone and lower levels of androstenedione and 5α-androstanedione from [4-^{14}C]progesterone on the day before proestrus, and this remained constant until the afternoon of proestrus when LH secretion became elevated; the levels of all three steroids then declined rapidly. Coincident with the LH surge and the decline in the production of androsterone, there was a marked increase in the production of 3α-hydroxy-5α-pregnan-20-one. These data suggest that LH may act not only to inhibit the side-chain cleavage of progesterone, but also to inhibit the cleavage of the ring A-reduced metabolites (e.g., 3α-hydroxy-5α-pregnane-20-one). The nature of this inhibition requires further investigation.

The isolated follicle is a multicomponent system, however, and the question of the relative contribution of the thecal cells and the granulosa cells to the overall steroid pattern can only be answered after study of the separate cell types. Of the two cell types, the theca interna cells possess the ultrastructural features which are generally associated with steroid synthesis. These characteristics include abundant smooth endoplasmic reticulum, a dispersed Golgi complex, and cytoplasmic lipid droplets. Morphological (Christensen and Gillim, 1969) and histochemical (Deane and Rubin, 1965) studies indicate that the primordial and early secondary follicles do not synthesize steroids; this capacity develops as the follicle matures. In contrast, the nonluteinized granulosa cells have predominantly rough endoplasmic reticulum and large Golgi complex, these features being characteristic of active protein synthesis.

It has been considered that the thecal cell component is the site of estrogen synthesis in the preovulatory follicle (Channing, 1969a; YoungLai and Short, 1970). Over the years, however, a considerable amount of evidence has accumulated which shows quite clearly that granulosa cells are involved in effecting maximum estrogen synthesis. The "two-cell type (theca and granulosa) theory" to explain the synthesis of estrogen in the follicle originated from a series of elegant studies by Falck (1959) in which granulosa and thecal cells were separated from the rat follicle and transplanted into the anterior chamber of the rat's eye, together with a piece of vaginal mucosa. Thecal or granulosa cells alone failed to produce sufficient estrogen to cause cornification of the vaginal transplant, but a combination of thecal and granulosa cells resulted in vaginal cornification. Since that time, several other investigators have demonstrated synergism between thecal and granulosa cells in the synthesis of estrogens (Ryan et al., 1968; YoungLai, 1973; Makris and Ryan, 1975). What can be the nature of this cooperation between these two types and how can this be explained in biochemical terms? In an attempt to answer this question the enzyme capacities of separated thecal and granulosa cells from human (Ryan and Petro, 1966; Ryan et al., 1968; Channing, 1969c), mare (Ryan and Short, 1965, 1966; Channing and Grieves, 1969), sow (Bjersing and Carstensen, 1964, 1967), rabbit (YoungLai, 1972), and hamster (Makris and Ryan, 1975) follicles have been analyzed. In each case, granulosa cells were obtained by puncturing and then scraping or washing the follicle wall; the remaining follicle wall was the source of thecal cells. In these experiments the cell populations are described as being "essentially" homogeneous under the light microscope. Careful characterization of the cell preparations is necessary before the precise degree of contamination of the cell populations is known. It is obvious that a small percentage of a contaminating cell type which contains a high activity of a particular enzyme could give misleading results, particularly if precursor–product interplay exists between these cell types.

In spite of the reservations, however, the capacity of granulosa cells appear to be quite clear-cut. Progesterone was the main steroid formed from [17α-^{3}H]pregnenolone by granulosa cells isolated from sow (Schomberg, 1969), human (Ryan and Petro, 1966), and mare (Channing, 1969b) follicles. Mare granulosa cells synthesized significant amounts of 20α-hydroxypregn-4-en-3-one but further metabolism of progesterone was minimal in granulosa cells from other species and only trace amounts of 17α-hydroxyprogesterone were detected.

Cultured granulosa cells from hamster preovulatory and medium-sized antral follicles (Makris and Ryan, 1975) and from rabbit preovulatory follicles (Erickson and Ryan, 1975) synthesized large amounts of progesterone (measured by radioimmunoassay). Further conversion to androstenedione or testosterone was low or undetectable in granulosa cells from the hamster (Makris and Ryan, 1975). Little or no 17-hydroxylase and 17–20 lyase activities have been found in equine (Channing, 1969b), porcine (Bjersing and Carstensen, 1967), and rabbit (YoungLai, 1973) granulosa cells. Granulosa cells, therefore, have a limited capacity for *de novo* androgen synthesis and, as a consequence of inadequate supplies of substrate for aromatization, estrogen synthesis is restricted. However, when equine (Channing, 1969b; Ryan and Short, 1965), porcine (Bjersing and Carstensen, 1967), rabbit (Erickson and Ryan, 1975), or rat (Dorrington *et al.,* 1975) granulosa cells were incubated with exogenous aromatizable substrate, either androstenedione or testosterone, significant amounts of estradiol-17β were synthesized, demonstrating that granulosa cells do contain an active aromatization system. Estradiol synthesis *in vitro* is short-lived, however, even in the presence of substrate, unless hormones are added. Human postmenopausal gonadotropin (Pergonal, which contains both FSH and LH activities) stimulated estrogen synthesis by rabbit granulosa cells (Erickson and Ryan, 1975), and highly purified FSH, but not LH, stimulated estrogen synthesis from testosterone by rat granulosa cells (Dorrington *et al.,* 1975).

In contrast, thecal cell preparations synthesize progestational steroids, androgens and estrogens. YoungLai (1973) has shown that rabbit follicle slices and thecal preparations converted pregnenolone and progesterone to androgens. Similarly, thecal cells from human follicles synthesized more androstenedione relative to progesterone than did the granulosa cells from the same follicles (Channing, 1969c). Recent studies on isolated thecal tissue from the hamster (Makris and Ryan, 1975) and the rabbit (Erickson and Ryan, 1975) have demonstrated that this follicular component has the capacity to synthesize significant amounts of androgens. The aromatization enzyme system in thecal cells isolated from preovulatory follicles of the mare (Ryan and Short, 1965) and the hamster (see Makris and Ryan, 1975) is not as active as in the granulosa cells from the same follicles. Because of the unknown degree of contamination of granulosa cells in the thecal tissue preparations, it is difficult to assess the capacity of thecal cells alone to aromatize C_{19} steroids, and the values may be overestimated.

Distinct differences in the enzymatic activities of the two cell types, thecal and granulosa, emerge. Granulosa cells have a weak 17α-hydroxylating system and little or no 17–20 lyase activity, but have the capacity to actively aromatize C_{19} steroids. In contrast, the thecal cells have 17α-hydroxylase and 17–20 lyase activities and are the main sites of androgen synthesis in the follicle. Together, therefore, the two cell types contain the full complement of enzymes required for estrogen synthesis. From the evidence cited above, it is possible to explain the nature of the synergism between the two cell types in effecting maximum estrogen biosynthesis; the thecal cells provide the substrate, androgen, which is transported to the granulosa cells for aromatization (Makris and Ryan, 1975; Dorrington *et al.*, 1975). If such a mechanism exists for the synthesis of estrogens *in vivo* then this type of precursor–product interplay may also provide a means of communication between the two cell types.

At this point, it is relevant to note the parallels which have been drawn between the granulosa cells of the ovary and the Sertoli cells of the testis, since they have the same embryonic origin. Sertoli cells synthesize estradiol-17β in the presence of exogenous testosterone and FSH (Dorrington and Armstrong, 1975). Sertoli cells cannot synthesize testosterone *de novo* and rely upon Leydig cells for their source of androgen (discussed by Dorrington and Armstrong, 1975). An analogous mechanism involving the cooperation of two cell types for the synthesis of estrogens may exist in both the ovary and the testis.

The studies on isolated granulosa cells and Sertoli cells do not, of course, exclude the possibility that the thecal cells and the Leydig cells may also synthesize estrogens. However, M. L. Dufau (personal communication) found that Leydig cells did not synthesize estrogen under conditions in which they synthesized large amounts of testosterone. Better techniques are required for the isolation of thecal cells to determine their capacity to synthesize estrogens in the absence of granulosa cells. If pure populations of thecal cells synthesize estrogens, then it is important to ascertain (1) if LH stimulates these cells and influences aromatization (since thecal cells contain LH receptors but not FSH receptors: Zeleznik *et al.*, 1974); and (2) their capacity at different stages of differentiation, since the two cell types (granulosa and thecal) may cooperate to give maximum estrogen synthesis only at certain stages of differentiation of the follicle.

Human granulosa cells and thecal cells did not appear to interact to synthesize estrogens in the experiments of Channing (1969c), but the cells had been in culture for 3 to 6 days and they were well established and luteinized. Estrogen synthesis by hamster granulosa cells is short-lived, since Makris and Ryan (1975) could demonstrate synergism between thecal and granulosa cells only during the first 48 hours after plating the cells. As mentioned earlier, estrogen synthesis by rat granulosa cells occurs for only a short time unless FSH is present, in which case estrogen synthesis from testosterone is stimulated (Dorrington *et al.*, 1975). Granulosa cells luteinize spontaneously in culture, and it is possible that the luteinized cell has a reduced capacity to synthesize estrogen compared with granulosa cells from the preovulatory follicle (Makris and Ryan, 1975).

C. Hormonal Control of Follicular Steroid Synthesis

Early in the 1940's Fevold (1941) and Greep *et al.* (1942) showed that FSH and LH were required for estrogen synthesis (as judged by uterine growth) by the ovary of immature hypophysectomized rats; FSH alone did not stimulate ovarian estrogen secretion. Lostroh and Johnson (1966) confirmed these results using partly purified FSH and ascertained that with FSH low levels of LH were essential for estrogen secretion and follicular development. Furthermore, Ryle *et al.* (1974) demonstrated a synergistic effect of FSH and LH on estrogen production by immature mouse ovaries in culture. For over 30 years, therefore, experiments *in vivo* and *in vitro* have emphasized the possible conjoint action of LH and FSH on estrogen biosynthesis, but very few studies have attempted to define the nature of this synergism in biochemical terms.

In 1958, Hollander and Hollander first demonstrated that FSH administered *in vivo* or *in vitro* stimulated the *in vitro* conversion of [^{14}C]testosterone to [^{14}C]estradiol-17β by canine ovarian slices, and suggested that a rate-limiting reaction in the aromatase system was regulated by FSH. In these experiments, however, high concentrations of impure FSH were used and HCG produced a similar stimulation of activity. Later, Mills *et al.* (1971) showed that LH acutely stimulated the biosynthesis of estrogens by the isolated rabbit follicle; FSH was without effect. From these and other studies on isolated follicles from a variety of species (reviewed above), the concept developed that LH was the steroidogenic hormone in the follicle, stimulating steroid biosynthesis at a site common to all gonadal steroids, i.e., between cholesterol and pregnenolone.

FSH, on the other hand, was known to promote the growth and maturation of the follicle (Kraiem and Samuels, 1974; Eshkol and Lunenfeld, 1972) to the stage at which it became responsive to the steroidogenic action of LH, but, in general, was not credited with steroidogenic properties itself

(Schwartz and McCormack, 1972). The demonstration of a specific effect of FSH on steroid biosynthesis (at a site between testosterone and estrogen) came from studies on Sertoli cells (Dorrington and Armstrong, 1975), and prompted the reexamination of the possible role of FSH in stimulating ovarian estradiol biosynthesis. Ovaries of immature rats explanted into organ culture 2–3 days after hypophysectomy responded to highly purified FSH, but not to LH, with an increase in estrogen production from testosterone (Moon *et al.*, 1975). Granulosa cells isolated from the same ovaries or from ovaries of hypophysectomized immature rats treated with estrogen for 5 days (to obtain increased yields of cells) and grown in monolayer culture also synthesized estrogen from testosterone in the presence of FSH but not LH (Dorrington *et al.*, 1975). FSH stimulated estrogen production in the presence, but not in the absence, of an aromatizable substrate. These studies defined both the cellular and biochemical sites of action of FSH on estrogen synthesis, and furthermore suggested a mechanism by which the two gonadotropins cooperate to regulate ovarian estrogen synthesis. Thus it is possible that thecal and/or interstitial cells, under the influence of LH, secrete androgen which is transported to the granulosa cells for aromatization under the influence of FSH (Fig. 1).

The literature reviewed above is in keeping with this theory: in freshly isolated follicles or granulosa cells, the aromatase would be active due to

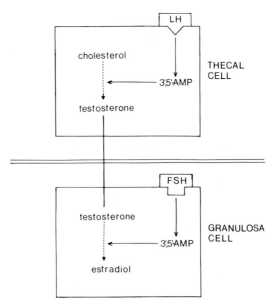

Fig. 1.   Model of the two-cell, two-gonadotropin hypothesis for the synthesis of follicular estradiol-17β.

prior exposure to FSH *in vivo*, and availability of substrate would be rate-limiting. Consequently, any agent (e.g., LH) which increases the level of testosterone would also cause an increase in estrogen production. Follicles deprived of hormone, for example, after hypophysectomy, would require LH and FSH (or testosterone and FSH) to synthesize estrogen.

Autoradiography studies indicated that labeled FSH binds only to granulosa cells, while HCG binds to thecal and interstitial cells, and variably to granulosa cells (Zeleznik *et al.*, 1974). These observations are consistent with the two-cell, two-gonadotropin theory.

III. THE CORPUS LUTEUM

A. Formation and Development

The corpus luteum is formed principally from the granulosa cells of the follicle following ovulation and luteinization. During the process of luteinization, the follicle, which secretes mainly estrogens, is transformed into the corpus luteum which synthesizes large amounts of progesterone. Morphological changes which accompany luteinization of granulosa cells include an increase in size and cytoplasm:nuclear ratio, the appearance of abundant smooth endoplasmic reticulum, cytoplasmic lipid droplets and eosinophilic granules, and dispersed Golgi elements.

The culture of granulosa cells has proved to be a valuable system in which to study the process of luteinization. The hormonal environment of the cells *in vivo* at the time of isolation is important in determining subsequent luteinization in culture. Granulosa cells isolated from mature Graafian follicles of a number of species (human, rhesus monkey, pig, and mare) will luteinize in culture in the absence of added pituitary gonadotropins (Channing, 1969a, 1970). Granulosa cells harvested from medium-sized follicles of rhesus monkeys (from day 7 to 12 of the cycle) required LH or FSH for full luteinization. In contrast, granulosa cells from small follicles underwent morphological luteinization and secreted increased amounts of progesterone only when both gonadotropins or dibutyryl $3',5'$-AMP were added (Channing, 1970). Channing (1970) therefore concluded that "the complete luteinization process will occur only after ovulation at the normal time or in culture after pre-stimulation with LH and FSH."

Other investigators have used a combined *in vitro–in vivo* approach to study luteinization. Follicles removed from the rabbit after mating, i.e., after exposure to elevated levels of LH *in vivo,* luteinized when transplanted to the kidney capsule (Keyes and Armstrong, 1969). Mature follicles were transformed to functional corpora lutea if they were incubated *in vitro* for

40 minutes in the presence of LH before transplantation (Keyes, 1969). Follicles incubated in buffer plus glucose alone, or with FSH, degenerated upon transplantation. Similarly, rat follicles removed during proestrus luteinized when transplanted to the kidney capsule of an hypophysectomized rat. Follicles removed at diestrus luteinized only when the follicles were incubated with LH or dibutyryl 3′,5′-AMP before transplantation (Ellsworth and Armstrong, 1971a,b). These experiments established the role of LH in initiating the process of luteinization and the formation of the corpus luteum.

LH-induced luteinization is characterized by a decrease in the level of estrogen and an increase in progesterone synthesis; the mitotic activity of granulosa cells is inhibited (McNatty and Sawers, 1975); the cumulus oophorus cells withdraw their cytoplasmic microvilli in contact with the oocyte (Zamboni, 1972); and the luteinizing granulosa cell layer becomes vascularized. Concomitant with these changes in the granulosa cells is the resumption of meiosis in the oocyte.

Once ovulation and luteinization have been induced by LH, striking species differences are evident in the hormonal requirements for the subsequent process of luteal differentiation. The rat has been studied most extensively: the onset of luteinization is marked by a loss of FSH receptors, a decrease in estrogen and LH receptors, and the appearance of prolactin receptors. Antiserum neutralization of endogenous LH on day 6 or 7 of pregnancy produced little or no adverse effects (Morishige and Rothchild, 1974). Neither LH nor FSH and LH maintained pregnancy in rats hypophysectomized from days 1 to 7 (Ahmad *et al.,* 1969; Lyons and Ahmad, 1973). On the other hand, suppression of the prolactin surges by the administration of ergot alkaloids terminates pregnancy before, but not after, the eighth day (Shelesnyak, 1957; Morishige and Rothchild, 1974). LH does not appear, therefore, to be required for the maintenance of luteal function from days 1 to 7, but rather prolactin seems to be the important luteotropic hormone during this period. Prolactin receptor activity is high in luteal tissue during the early stages of pregnancy, but falls at the time when the placental luteotropin becomes detectable (i.e., about day 8) (Richards and Midgley, 1976).

At day 12 of pregnancy the corpus luteum enters a new phase of development requiring a different hormonal environment. This is clear from the observations that antiserum to LH did not interrupt pregnancy (Raj and Moudgal, 1970), and hypophysectomy did not influence plasma and luteal levels of progesterone (Takayama and Greenwald, 1973). Hypophysectomy and hysterectomy at day 12, however, led to an abrupt decline in progesterone levels by day 16 and the effect could not be reversed by prolactin, FSH, and LH, separately or in combination (Takayama and

Greenwald, 1973). Daily administration of large amounts of estrone restored luteal and plasma levels of progesterone, suggesting a direct luteotropic effect of estrogen from day 12 of pregnancy in the rat.

Rat luteal cell cytosol and nuclei bind estradiol-17β specifically, supporting the view that the rat corpus luteum is a target tissue for estrogen (Richards, 1974). Richards (1975) has reported changes in total content and intracellular distribution of estradiol-17β receptors in rat corpora lutea during pregnancy. Estradiol-17β binding to cytosol was high between days 3 and 11 of pregnancy after which time it declined and remained low to the end of pregnancy. Nuclear binding, on the other hand, remained low until day 10 of pregnancy, increased from day 10 to 15, remained elevated until day 18, and then declined dramatically. There was an increase in the total amount of receptor during the second half of pregnancy until day 18 when the total estradiol-17β receptor content decreased. Treatment with estradiol-17β early in pregnancy stimulated an increase in nuclear receptor content and it was suggested that in the presence of elevated levels of estrogen, estradiol-17β binding components in the cytosol are capable of translocation to the nucleus (Richards, 1975). These data are provocative since they suggest that the concentration of estrogen and the availability of estrogen receptors may be important factors in the control of luteal cell function.

B. Hormonal Control of Luteal Steroidogenesis

It is clear from the foregoing that the hormonal requirements for the maintenance of progesterone synthesis by the rat corpus luteum differ depending upon the stage of development. Both LH and prolactin are intimately involved in sustaining progesterone synthesis, at least during the first half of pregnancy, and any consideration of control mechanisms must take into account the luteotropic activities of both of these hormones. While both hormones can increase progesterone output, the two hormones have quite different amino acid sequences and influence different steps in the biosynthetic pathway. LH administration *in vivo* or *in vitro* stimulated the *in vitro* synthesis of progesterone by slices of rat luteinized ovarian tissue (Huang and Pearlman, 1962; Armstrong *et al.*, 1964), and this was accomplished by an increased rate of conversion of cholesterol to pregnenolone (reviewed in Section VI,A).

The luteotropic action of prolactin in the rat can be attributed in part to an inhibitory effect on the further metabolism of progesterone to the progestationally inactive form, 20α-hydroxypregn-4-en-3-one (Hashimoto and Wiest, 1969; Armstrong *et al.*, 1969b, 1970) and to 5α-reduced metabolites (Zmigrod *et al.*, 1972). Armstrong *et al.* (1970) demonstrated that total progestin synthesis *in vitro* by corpora lutea isolated from hypophysectomized and intact pseudopregnant rats was similar. However, the 20α-hydroxypregn-4-en-3-one:progesterone ratio was higher in the corpora lutea from the hypophysectomized animals. Treatment of hypophysectomized rats for 4 days with prolactin did not influence total progestin synthesis *in vitro* but inhibited 20α-hydroxysteroid dehydrogenase resulting in a considerable reduction in the 20α-hydroxypregn-4-en-3-one:progesterone ratio. On the basis of these experiments, Armstrong *et al.* (1970) concluded that the synthesis of steroids by the rat corpus luteum was autonomous for at least 7 days of pseudopregnancy, and that prolactin was necessary to prevent progesterone from being converted to 20α-hydroxypregn-4-en-3-one.

A functional relationship between prolactin and LH in the regulation of cholesterol synthesis and storage in the rat corpus luteum was suggested by the histochemical studies of Everett (1947), and has been confirmed biochemically. Hypophysectomy of pseudopregnant rats caused little change in the content of free and esterified cholesterol in corpora lutea isolated 5 days after hypophysectomy, but replacement therapy with prolactin significantly increased the concentration of both forms of cholesterol (Armstrong *et al.*, 1970). In contrast, the administration of prolactin *in vivo* to intact rats caused a marked decrease in esterified cholesterol and a slight decrease in free cholesterol (Armstrong *et al.*, 1969b). Prolactin, therefore, promotes the storage of esterified cholesterol in the absence of the pituitary, but in the presence of the pituitary (the important factor is probably LH) the utilization of cholesterol is favored (Armstrong *et al.*, 1970). Prolactin also appears to act synergistically with a hypophyseal factor to stimulate the incorporation *in vitro* of [1-^{14}C]acetate into the free and esterified cholesterol pools of the rat corpus luteum, since the effect of prolactin was considerably greater in corpora lutea from intact rats than from hypophysectomized animals. Prolactin also stimulated the synthesis of long-chain fatty acids from [1-^{14}C]acetate, but this occurred in the absence of the pituitary, indicating that other pituitary hormones were not essential (Armstrong *et al.*, 1969b, 1970).

A possible mechanism by which prolactin controls the supply of cholesterol available for progesterone synthesis was suggested by experiments of Behrman *et al.* (1970). These workers showed that the levels of the enzymes involved in the turnover of cholesterol esters, i.e., sterol acyltransferase, which catalyzes the synthesis of cholesterol esters from fatty acids and cholesterol, and sterol esterase, which hydrolyzes cholesterol esters, were both reduced 3 days after hypophysectomy. The administration of purified rat prolactin prevented the rapid decline of both of the above enzymes following hypophysectomy. LH, progesterone, and progesterone plus estrogen were unable to maintain the enzyme levels (Behrman *et al.,* 1970). LH stimulation caused a depletion of the pool of esterified cholesterol, whereas the level of free cholesterol remained essentially unaffected. One means by which LH caused this effect was by stimulating cholesterol esterase activity (Behrman and Armstrong, 1969). Stimulation of cholesterol esterase activity by LH was retained in the presence of aminoglutethimide which inhibits cholesterol side-chain cleavage (Behrman *et al.,* 1970). Therefore, it appears that cholesterol ester depletion in the presence of LH is not the result of increased utilization of free cholesterol for steroid synthesis but represents a separate and independent action of LH. LH also inhibited the *in vitro* incorporation of [1-^{14}C]acetate into long-chain fatty acids esterified as sterol esters by luteinized rat ovaries. In contrast, LH had no effect on the incorporation of [1-^{14}C]acetate into fatty acids esterified as triglycerides.

Thus, progesterone synthesis by rat luteal tissue is influenced by prolactin and LH. Prolactin promotes the synthesis of long-chain fatty acids and free esterified cholesterol, and maintains the enzymes involved in the turnover of cholesterol esters. LH activates cholesterol esterase and inhibits the esterification of fatty acids, thereby increasing the amount of free cholesterol available for conversion to pregnenolone and progesterone.

Pituitary hormones are required for the preservation of structural integrity of the corpus luteum and for progesterone synthesis in the rabbit. On the basis of morphological studies on luteal and endometrial cells, Kilpatrick *et al.* (1964) concluded that LH served as a luteotropin in the rabbit, since administration of this hormone partly maintained corpora lutea for at least 7 days. Later, Hilliard *et al.* (1971) were unable to repeat these experiments. In their hands, LH consistently failed to maintain the structure and function of corpora lutea, but achieved some degree of maintenance when administered with FSH. Estrogen is even more effective as a luteotropin in the rabbit. There is overwhelming evidence that estrogen is essential for the maintenance and continued function of the corpus luteum, and that the effects of LH (and FSH) are mediated via estrogens. Early studies demonstrated that injections of estrogen could maintain the normal morphological appearance of corpora lutea in hypophysectomized rabbits and could maintain pregnancy in these animals (Robson, 1937). Chips or cylinders of estrogen implanted adjacent to the corpora lutea in pseudopregnant and pregnant rabbits maintained the morphological structure of the corpus luteum beyond its normal life span (Hammond and Robson, 1951). More recently, the direct luteotropic action of estrogens on the corpus luteum has been demonstrated in rabbits treated with anti-LH serum (Spies and Quadri, 1967), in hypophysectomized rabbits (Spies *et al.,* 1968), and rabbits with X-irradiated ovaries (Keyes and Nalbandov, 1967). Corpora lutea, which were autotransplanted under the kidney capsule of ovariectomized rabbits, developed and secreted progesterone for approximately 5 days with or without estradiol treatment. Following this, estradiol treatment was essential for further development and continued progesterone synthesis of the corpus luteum (Miller and Keyes, 1975).

The follicles are the sites of estrogen synthesis and must be present for the maintenance of the rabbit corpus luteum (Keyes and Armstrong, 1968). Estrogen administration to rabbits in which the follicles had been eliminated by X-irradiation maintained pregnancy, and progesterone was detectable in the corpora lutea and plasma. Withdrawal of estrogen resulted in abortion and the levels of progesterone were undetectable (Keyes and Nalbandov, 1967). LH failed to maintain the corpus luteum and pregnancy in the absence of follicles, and it was suggested that LH acts on the follicle of the pregnant rabbit to stimulate the synthesis of estrogen, which is the actual luteotropic hormone (Keyes and Nalbandov, 1967; Keyes and Armstrong, 1968). In Section II,C the "two-cell type-two-gonadotropin" theory was presented to explain the control of estrogen synthesis in the follicle. If this is the case, then LH alone would be expected to increase testosterone levels in the follicle but not influence the aromatization enzyme system (Dorrington *et al.*, 1975). It is possible that in the *in vivo* experiments (e.g., Kilpatrick *et al.*, 1964) there was sufficient FSH contamination in the LH preparation to stimulate the aromatization system. The requirements for both LH and FSH for estrogen synthesis by the follicle and the preservation of the corpus luteum is consistent with the experiments of Hilliard *et al.* (1971). Alternatively, it is possible that the follicle synthesizes testosterone under the influence of LH and then the testosterone is converted to estrogen by the luteal cells.

Rabbit corpora lutea have been isolated and steroidogenesis studied *in vitro*. Corpora lutea from pseudopregnant (Dorrington and Kilpatrick, 1966; Gorski and Padnos, 1966) and pregnant rabbits (Telegdy and Savard, 1966) synthesized more progesterone than 20α-hydroxypregn-4-en-3-one. This contrasted with whole ovarian and interstitial tissue incubated under the same conditions in which 20α-hydroxypregn-4-en-3-one was the major steroid synthesized (Dorrington and Kilpatrick, 1966). Corpora lutea from pregnant rabbits incubated with radioactive acetate also synthesized small amounts of labeled 20β-hydroxypregn-4-en-3-one, 17α-hydroxypregn-4-en-3-one, dehydroepiandrosterone, androstenedione, and testosterone; no estrogen was detected (Telegdy and Savard, 1966). Addition of LH to rabbit corpora lutea *in vitro* produced only a slight increase in progesterone synthesis (Dorrington and Kilpatrick, 1966). Estrogen also stimulated progesterone and 20α-hydroxypregn-4-en-3-one synthesis *in vitro*, but again the effects were small (Fuller and Hansel, 1971). Additional support for the concept that rabbit corpus luteum is an estrogen target tissue was provided by the isolation of a receptor which strongly bound estradiol-17β *in vitro* (Lee *et al.*, 1971; Scott and Rennie, 1971).

Savard *et al.* (1965) divided corpora lutea from different species into two classes; those which synthesize progestational steroids and those which synthesize estrogens in addition to progestational steroids. It is possible that the amount of estrogen produced may be determined by the number of theca interna cells present in the corpus luteum, since these cells can provide the substrate for aromatization (Brambell, 1956; Short, 1962). The human corpus luteum falls into the second category. Zander (1958) analyzed the steroid content of the human corpus luteum and estimated the concentration of estrone and estradiol to be approximately 0.3 μg/gm. Human corpus luteum slices incubated in the presence of [1-^{14}C]acetate synthesized progesterone as the major labeled product, together with smaller amounts of 17α-hydroxyprogesterone, androstenedione, and estrogen (Hammerstein *et al.*, 1964; Rice *et al.*, 1964; Savard *et al.*, 1969). Radioactive testosterone was detected after 5 hours of incubation (Savard *et al.*, 1969). The pattern of steroids synthesized *in vitro* was comparable to the steroid content of ovarian venous blood collected during the luteal stage of the menstrual cycle (Mikhail *et al.*, 1963).

The addition of LH or HCG to human corpus luteum slices increased the incorporation of [1-^{14}C]acetate into all the steroids, but the relative distribution of the ^{14}C in the products was not significantly altered (Savard *et al.*, 1969). Human corpora lutea of the menstrual cycle had a greater capacity to respond to HCG, as judged by 3′,5′-AMP accumulation and steroid synthesis, than did the corpora lutea of pregnancy (LeMaire *et al.*, 1971; Hermier *et al.*, 1972). The effect of 3′,5′-AMP was greater than that of HCG on the corpus luteum of ectopic pregnancy, indicating that the enzyme potential for steroidogenesis was not realized in the presence of HCG, perhaps due to restricted availability of receptors, reduced efficiency of the receptor–adenylate cyclase coupling mechanism, or inactivation of the adenylate cyclase system.

The bovine corpus luteum synthesizes only progestational steroids. Progesterone is the major steroid synthesized and the cow is unusual in producing 20β-hydroxypregn-4-en-3-one. Slices of bovine corpus luteum *in vitro* were unable to synthesize estrogen from labeled acetate, progesterone, or testosterone (Savard *et al.*, 1965). In general, corpora lutea have little or no 17-hydroxylase and 17–20-lyase activities; however, luteal tissue from the mare (Mahajan and Samuels, 1963) and the pig (Bjersing and Carstensen, 1967) have the capacity to aromatize C_{19} steroids. In contrast, the bovine corpus luteum is deficient in the 17-hydroxylating and the aromatizing systems.

Slices of bovine corpus luteum respond to LH with an increase in progesterone biosynthesis *in vitro* (Mason *et al.,* 1962; Armstrong and Black, 1966, 1968; Cook *et al.,* 1967). Homogenates of corpora lutea also synthesize progesterone but will not respond to LH (Hafs and Armstrong, 1968). Prolactin and peroxide-inactivated LH were ineffective in stimulating progesterone synthesis and the stimulatory effect of FSH was attributed to the LH contamination (Savard *et al.,* 1965). Progesterone synthesis per gram of luteal tissue did not change significantly until 13 days after estrus when the production rate decreased. At day 18 there was a dramatic decrease in the rate of production of progesterone and the levels became undetectable (Armstrong and Black, 1966). LH consistently stimulated progesterone synthesis by corpora lutea obtained on days 1 to 18 after estrus. Corpora lutea obtained on days 19 and 20 either failed to respond to LH or responded only minimally. The corpora lutea which were unresponsive contained normal levels of unesterified sterol and retained the capacity to convert pregnenolone to progesterone. The rate of synthesis of unesterified sterol from acetate was much lower in the inactive corpora lutea (19–23 days old) than in the active corpora lutea (3–18 days old). As suggested by Armstrong and Black (1966), the inability of corpora lutea to synthesize progesterone after 19 days may be due to a decrease in the enzymes responsible for the conversion of cholesterol to pregnenolone, or to inadequate supplies of newly synthesized cholesterol which is utilized preferentially for steroidogenesis (Armstrong and Black, 1966). Alternatively, there may be a loss of LH receptors in the inactive corpora lutea.

A method has been devised for preparing single cell suspensions from the bovine corpus luteum (Gospodarowicz and Gospodarowicz, 1972). The isolated cells secreted progestins for up to 4 hours, and LH increased the rate of release of progestins. Subsequent studies on the isolated luteal cells in culture indicated that dibutyryl $3',5'$-AMP as well as LH could maintain progesterone synthesis. In the presence of either stimulant the morphological appearance of the cells changed from fibroblastic to epithelial. The cells reverted to a fibroblastic appearance after withdrawal of LH, steroid synthesis declined, and the cells divided more rapidly. It was suggested that LH acted to regulate the level of intracellular $3',5'$-AMP which was responsible for the maintenance of differentiation, i.e., a high level of progesterone secretion, epithelial appearance of the cells, a slow rate of cell division, and contact inhibition (Gospodarowicz and Gospodarowicz, 1975).

The maintenance of the corpus luteum of the pig is independent of the pituitary since it functions normally after hypophysectomy (Anderson *et al.,* 1967), and the gonadotropin present at the time of ovulation seems to be adequate to program the function of the corpus luteum for its normal lifespan. Slices of porcine corpora lutea synthesize progesterone and 20α-hydroxypregn-4-en-3-one *in vitro* (Cook *et al.,* 1967). Duncan *et al.* (1960)

were unable to stimulate steroidogenesis by the addition of gonadotropins to slices of porcine luteal tissue. However, in a more detailed study, LH did stimulate progesterone production *in vitro* but the response was small (Cook *et al.,* 1967). HCG caused morphological luteinization of porcine granulosa cells in culture (van Thiel *et al.,* 1971) and also significantly increased progesterone synthesis (Goldenberg *et al.,* 1972a). The finding that diethylstilbestrol (or estradiol) consistently produced a greater increase in progesterone levels than HCG was surprising, even though very high levels of estrogen were required. Maximal production of progesterone was achieved following sequential treatment of the cultured cells, first, with HCG and then with estrogen (Goldenberg *et al.,* 1972a).

Duncan *et al.* (1960) found that the concentration of progesterone in porcine luteal tissue increased from day 4 to 16 of the cycle, followed by a dramatic decrease such that at day 18 there was no detectable progesterone. Du Mesnil du Buisson (1961) first obtained evidence that a substance produced by the uterus may cause luteal regression. Using an *in vitro* system, Duncan *et al.* (1961) observed that uterine extracts obtained from days 16 to 18 of the cycle inhibited progesterone synthesis. Later, Schomberg (1969) found that uterine flushings obtained at days 12 to 18 of the cycle contained a substance which caused luteolysis of porcine granulosa cells in culture.

In summary, the corpus luteum is amenable to studies *in vitro* since it can be readily dissected free from other tissue components. Luteal tissue isolated from a number of species will continue to synthesize progesterone *in vitro* and experiments *in vivo* and *in vitro* have indicated that in many species LH is involved in the control process. LH appears to be the major luteotropin in some species (e.g., cow), while in others LH constitutes a component of the "luteotropic complex" which also includes prolactin and estrogen. Once formed under the influence of LH, the ability of the corpus luteum to synthesize steroids is autonomous during early development in some species (e.g., rat, rabbit, and pig); in the case of the rat, however, prolactin is required to prevent progesterone from being converted to progestationally inactive forms. The progestin levels of corpora lutea from the rabbit and the pig were higher when incubated in the presence of estrogen, but the mechanism by which this was effected is unknown.

IV. THE INTERSTITIAL TISSUE

The amount of ovarian interstitial tissue varies considerably in different species. The interstitial tissue in the mature rabbit ovary is particularly well developed, and contains considerable stores of esterified cholesterol, which are mobilized after coitus, or following the administration of exogenous gonadotropins (Claesson *et al.*, 1948). Hilliard *et al.* (1963) demonstrated that, in the estrous rabbit, various gonadotropins which contain LH activity caused an increase in the levels of progesterone and 20α-hydroxypregn-4-en-3-one, the latter being the predominant steroid secreted *in vivo*. These findings were confirmed by Dorrington and Kilpatrick (1966).

Ovarian tissue from pseudopregnant rabbits was sensitive to LH *in vitro*; a detectable increase in steroid synthesis was produced by 0.05 μg LH/ml. 20α-Hydroxypregn-4-en-3-one was the major steroid synthesized in the control and LH-treated tissue (Dorrington and Kilpatrick, 1966). Experiments *in vivo* indicated that the response to LH was unrelated to the presence of corpora lutea (Hilliard *et al.*, 1963), and this was confirmed by *in vitro* studies in which the effects of LH on separated corpora lutea and interstitial tissue were compared. The isolated corpora lutea responded only minimally to LH, in contrast to the increase in the levels of progesterone and 20α-hydroxypregn-4-en-3-one in the interstitial tissue free of corpora lutea (Dorrington and Kilpatrick, 1966). The elimination of mature follicles has been effected by cautery (Hilliard *et al.*, 1963) and by hypophysectomy (Dorrington and Kilpatrick, 1969) and has clearly shown that the mature follicles do not contribute significantly to the synthesis of progestational steroids in the rabbit ovary. Therefore, the rabbit ovary synthesizes and releases progestational steroids in the absence of corpora lutea and also in the absence of mature follicles, indicating that the interstitial cells are the main sites of progestational steroid synthesis. The probable significance of the large amounts of 20α-hydroxypregn-4-en-3-one synthesized by the interstitial tissue was discussed by Hilliard *et al.* (1967), who proposed that coitus caused a rise in the level of LH sufficient to produce an elevated 20α-hydroxypregn-4-en-3-one output, which, in turn, had a positive feedback action on the pituitary to maintain the elevated level of gonadotropin necessary for ovulation.

In addition to synthesizing progestins, the interstitial tissue may also be a source of testosterone (Hilliard *et al.*, 1974). In intact rabbits, exogenous LH or synthetic LH-RH increases the concentration of testosterone and estrogens. As discussed above, the Graafian follicle has the capacity to synthesize testosterone and estradiol, and the destruction of Graafian follicles by thermocautery caused a marked reduction in the level of both steroids in the ovarian venous plasma. However, systemic injections of LH-RH markedly enhanced testosterone secretion but failed to stimulate estrogen release from the cauterized ovary (Hilliard *et al.*, 1974). This suggests that the interstitial tissue in the rabbit may act as a supplementary source of testosterone but, in contrast to the follicle, the interstitial tissue cannot convert testosterone to estrogens.

Interstitial cells with characteristic ultrastructural features associated with steroidogenic activity were identified in ovarian tissue from human fetuses aged 12 to 20 weeks, i.e., before the development of the theca interna and before the granulosa cells acquire similar ultrastructural features (Gondos and Hobel, 1973). Rice and Savard (1966) reported that androgens were the principal steroids secreted by the human ovarian interstitial cells during the follicular phase of the cycle and during pregnancy.

In the rat ovary, the interstitial tissue is recognizable by the end of the first week of life and until the third week is the only ovarian component with the structural characteristics associated with steroidogenesis (Quattropani and Weisz, 1973). During the neonatal development of the ovary, therefore, the interstitial cells may be the main sites of androgen biosynthesis (Dohler and Wuttke, 1975). The very interesting work of Louvet et al. (1975a,b) has suggested that androgen production by interstitial cells may play an important role in the control of follicular maturation. These workers showed that low doses of HCG (or LH) inhibited the ovarian weight response to estrogen in hypophysectomized immature rats. In the presence of estrogen alone there was granulosa cell proliferation and little follicular atresia, but in the presence of estrogen and HCG, there was no follicular enlargement, many atretic follicles, and interstitial cell stimulation. The effect of HCG on ovarian weight response and follicular atresia was inhibited by antiandrogens, suggesting that these events were mediated via androgen production, most probably as a result of interstitial cell stimulation (Louvet et al., 1975b). The maintenance of a correct balance between the two opposing factors, androgens (probably dihydrotestosterone) and estrogens, may be a key factor in determining the ultimate fate of a follicle, i.e., atresia or enlargement and ovulation.

V. CHOLESTEROL

A. Role as Precursor

Cholesterol is widely distributed throughout the body. Histochemical studies of the ovary revealed fluctuations in the level of stored cholesterol, found mainly in the esterified form, at various stages of the ovarian cycle, gestation, and after coitus (Everett, 1945; Claesson and Hillarp, 1947; Claesson et al., 1948). Everett (1945) found that there was an inverse relationship between the amount of cholesterol stored in the corpus luteum and the level of progesterone secreted and suggested that cholesterol was a precursor of progesterone in the rat.

The conversion of cholesterol to progesterone was demonstrated unequivocally by Tamaoki and Pincus (1961), but the role of cholesterol as an obligatory intermediate in the biosynthesis of progesterone from acetate has been controversial. If steroids are synthesized from acetate by a single pathway, and assuming equilibration of labeled intermediates, newly synthesized from [1-^{14}C]acetate with the existing unlabeled material within the cell, then the specific activities of the intermediates and products should be the same. The results obtained, however, were not consistent with this prediction, as initially indicated by Hechter *et al.* (1953). These workers perfused adrenal glands with labeled acetate and observed that the specific activity of the newly synthesized steroids was higher than that of the isolated cholesterol. Subsequently, this phenomenon was demonstrated in the testis (Mason and Samuels, 1961; Hall *et al.*, 1963) and the ovary. Slices of bovine corpus luteum (Savard *et al.*, 1965; Koritz and Hall, 1965), luteinized rat ovary (Armstrong *et al.*, 1964), and rabbit interstitial tissue (Armstrong *et al.*, 1969a) incubated in the presence of [1-^{14}C]acetate all synthesized progesterone which had a much higher specific activity than the tissue cholesterol. In the light of these observations the following explanations have been proposed: (1) the existence of an alternative pathway for the synthesis of progesterone which does not involve cholesterol as an intermediate; (2) cholesterol is an obligatory intermediate, but not all of the entire pool of cholesterol in the cell is available for steroidogenesis (Hechter *et al.*, 1953; Stone and Hechter, 1954; Savard *et al.*, 1965).

To differentiate between these possibilities, Major *et al.* (1967) administered labeled cholesterol to rats 24 hours before removal of the ovaries for incubation *in vitro* to allow equilibration of the exogenous cholesterol with intracellular cholesterol. Under these conditions, the specific activities of the progestational steroids and the free and esterified cholesterol were similar in the presence and absence of LH. These experiments clearly emphasized the precursor role of cholesterol and there was no indication of the involvement of an alternative pathway. This conclusion was supported by Wilks *et al.* (1970) who showed that aminoglutethimide, which competitively inhibits the 20α-hydroxylation of cholesterol, completely inhibited the biosynthesis of progesterone and the incorporation of [1-^{14}C]acetate into progesterone by luteinized rat ovaries *in vitro*. This inhibition was not overcome by LH. Aminoglutethimide was unable to effect complete inhibition of the conversion of cholesterol to pregnenolone by bovine corpus luteum slices *in vitro*; nevertheless, a net loss of progesterone resulted. It was concluded that the primary pathway for the biosynthesis of progesterone involves cholesterol as an intermediate.

Since there is no experimental evidence to support the participation of an alternative pathway for the synthesis of steroids, the concept has developed that separate pools of cholesterol exist within the cell and only a portion is readily available for steroidogenesis. The remaining cholesterol cannot be utilized for steroidogenesis, and is not responsive to LH (Armstrong et al., 1964; Herbst, 1967). The cholesterol newly synthesized from [1-^{14}C]acetate is incorporated into the steroidogenic pool and equilibrates very slowly with the other pools of cholesterol (Armstrong and Black, 1968; Flint and Armstrong, 1971a,b). Following tissue homogenization, the separate cholesterol pools are mixed and the measured specific activity is that of the total intracellular cholesterol. Consequently, the value obtained is much less than that of the newly synthesized progesterone.

There are three major sources of free cholesterol in the ovary: (1) synthesis de novo from acetate; (2) uptake of preformed cholesterol by the ovary from the blood, and (3) from the hydrolysis of esterified cholesterol. Cholesterol is synthesized from acetate in the ovary by the pathways which were established in the liver by the classical studies of Popjak and Cornforth (1960). This was ascertained in the bovine corpus luteum by studying the kinetics of the incorporation of [1-^{14}C]acetate into selected intermediates along the biosynthetic pathway to cholesterol. Squalene was labeled rapidly (after 7.5 minutes), followed by lanosterol (15 minutes), C_{29} and C_{28} sterols (15–30 minutes), and cholesterol (30 minutes) (Hellig and Savard, 1966). Free cholesterol can diffuse readily through cell membranes and labeled cholesterol taken up from the plasma equilibrates rapidly with the existing pools of free cholesterol within the cell (Flint and Armstrong, 1971b). Equilibration between free and esterified cholesterol proceeds much more slowly but is attained 24 hours after the administration of labeled cholesterol in vivo (Major et al., 1967).

In most species, a large proportion of ovarian cholesterol is esterified to long-chain fatty acids and is located in intracellular lipid droplets (Claesson, 1954; Major et al., 1967; Herbst, 1967). However, there are considerable species differences both in the total cholesterol content and in the proportion which is esterified. The luteinized rat ovary contains large amounts of cholesterol (10–40 mg/gm tissue), most of which is in the esterified form (Armstrong et al., 1964). The rabbit ovary also contains appreciable stores of esterified cholesterol (Claesson et al., 1948). In contrast, the free and esterified cholesterol content of the porcine corpus luteum is low (0.2 mg and 0.4 mg/gm of tissue, respectively: Cook et al., 1967). The stores of esterified cholesterol in the bovine corpus luteum are also low (0.2–0.4 mg/gm tissue), but this species is an exception in that most of the total intracellular cholesterol is unesterified (2–3 mg/gm tissue: Armstrong and Black, 1966). Administration of LH to immature rats causes a depletion of total cholesterol, the greatest changes occurring in the cholesterol ester fraction (Levin and Jailer, 1948; Herbst, 1967). The esterified cholesterol fraction reaches a minimum 3–5 hours after the LH treatment, and reaccumulation occurs after 8 to 18 hours (Herbst, 1967). The cholesterol was not completely depleted even after administration of high doses of LH, supporting the concept that only a portion of the total cholesterol is utilized for steroid synthesis (Herbst, 1967).

Free cholesterol, as opposed to esterified cholesterol, is the immediate precursor for steroidogenesis, as shown by Flint and Armstrong (1971b). In this study, the specific activities of steroids and free and esterified cholesterol at different times after the administration of [^{14}C]cholesterol revealed that the specific activity of the steroids (progesterone and 20α-hydroxypregn-4-en-3-one) followed that of the free cholesterol. The specific activity of the esterified cholesterol at all times up to 8 hours after the administration of [^{14}C]cholesterol was considerably lower than that of the free cholesterol.

B. Cholesterol Side-Chain Cleavage Enzyme System

The conversion of cholesterol to pregnenolone is a complex process and details of the reaction mechanism and the components of the cholesterol side-chain cleavage enzyme complex have been extensively reviewed (Sulimovici and Boyd, 1969).

In the bovine corpus luteum (Hall and Koritz, 1964; Yago et al., 1967b) and the immature rat ovary, cholesterol side-chain cleavage activity was localized exclusively in the mitochondrial fraction. The conversion of cholesterol to pregnenolone is the rate-limiting step in the synthesis of steroids and the step in the pathway at which LH influences the rate of production of steroids. However, LH did not act directly upon active powders of mitochondria (Ichii et al., 1963; Hall and Koritz, 1964; Yago et al., 1967a,b) and neither LH nor 3′,5′-AMP acted upon the isolated mitochondrion to stimulate side-chain cleavage (Jackanicz and Armstrong, 1968).

The side-chain cleavage enzyme system has an absolute requirement for NADPH and molecular oxygen, as clearly demonstrated using suspensions of acetone-dried preparations of mitochondria (Hall and Koritz, 1964; Yago et al., 1967a,b). Nevertheless, the side-chain activity of freshly prepared intact mitochondria was not influenced by exogenous NADPH, and NADH and NADPH were not oxidized by the respiratory chain, indicating that the mitochondrial membrane acts as a permeability barrier to these cofactors (Yago et al., 1967b). In the presence of Ca^{2+}, which causes mitochondrial swelling, the permeability barrier to exogenous NADPH is lost, and steroid synthesis is stimulated (McIntosh et al., 1971). As will be discussed later, the inaccessibility of NADPH to intact mitochondria is an important factor in evaluating the theories of hormone action which have suggested that NADPH produced outside the mitochondria is made available for cholesterol side-chain cleavage.

VI. LUTEINIZING HORMONE

A. Site of Action

Early histochemical studies on the depletion of ovarian cholesterol elicited by LH suggested that this gonadotropin exerted its effect on steroidogenesis at a step subsequent to cholesterol (Levin and Jailer, 1948; Claesson et al., 1948). More definite evidence concerning the site of action of LH has been obtained by labeling the intracellular pools of cholesterol and studying their conversion to progesterone. However, the in vitro approach to the study of cholesterol has been hazardous. One problem has been the insolubility of cholesterol in the incubation medium. Mason and Savard (1964) added albumin to the incubation medium to help solubilize the radioactive cholesterol, and thus aid the entry of the precursor into bovine corpus luteum slices. LH increased the amount of progesterone synthesized de novo, but this increase was not paralleled by the conversion of labeled cholesterol to progesterone. The specific activity of the newly synthesized progesterone was lower in the presence of LH than in controls. The addition of NADP plus glucose-6-phosphate increased the amount of progesterone synthesized de novo, but the specific activity of the progesterone was higher than that isolated from LH-treated and control tissues. In the light of these observations, several explanations were put forward by Savard et al. (1965). They favored the concept that cholesterol is an intermediate in progesterone synthesis and that LH influences at least two steps in the biosynthetic pathway: (1) the conversion of cholesterol to progesterone, and (2) the synthesis of cholesterol from acetate. According to this proposal, the newly synthesized unlabeled cholesterol, produced at a higher rate in the presence of LH, was incorporated into the labeled cholesterol pool, thereby reducing the specific activity of the cholesterol and consequently of the progesterone. In addition, it was concluded that LH and NADPH influenced progesterone synthesis by different and independent mechanisms; NADPH stimulated steroidogenesis at a step between cholesterol and progesterone whereas LH exerted its effects upon steroidogenesis primarily by stimulating the synthesis of cholesterol.

Other interpretations of these data (Mason and Savard, 1964; Savard et al., 1965) are possible. Armstrong et al. (1964) demonstrated the poor penetration of exogenous [^{14}C]cholesterol into slices of luteinized rat ovaries by showing that the specific activity of the tissue cholesterol was much lower than that of the combined tissue and medium. The small proportion of [^{14}C]cholesterol which did enter the cell failed to equilibrate with the pool of cholesterol utilized for progesterone synthesis, and tended to remain in the unesterified form. Consequently, the decreased specific activity of progesterone synthesized after stimulation by LH, observed by Mason and Savard (1964), could be explained by inadequate equilibration.

The solubility of cholesterol in the incubation medium has also been increased by the addition of a detergent, Tween 80, and, under these conditions, LH stimulated the conversion of labeled cholesterol to progesterone by slices of bovine corpus luteum (Hall and Koritz, 1965). Furthermore, the radioactivity incorporated into cholesterol from [1-^{14}C]acetate by bovine corpus luteum slices was consistently less in the presence of LH than in control incubations. However, the mass of cholesterol was not significantly altered, and consequently, the specific activity of the cholesterol was lower in the presence of LH (Koritz and Hall, 1965). This observation is not in keeping with a significant effect of LH on the biosynthetic pathway from acetate to cholesterol, at least during the early stages of increased steroidogenesis, and Hall and Koritz (1965) and Koritz and Hall (1965) concluded that the principal site of action of LH is at a point in the biosynthetic pathway between cholesterol and progesterone.

Even in the presence of albumin or Tween 80, incomplete equilibration of the exogenous labeled cholesterol with the relatively large amounts of endogenous cholesterol remains a problem, as emphasized by Savard *et al.* (1969). These workers showed that the effects of LH on steroidogenesis by bovine corpus luteum slices *in vitro* were detectable within 30 minutes whereas only 2.3% of the labeled cholesterol had entered the tissue slice at that time, although albumin was present in the incubation medium. In view of this time discrepancy between the rapid effect of LH on progesterone biosynthesis and the slow entry of exogenous labeled cholesterol, this experimental design is particularly unsatisfactory for the study of hormone action.

The leakage of enzymes from the tissue slice into the incubation medium is another area of concern. The cholesterol side-chain cleavage enzyme system has been detected in the incubation medium (Armstrong, 1966). Therefore, in the presence of NADPH or a NADPH-generating system, labeled cholesterol can be converted extracellularly to pregnenolone. This could explain the high specific activity of progesterone obtained by Mason and Savard (1964) when they incubated bovine corpus luteum slices with labeled cholesterol in the presence of NADPH (Armstrong, 1966; Armstrong and Black, 1968).

To circumvent the problems encountered in the use of labeled cholesterol *in vitro,* cholesterol pools have been labeled in rats (Major *et al.,* 1967) and rabbits (Solod *et al.,* 1966) by the intravenous administration of [7α-^{3}H]cholesterol 24 hours before experimentation. The *in vitro* synthesis of progesterone by luteinized rat ovaries was increased by LH, whether or not the hormone was added directly to the incubation medium or administered intravenously 30 minutes before killing. The increase in the mass of progesterone synthesized after treatment with LH was paralleled by an increase in the incorporation of radioactive cholesterol into progesterone; thus the specific activity of progesterone was similar in control and LH-stimulated tissues. If LH stimulated the synthesis of cholesterol from un-labeled precursors, then the specific activity of the steroid would have been less in the presence of LH than in its absence. Since this was not the case, Major *et al.* (1967) concluded that cholesterol is an obligatory intermediate in the synthesis of progesterone and that LH exerts its effects on steroidogenesis in the luteinized rat ovary by stimulating the conversion of cholesterol to progesterone.

Similar experiments with rabbit ovaries revealed that LH treatment *in vivo* or *in vitro* caused an increase in the conversion of prelabeled choles-terol to progesterone and 20α-hydroxypregn-4-en-3-one (Solod *et al.,* 1966). The results were not as clear as those obtained using luteinized rat ovaries because the specific activity of progesterone synthesized in the presence of LH was less than in unstimulated tissue. Thus, the possibility remains that in this species LH stimulates steroid synthesis, at least to some extent, by a primary action on the synthesis of cholesterol (Solod *et al.,* 1966). Pursuing this possibility further, steroidogenesis in rabbit interstitial tissue (Armstrong, 1967) and bovine corpus luteum tissue (Armstrong, 1968) was evaluated in the presence of AY9944 [*trans*-1,4-*bis*-2-(chlorobenzyl-aminomethyl)cyclohexane dihydrochloride] which decreases the synthesis of cholesterol by inhibiting 7-dehydrocholesterol Δ^7-reductase. AY9944 has previously been shown to interfere with adrenal (Givner and Dvornik, 1964) and testicular (Menon *et al.,* 1965) cholesterogenesis *in vitro.* This inhibitor greatly reduced the incorporation of [1-^{14}C]acetate into free and esterified cholesterol and steroids by rabbit interstitial tissue and bovine corpus luteum tissue, but this effect did not significantly influence the effectiveness of LH in stimulating steroid synthesis.

The bulk of the evidence now available suggests that, as far as the acute *in vitro* stimulation of progesterone production is concerned, the conversion of cholesterol to progesterone is the principal site of action of LH. If an increase in cholesterol synthesis occurs in response to LH treatment, then the experiments with AY9944 suggest that this increase does not contribute significantly to the increased amount of progesterone synthesized in short-term experiments *in vitro*. After LH treatment, the cholesterol stores become depleted. Reaccumulation of ovarian cholesterol may be accomplished by the uptake of cholesterol from the plasma or by *de novo* synthesis of cholesterol from acetate. There is a marked increase in the specific activity of progesterone synthesized from [1-^{14}C]acetate *in vitro* by luteinized rat ovaries from 2 to 5 hours after the administration of LH *in vivo*, but during this time progesterone synthesis declines. This strongly suggests that increased conversion of acetate to cholesterol is a consequence of increased steroidogenesis and is a mechanism whereby the stores of cholesterol are reaccumulated (Armstrong, 1968).

B. Mechanism of Action

1. Cofactor Availability

The synthesis of progesterone proceeds by the same pathway in the ovary, the adrenal, and the testis; LH and ACTH act upon this pathway in their respective target tissues at a site between cholesterol and pregnenolone (Hall and Young, 1968). NADPH is essential for cholesterol side-chain cleavage and the control of NADPH availability has formed the basis of a number of theories of hormonal control of steroidogenesis (Haynes and Berthet, 1957; Haynes *et al.*, 1960; McKerns, 1969). Briefly, Haynes and Berthet (1957) and Haynes *et al.* 1960) proposed that ACTH stimulated the synthesis of 3′,5′-AMP which activated phosphorylase, thereby increasing the amount of glucose-6-phosphate entering the pentose phosphate pathway, and thus producing increased levels of NADPH.

The possibility that the Haynes–Berthet theory may also explain the action of LH on the ovary has received considerable attention. LH was shown to increase phosphorylase activity *in vitro* in bovine corpus luteum slices (Marsh and Savard, 1964) and in prepubertal rat ovaries (Selstam and Ahrén, 1971) and this correlated with an increase in progesterone synthesis. On the other hand 3′,5′-AMP stimulated progesterone synthesis without an effect on phosphorylase activity, indicating that phosphorylase activation is not obligatory for increased steroid production (Savard *et al.*, 1965). The administration of LH to pseudopregnant rats increased ovarian phosphorylase activity and caused a slight decrease in total glycogen (Stansfield and Robinson, 1965). However, the level of glycogen in this tissue was so low (0.29 μg/mg tissue) that it was considered unlikely that glycogen breakdown was a significant factor in the stimulation of hormone production.

Steroid hormone production may be regulated by virtue of an increased supply of NADPH as a result of an increase in the rate of transport of glucose across the cell membrane. Some support for this theory came from the observation that LH administration *in vivo* stimulated the subsequent uptake of glucose and the production of lactic acid *in vitro* by whole prepubertal rat ovaries (Armstrong *et al.,* 1963; Ahrén and Kostyo, 1963; Hamberger and Ahrén, 1967) and by slices (Armstrong *et al.,* 1963) and minces (Channing and Villee, 1966) of luteinized rat ovaries. Administration of LH *in vivo* influenced glucose uptake, but LH had no effect on glucose metabolism *in vitro,* although progesterone synthesis was stimulated under the same conditions. These experiments demonstrated that the effects of LH on progesterone synthesis and glycolysis could be dissociated (Armstrong, 1968). Other observations were inconsistent with the theory that increased glucose uptake and increased NADPH production, via the metabolism of glucose-6-phosphate by the pentose phosphate pathway, was essential for an increase in steroidogenesis. These included the demonstration that the stimulatory effect of LH on steroid production was independent of the presence of exogenous glucose. Moreover, in the luteinized rat ovary, glucose is metabolized by the glycolytic pathway rather than by the pentose phosphate pathway (Armstrong and Greep, 1962; Flint and Denton, 1969).

Many of the above considerations are relevant to the evaluation of McKerns' theory which proposed that the initial action of gonadotropins was to stimulate glucose-6-phosphate dehydrogenase activity and, by this means, increase the level of NADPH available for steroidogenesis (McKerns, 1965, 1969). Studies *in vivo* clearly demonstrated that glucose-6-phosphate dehydrogenase activity in rat ovaries increased in response to gonadotropins (Wiest and Kidwell, 1969; Kidwell *et al.,* 1966; McKerns, 1965). Also, there was a direct relationship between the steroidogenic capacity of an ovarian component and glucose-6-phosphate dehydrogenase activity. However, Wiest and Kidwell (1969) emphasized that there is no evidence to suggest that the activity of glucose-6-phosphate dehydrogenase is a rate-limiting factor in steroidogenesis, and its activity may be secondary to elevated steroid synthesis.

The theories of Haynes and Berthet (1957) and McKerns (1969) of hormone action on steroidogenesis both suggest that the NADPH required for cholesterol side-chain cleavage in mitochondria is produced in the cytoplasm by the oxidation of glucose-6-phosphate. Flint and Denton (1970) have compared the rates of steroid synthesis with the intracellular concentration of oxidized and reduced pyridine nucleotides. In the presence of glucose the total intracellular content of NADPH and NADH increased, but no change in the rate of steroidogenesis was detectable. Pretreatment with LH increased steroid synthesis in the presence of glucose but did not change the NADPH:NADP ratio. It was concluded that "a change in the concentration of cytoplasmic NADPH is not involved in the primary steroidogenic response of rat luteum tissue to luteinising hormone" (Flint and Denton, 1970).

Hall (1971) criticized the above interpretation and pointed out that an increase in NADPH production may occur after treatment with LH, but if this was paralleled by an increase in rate of utilization of NADPH for steroid synthesis, then the NADPH:NADP ratio would remain constant. NADPH will not penetrate intact ovarian mitochondria. Therefore, if a NADPH-generating system outside the mitochondria provides reducing equivalents for intramitochondrial steroid synthesis, then there must be a means of transporting NADPH across the mitochondrial membrane. Hall (1971) considered that an increase in the total cell content of NADPH elicited by glucose, as observed by Flint and Denton (1970), may be ineffective in stimulating steroid synthesis because the transport of reducing equivalents across mitochondrial membranes may not be facilitated under these conditions.

There are a number of ways by which intramitochondrial NADPH may be increased: (1) by the transfer of cytoplasmic reducing equivalents to the mitochondrial NADPH pool via the malate shuttle (Simpson and Estabrook, 1968); (2) by reversed electron transport coupled to a transhydrogenase (Harding et al., 1965); (3) by malate and isocitrate dehydrogenase activity; or (4) by oxidation of fatty acids to produce NADH followed by transhydrogenation. The observations of McIntosh et al. (1971) and Uzgiris et al. (1971) support the second possibility. Mitochondria isolated from bovine corpora lutea readily converted cholesterol to progesterone and 20β-hydroxypregn-4-en-3-one when incubated with various Krebs cycle intermediates (isocitrate, malate or succinate) or with NADPH and $CaCl_2$. However, cholesterol side-chain cleavage activity, supported by the citric acid cycle intermediates, was blocked by rotenone, antimycin, and amobarbital; NADPH-supported side-chain cleavage was only affected by amobarbital. It appeared, therefore, that NADPH production by mitochondria from bovine corpus luteum occurred almost exclusively by reversed electron transport coupled to an energy-dependent transhydrogenase (Uzgiris et al., 1971).

2. Cholesterol Availability and Transport

The availability of cholesterol to the side-chain cleavage system may be a critical factor in the control of steroidogenesis. As reviewed earlier, the hydrolysis of cholesterol esters which are stored in cytoplasmic lipid droplets is stimulated by LH and $3',5'$-AMP (Behrman and Armstrong, 1969; Goldstein and Marsh, 1973). LH also influences the concentration of free cholesterol in slices of rabbit interstitial tissue by inhibiting the cholesterol ester synthetase activity (Flint *et al.*, 1973). This inhibitory effect of LH was mimicked by 3′,5′-AMP *in vitro,* but was prevented by cycloheximide and aminoglutethimide phosphate, both of which inhibit the synthesis of steroids from cholesterol. That this effect of LH was due to elevated levels of steroids was suggested by the inhibitory effects of progesterone and 20α-hydroxypregn-4-en-3-one on cholesterol ester synthetase activity in cell-free extracts of rabbit interstitial tissue. LH, therefore, increases the concentration of free cholesterol by activating cholesterol esterase and by inhibiting cholesterol ester synthetase. The physiological significance of the increased level of free cholesterol at the expense of esterified cholesterol is uncertain, particularly since free cholesterol is available from the plasma. At least one of the effects, i.e., cholesterol ester synthetase inhibition, appears to be a consequence of elevated steroid levels and therefore could be a means by which elevated steroid synthesis is sustained.

Since cholesterol is synthesized and stored outside the mitochondria and converted to pregnenolone within, the means by which cholesterol is transported to the side-chain cleavage system is another possible site of control. Cholesterol transport has not been studied thoroughly in the ovary, but, in a series of elegant studies on adrenal tissue, Garren and his associates have implicated this process in the control of steroidogenesis by ACTH (Garren *et al.*, 1965, 1971). These investigators showed that the administration of cycloheximide to hypophysectomized rats maximally stimulated with ACTH resulted in a rapid decay of steroid output with a half-life of several minutes. In contrast, actinomycin D did not affect ACTH-stimulated steroidogenesis, suggesting that the messenger RNA was stable and that the control was at the level of translation. The inhibition of ACTH action by cycloheximide was not due to a decrease in the levels of the enzymes involved in the synthesis of steroids (Koritz and Kumar, 1970); the incorporation of [³H]acetate into adrenal cholesterol was unaffected (Garren *et al.*, 1971). The stimulation of the hydrolysis of cholesterol esters to free cholesterol by ACTH in the cytoplasmic lipid droplets did not require protein synthesis, and in the presence of ACTH and cycloheximide there was a marked increase in the amount of free cholesterol in the lipid droplets, but *not* in the mitochondria. Garren *et al.* (1971) concluded from these observations that cycloheximide blocked the effects of ACTH on the conversion of cholesterol to pregnenolone by inhibiting the synthesis of a regular protein which was required for the transfer of cholesterol from the lipid droplets to the mitochondria. A cholesterol-binding protein which stimulates the cholesterol side-chain cleavage activity of adrenal mitochondrial preparations has been isolated from rat adrenals (Kan *et al.*, 1972). However, the level of this protein was unaffected by ACTH (Ungar *et al.*,

1973), and its relationship to the hypothetical regulator protein required for the transfer of cholesterol from the lipid droplets is not clear.

It is apparent from the foregoing that the mechanism by which LH activates the side-chain cleavage of cholesterol is not understood. LH may accelerate this reaction by increasing (1) the levels of NADPH, (2) substrate levels, and/or (3) substrate transport. Other possibilities exist, such as a direct action on the activity of the side-chain cleavage system, or an increase in the transport of pregnenolone, an inhibitor of cholesterol side-chain cleavage, from the mitochondria (Koritz and Kumar, 1970).

In an attempt to determine the point at which ACTH stimulates cholesterol side-chain cleavage *in vivo,* Bell and Harding (1974) studied the catalytic properties of the enzyme system in mitochondria obtained from rat adrenals, using cholesterol–lecithin emulsion as a substrate, and presented some interesting results. In mitochondria from normal rats, hypophysectomized rats, and hypophysectomized rats treated with ACTH, the cholesterol and 20α-hydroxycholesterol side-chain cleavage activities were high and were not influenced by ACTH. ACTH stimulation of hypophysectomized rats increased the levels of cytochrome P450-bound cholesterol in rat mitochondria, and cholesterol accumulation occurred at the enzyme site only when hydroxylation was inhibited by anoxia. Pregnenolone was rapidly metabolized in the presence of added microsomes and did not inhibit cholesterol side-chain cleavage except under abnormal conditions, suggesting that pregnenolone exit is not restricted and does not act as an inhibitor of cholesterol metabolism. Bell and Harding (1974) concluded that ACTH did not increase the rate of cholesterol side-chain cleavage *per se,* but stimulated the transport of cholesterol to the cholesterol side-chain cleavage cytochrome P450 enzyme, and that the transported cholesterol was rapidly metabolized.

3. Protein Synthesis

Ongoing protein synthesis is required for the stimulation of steroidogenesis by cyclic AMP and tropic hormones (reviewed by Wicks, 1974). Puromycin or cycloheximide inhibited the effects of LH on steroidogenesis in the bovine corpus luteum (Savard *et al.,* 1965; Marsh *et al.,* 1966), rat luteinized ovarian tissue (Hermier *et al.,* 1971), rabbit ovarian tissue (Gorski and Padnos, 1966), and rabbit (YoungLai, 1975) and rat follicles (Lieberman *et al.,* 1975).

In the ovary, as in the adrenal (Garren *et al.,* 1971), a labile protein has been implicated in the regulation of steroidogenesis (Hermier *et al.,* 1971; Lieberman *et al.,* 1975). In the first of these studies, cycloheximide was shown to inhibit progesterone biosynthesis within 30 minutes when added to luteinized rat ovarian tissue already stimulated by LH or 3′,5′-AMP. Consequently, the protein mediating the effects of LH on steroidogenesis has a half-life of approximately 10 minutes. Hermier *et al.* (1971) demonstrated that the rapidly turning over protein was not required for cholesterol side-chain cleavage but was involved at an earlier step. Oxygen deprivation inhibited the synthesis of progesterone in the presence of LH or 3′,5′-AMP, presumably because oxygen is required for side-chain cleavage. Subsequent oxygenation resulted in the rapid synthesis of progesterone, even in the presence of cycloheximide. Hermier *et al.* suggested that, in an oxygen-deprived medium, LH stimulated the synthesis of a regulator protein which caused the accumulation of an intermediate involved in steroid synthesis. In the presence of oxygen this intermediate was rapidly consumed for the synthesis of progesterone, a process which did not require *de novo* protein synthesis. They suggested further that the intermediate may be intramitochondrial free cholesterol and that the regulator protein may be involved in the translocation of cholesterol to interact with the mitochondrial hydroxylase system (Hermier *et al.*, 1971).

In the rat Graafian follicle the LH-induced protein required for the synthesis of progesterone appeared to be short-lived, since inhibition of protein synthesis at any time after LH treatment prevented further steroid synthesis within 0.5 to 1 hour (Lieberman *et al.*, 1975). Whether the labile proteins described by Lieberman *et al.* (1975), Hermier *et al.* (1971), and Garren *et al.* (1971) are similar and influence steroidogenesis by the same mechanism remains to be investigated.

To explain the effect of inhibitors of protein synthesis on hormone-induced steroidogenesis in both the adrenal and the ovary, the existence of a regulator protein has been proposed. In both tissues this hypothetical protein is thought to exert its effects before cholesterol side-chain cleavage, possibly at the level of cholesterol transport to the catalytic site. Further studies on this putative protein and the mechanism by which cholesterol is translocated could provide new insight into the control of steroidogenesis.

The scheme shown in Fig. 2 to explain the mode of action of tropic hormones on steroidogenesis has been constructed with the liberal use of Occam's razor, i.e., it is the simplest scheme possible for the combined results from studies on the gonads and the adrenal.

C. Role of Adenosine 3′,5′-Monophosphate

The impact of the discovery of 3′,5′-AMP has been felt in almost every area of endocrinology, and it is clear that this nucleotide is an important

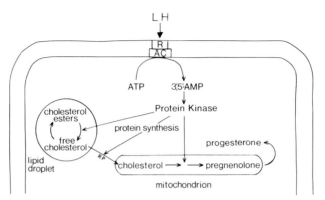

Fig. 2:   Possible sites of action of LH and 3′,5′-AMP on steroidogenesis. R, the hormone receptor; AC, adenylate cyclase; and RP, the regulator protein.

agent in the regulation of cellular function in a wide variety of tissues
(reviewed in *Advan. Cyclic Nucleotide Res.* Vols. 1–6).

The involvement of 3′,5′-AMP in the regulation of steroidogenesis was
first indicated by Haynes (1958) who found that the steroidogenic response
of bovine adrenal tissue to ACTH was associated with elevated levels of
3′,5′-AMP. Subsequently, Haynes *et al.* (1959) showed that exogenous
3′,5′-AMP stimulated corticosteroid synthesis by rat adrenals *in vitro*. In
view of the findings, Haynes *et al.* (1960) proposed that 3′,5′-AMP
mediated these steroidogenic effect of ACTh. Similar studies on the ovary
and the testis showed that the stimulatory effect of 3′,5′-AMP on
steroidogenesis was not peculiar to the adrenal. Steroidogenesis by ovarian
tissue components from a variety of species was stimulated *in vitro* by the
addition of 3′,5′-AMP. Corpora lutea isolated from cows (Marsh and
Savard, 1964, 1966), women (LeMaire *et al.,* 1971; Hermier *et al.,* 1972),
rats (Hermier and Jutisz, 1969) and rabbits (Dorrington and Kilpatrick,
1967) responded to 3′,5′-AMP with an increase in progesterone synthesis.
The magnitude of the response to 3′,5′-AMP varied considerably from one
species to another, but in each case, maximum stimulation by 3′,5′-AMP
produced a response which was comparable to that produced by saturating
concentrations of LH (or HCG), both in terms of total mass of steroid and
relative proportions of various steroids produced. Similarly, the rabbit
preovulatory follicle (Mills, 1975) and interstitial tissue (Dorrington and
Kilpatrick, 1967, 1969) responded to 3′,5′-AMP with increases in steroid
synthesis which were comparable to those produced by LH. The addition of
3′-AMP, 5′-AMP, ADP, and ATP did not significantly affect
steroidogenesis under conditions in which the same concentrations of 3′,5′-
AMP produced a marked stimulation (Marsh and Savard, 1966; Dorrington and Kilpatrick, 1969). Addition of 3′,5′-AMP to corpora lutea

(Marsh and Savard, 1966) and rabbit ovarian tissue (Dorrington and Kilpatrick, 1967), which were maximally stimulated by LH, did not cause any additional response. The simultaneous addition of submaximal concentrations of LH and 3′,5′-AMP to rabbit ovarian tissue produced an additive effect both on 20α-hydroxypregn-4-en-3-one and progesterone levels. Furthermore, kinetic studies showed that the rates of steroidogenesis induced by LH and 3′,5′-AMP were the same for up to 10 hours (Dorrington and Kilpatrick, 1969).

Therefore, 3′,5′-AMP reproduces the effects of LH qualitatively and quantitatively and all the above data are consistent with the proposal that 3′,5′-AMP is the intracellular mediator of LH action on steroid synthesis in the ovary. However, the exogenous concentrations of 3′,5′-AMP required to stimulate steroidogenesis in preparations from the ovary (2–20 μmoles/ml) were high compared with the endogenous levels found in these tissues. The requirement for high concentrations of 3′,5′-AMP could be due to two factors: poor penetration of 3′,5′-AMP through cell membranes and the breakdown of 3′,5′-AMP by phosphodiesterase, intracellularly or extracellularly due to leakage of enzyme from damaged cells. Theophylline (10 mM), which inhibits phosphodiesterase (Butcher and Sutherland, 1959), caused a small but significant increase in steroid production by rabbit interstitial tissue *in vitro*. Theophylline potentiated the ovarian response to submaximal amounts of 3′,5′-AMP and LH (Dorrington and Kilpatrick, 1967, 1969). The effect of HCG on progesterone synthesis by human corpora lutea was potentiated by theophylline whereas imidazole, a stimulator of phosphodiesterase, reduced progesterone synthesis.

In 1966, Marsh *et al.* provided clear evidence that 3′,5′-AMP was present in bovine and human corpora lutea and showed that the level was dramatically increased in the presence of LH (or HCG). The effect of LH on the accumulation of 3′,5′-AMP in the bovine corpus luteum was rapid and preceded the increase in steroidogenesis. Prolactin, ACTH, glucagon, and LH inactivated by treatment with hydrogen peroxide did not influence 3′,5′-AMP levels or stimulate steroidogenesis (Mason et al., 1962; Marsh et al., 1966). Corpora lutea from luteinized rat ovaries or pregnant rats also responded to LH *in vitro* with an increase in 3′,5′-AMP accumulation, but the response appeared to decline with increasing age of the corpus luteum (Lamprecht et al., 1973; Mason et al., 1973; Ahrén et al., 1974). LH caused a rapid accumulation of 3′,5′-AMP in isolated rat follicles (Lindner et al., 1974; Tsafriri et al., 1972; Lamprecht et al., 1973; Ahrén et al., 1974) and porcine granulosa cells (Kolena and Channing, 1971, 1972).

Although there is overwhelming supporting evidence for the role of 3′,5′-AMP as a mediator of tropic hormone stimulation of steroidogenesis, this theory has been carefully reconsidered in recent years, largely as a result of the kinetic analyses of LH and ACTH effects on Leydig and adrenal cells, respectively (Catt and Dufau, 1973; Moyle and Ramachandran, 1973; Mendelson et al., 1975; Beall and Sayers, 1972). These studies showed that low concentrations of hormones could stimulate steroidogenesis without causing detectable increases in $3',5'$-AMP. Thus, even though tropic hormone-induced steroid synthesis is often accompanied by elevated $3',5'$-AMP levels, there is a dissociation between the steroidogenic and $3',5'$-AMP responses to low levels of hormone. From a consideration of the above papers, two possibilities emerge. Either $3',5'$-AMP mediates the effect of low concentrations of hormone on steroidogenesis, and the amount produced is below the level of detectability of the present assay techniques, or the stimulation of steroidogenesis proceeds by mechanisms other than $3',5'$-AMP formation. In the latter case, $3',5'$-AMP may be involved in prolonging the steroidogenic response, and other membrane-associated events may be crucial during the early stages of tropic hormone stimulation, as suggested by Mendelson et al. (1975). The development of more sensitive techniques for the measurement of $3',5'$-AMP would help to resolve this issue.

1. Adenylate Cyclase

The intracellular concentration of $3',5'$-AMP in the ovary may be controlled by LH either by activating adenylate cyclase or by inhibiting phosphodiesterase. To investigate which control mechanism was operating, adenylate cyclase and phosphodiesterase activities were evaluated simultaneously in homogenates of rabbit ovarian tissue by following the rate of formation of [^{14}C]$3',5'$-AMP from [^{14}C]ATP and also the rate of destruction of [^{3}H]$3',5'$-AMP (Baggett and Dorrington, 1969; Dorrington and Baggett, 1969; Dorrington and Kilpatrick, 1969). LH had no effect on phosphodiesterase activity, either in the presence or absence of caffeine or theophylline. In contrast, a consistent effect of LH on the activity of adenylate cyclase was demonstrated. A significant increase in the level of [^{14}C]$3',5'$-AMP was apparent 2 minutes after the addition of LH to the homogenate (Dorrington and Baggett, 1969). Marsh (1970a,b) also concluded that LH increased the level of $3',5'$-AMP in homogenates of bovine corpora lutea by activating adenylate cyclase rather than by inhibiting phosphodiesterase.

D. Role of Prostaglandins

Most tissues have the capacity to synthesize prostaglandins from arachidonic acid and those highly potent substances are able to elicit a wide variety of biochemical effects depending upon the cell type which is stimulated. Many of the effects of prostaglandins have been observed in tissues in which $3',5'$-AMP acts as the intracellular mediator of hormone action, and these effects are similar to those produced by $3',5'$-AMP. The possibility that prostaglandins may be mediators of LH action was realized when it was observed that prostaglandins stimulated progesterone synthesis in luteinized rat ovaries *in vitro* (Pharriss *et al.*, 1968). Later, Speroff and Ramwell (1970) showed that several prostaglandins (PGE$_2$, PGE$_1$, PGF$_2$, and PGA) stimulated the synthesis of progesterone by slices of bovine corpus luteum. The effects of the prostaglandins and LH were blocked by cycloheximide. Marsh (1970b) confirmed that PGE$_2$, as well as LH, increased the mass of steroid synthesized and the incorporation of $[1-^{14}C]$acetate into steroid by the bovine corpus luteum *in vitro*. Both PGE$_2$ and LH increased the conversion of $[^{32}P]ATP$ to $[^{32}P]3',5'$-AMP by homogenates of bovine corpus luteum, and the increases in adenylate cyclase activity correlated with the increases in steroid production caused by PGE$_1$ and LH in slices of the same tissue. PGE$_1$ and LH increased $3',5'$-AMP accumulation in homogenates in the presence of theophylline, suggesting that both substances exerted their effects by activating adenylate cyclase (Marsh, 1970a,b).

Since prostaglandins can mimic many of the actions of LH in the ovary including those related to ovulation, luteinization, and stimulation of steroidogenesis, the question was raised whether prostaglandins are essential mediators of LH action, coupling the initial interaction of hormone and receptor with the adenylate cyclase system. Some support for this theory came from the report that 7-oxa-13-prostynoic acid, a competitive inhibitor of prostaglandin synthesis, blocked the effects of PGE on $3',5'$-AMP accumulation in rat ovaries, but more significantly that it blocked the effect of LH on cyclic AMP accumulation (Kuehl *et al.*, 1970). Later, however, a body of evidence accumulated which was inconsistent with the above theory, and rather favored the view that LH and prostaglandins acted by different mechanisms. First, the combined effects of saturating concentrations of LH and PGE$_2$ or adenylate cyclase activity and $3',5'$-AMP accumulation were greater than either stimulant alone (Marsh, 1971; Lindner *et al.*, 1974). Second, the effect of LH on $3',5'$-AMP was rapid, whereas LH failed to stimulate prostaglandin synthetase when rat ovaries were incubated for 1 hour (Bauminger *et al.*, 1973). Furthermore, the ovarian and follicular content of PGE$_2$ did not increase for several hours after LH administration (LeMaire *et al.*, 1973; Lindner *et al.*, 1974). Dibutyryl $3',5'$-AMP stimulated prostaglandin synthesis in mouse ovaries (Kuehl *et al.*, 1973), suggesting that the sequence of events was LH activation of adenylate cyclase followed by prostaglandin synthesis.

Inhibitors of prostaglandin synthesis, such as flufenamic acid, indomethacin, and aspirin, effectively inhibited the effect of LH on ovarian prostaglandin synthesis, but did not block the effect of LH on $3',5'$-AMP accumulation or upon steroid synthesis (Grinwich *et al.*, 1972; Zor *et al.*, 1973). Even though indomethacin did not block the effect of LH on $3',5'$-AMP and steroid synthesis and ovum maturation, it did block ovulation in rats and rabbits (Armstrong and Grinwich, 1972).

Prostaglandins do not therefore appear to play an obligatory role in the action of LH on $3',5'$-AMP accumulation and steroid synthesis. They may, however, participate in later events leading to follicular rupture, perhaps by influencing the contraction of the smooth muscle cells in the follicular wall (Armstrong *et al.*, 1974).

E. Protein Kinase

The many diverse responses elicited by $3',5'$-AMP in different tissues may involve the activation of protein kinases (Kuo and Greengard, 1969). The protein kinases, in turn, could, by the process of phosphorylation, activate proteins important in the control of cell function. In the case of epinephrine action on glycogenolysis, for example, the primary effect of $3',5'$-AMP is activation of protein kinase (Robison *et al.*, 1971). Using a genetic approach, Insel *et al.* (1975) have provided compelling evidence for the "pivotal" role of protein kinases in $3',5'$-AMP regulation of enzyme induction and growth of S49 cells. These authors demonstrated parallelism between $3',5'$-AMP-dependent protein kinase activity and the cellular responses to dibutyryl $3',5'$-AMP in wild-type cells and three classes of clones resistant to $3',5'$-AMP, indicating that the cellular content of $3',5'$-AMP-dependent kinase determines the responsiveness to $3',5'$-AMP.

Protein kinase consists of a regulatory and a catalytic subunit (Walsh and Ashby, 1973). When $3',5'$-AMP binds to the regulatory subunit, the catalytic subunit dissociates and is no longer $3',5'$-AMP-dependent. Protein kinase activity has been found in a wide variety of different tissues, including the ovary (Lamprecht *et al.*, 1973; Lindner *et al.*, 1974; Menon, 1973). LH and PGE stimulated protein kinase activity in intact ovaries from 28-day-old rats and, concomitantly, there was a decrease in the binding of exogenous [^{3}H]$3',5'$-AMP. The decrease in binding presumably reflects saturation of the regulatory subunit by endogenous cyclic nucleotide. The ability of ovarian protein kinase to respond to $3',5'$-AMP and of adenylate cyclase to respond to LH are both acquired during the second week of postnatal life (Lamprecht *et al.*, 1973).

Vaitukaitis *et al.* (1975) have presented evidence for an acute role of protein kinase in HCG action. The $3',5'$-AMP-dependent protein kinase activity of pseudopregnant rat ovaries decreased fivefold within 5 to 30 minutes after treatment with HCG. Inhibitors of protein synthesis did not influence the HCG-induced changes in the tissue concentration of $3',5'$-AMP-dependent protein kinase activity, but did suppress the recovery of the $3',5'$-AMP-dependent protein kinase activity.

VII. MECHANISM OF ACTION OF FSH

FSH binds only to granulosa cells in the rat ovary (Zeleznik *et al.*, 1974). Granulosa cells in some large follicles bind ^{125}I-labeled FSH and ^{125}I-labeled HCG whereas other follicles bind only ^{125}I-labeled FSH (Midgley, 1973). The most rapid biochemical event after exposure of porcine granulosa cells to FSH was an increase in $3',5'$-AMP levels (Kolena and Channing, 1972); the stimulatory effect of LH was greater than that of FSH, and maximal doses of LH and FSH were not additive.

FSH stimulated $3',5'$-AMP accumulation within 20 minutes in cultured rat Graafian follicles (Lindner *et al.*, 1974; Ahrén *et al.*, 1974). Stimulation of adenylate cyclase was retained when the FSH preparation was preincubated with an antiserum to the β-subunit of LH, indicating that the effect was an inherent property of FSH and not due to the LH contamination. In similar experiments on rabbit Graafian follicles, however, the effect of FSH was lost in the presence of LH antiserum (Marsh *et al.*, 1972). This may reflect a species difference.

Rat ovaries, particularly from prepubertal animals, responded to FSH, and the maximal stimulation obtained was less than that found with LH (Koch *et al.*, 1973; Mason *et al.*, 1973). The amount of LH in the FSH preparation could not account for the stimulation (Mason *et al.*, 1973) and the effect of FSH was not impaired by previous exposure of the hormone preparation to antiserum to the β-subunit of LH. These findings are in agreement with those of Fontaine *et al.* (1971), who found that FSH and LH increased adenylate cyclase activity in homogenates of prepubertal rat ovaries. The stimulatory effect of FSH on $3',5'$-AMP formation by the ovary, therefore, seems to represent an intrinsic property of the FSH molecule. Similar effects of FSH on $3',5'$-AMP levels of the testis have been reported (Dorrington and Fritz, 1974; Dorrington *et al.*, 1975).

Both LH and FSH can stimulate progestin secretion as well as cause morphological luteinization of cultured granulosa cells (Channing, 1974). The requirement for LH and FSH for the differentiation and subsequent luteinization of granulosa cells can be satisfied by $3',5'$-AMP or dibutyryl $3',5'$-AMP (Channing, 1970). FSH can also stimulate progesterone accumulation by cultured follicles, even after treatment of the hormone with antiserum to the LH β-subunit. FSH can induce ovum maturation and ovulation, and these properties could not be attributed to LH contamination. Presumably, FSH can stimulate progesterone synthesis and the resumption of meiosis by virtue of its ability to augment $3',5'$-AMP production in granulosa cells in a manner similar to that of LH. However, it is important to note that in addition to LH and FSH, other agents, such as prostaglandins and enterotoxin, which stimulate adenylate cyclase activity, can also induce ovum maturation. The physiological significance of the intrinsic capability of FSH to mimic the effects of LH on the preovulatory follicle is not clear. It is possible that in the intact rat FSH may cooperate with LH to effect these events.

FSH specifically influences at least two proteins: presumptive receptors for LH (HCG) and aromatase. Goldenberg *et al.* (1972b) found that FSH treatment *in vivo* caused an increase in the ovarian uptake of HCG in hypophysectomized rats. Granulosa cells obtained from intact or hypophysectomized immature rats did not bind ^{125}I-labeled HCG, whereas binding did occur in granulosa cells from animals treated with FSH *in vivo* for 2 days (Zeleznik *et al.*, 1974). This effect of FSH was probably not mediated via estrogens since diethylstilbestrol did not influence binding of ^{125}I-labeled HCG, and this effect of FSH was manifested in the presence of cyanoketone (Midgley *et al.*, 1974). 3β-Hydroxysteroid dehydrogenase activity was also apparent in granulosa cells from FSH-stimulated animals, but it was not clear if this was a direct effect of FSH or a consequence of the appearance of LH binding sites.

As discussed earlier, the aromatase activity of granulosa cells is also regulated by FSH. Cultured granulosa cells isolated from hypophysectomized rats could not convert testosterone to estrogen. However, when FSH and testosterone were added to the culture medium, estrogen synthesis and secretion were stimulated (Dorrington *et al.*, 1975). The mechanisms by which FSH influences the appearance of LH binding sites and aromatase activity is not known, but may involve activation of preexisting proteins or induction of protein synthesis (Fig. 3).

The synthesis of androgen-binding protein (ABP), a protein with a high affinity for testosterone and dihydrotestosterone, by Sertoli cells is regulated by FSH (Fritz *et al.*, 1974, 1976). It would be of interest to determine if this protein is also synthesized and secreted by granulosa cells, particularly in view of the fact that testosterone is retained in follicular fluid.

FSH stimulated the incorporation of amino acids into total ovarian protein *in vivo* and *in vitro* (Reel and Gorski, 1968; Ahrén *et al.*, 1969).

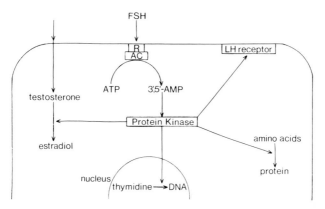

Fig. 3. Possible sites of action of FSH and 3′,5′-AMP on the metabolism of granulosa cells. R, hormone receptor; AC, adenylate cyclase.

Whether this reflects the growth-promoting effect of FSH and/or the effect of FSH on the synthesis of specific proteins remains to be determined.

The past decade has witnessed major advances in the elucidation of steroid biosynthetic pathways and the sites and mechanisms of actions of hormones at the cellular and subcellular levels. Many questions remain unanswered, some of which have been mentioned in this review. Hopefully, the future will bring an understanding of the molecular basis of gonadotropic action, information which will contribute not only to basic endocrinology, but also to clinical endocrinology and population control.

REFERENCES

Ahmad, N., Lyons, W. R., and Papkoff, H. (1969). Maintenance of gestation in hypophysectomised rats with highly purified pituitary hormones. *Anat. Rec.* **164,** 291–303.

Ahrén, K. E. B., and Kostyo, J. L. (1963). Acute effects of pituitary gonadotrophins on the metabolism of isolated rat ovaries. *Endocrinology* **73,** 81–91.

Ahrén, K. E. B., Hamberger, L., and Rubinstein, L. (1969). Acute *in vivo* and *in vitro* effects of gonadotrophins on the metabolism of the rat ovary. *In* "The Gonads" (K. W. McKerns, ed.), pp. 327–354. Appleton, New York.

Ahrén, K. E. B., Herlitz, H., Nilsson, L., Perklev, T., Rosberg, S., and Selstam, G. (1974). Gonadotropins and cyclic AMP in various compartments of the rat ovary. *In* "Symposium on Advances in Chemistry, Biology and Immunology of Gonadotropins" (N. R. Moudgal, ed.), pp. 364–375. Academic Press, New York.

Anderson, L. L., Dyk, G. W., Mori, H., Henricks, D. M., and Melampy, R. M. (1967). Ovarian function in pigs following hypophysial stalk transection or hypophysectomy. *Am. J. Physiol.* **212,** 1188–1194.

Armstrong, D. T. (1966). Comparative study of the action of luteinizing hormone upon steroidogenesis. *J. Reprod. Fertil., Suppl.* **1**, 101–112.

Armstrong, D. T. (1967). On the site of action of luteinizing hormone. *Nature (London)* **213**, 633–634.

Armstrong, D. T. (1968). Gonadotropins, ovarian metabolism, and steroid biosynthesis. *Recent Progr. Horm. Res.* **24**, 255–319.

Armstrong, D. T. (1974). Inhibition of androgen synthesis in prepubertal rat ovaries by exogenous and endogenous luteinizing hormone. *J. Steroid Biochem.* **5**, 385.

Armstrong, D. T., and Black, D. L. (1966). Influence of luteinising hormone on corpus luteum metabolism and progesterone biosynthesis throughout the bovine estrous cycle. *Endocrinology* **78**, 937–944.

Armstrong, D. T., and Black, D. L. (1968). Control of progesterone biosynthesis in the bovine corpus luteum; effects of luteinizing hormone and of reduced nicotinamide-adenine dinucleotide phosphate *in vitro*. *Can. J. Biochem.* **46**, 1137–1145.

Armstrong, D. T., and Greep, R. O. (1962). Effect of gonadotrophic hormones on glucose metabolism by luteinized rat ovaries. *Endocrinology* **70**, 701–710.

Armstrong, D. T., and Grinwich, D. L. (1972). Blockade of spontaneous and LH-induced ovulation in rats by indomethacin, an inhibitor of prostaglandin biosynthesis. *Prostaglandins* **1**, 21–26.

Armstrong, D. T., Kilpatrick, R., and Greep, R. O. (1963). *In vitro* and *in vivo* stimulation of glycolysis in prepubertal rat ovary by luteinizing hormone. *Endocrinology* **73**, 165–169.

Armstrong, D. T., O'Brien, J., and Greep, R. O. (1964). Effects of luteinizing hormone on progestin biosynthesis in the luteinized rat ovary. *Endocrinology* **75**, 488–500.

Armstrong, D. T., Jackanicz, T. M., and Keyes, P. L. (1969a). Regulation of steroidogenesis in rabbit ovary. *In* "The Gonads" (K. W. McKerns, ed.), pp. 3–25. Appleton, New York.

Armstrong, D. T., Miller, L. S., and Knudsen, K. A. (1969b). Regulation of lipid metabolism and progesterone production rat corpora lutea and ovarian interstitial elements by prolactin and luteinizing hormone. *Endocrinology* **85**, 393–401.

Armstrong, D. T., Knudsen, K. A., and Miller, L. S. (1970) Effects of prolactin upon cholesterol metabolism and progesterone biosynthesis in corpora lutea of rats hypophysectomized during pseudopregnancy. *Endocrinology* **86**, 634–641.

Armstrong, D. T., Moon, Y. S., and Zamecnik, J. (1974). Evidence for the role of ovarian prostaglandins in ovulation. *In* "Gonadotropins and Gonadal Function" (N. R. Moudgal, ed.), pp. 345–356. Academic Press, New York.

Baggett, B., and Dorrington, J. H. (1969). Adenyl cyclase activity in the rabbit ovary. *Prog. Endocrinol., Proc. Int. Congr. Endocrinol., 3rd, 1968* Excerpta Med. Found. Int. Congr. Ser. No. 157, Abstract 321.

Bauminger, S., Zor, V., and Lindner, H. R. (1973). Radioimmunological assay of prostaglandin synthetase activity. *Prostaglandins* **4**, 313–324.

Beall, R. J., and Sayers, G. (1972). Isolated adrenal cells: steroidogenesis and cyclic AMP accumulation in response to ACTH. *Archs. Biochem. Biophys.* **148**, 70–76.

Behrman, H. R., and Armstrong, D. T. (1969). Cholesterol esterase stimulation by luteinizing hormone in luteinized rat ovaries. *Endocrinology* **85**, 474–480.

Behrman, H. R., Orczyk, G. P., MacDonald, G. J., and Greep, R. O. (1970).

Prolactin induction of enzymes controlling luteal cholesterol ester turnover. *Endocrinology* **87**, 1251–1256.

Bell, J. J., and Harding, B. W. (1974) The acute action of adrenocorticotropic hormone on adrenal steroidogenesis. *Biochim. Biophys. Acta* **348**, 285–298.

Bjersing, L., and Carstensen, H. (1964). The role of the granulosa cell in the biosynthesis of ovarian steroid hormones. *Biochim. Biophys. Acta* **86**, 639–640.

Bjersing, L., and Carstensen, H. (1967). Biosynthesis of granulosa cells of the porcine ovary *in vitro*. *J. Reprod. Fertil.* **14**, 101–111.

Brambell, F. W. R. (1956). Ovarian changes. *In* "Marshall's Physiology of Reproduction." (A. S. Parkes, ed.), 3rd ed., Vol. 1, Part 1, pp. 397–542. Longmans Green, London and New York.

Butcher, R. W., and Sutherland, E. W. (1959). Enzymatic inactivation of adenosine-3′,5′-phosphate by preparations from heart. *Pharmacologist* **1**, 63.

Catt, K. J., and Dufau, M. L. (1973). Spare gonadotrophin receptors in rat testis. *Nature (London), New Biol.* **244**, 219–221.

Channing, C. P. (1969a). The use of tissue culture of granulosa cells as a method of studying the mechanism of luteinization. *In* "The Gonads" (K. W. McKerns, ed.), pp. 245–275. Appleton, New York.

Channing, C. P. (1969b). Studies on tissue culture of equine ovarian cell types. Pathways of steroidogenesis. *J. Endocrinol.* **43**, 403–414.

Channing, C. P. (1969c). Steroidogenesis and morphology of human ovarian cell types in tissue culture. *J. Endocrinol.* **45**, 297–308.

Channing, C. P. (1970). Influences of *in vitro* and *in vivo* hormonal environment upon luteinization of granulosa cells in tissue culture *Recent Progr. Horm. Res.* **26**, 589–622.

Channing, C. P. (1974). The use of granulosa cell cultures and short-term incubations in the assay for gonadotropins. *In* "Gonadotropins and Gonadal Function" (N. R. Moudgal, ed.), pp. 185–198. Academic Press, New York.

Channing, C. P., and Grieves, S. A. (1969). Studies on tissue culture of equine ovarian cell types: steroidogenesis. *J. Endocrinol.* **43**, 391–402.

Channing, C. P., and Kammerman, S. (1973). Characteristics of gonadotropin receptors of porcine granulosa cells during follicle maturation. *Endocrinology* **92**, 531–540.

Channing, C. P., and Villee, C. A. (1966). Luteinizing hormone: effects on uptake and metabolism of hexoses by luteinized rat ovaries. *Biochim. Biophys. Acta* **115**, 205–218.

Christensen, A. K., and Gillim, S. W. (1969). The correlation of fine structure and function in steroid-secreting cells, with emphasis on those of the gonads. *In* "The Gonads" (K. W. McKerns, ed.), pp. 415–488. Appleton, New York.

Claesson, L. (1954). The intracellular localization of the esterified cholesterol in the living interstitial gland cell of the rabbit ovary. *Acta Physiol. Scand.* **31**, Suppl. **113**, 53–78.

Claesson, L., and Hillarp, N. (1947). The formation mechanism of oestrogenic hormones. 1. The presence of an oestrogen precursor in the rabbit ovary. *Acta Physiol. Scand.* **13**, 115–129.

Claesson, L., Diczfalusy, E., Hillarp, N., and Hogeberg, B. (1948). The formation mechanism of oestrogenic hormones, lipids of the pregnant rabbit ovary, and their changes on gonadotrophic stimulation. *Acta Physiol. Scand.* **16**, 183–200.

Cook, B., Kaltenbach, C. C., Norton, H. W., and Nalbandov, A. V. (1967). Synthesis of progesterone *in vitro* by porcine corpora lutea. *Endocrinology* **81**, 573–584.

Deane, H. W., and Rubin, B. L. (1965). Identification and control of cells that synthesize steroid hormones in the adrenal glands, gonads and placentae of various mammalian species. *Arch. Anat. Microsc. Morphol. Exp.* **54**, 49–64.

Dohler, K. D., and Wuttke, W. (1975). Changes with age in levels of serum gonadotrophins, prolactin, and gonadal steroids in prepubertal male and female rats. *Endocrinology* **97**, 898–907.

Dorrington, J. H., and Armstrong, D. T. (1975). Follicle-stimulating hormone stimulates estradiol-17β synthesis in cultured Sertoli cells. *Proc. Natl. Acad. Sci. U.S.A.* **72**, 2677–2681.

Dorrington, J. H., and Baggett, B. (1969). Adenyl cyclase activity in the rabbit ovary. *Endocrinology* **84**, 989–996.

Dorrington, J. H., and Fritz, I. B. (1974). Effect of gonadotrophins on cyclic AMP production by isolated seminiferous tubules and interstitial cell preparations. *Endocrinology* **94**, 395–403.

Dorrington, J. H., and Kilpatrick, R. (1966). Effects of pituitary hormones on progestational hormone production by the rabbit ovary *in vivo* and *in vitro*. *J. Endocrinol.* **35**, 53–64.

Dorrington, J. H., and Kilpatrick, R. (1967). Effect of adenosine 3′,5′-monophosphate on the synthesis of progestational steroids by the rabbit ovarian tissue *in vitro*. *Biochem. J.* **104**, 725–730.

Dorrington, J. H., and Kilpatrick, R. (1969). The synthesis of progestational steroids by the rabbit ovary. *In* "The Gonads" (K. W. McKerns, ed.), pp. 27–54. Appleton, New York.

Dorrington, J. H., Moon, Y. S., and Armstrong, D. T. (1975). Estradiol-17β biosynthesis in cultured granulosa cells from hypophysectomized immature rats; stimulation by follicle stimulating hormone. *Endocrinology* **97**, 1328–1331.

du Mesnil du Buisson, F. (1961). Regression unilatérale des corps jaunes après hystérectomie partielle chez la truie. *Ann. Biol. Anim., Biochim., Biophys.* **1**, 105–112.

Duncan, G. W., Bowerman, A. M., Hearn, W. R., and Melampy, R. M. (1960). *In vitro* synthesis of progesterone by swine corpora lutea. *Proc. Soc. Exp. Biol. Med.* **104**, 17–19.

Duncan, G. W., Bowerman, A. M., Anderson, L. L., Hearn, W. R., and Melampy, R. M. (1961). Factors influencing *in vitro* synthesis of progesterone. *Endocrinology* **68**, 199–207.

Eaton, L. W., and Hilliard, J. (1971). Estradiol-17β, progesterone and 20α-hydroxypregn-4-en-3-one in rabbit ovarian venous plasma. I. Steroid secretion from paired ovaries with and without corpora lutea; effect of LH. *Endocrinology* **89**, 105–111.

Ellsworth, L. R., and Armstrong, D. T. (1971a). Effect of LH on luteinization of ovarian follicles transplanted under the kidney capsule in rats. *Endocrinology* **88**, 755–762.

Ellsworth, L. R., and Armstrong, D. T. (1971b). Initiation by dibutyryl cyclic AMP of luteinization in transplanted ovarian follicles. *Proc. Can. Fed. Biol. Soc.* **14**, 65.

Erickson, G. F., and Ryan, K. J. (1975). The effect of LH/FSH, dibutyryl cyclic AMP, and prostaglandins on the production of estrogens by rabbit granulosa cells *in vitro*. *Endocrinology* **97**, 108–113.

Eshkol, A., and Lunenfeld, B. (1972). Gonadotrophic regulation of ovarian development in mice during infancy. *In* "Gonadotropins" (B. B. Saxena, G. G. Beling, and H. M. Gandy, eds.), pp. 335–346. Wiley, New York.

Everett, J. W. (1945). The microscopically demonstrable lipids of cyclic corpora lutea in the rat. *Am. J. Anat.* **77**, 293–323.

Everett, J. W. (1947). Hormonal factors responsible for deposition of cholesterol in the corpus luteum of the rat. *Endocrinology* **41**, 364–377.

Falck, B. (1959). Site of production of oestrogen in rat ovary as studied in microtransplants. *Acta Physiol. Scand.* **47**, Suppl.163,1–101.

Fevold, H. L. (1941). Synergism of follicle stimulating and luteinising hormone in producing estrogen secretion. *Endocrinology* **28**, 33–36.

Flint, A. P. F., and Armstrong, D. T. (1971a). The compartmentation of non-esterified and esterified cholesterol in the superovulated rat ovary. *Biochem. J.* **123**, 143–152.

Flint, A. P. F., and Armstrong, D. T. (1971b). Cholesterol side-chain cleavage enzyme: intracellular localization in corpora lutea of cow and rat. *Nature (London), New Biol.* **231**, 60–61.

Flint, A. P. F., and Denton, R. M. (1969). Glucose metabolism in the superovulated rat ovary *in vitro*. *Biochem. J.* **112**, 243–254.

Flint, A. P. F., and Denton, R. M. (1970). Mechanism of action of luteinizing hormone. *Nature (London)* **228**, 376–377.

Flint, A. P. F., Grinwich, D. L., and Armstrong, D. T. (1973). Control of ovarian cholesterol ester biosynthesis. *Biochem. J.* **132**, 313–321.

Fontaine, Y. A., Fontaine-Bertrand, E., Salmon, C., and Delerue-Lebelle, N. (1971). Stimulation *in vitro* par les deux hormones gonadotropes hypophysaires (LH et FSH) et l'activité adenylcyclasique de l'ovarie chez la ratte prépubère. *C. R. Acad. Sci., Ser. D* **272**, 1137–1140.

Fritz, I. B., Kopec, B., Lam, K., and Vernon, R. G. (1974). Effects of FSH on levels of adrogen-binding protein in the testis. *In* "Hormone Binding and Target Cell Activation in the Testis" (M. L. Dufau and A. R. Means, eds.), pp. 311–327. Plenum, New York.

Fritz, I. B., Rommerts, F. G., Louis, B. G., and Dorrington, J. H. (1976). Regulation by FSH and dibutyryl cyclic AMP of androgen-binding protein formation by Sertoli cell-enriched cultures. *J. Reprod. Fertil.* **46**, 17–24.

Fuller, G. B., and Hansel, W. (1971). Estrogen-stimulated progesterone synthesis by rabbit corpora lutea *in vitro*. *Proc. Soc. Exp. Biol. Med.* **137**, 539–542.

Garren, L. D., Ney, R. L., and Davis, W. W. (1965). Studies on the role of protein synthesis in the regulation of corticosterone production by adrenocorticotropic hormone *in vivo*. *Proc. Natl. Acad. Sci. U.S.A.* **53**, 1443–1450.

Garren, L. D., Gill, G. N., Masui, H., and Walton, G. M. (1971). On the mechanism of action of ACTH. *Recent Progr. Horm. Res.* **27**, 433–478.

Givner, M. L., and Dvornik, D. (1964). Inhibition of adrenal cholesterogenesis *in vitro* by AY-9944, 22,25-diazocholestanol and by triparanol. *Proc. Soc. Exp. Biol. Med.* **117**, 3–6.

Goldenberg, R. L., Bridson, W. E., and Kohler, P. O. (1972a). Estrogen stimulation of progesterone synthesis by porcine granulosa cells in culture. *Biochem. Biophys. Res. Commun.* **48**, 101–107.

Goldenberg, R. L., Vaitukaitis, J. L., and Ross G. T. (1972b). Estrogen and follicle stimulating hormone interactions on follicle growth in rats. *Endocrinology* **90**, 1492–1498.

Goldstein, S., and Marsh, J. M. (1973). Protein kinase in the bovine corpus luteum. *In* "Protein Phosphorylation in Control Mechanisms" (F. Huijing and E. Y. C. Lee, eds.), pp. 123–144. Academic Press, New York.

Gondos, B., and Hobel, C. J. (1973). Interstitial cells of the human fetal ovary. *Endocrinology* **93**, 736–739.

Gorski, J., and Padnos, D. (1966). Translational control of protein synthesis and the control of steroidogenesis in the rabbit ovary. *Arch. Biochem. Biophys.* **113**, 100–106.

Gospodarowicz, D., and Gospodarowicz, F. (1972). A technique for the isolation of bovine luteal cells and its application to metabolic studies of luteal cells *in vitro*. *Endocrinology* **90**, 1427–1434.

Gospodarowicz, D., and Gospodarowicz, F. (1975). The morphological transformation and inhibition of growth of bovine luteal cells in the tissue culture induced by luteinizing hormone and dibutyryl cyclic AMP. *Endocrinology* **96**, 458–467.

Greep, R. O., Van Dyke, H. B., and Chow, B. F. (1942). Gonadotropins of the swine pituitary. 1. Various biological effects of purified thylakentrin (FSH) and pure metakentrin (ICSH). *Endocrinology* **30**, 635–649.

Grinwich, D. L., Kennedy, T. G., and Armstrong, D. T. (1972). Dissociation of ovulatory and steroidogenic actions of luteinizing hormone in rabbits with indomethacin, an inhibitor of prostaglandin biosynthesis. *Prostaglandins* **1**, 89–96.

Hafs, H. P., and Armstrong, D. T. (1968). Corpus luteum growth and progesterone synthesis during the bovine estrous cycle. *J. Anim. Sci.* **27**, 134–141.

Hall, P. F. (1971). Role of reduced TPH in the response to interstitial cell stimulatory hormone. *Nature (London)* **232**, 631–632.

Hall, P. F., and Koritz, S. B. (1964). The conversion of cholesterol and 20α-hydroxycholesterol to steroids by acetone powder of particles from bovine corpus luteum. *Biochemistry* **3**, 129–134.

Hall, P. F., and Koritz, S. B. (1965). Influence of interstitial cell-stimulating hormone on the conversion of cholesterol to progesterone by bovine corpus luteum. *Biochemistry* **4**, 1037–1043.

Hall, P. F., and Young, D. E. (1968). Site of action of trophic hormones upon the biosynthetic pathways to steroid hormones. *Endocrinology* **82**, 559–568.

Hall, P. F., Nishizawa, E. E., and Eik-Nes, K. B. (1963). Biosynthesis of testosterone by rabbit testis. Homogenate vs. slices. *Proc. Soc. Exp. Biol. Med.* **144**, 791–794.

Hamberger, L. A., and Ahrén, K. E. B. (1967). Effects of gonadotropins *in vitro* on glucose uptake and lactic acid production of ovaries from prepubertal and hypophysectomized rats. *Endocrinology* **81**, 93–100.

Hammerstein, J., Rice, B. F., and Savard, K. (1964). Steroid formation in the human ovary. 1. Identification of steroids formed *in vitro* from acetate-1-^{14}C in the corpus luteum. *J. Clin. Endocrinol. Metab.* **24**, 597–605.

Hammond, J., Jr., and Robson, J. M. (1951). Local maintenance of the rabbit corpus luteum with oestrogen. *Endocrinology* **49**, 384–389.

Harding, B. W., Wilson, L. D., Wong, S. H., and Nelson, D. H. (1965). Electron carriers of the rat adrenal and the 11β-hydroxylating system. *Steroids, Suppl.* **11**, 51–75.

Harman, S. M., Louvet, J. P., and Ross, G. T. (1975). Interaction of estrogen and gonadotrophins on follicular atresia. *Endocrinology* **96**, 1145–1152.

Hashimoto, I., and Wiest, W. G. (1969). Luteotrophic and luteolytic mechanisms in rat corpora lutea. *Endocrinology* **84**, 886–892.

Haynes, R. C. (1958). Activation of adrenal phosphorylase by adrenocorticotrophic hormone. *J. Biol. Chem.* **233**, 1220-1222.

Haynes, R. C., and Berthet, L. (1957). Studies on the mechanism of action of adrenocorticotrophic hormone. *J. Biol. Chem.* **225**, 115-124.

Haynes, R. C., Koritz, S. B., and Peron, F. G. (1959). Influence of adenosine 3′,5′-monophosphate on corticoid production by rat adrenal glands. *J. Biol. Chem.* **234**, 1421-1423.

Haynes, R. C., Sutherland, E. W., and Rall, T. W. (1960). The role of cyclic adenylic acid in hormone action. *Recent Progr. Horm. Res.* **16**, 121-138.

Hechter, O., Solomon, M. M., Zaffaroni, A., and Pincus, G. (1953). Transformation of cholesterol and acetate to adrenal cortical hormones. *Arch. Biochem.* **46**, 201-207.

Hellig, H., and Savard, K. (1966). Sterol biosynthesis in the bovine corpus luteum *in vitro. Biochemistry* **5**, 2944-2956.

Herbst, A. L. (1967). Response of rat ovarian cholesterol to gonadotropins and anterior pituitary hormones. *Endocrinology* **81**, 54-60.

Hermier, C., and Jutisz, M. (1969). Biosynthèse de la progestérone *in vitro* dans le corps jaune de la ratte pseudo-gestante. *Biochim. Biophys. Acta* **192**, 96-105.

Hermier, C., Combarnous, Y., and Jutisz, M. (1971). Role of a regulating protein and molecular oxygen in the mechanism of action of luteinizing hormone. *Biochim. Biophys. Acta* **244**, 625-633.

Hermier, C., Santos, A. A., Wisnewsky, C., Netter, A., and Jutisz, M. (1972). Rôle de l'AMPc et d'une protéïne regulatrice dans le action *in vitro* de la gonadotropine choriale humaine (HCG) sur le corps jaune humain. *C. R. Acad. Sci., Ser. D* **275**, 1415-1418.

Hilliard, J., Penardi, R., and Sawyer, C. H. (1967). A functional role of 20α-hydroxypregn-4-en-3-one in the rabbit. *Endocrinology* **80**, 901-909.

Hilliard, J., Saldarini, R. J., Spies, H. G., and Sawyer, C. H. (1971). Luteotrophic and luteolytic actions of LH in hypophysectomised pseudopregnant rats. *Endocrinology* **89**, 513-521.

Hilliard, J., Scaramuzzi, R. J., Pang, C., Penardi, R., and Sawyer, C. H. (1974). Testosterone secretion by rabbit ovary *in vivo. Endocrinology* **94**, 267-271.

Hilliard, J., Archibald, D., and Sawyer, C. H. (1963). Gonadotropic activation of preovulatory synthesis and release of progestin in the rabbit. *Endocrinology* **72**, 59-66.

Hollander, N., and Hollander, V. P. (1958). The effect of follicle-stimulating hormone on the biosynthesis *in vitro* of estradiol-17β from acetate-1-¹⁴C and testosterone-4-¹⁴C. *J. Biol. Chem.* **233**, 1097-1099.

Huang, W. D., and Pearlman, W. H. (1962). The corpus luteum and steroid formation. 1. Studies on luteinized rat ovarian tissue *in vitro. J. Biol. Chem.* **237**, 1060-1065.

Ichii, W., Forchielli, F., and Dorfman, R. I. (1963). *In vitro* effect of gonadotropins on the soluble cholesterol side-chain cleavage enzyme system of bovine corpus luteum. *Steroids* **2**, 631-656.

Insel, P. A., Bourne, H. R., Coffino, P., and Tomkins, G. M. (1975). Cyclic AMP-dependent protein kinase: pivotal role in regulation of enzyme induction and growth. *Science* **190**, 896-898.

Jackanicz, T. M., and Armstrong, D. T. (1968). Progesterone biosynthesis in rabbit ovarian interstitial tissue mitochondria. *Endocrinology* **83**, 769-776.

Kan, K. W., Riter, M. C., Ungar, F., and Dempsey, M. E. (1972). The role of a carrier protein in cholesterol and steroid hormone synthesis by adrenal enzymes. *Biochem. Biophys. Res. Commun.* **48**, 423–429.

Keyes, P. L. (1969). Luteinizing hormone; action on the Graafian follicle *in vitro*. *Science* **164**, 846–847.

Keyes, P. L., and Armstrong, D. T. (1968). Endocrine role of follicles in the regulation of corpus luteum function in the rabbit. *Endocrinology* **83**, 509–515.

Keyes, P. L., and Armstrong, D. T. (1969). Development of corpora lutea from follicles autotransplanted under the kidney capsule in rabbits. *Endocrinology* **85**, 423–427.

Keyes, P. L., and Nalbandov, A. V. (1967). Maintenance and function of corpora lutea in rabbits depend on estrogen. *Endocrinology* **80**, 938–946.

Kidwell, W. R., Balogh, K., Jr., and Wiest, W. G. (1966). Effects of luteinizing hormone on glucose-6-phosphate and 20α-hydroxysteroid dehydrogenase activities in superovulated rat ovaries. *Endocrinology* **79**, 352–361.

Kilpatrick, R., Armstrong, D. T., and Greep, R. O. (1964). Maintenance of the corpus luteum by gonadotropins in the hypophysectomized rabbit. *Endocrinology* **74**, 453–461.

Koch, Y., Zor, U., Pomerantz, S., Chobsieng, P., and Lindner, H. R. (1973). Intrinsic stimulatory action of follicle-stimulating hormone on ovarian adenylate cyclase. *J. Endocrinol.* **58**, 677–678.

Kolena, J., and Channing, C. P. (1971). Stimulatory effects of gonadotrophins on the formation of cyclic adenosine 3,5-monophosphate by porcine granulosa cells. *Biochim. Biophys. Acta* **252**, 601–606.

Kolena, J., and Channing, C. P. (1972). Stimulatory effects of LH, FSH and prostaglandins upon cyclic 3,5-AMP levels in porcine granulosa cells. *Endocrinology* **90**, 1543–1550.

Koritz, S. B., and Hall, P. F. (1965). Further studies on the locus of action of interstitial cell-stimulating hormone on the biosynthesis of progesterone by bovine corpus luteum. *Biochemistry* **4**, 2740–2747.

Koritz, S. B., and Kumar, A. M. (1970). On the mechanism of action of adrenocorticotropic hormone. *J. Biol. Chem.* **245**, 152–159.

Kraiem, Z., and Samuels, L. T. (1974). The influence of FSH and FSH + LH on steroidogenic enzymes in the immature mouse ovary. *Endocrinology* **95**, 660–668.

Kuehl, F. A., Humes, J. L., Tarnoff, J., Cirillo, V. J., and Ham, E. A. (1970). Prostaglandin receptor site. Evidence for an essential role in the action of luteinizing hormone. *Science* **169**, 883–885.

Kuehl, F. A., Cirillo, V. J., Ham, E. A., and Humes, J. L. (1973). The regulatory role of the prostaglandins on the cyclic AMP system. *Adv. Biosci.* **9**, 155–172.

Kuo, J. F., and Greengard, P. (1969). Cyclic nucleotide dependent protein kinases. IV. Widespread occurrence of adenosine 3,5-monophosphate-dependent protein kinases in various tissues and phyla of the animal kingdom. *Proc. Natl. Acad. Sci. U.S.A.* **64**, 1349–1355.

Lamprecht, S. A., Zor, U., Tsafriri, A., and Lindner, H. R. (1973). Action of prostaglandin E_2 and of luteinizing hormone on ovarian adenylate cyclase, protein kinase and ornithine decarboxylase activity during postnatal development and maturity in the rat. *J. Endocrinol.* **57**, 217–233.

Lee, C., Keyes, P. L., and Jacobson, H. I. (1971). Estrogen receptor in the rabbit corpus luteum. *Science* **173**, 1032–1033.

LeMaire, W. J., Askari, H., and Savard, K. (1971). Steroid hormone formation in the human ovary. *Steroids* 17, 65–84.

LeMaire, W. J., Yang, N. S. T., Behrman, H. R., and Marsh, J. M. (1973). Preovulatory changes in the concentration of prostaglandins in rabbit Graafian follicles. *Prostaglandins* 3, 367–376.

Levin, L., and Jailer, J. W. (1948). The effect of induced secretory activity on the cholesterol content of the immature rat ovary. *Endocrinology* 43, 154–166.

Lieberman, M. E., Barnea, A., Bauminger, S., Tsafriri, A., Collins, W. P., and Lindner, H. R. (1975). LH effect on the pattern of steroidogenesis in cultured Graafian follicles of the rat. *Endocrinology* 96, 1533–1542.

Lindner, H. R., Tsafriri, A., Lieberman, M. E., Zor, U., Koch, Y., Bauminger, S., and Barnea, A. (1974). Gonadotropin action on cultured Graafian follicles: induction of maturation division of the mammalian oocyte and differentiation of the luteal cell. *Recent Progr. Horm. Res.* 30, 79–126.

Lipner, H., and Greep, R. O. (1971). Inhibition of steroidogenesis at various sites in the biosynthetic pathway in relation to induced ovulation. *Endocrinology* 88, 602–607.

Lostroh, A., and Johnson, R. E. (1966). Amounts of interstitial-cell stimulating hormone and follicle-stimulating hormone required for follicular development, uterine growth and ovulation in the hypophysectomized rat. *Endocrinology* 79, 991–996.

Louvet, J. P., Harman, S. M., and Ross, G. T. (1975a). Effects of human chorionic gonadotropin, human interstitial cell stimulating hormone and human follicle-stimulating hormone on ovarian weights in estrogen-primed hypophysectomized immature female rats. *Endocrinology* 96, 1179–1186.

Louvet, J. P., Harman, S. M., Schreiber, J. R., and Ross, G. T. (1975b). Evidence for a role of androgens in follicular maturation. *Endocrinology* 97, 366–372.

Lyons, W. R., and Ahmad, N. (1973). Hormonal maintenance of pregnancy in hypophysectomized rats. *Proc. Soc. Exp. Biol. Med.* 142, 198–202.

McIntosh, E. N., Uzgiris, V. I., Alonso, C., and Salhanick, H. A. (1971). Special properties, respiratory activity and enzyme systems of bovine corpus luteum mitochondria. *Biochemistry* 10, 2909–2916.

McKerns, K. W. (1965). Gonadotropin regulation of the activities of dehydrogenase enzymes of the ovary. *Biochim. Biophys. Acta* 97, 542–550.

McKerns, K. W. (1969). Studies on the regulation of ovarian function by gonadotropins. *In* "The Gonads" (K. W. McKerns, ed.), pp. 137–173. Appleton, New York.

McNatty, K. P., and Sawers, R. S. (1975). Relationship between the endocrine environment within the Graafian follicle and the subsequent rate of progesterone secretion by human granulosa cells *in vitro*. *J. Endocrinol.* 66, 391–400.

Mahajan, D. K., and Samuels, L. T. (1963). Biosynthesis of steroids by different ovarian tissues. *Fed. Proc. Fed. Am. Soc. Exp. Biol.* 22, 531.

Mahesh, V. B., and Greenblatt, R. B. (1964). Steroid secretions of the normal and polycystic ovary. *Recent Progr. Horm. Res.* 20, 341–379.

Major, P. W., Armstrong, D. T., and Greep, R. O. (1967). Effects of luteinizing hormone *in vivo* and *in vitro* on cholesterol conversion to progestins in rat corpus luteum tissue. *Endocrinology* 81, 19–28.

Makris, A., and Ryan, K. J. (1975). Progesterone, androstenedione, testosterone, estrone, and estradiol synthesis in hamster ovarian follicle cells. *Endocrinology* 96, 694–701.

Marsh, J. M. (1970a). The stimulatory effect of luteinizing hormone on adenyl cyclase in the bovine corpus luteum. *J. Biol. Chem.* **245**, 1596–1603.

Marsh, J. M. (1970b). The stimulatory effect of prostaglandin E_2 on adenyl cyclase in the bovine corpus luteum. *FEBS Lett.* **7**, 283–286.

Marsh, J. M. (1971). The effect of prostaglandins on the adenyl cyclase of the bovine corpus luteum. *Ann. N.Y. Acad. Sci.* **180**, 416–425.

Marsh, J. M., and Savard, K. (1964). The activation of luteal phosphorylase by luteinising hormone. *J. Biol. Chem.* **239**, 1–7.

Marsh, J. M., and Savard, K. (1966). The stimulation of progesterone synthesis in bovine corpora lutea by adenosine 3′,5′-monophosphate. *Steroids* **8**, 133–148.

Marsh, J. M., Butcher, R. W., Savard, K., and Sutherland, E. W. (1966). The stimulatory effect of luteinizing hormone on adenosine 3′,5′-monophosphate accumulation in corpus luteum slices. *J. Biol. Chem.* **241**, 5436–5440.

Marsh, J. M., Mills, T. M., and LeMaire, W. J. (1972). Cyclic AMP synthesis in rabbit Graafian follicles and the effect of luteinizing hormone. *Biochim. Biophys. Acta* **273**, 389–394.

Mason, N. R., and Samuels, L. T. (1961). Incorporation of acetate-1-^{14}C into testosterone and 3β-hydroxysterols by the canine testis. *Endocrinology* **68**, 899–907.

Mason, N. R., and Savard, K. (1964). Conversion of cholesterol to progesterone by corpus luteum slices. *Endocrinology* **75**, 215–221.

Mason, N. R., Marsh, J. M., and Savard, K. (1962). An action of gonadotrophin *in vitro. J. Biol. Chem.* **237**, 1801–1806.

Mason, N. R., Schaffer, R. J., and Toomey, R. E. (1973). Stimulation of cyclic AMP accumulation in rat ovaries *in vitro. Endocrinology* **93**, 34–41.

Mendelson, C., Dufau, M. L., and Catt, K. J. (1975). Gonadotropin binding and stimulation of cyclic adenosine 3′,5′-monophosphate and testosterone production in isolated Leydig cells. *J. Biol. Chem.* **250**, 8818–8823.

Menon, K. M. J. (1973). Purification and properties of a protein kinase from bovine corpus luteum that is stimulated by cyclic adenosine 3′,5′-monophosphate and luteinizing hormone. *J. Biol. Chem.* **248**, 494–501.

Menon, K. M. J., Dorfman, R. I., and Forchielli, F. (1965). The obligatory nature of cholesterol in the biosynthesis of testosterone in rabbit testis slices. *Steroids, Suppl.* **2**, 165–175.

Midgley, A. R. (1973). Autoradiographic analysis of gonadotropin binding to rat ovarian tissue sections. *Adv. Exp. Med. Biol.* **36**, 365–378.

Midgley, A. R., Zeleznik, A. J., Rajaniemi, H. J., Richards, J. S., and Reichert, L. E. (1974). Gonadotropin-receptor activity and granulosa luteal cell differentiation. *In* "Gonadotropins and Gonadal Function" (N. R. Moudgal, ed.), pp. 416–429. Academic Press, New York.

Mikhail, E., Zander, J., and Allen, W. M. (1963). Steroids in human ovarian blood. *J. Clin. Endocr. Metab.* **23**, 1267–1270.

Miller, J. B., and Keyes, P. L. (1975). Progesterone synthesis in developing rabbit corpora lutea in the absence of follicular estrogen. *Endocrinology* **97**, 83–90.

Mills, T. M. (1975). Effect of luteinizing hormone and cyclic adenosine 3′,5′-monophosphate on steroidogenesis in the ovarian follicle of rabbit. *Endocrinology* **96**, 440–445.

Mills, T. M., and Savard, K. (1973). Steroidogenesis in ovarian follicles isolated from rabbits before and after mating. *Endocrinology* **92**, 788–791.

Mills, T. M., Davies, P. J. A., and Savard, K. (1971). Stimulation of estrogen synthesis in rabbit follicles by luteinizing hormone. *Endocrinology* **88**, 857–862.

Moon, Y. S., Dorrington, J. H., and Armstrong, D. T. (1975). Stimulatory action of follicle-stimulating hormone on estradiol-17β secretion by hypophysectomized rat ovaries in organ culture. *Endocrinology* **97**, 244–247.

Morishige, W. K., and Rothchild, I. (1974). Temporal aspects of the regulation of corpus luteum function by luteinizing hormone, prolactin, and placental luteotrophin during the first half of pregnancy in the rat. *Endocrinology* **95**, 260–274.

Moyle, W. R., and Ramachandran, J. (1973). Effect of LH on steroidogenesis and cyclic AMP accumulation in rat Leydig cell preparations and mouse tumor Leydig cells. *Endocrinology* **93**, 127–134.

Pedersen, T. (1970). Follicle kinetics in the ovary of the cyclic mouse. *Acta Endocrinol. (Copenhagen)* **64**, 304–323.

Pharriss, B. B., Wyngarden, L. J., and Gutnecht, G. D. (1968). Biological interactions between prostaglandins and luteotropins in the rat. *In* "Gonadotropins" (E. Rosemberg, ed.), pp. 121–129. Geron-X Inc., Los Altos, California.

Popjak, G., and Cornforth, J. W. (1960). The biosynthesis of cholesterol. *Adv. Enzymol.* **22**, 281–335.

Quattropani, S. L., and Weisz, J. (1973). Conversion of progesterone to estrone and estradiol *in vitro* by the ovary of the infantile rat in relation to the development of its interstitial tissue. *Endocrinology* **53**, 1269–1276.

Raj, H. G. M., and Moudgal, N. R. (1970). Hormonal control of gestation in the intact rat. *Endocrinology* **86**, 874–889.

Reel, J. R., and Gorski, J. (1968). Gonadotropic regulation of precursor incorporation into ovarian RNA, protein, and acid-soluble fractions. *Endocrinology* **83**, 1083–1091.

Rice, B. F., and Savard, K. (1966). Steroid hormone formation in the human ovary. IV. Ovarian stromal compartment; formation of radioactive steroids from acetate-1-^{14}C and action of gonadotropins. *J. Clin. Endocrinol. Metab.* **26**, 593–609.

Rice, B. F., Hammerstein, J., and Savard, K. (1964). Steroid hormone formation in the human ovary. II. Action of gonadotrophins *in vitro* in the corpus luteum. *J. Clin. Endocrinol. Metab.* **24**, 606–615.

Richards, J. S. (1974). Estradiol binding to rat corpora lutea during pregnancy. *Endocrinology* **95**, 1018–1053.

Richards, J. S. (1975). Content of nuclear estradiol receptor complex in rat corpora lutea during pregnancy: relationship to estrogen concentrations and cytosol receptor availability. *Endocrinology* **96**, 227–230.

Richards, J. S., and Midgley, A. R. (1976). Protein hormone action: a key to understanding ovarian follicular and luteal cell development. *Biol. Reprod.* **14**, 82–94.

Robison, E. A., Butcher, R. W., and Sutherland, E. W. (1971). "Cyclic AMP." Academic Press, New York.

Robson, J. M. (1937). Maintenance by oestrin of the luteal function in hypophysectomized rabbits. *J. Physiol. (London)* **90**, 435–439.

Ryan, K. J., and Petro, Z. (1966). Steroid biosynthesis by human ovarian granulosa and thecal cells. *J. Endocrinol. Metab.* **26**, 46–52.

Ryan, K. J., and Short, R. V. (1965). Formation of estradiol by granulosa and theca cells of the equine ovarian follicle. *Endocrinology* **76**, 108–114.

Ryan, K. J., and Short, R. V. (1966). Cholesterol formation by granulosa and thecal cells of the equine follicle. *Endocrinology* **78**, 214–216.

Ryan, K. J., and Smith, O. W. (1961a). Biogenesis of estrogens by the human ovary.

1. Conversion of acetate-1-C^{14} to estrone and estradiol. *J. Biol. Chem.* **236**, 705-709.

Ryan, K. J., and Smith, O. W. (1961b). Biogenesis of estrogens by the human ovary. II. Conversion of progesterone-4-C^{14} to estrone and estradiol. *J. Biol. Chem.* **236**, 710-714.

Ryan, K. J., and Smith, O. W. (1961c). Biogenesis of estrogens by the human ovary. III. Conversion of cholesterol-4-C^{14} to estrone. *J. Biol. Chem.* **236**, 2204-2206.

Ryan, K. J., and Smith, O. W. (1961d). Biogenesis of estrogens by the human ovary. IV. Formation of neutral steroid intermediates. *J. Biol. Chem.* **236**, 2207-2212.

Ryan, K. J., and Smith, O. W. (1965). Biogenesis of steroid hormones in the human ovary. *Recent Progr. Horm. Res.* **21**, 367-403.

Ryan, K. J., Petro, Z., and Kaiser, J. (1968). Steroid formation by isolated and recombined ovarian granulosa and thecal cells. *J. Clin. Endocrinol. Metab.* **28**, 355-358.

Ryle, M. (1969). Morphological responses to pituitary gonadotrophins by mouse ovaries *in vitro. J. Reprod. Fertil.* **20**, 307-312.

Ryle, M. (1971). The activity of human follicle-stimulating hormone preparations as measured by a response *in vitro. J. Endocrinol.* **51**, 97-107.

Ryle, M., Court, J., and Morris, R. (1974). Synergistic action of follicle-stimulating hormone and luteinizing hormone on oestrogen production by mouse ovaries in culture. *J. Endocrinol.* **61**, xxiii.

Savard, K., Marsh, J. M., and Rice, B. F. (1965). Gonadotropins and ovarian steroidogenesis. *Recent Progr. Horm. Res.* **21**, 285-365.

Savard, K., LeMaire, W. J., and Kumari, L. (1969). Progesterone synthesis from labeled precursors in the corpus luteum. *In* "The Gonads" (K. W. McKerns, ed.), pp. 119-136. Appleton, New York.

Schomberg, D. W. (1969). The concept of a uterine luteolytic hormone. *In* "The Gonads" (K. W. McKerns, ed.), pp. 383-414. Appleton, New York.

Schwartz, N. B., and McCormack, C. E. (1972). Reproduction: gonadal function and its regulation. *Annu. Rev. Physiol.* **34**, 425-472.

Scott, R. S., and Rennie, P. I. C. (1971). An estrogen receptor in the corpora lutea of the pseudopregnant rabbit. *Endocrinology* **89**, 297-301.

Selstam, G., and Ahrén, K. (1971). Effects of gonadotrophins on ovarian phosphorylase activity. *Acta Endocrinol. (Copenhagen), Suppl.* **155**, 55.

Shelesnyak, M. (1957). Aspects of reproduction. Some experimental studies on the mechanisms of ova-implantation in the rat. *Recent Progr. Horm. Res.* **13**, 269-322.

Short, R. V. (1962). Steroids in the follicular fluid and the corpus luteum of the mare. A "two cell type" theory of ovarian steroid synthesis. *J. Endocrinol.* **24**, 59-63.

Short, R. V. (1964). Ovarian steroid synthesis and secretion *in vivo. Recent Progr. Horm. Res.* **20**, 303-340.

Simpson, E. R., and Estabrook, R. W. (1968). A possible mechanism for the transfer of cytosol-generated NADPH to the mitochondrial mixed-function oxidases in bovine adrenal cortex; a malate shuttle. *Arch. Biochem. Biophys.* **126**, 977-978.

Smith, O. W., and Ryan, K. J. (1961). Biogenesis of estrogens by the human ovary. Formation of neutral steroid intermediates from progesterone-4-C^{14} androstenedione-4-C^{14} and cholesterol-4-^{14}C. *Endocrinology* **69**, 970-983.

Solod, E. A., Armstrong, D. T., and Greep, R. O. (1966). Action of luteinizing hormone on conversion of ovarian cholesterol stores to steroids secreted *in vivo* and synthesised *in vitro* by the pseudopregnant rabbit ovary. *Steroids* **7,** 607–620.

Speroff, L., and Ramwell, P. W. (1970). Prostaglandin stimulation of *in vitro* progesterone synthesis. *J. Clin. Endocrinol. Metab.* **30,** 345.

Spies, H. E., and Quadri, S. K. (1967). Regression of corpora lutea and interception of pregnancy in rabbits following treatment with rabbit serum to ovine LH. *Endocrinology* **80,** 1127–1132.

Spies, H. G., Hilliard, J., and Sawyer, C. H. (1968). Maintenance of corpora lutea and pregnancy in hypophysectomized rabbits. *Endocrinology* **83,** 354–367.

Stansfield, D. A., and Robinson, J. W. (1965). Glycogen and phosphorylase in bovine and rat corpora lutea and the effect of luteinizing hormone. *Endocrinology* **76,** 390–395.

Stoklosowa, S., and Nalbandov, A. V. (1972). Luteinization and steroidogenic activity of rat ovarian follicles cultured *in vitro. Endocrinology* **91,** 25–32.

Stone, D., and Hechter, O. (1954). Studies on ACTH action in perfused bovine adrenals: the site of action of ACTH in corticosteroidogenesis. *Arch. Biochem. Biophys.* **51,** 457–469.

Sulimovici, S. I., and Boyd, E. S. (1969). The cholesterol side chain cleavage enzymes in steroid hormone-producing tissues. *Vitam. Horm. (N.Y.)* **27,** 199–234.

Takayama, M., and Greenwald, G. S. (1973). Direct luteotropic action of estrogen in the hypophysectomised-hysterectomized rat. *Endocrinology* **92,** 1405–1413.

Tamaoki, B., and Pincus, G. (1961). Biogenesis of progesterone in ovarian tissues. *Endocrinology* **69,** 527–533.

Telegdy, G., and Savard, K. (1966). Steroid formation *in vitro* in rabbit ovary. *Steroids* **8,** 685–694.

Tsafriri, A., Lindner, H. R., Zor, V., and Lamprecht, S. A. (1972). Physiological role of prostaglandins in the induction of ovulation. *Prostaglandins* **2,** 1–10.

Tsafriri, A., Lieberman, M. E., Barnea, A., Bauminger, S., and Lindner, H. R. (1973). Induction by luteinizing hormone of ovum maturation and of steroidogenesis in isolated Graafian follicles of the rat: role of RNA and protein synthesis. *Endocrinology* **93,** 1378–1386.

Ungar, F., Kan, K. W., and McCoy, K. E. (1973). Activator and inhibitor factors in cholesterol side-chain cleavage. *Ann. N.Y. Acad. Sci.* **212,** 276–289.

Uzgiris, V. I., McIntosh, E. N., Alonso, C., and Salhanick, H. A. (1971). Role of reversed electron transport in bovine corpus luteum mitochondrial steroid synthesis. *Biochemistry* **10,** 2916–2922.

Vaitukaitis, J. L., Lee, C. Y., Ebersole, E. R., and Lerario, A. C. (1975). New evidence for an acute role of protein kinase in hCG action. *Endocrinology* **97,** 215–222.

van Thiel, D. H., Bridson, W. E., and Kohler, P. O. (1971). Induction of "luteinization" in cultures of granulosa cells by chorionic gonadotropins. *Endocrinology* **89,** 622–628.

Villee, C. A., Channing, C. P., Eckstein, B., and Sulovic, V. (1969). Effects of gonadotropins on the metabolism of luteinized and follicular ovarian tissues. *In* "The Gonads" (K. W. McKerns, ed.), pp. 277–291. Appleton, New York.

Walsh, D. A., and Ashby, C. D. (1973). Protein kinases: aspects of their regulation and diversity. *Recent Prog. Horm. Res.* **29,** 329–353.

Wicks, W. D. (1974). Regulation of protein synthesis by cyclic AMP. *Adv. Cyclic Nucleotide Res.* **4**, 335 438.

Wiest, W. G., and Kidwell, W. R. (1969). The regulation of progesterone secretion by ovarian dehydrogenases. *In* "The Gonads" (K. W. McKerns, ed.), pp. 295–325. Appleton, New York.

Wilks, J. W., Fuller, G. B., and Hansel, W. (1970). Role of cholesterol as a progestin precursor in the rat, rabbit and bovine luteal tissue. *Endocrinology* **87**, 581–587.

Yago, N., Nightingale, M. S., Dorfman, R. I., and Forchielli, E. (1967a). NADPH inhibition and *in vitro* effects of gonadotrophins on the cholesterol side-chain cleavage enzyme system in bovine corpora lutea. *J. Biochem. (Tokyo)* **62**, 274–275.

Yago, N., Dorfman, R. I., and Forchielli, E. (1967b). NADPH and cholesterol side-chain cleavage enzyme system in the heavy mitochondria of bovine corpora lutea. *J. Biochem. (Tokyo)* **62**, 345–352.

YoungLai, E. (1972). Steroid secretion and the aromatization of androgens by rabbit thecal cells. *Endocrinology* **91**, 1267–1272.

YoungLai, E. (1973). Biotransformation of pregnenolone and progesterone by rabbit ovarian follicles and corpora lutea. *Acta Endocrinol. (Copenhagen)* **74**, 755–782.

YoungLai, E. (1975). Steroid production by the isolated rabbit ovarian follicle. III. Actinomycin D-insensitive stimulation of steroidogenesis by LH. *Endocrinology* **96**, 468–474.

YoungLai, E. V., and Short, R. V. (1970). Pathways of steroid biosynthesis in the intact Graafian follicle of mares in oestrus. *J. Endocrinol.* **47**, 321–331.

Zamboni, L. (1972). Comparative studies on the ultrastructure of mammalian oocytes. *In* "Oogenesis" (J. D. Biggers and A. W. Schuetz, eds.), pp. 2–45. Univ. Park Press, Baltimore, Maryland.

Zander, J. (1958). Steroids in the human ovary. *J. Biol. Chem.* **232**, 117–122.

Zeleznik, A. J., Midgley, A. R., and Reichert, L. E. (1974). Granulosa cell maturation in the rat: increased binding of human chorionic gonadotropin following treatment with follicle-stimulating hormone *in vivo*. *Endocrinology* **95**, 818–825.

Zmigrod, A., Lindner, H. R., and Lamprecht, S. A. (1972). Reductive pathways in progesterone metabolism in the rat ovary. *Acta Endocrinol. (Copenhagen)* **69**, 141–146.

Zor, U., Bauminger, S., Lamprecht, S. A., Koch, Y., Chobsieng, P., and Lindner, H. R. (1973). Stimulation of cyclic AMP production in the rat ovary by luteinizing hormone: independence of prostaglandin mediation. *Prostaglandins* **4**, 499–507.

Author Index

Numbers in italics refer to the pages on which the complete references are listed.

413

Subject Index

A